Banana Breeding

Banana Breeding
Progress and Challenges

Edited by
Michael Pillay
Abdou Tenkouano

CRC Press
Taylor & Francis Group
Boca Raton London New York

CRC Press is an imprint of the
Taylor & Francis Group, an **informa** business

The picture of male flowers and seeds are courtesy of Mauricio Guzman, Corbana, Costa Rica. The female flowers are courtesy of Moses Nyine, Iita, Uganda.

CRC Press
Taylor & Francis Group
6000 Broken Sound Parkway NW, Suite 300
Boca Raton, FL 33487-2742

International Standard Book Number: 978-1-4398-0017-1 (Hardback)

Visit the Taylor & Francis Web site at
http://www.taylorandfrancis.com

and the CRC Press Web site at
http://www.crcpress.com

This book is dedicated to our families, who enrich our lives, and to our late colleagues Dirk Vuylsteke, Paul Speijer, John Hartman (IITA), and Phil Rowe (FHIA) for their contributions to banana breeding.

Contents

Foreword

Food production must be increased year after year to keep pace with population growth. At the current population growth rate of 1.2%, world population is expected to reach 9 to 10 billion by 2050. On the basis of this estimate, food production would need to be doubled in the next 30 years or so and tripled in the next 50 years. We will need to do this without causing ecological damage to our natural resources and the environment.

Although cereals are expected to continue to be the most important calorie providers in the world, crops like bananas will also remain important calorie-providing staples throughout the world, especially in developing tropical countries. Because of the importance of banana in some parts of the world, especially Africa, attention must be focused on it as a staple food.

Several previously published books have included chapters on bananas, but there is no recent book that provides in-depth coverage of all aspects of banana breeding and genetics, including biotechnology. In some recently published books, individual chapters can be found on banana improvement, but all aspects of banana breeding, genetics, biotechnology, genetic resources, and morphology have not received treatment in sufficient detail, especially in light of the fact that major advances have occurred in modern methods of banana breeding and related aspects during the past couple of decades. Thus, there was a need to bring together these advances in a single title. The current book, *Banana Breeding: Progress and Challenges*, edited by Michael Pillay and Abdou Tenkouano, fills this need.

The book is a wide-ranging compilation of chapters by various experts. The book begins with a chapter on the general plant morphology of *Musa*. Subsequently, chapters such as Evolution and Genetic Relationships in Bananas and Plantains, Genetic Resources for Banana Improvement, Genomes, Cytogenetics, and Flow Cytometry of *Musa*, and Genetics of Important Traits in *Musa* are included. Two chapters cover the major diseases and pests of banana. Five chapters cover the central focus of the book, including Reproductive Biology, Breeding Techniques, Mutations and Cultivar Development of Banana, Biotechnology in *Musa* Improvement, and Genotype by Environment Interaction and *Musa* Improvement. The latter chapter should help provide tools for selecting both narrowly adapted and broadly adapted cultivars. The chapters on quality improvement of cultivated *Musa* and postharvest processed products provide researchers and teachers with information to improve quality aspects of banana and how to reduce postharvest losses, respectively.

Because of its vegetative reproduction, the chapter on propagation methods in *Musa* should prove valuable to small-scale farmers in providing enough planting material. In the chapter Molecular Breeding of Other Vegetatively Propagated Crops: Lessons for Banana, the book draws on the strengths and weaknesses of other vegetatively propagated crops to avoid mistakes and to make achievement of success more certain.

The editors and other authors have vast experience in banana breeding, genetics, biotechnology, molecular-marker technology, tissue culture, and other areas. Thus, all the chapters are authoritative contributions. I must congratulate all of them for generating a first-class publication that should be useful to researchers, teachers, and extension personnel around the world. This book is expected to have a major impact on banana research and teaching. This comprehensive book should serve as a ready reference for all researchers and teachers interested in banana breeding and production.

Manjit S. Kang

Introduction

Finally we have a book centering on banana breeding. It is even entitled *Banana Breeding*. This is a most welcome addition to the classical book by N. W. Simmonds, *The Evolution of the Bananas*, now some 50 years old.

This book is a wide-ranging compilation of chapters by various authors covering plant morphology, origin, genetic resources, reproductive biology, diseases, pests, quality improvement, propagation, and distribution to farmers, as well as the central focus chapters covering breeding, genetics, and biotechnology. There is even a chapter comparing breeding with three other major clonal tropical crops.

When one considers that banana breeding based on modern science is now some 90 years old and yet farmers are still mainly growing the diverse clones selected by villagers thousands of years ago from their natural environment, one must ask why. How could so many diverse and excellent clones have arisen by purely natural events of pollination, seed set, seed germination, and then selection?

The answers are complex, but things must have been very different then. Indeed, in an excellent chapter by Fortescue and Turner on reproductive biology, it is made very clear that low seed set and germination today are major detriments to breeding progress and that little research has been applied to understand the physiological/biochemical reasons for ovule abortion/low seed set. When early man first found and brought into cultivation parthenocarpic plants, they were probably quite fertile. As plants were moved about, encountering genetically diverse bananas, seed set was probably abundant. With time, seed set was negatively selected and, hence, has led to present-day sterility. Some parthenocarpic clones are still quite fertile, but exploration and evaluation for this character has been neglected. Even our understanding of parthenocarpy itself and its inheritance go back to work now 50 years old.

But the breeding programs themselves have waxed and waned as support varied and objectives changed. There are six existing breeding programs outside the center of origin of *Musa*, and only two minor efforts in India and none at all in the Southeast Asia center of diversity, where wild bananas still exist and clones are still being domesticated. Research on coevolved pathosystems is neglected, and no feedback from the natural systems into breeding exists.

Many chapters in this book reveal the enormity of molecular research applied to bananas and the attempts to apply molecular techniques to banana breeding itself. Molecular-assisted selection is still in its infancy and much dynamism continues in the molecular field.

Breeding techniques and breeding philosophies are expertly detailed in a chapter by Tenkouano et al. It is clear that much has been learned to direct the future of breeding. Excellent bibliographies in many chapters provide a valuable documentation of the diverse and enormous scattered research activity on bananas of the last 50 years.

Breeding bananas started with the simple objective of a Fusarium wilt-resistant 'Gros Michel.' Breeding objectives changed and proliferated as new programs started and local farmers' needs were addressed. Objectives are now very diverse and complex, and they differ in different regions. Much has been learned of banana evolution through molecular science. Yet breeders still use only a very limited pool of parents compared with the great natural diversity existing. It is clear that much research is still needed to assess and to reduce reproductive barriers.

A perusal of these chapters with the literature and an examination of the experience on which they are based reveal a wealth of knowledge and views not readily available elsewhere. It is an excellent new resource on bananas and banana breeding.

Ivan W. Buddenhagen

Preface

This book comprises a collection of chapters written by experts in banana research. Banana is one of the most important agricultural crops, providing food, income, and employment for millions of people, especially in the tropics. The crop is threatened by various diseases and pests that are being compounded by environmental change. Breeding resistant cultivars appears to be the only sustainable solution. The crop has been largely neglected and there are a few isolated breeding programs in the world. The last major book covering a diverse range of topics in bananas and plantains was written over 10 years ago. Since then, a large body of new information on banana has emerged, and this is reflected in the number of publications in various research areas. We believe that there is a need for updated information in banana breeding and new responses to old challenges facing the crop. The purpose of this book is to portray our personal perspectives on the challenges facing the crop. To enable us to cover a wider range of topics, we have enlisted the ideas of leading experts, including agronomists, biologists, biotechnologists, breeders, crop improvement and integrated pest management (IPM) specialists, plant pathologists, and taxonomists. In this book our aim is to concentrate the current information and provide an accessible source of information to those interested in banana research, especially the development of new disease- and pest-resistant cultivars. This book provides basic as well as advanced information for those interested in learning more about banana as well for those pursuing further research in the crop.

Chapter 1, written by Deborah Karamura, Eldad Karamura, and Guy Blomme, individuals with vast experience in bananas and plantains, provides a detailed botanical description of the plant. In addition to describing the aerial shoot corm and root systems, they also address the role of morphology in classifying banana and the confounding effects of mutations and genotype x environment (GxE) interactions. They emphasize the importance of distinguishing cultivars, even those produced by breeding programs, with regards to breeders' rights and the fact that traders and buyers can select the cultivar of their choice.

In Chapter 2, Uma Subbaraya, Marimuthu S. Saraswathi, and Michael Pillay draw on their personal experiences to present a comprehensive treatment of the evolution, diversification, and molecular genetic relationships in bananas and plantains. The early recordings of banana in India and indigenous knowledge of the uses of banana is a unique aspect of this chapter. The authors also indicate the various research needs in banana with regards to molecular taxonomy, especially of unique germplasm available in India.

Markku Häkkinen, a highly respected field botanist, and Richard Wallace provide information in Chapter 3 on some of the new banana germplasm that have been recently identified. They report on members of the five sections and their potential usefulness for conventional breeding. They conclude by stating that "The incorporation of genetic traits (disease and pest resistance, drought and cold tolerance, and so forth) from these sections will play an important role in the development of new, improved hybrid bananas for use by future generations."

In Chapter 4, Michael Pillay and Abdou Tenkouano provide a comprehensive treatise of the genomes in *Musa*, the part played by molecular cytogenetics in identifying the genomes, and the role of genomes in *Musa* classification. They formulate from their personal and practical experiences the opinion that conventional and molecular cytogenetics have played and will continue to play a vital role in breeding of banana. Various biochemical markers to identify the different genomes are discussed. The importance of flow cytometry, especially in ploidy identification, is highlighted. Recent information on cytogenetical aspects of fertility is addressed briefly.

The genetics of important traits are discussed in Chapter 5 by Eli Khayat and Rodomiro Ortiz, two individuals with vast experience and knowledge in banana genetics. The chapter presents new perspectives on the genetics of plant architecture, fruit parthenocarpy, fruit ripening and senescence, nematode resistance, and resistance to black leaf streak disease.

In Chapter 6, Guy Blomme, Simon Eden-Green, Mohammed Mustaffa, Bartholemew Nwauzoma, and Raman Thangavelu use their firsthand field experiences to provide an extensive review of the Sigatoka diseases, Fusarium wilt, Xanthomonas wilt, and viral diseases of banana. This is perhaps the first comprehensive treatise of Xanthomonas wilt of banana.

Thomas Dubois and Daniel Coyne present an excellent overview of integrated pest management of banana in Chapter 7. In addition to the major pests—nematodes and the banana weevil—that generally receive the most attention, they outline a wide range of pests associated with banana cultivation. They define integrated pest management (IPM) and present an important assessment on how IPM principles can be applied to various banana cultivation systems.

Chapter 8 is a comprehensive treatment of the reproductive biology of banana by Jeanie Fortescue and David Turner. The authors' probing writing style and personal experiences make this chapter one of the best treatments written on this topic. The floral biology, breeding systems, pollen and seed production, and reproductive systems are fully discussed. This is excellent reading for researchers wishing to start banana breeding programs.

Chapter 9 by Abdou Tenkouano, Michael Pillay, and Rodomiro Ortiz highlights the experience of these banana breeders with their work in Africa and presents some of their own research findings. A short history of banana breeding is followed by the main reasons for breeding in the crop. The chapter outlines the objectives and difficulties of banana breeding and the progress made in this field. Future breeding goals are elaborated.

Shri Mohan Jain, Bradley Till, Prasnna Suprasanna, and Nicolas Roux, specialists in mutation research, discuss the value of mutations in producing new banana cultivars in Chapter 10. The chapter is ideal for anyone interested in using mutations in banana since it addresses the best methods for inducing mutations, the types of materials to use, and the steps to follow after mutation induction. The value of targeting-induced local lesions in genomes (TILLING) in banana is also introduced.

The value of the many facets of biotechnology in *Musa* improvement is addressed by Leena Tripathi in Chapter 11. The application of tissue culture in *Musa* research is discussed. The role of genomics and transgenic technology for *Musa* genetic improvement is highlighted, with examples of genes that will be useful for developing transgenic banana. The challenges facing researchers in the development of transgenics, especially in less-developed countries, are elaborated.

In Chapter 12, Rodomiro Ortiz and Abdou Tenkouano give an overview of the progress made in research related to GxE interactions in banana. The components of phenotypic stability for some traits are discussed. The chapter reviews how to manage GxE to have efficient selection schemes and multilocation testing before release of improved cultivars. Broad-sense heritability (H^2) and repeatability (R) estimates for growth, bunch, and fruit traits in triploid *Musa* germplasm are provided. The authors discuss new ways of using GxE information, inclusive of market-related information, to develop end-use approaches to breeding and to target new cultivars to areas where they are most likely to add value.

In Chapter 13, Edson P. Amorim, Sebastião de Oliveira e Silva, Vanusia B. de Oliveira Amorim, and Michael Pillay address the nutritional value of banana. Breeding objectives for quality improvement are discussed. The role of biofortification and breeding strategies for developing biofortified cultivars with improved nutritional quality is highlighted.

Cherukatu K. Narayana and Michael Pillay list some of the various postharvest products obtained from banana in Chapter 14. The need for new products from banana to reduce large postharvest losses is discussed. The authors conclude by stating that bananas represent a great potential raw material for food and nonfood processing industries.

Chapter 15 by Michael Pillay, Christopher A. Cullis, David Talengera, and Leena Tripathi reviews various ways of propagating banana. Embryo culture of hybrid seeds from breeding programs is discussed. The role of micropropagation is reviewed. Low-cost techniques to rapidly multiply banana seedlings are reviewed. Since micropropagation is usually associated with somaclonal variation, molecular methods to detect somaclonal variants are outlined.

The success of plant breeding programs is measured by the extent to which the breeding products are adopted and used by the growers, profitably and durably. Beyond the genetic products, perhaps the more challenging task for the breeders is to understand and help put in place the complex battery of institutional and transactional measures that create a conducive delivery framework. These issues are discussed by Abdou Tenkouano, Michael Pillay, and Ousmane Coulibaly in Chapter 16. The authors also elaborate on the asymmetric nature of cultivar-based transactions in an evolving market within domestic, regional, and international contexts.

Molecular breeding in banana has not been as progressive as that of other crops. In Chapter 17 Michael Pillay, Abdou Tenkouano, and Rodomiro Ortiz review some aspects of molecular breeding in potato, cassava, and sugarcane. The breeding challenges, production constraints, and breeding objectives of these crops are similar to those of banana. Greater progress has been made in the search for molecular markers, mapping, and developing transgenics for a number of traits in potato, cassava, and sugarcane. These are useful lessons for banana scientists.

Chapter 18, by Rodomiro Ortiz, Michael Pillay, and Abdou Tenkouano, is a brief chapter on future prospects in *Musa* research.

We trust that those interested in banana and plantain will find the information in this book useful and stimulating. This book is intended for students, teachers, and banana breeders. We hope that academics throughout the world interested in developing tropical crops will also find use for this book.

We thank all the authors for their valuable contributions and for sharing their knowledge to make this book a success.

Editors

Michael Pillay (BSc, UHDE, BEd, BA, BSc [Hons] MS, PhD) is a professor in the Department of Biosciences at Vaal University of Technology, Vanderbijlpark, South Africa. He completed a BSc degree in botany and zoology and a university higher diploma in education at the University of Durban-Westville (now the University of KwaZulu–Natal), a BEd and BA from the University of South Africa (UNISA), a BSc (Hons) at the University of Durban-Westville, an MS (agronomy) at Louisiana State University, and a PhD at Virginia Polytechnic Institute and State University.

Dr. Pillay is the author and coauthor of many articles and book chapters. He served as an editor for the *American Journal of Agronomy* and is on the editorial board of the *Journal of Crop Improvement*. Dr. Pillay's research interests are biotechnology, breeding, molecular genetics, genetics, plant sciences, plant systematics, cytogenetics, tissue culture, and germplasm conservation of crop plants. Dr. Pillay started his career as an educator (1974–1984) before obtaining a scholarship to Louisiana State University, in the United States. After completing three postdoctoral fellowships in the United States, he joined the International Institute of Tropical Agriculture (IITA) in Nigeria as an associate scientist to work on cytogenetics and molecular biology of bananas and plantains. He was then transferred to Uganda as a scientist and banana breeder/molecular biologist.

He joined Vaal University of Technology in 2007 as an associate professor, was employed at UNISA as professor in 2009, and rejoined Vaal University of Technology in 2010, where his main interests are research and teaching.

Abdou Tenkouano (BSc, MSc, PhD) is the director of the Regional Center for Africa, AVRDC—The World Vegetable Center in Arusha, Tanzania, and formerly a banana and plantain breeder at the International Institute of Tropical Agriculture. Dr. Tenkouano obtained his MSc in plant breeding and his PhD in genetics from the Department of Soil and Crop Sciences, Texas A&M University, College Station, Texas, and his BSc in agronomy and *ingénieur agronome* in rural development (agricultural planning and resource management) from the University of Ouagadougou, Burkina Faso.

Dr. Tenkouano is the author of numerous articles and book chapters in a wide range of topics on banana and plantain. His research interests lie in genetics, breeding, and improvement of crop plants, with a particular focus on improving nutritional quality and host-plant reaction to biotic and abiotic stress factors.

Dr. Tenkouano worked as a researcher at the Institut de l'Environnement et de Recherches Agricoles (INERA) and as a lecturer at the University of Ouagadougou in Burkina Faso before joining the International Crops Research Institute for the Semi-Arid Tropics (ICRISAT, Mali) as coordinator, West and Central Africa Sorghum Research Network. He then started employment at the International Institute of Tropical Agriculture (IITA, Nigeria) where he was the research team coordinator, Plantain and Banana Research, Genetic Improvement Group, and thereafter became the program coordinator, Plantain and Banana Research.

He moved to Cameroon where he assumed the role of program leader, Diversification of Agricultural Systems in Humid and Sub-Humid Agro-Ecologies for IITA. He was elected councilor, Research for Development Council (RDC), at IITA. Dr. Tenkouano has been involved in the supervision of many MSc and PhD students.

Contributors

Edson Perito Amorim, PhD
Embrapa Cassava and Tropical Fruits
Cruz das Almas, Bahia, Brazil

Vanusia Batista de Oliveira Amorim, PhD
Embrapa Cassava and Tropical Fruits
Cruz das Almas, Bahia, Brazil

Guy Blomme, PhD
Bioversity International
Kampala, Uganda

Ousmane Coulibaly, PhD
International Institute of Tropical Agriculture (IITA)
Biological Control Center for Africa
Cotonou, Benin

Daniel Leigh Coyne, PhD
International Institute of Tropical Agriculture (IITA)
Dar es Salaam, Tanzania

Christopher Ashley Cullis, PhD
Department of Biology
Case Western Reserve University
Cleveland, Ohio, USA

Thomas Dubois, PhD
International Institute of Tropical Agriculture (IITA)
Kampala, Uganda

Simon Eden-Green, PhD
EG Consulting
Larkfield, Kent, UK

Jeanie Anne Fortescue, PhD
School of Plant Biology
Faculty of Natural and Agricultural Sciences
University of Western Australia
Crawley, Australia

Markku Anton Häkkinen, PhD
Finnish Museum of Natural History
Botanic Garden, University of Helsinki
Jyrängöntie, Finland

Shri Mohan Jain, PhD
Department of Agricultural Sciences
University of Helsinki
Helsinki, Finland

Deborah Karamura, PhD
Bioversity International
Kampala, Uganda

Eldad Karamura, PhD
Bioversity International
Kampala, Uganda

Eli Khayat, PhD
Department of R&D
Rahan Meristem (1998) Ltd.
Rosh Hanikra, Israel

Mohammed Mustaffa, PhD
National Research Centre for Banana (ICAR)
Tamil Nadu, India

Cherukatu Kalathil Narayana, PhD
Division of Post Harvest Technology
Indian Institute of Horticultural Research
Karnataka, India

Bartholomew Nwauzoma, PhD
Department of Applied & Environmental Biology
Faculty of Science
Rivers State University of Science & Technology
Port Harcourt, Nigeria

Rodomiro Ortiz, PhD
Sociedad Anomia
Lima, Peru

Michael Pillay, PhD
Biosciences Department
Vaal University of Technology
Vanderbijlpark, South Africa

Nicolas Roux, PhD
Commodities for Livelihoods Programme
Bioversity International
Montpellier, France

Sebastião de Oliveira e Silva, PhD
Embrapa Cassava and Tropical Fruits
Cruz das Almas, Bahia, Brazil

Marimuthu S. Saraswathi, PhD
National Research Center for Banana (ICAR)
Tamil Nadu, India

Uma Subbaraya, PhD
National Research Center for Banana (ICAR)
Tamil Nadu, India

Prasnna Suprasanna, PhD
Plant Cell Culture Technology Section
Nuclear Agriculture and Biotechnology Division
Bhabha Atomic Research Centre
Mumbai, India

David Talengera, MSc
International Institute of Tropical Agriculture (IITA)
Kampala, Uganda

Abdou Tenkouano, PhD
Regional Center for Africa, AVRDC—The World Vegetable Center
Duluti, Arusha, Tanzania

Raman Thangavelu, PhD, PDF
National Research Centre for Banana
Tamil Nadu, India

Bradley Till, PhD
Plant Breeding Unit
FAO/IAEA Agricultural and Biotechnology Laboratory
International Atomic Energy Agency
Vienna, Austria

Leena Tripathi, PhD
International Institute of Tropical Agriculture (IITA)
Kampala, Uganda

David William Turner, PhD
School of Plant Biology
Faculty of Natural and Agricultural Sciences
University of Western Australia
Crawley, Australia

Richard Henry Wallace, PhD
Department of Chemistry and Physics
Armstrong Atlantic State University
Savannah, Georgia, USA

1 General Plant Morphology of *Musa*

Deborah Karamura, Eldad Karamura, and Guy Blomme

CONTENTS

1.1 INTRODUCTION

In this chapter, *Musa* stands for the genus and musa for the crop that includes both bananas and plantains. This chapter describes the morphology of the musa plant and discusses the potential benefits of morphological diversity. The morphology of *Musa* has been extensively described by Cheesman (1947, 1948a, 1948b, 1950), Simmonds and Shepherd (1955), Champion (1961), De Langhe (1964), Simmonds (1962, 1966), Purseglove (1972), Stover and Simmonds (1987), and Karamura and Karamura (1995). These authors focused on the aerial parts of the plant, but an understanding of the root morphology is also critical, given its role in plant nutrition and anchorage. Root architecture

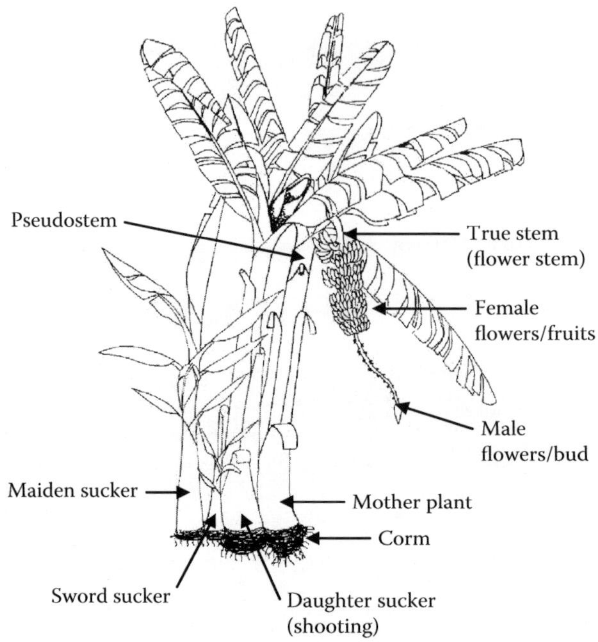

FIGURE 1.1 The morphology of the *Musa* plant: a mat or stool.

has been studied by Champion (1961, 1963), Swennen et al. (1984), Stover and Simmonds (1987), Price (1995), Blomme (2000), and Blomme et al. (2003).

The musa plants constitute some of the largest herbaceous perennial plants that range from 2–9 m in height in cultivated plants and 10–15 m in some wild species. Basically, the musa plant consists of a subterranean stem or corm, an aerial pseudostem, the leaves, and the inflorescence (Figure 1.1). The corm is the true stem to which are attached developing suckers (that perpetuate the life cycle of the plant) and roots; the corm supports the pseudostem, the leaves, and the inflorescence that bears the flowers and subsequently the fruit. The plants are monocarpic, that is, the shoots flower only once and die after fruiting. Together, the corm and attached structures form a stool or a mat.

1.2 BOTANICAL DESCRIPTION

1.2.1 The Aerial Shoot

The aerial shoot is the above-ground part of the musa plant, consisting of the pseudostem, leaves, and the inflorescence. The aerial shoot has the greatest utility and value.

1.2.2 The Pseudostem

The pseudostem is made of large overlapping leaf sheaths that are tightly rolled on each other to form a firm cylindrical structure (Purseglove, 1972; Stover and Simmonds, 1987). The aerial stem entirely depends on the leaf sheaths for mechanical support and it essentially provides vascular connection between the leaves, the roots, and the fruit (Stover and Simmonds, 1987). Considerable variation in height, color, and disposition of the pseudostem occurs and is used to distinguish banana cultivars. Thus, the pseudostems of *Musa* AB (e.g., 'Kisubi'), the AAB (plantains, 'Silk,' 'Mysore,' and 'Sukali Ndizi'), and the ABB ('Bluggoes' and 'Pisang Awak') groups are predominantly yellow-green with little or no pink pigmentation on the undersheaths; the AA ('Sucrier') and the AAA ('Gros Michel') groups are generally multicolored (Figure 1.2). The pseudostem of the

AAA-EA AAB Plantain ABB Bluggoe

FIGURE 1.2 Variation in color of the underlying pseudostem in the different *Musa* groups.

AA group tends to be rich chocolate-brown at the junction of the pseudostem–petiole region while the AAA has dull green to green-brown pink-flushed uppersheaths, and a rich mix of green, pink, and rust-brown background with black mottling along the pseudostem length. The East African Highland bananas (AAA-EA), also considered as the Lujugira-Mutika subgroup, display a very variable pseudostem color, the intensity of which varies with environmental conditions. Thus, in the Lujugira-Mutika subgroup, the pseudostem and leaf petioles tend to get darker as altitude increases. Pseudostem height also varies across cultivars and agro-ecological conditions, for example, from 4 m on the plains to 8 m in sheltered valleys for the AAA cultivar 'Gros Michel' (Purseglove, 1972). Likewise, Cavendish cultivars may be relatively tall in lowlands where conditions are ideal but shorter at higher altitudes. In the Lujugira-Mutika, the cultivar 'Nakyetengu' usually grows up to 2 m and is early maturing, but it can reach 3 m under some conditions and takes then longer to shoot. In general short-stemmed cultivars are susceptible to drought while the tall ones are tolerant (Stover and Simmonds, 1987). Most cultivars have a straight and erect pseudostem, but some cultivars of the Lujugira-Mutika subgroup, such as 'Mukazi-aranda,' distinctly display a "creeping" habit from which it derives its name (meaning "crawling lady" in Luganda dialect).

1.2.3 The Suckers

The suckers are lateral shoots that emerge from buds located opposite the base of leaves on the corm. Only 3–4 buds develop into shoots or suckers. Continuous emergence of new suckers perpetuates the corm's life, giving musa its perennial status. There are two types of suckers: the broad-leafed suckers from superficially placed buds (usually due to damaged corms) and the sword-leafed suckers from deep-lying buds. Thus, unless they are removed for propagation purposes, the aerial part of a plant consists of a large number of shoots forming a clump and the rate of production of shoots varies from a few to several tens, depending on the clone. The growth of suckers is also greatly influenced by the stage of maturity reached by the parent plant, and field operations like pruning or manure/mulch application may also affect sucker production.

1.2.4 The Leaves

The corm also consists of the apical meristem from which the leaves and the flowers are initiated. At emergence the leaf is more or less vertical, gradually becoming horizontal and drooping as it ages. The lamina develops as a rolled cylinder (hence the name *cigar leaf* or *heart leaf*) during its passage through the pseudostem, with the right half rolled upon itself and the left half rolled over the right and the midrib. Purseglove (1972) has attributed the unfurling process of the leaf and other diurnal lamina movements to growth and turgor changes of specific motor cells of the pulvinar band exerting pressure against rigid structures. The pulvinar bands are found where the two leaf blades join the midrib and appear as two pale lines. The leaf blade gradually expands into a large, oblong lamina with a pronounced supporting midrib and well-marked, pinnately arranged parallel veins. The whole process of leaf blade formation is completed when the leaf sheaths narrow on both sides to form the petiole (leaf stalk), which is rounded beneath and channeled above, retaining the crescent-shaped section of the sheath and enlarging at the tip to form the blade. The petiole is variously colored in many banana cultivars. In the *Musa acuminata* cultivars, the petiole coloration has a predominantly reddish to purplish background, with traces of green. Conversely, in cultivars of *M. balbisiana* the petioles are largely green and in many cases waxy. Hybrids have mixtures of green and purplish colors. However, due to somatic mutations, coloration and waxiness may vary even within the same stool.

Leaf disposition on the plant enables them to receive maximum light for photosynthesis. The blade is often torn by the wind and hangs in ribbons from the midrib. New leaves are continually forced up through the center of the pseudostem and expand at the top, where the leaf blades form a handsome crown or canopy that enables the giant herb to outcompete other herbaceous plants for light. Banana leaves are light green in color, smooth and glossy, and attain a very large size, often being used as a temporary shade for other crops. The final leaf to emerge through the pseudostem is much smaller than the rest and curves to protect the developing inflorescence from direct insulation. In some Lujugira-Mutika cultivars such as 'Nsowe' or 'Kataribwambuzi' (translated as "that which cannot be eaten by goats"), the sword suckers continue to produce scale leaves until the plant is taller than 2 meters. This enables the cultivar to grow without being eaten by herbivores that may devour the foliage of cultivars that unfurl their leaves at lower heights. The lamina is thickest near the midrib and thinnest at the edges. The veins of the lamina are parallel as they leave the midrib but become S-shaped as they approach the margins. Stover and Simmonds (1987), quoting Skutch (1930), have indicated that there are about 17,000 veins near the midrib in one lamina half of a large leaf. This venation offers little or no resistance to transverse tearing of the lamina. When this happens, commissural bundles and some of the ground tissue may be destroyed but the vascular connection between the midrib and margin remains intact.

Splitting of the lamina along the veins is a normal occurrence, which may increase the surface area for cooling the plant while reducing transpiration due to suberization of torn surfaces (Taylor, 1969). This results in higher photosynthetic rates in torn leaves because of favorable temperatures and carbon dioxide exchange (Taylor and Sexton, 1972). However, excessive tearing of the lamina may reduce photosynthetic rates and cause yield losses in highland agro-ecologies that are prone to hailstorms (Karamura and Karamura, 1995). Excessive tearing can be prevented with wind breaks, as is common in Uganda where *Ficus* trees are prominent on banana farms (Davis, 1995).

Purseglove (1972) observed that the number of functioning leaves remains approximately constant at 10–15, with a new leaf emerging every 7–10 days. Leaf emergence is influenced by cultivar and ecological conditions, with a pronounced seasonality effect whereby more leaves are produced in wet than dry seasons. Likewise, leaf retention is affected by prevailing soil fertility and soil moisture levels. Air temperature, day length, plantation age, plant density, and plant stature are also known to influence leaf emergence, notably in the Cavendish and Gros Michel subgroups (Allen et al., 1988). The posture of the leaf varies with age, gradually shifting from a tightly rolled vertical cylinder at emergence to an expanded horizontal posture before drooping as the petiole starts to collapse. The lamina eventually senesces and dies, preceding the decay and shedding of the leaf sheath.

Much variability exists in the color, disposition, and waxiness of musa leaves. In general, diploids tend to have more erect leaves, whereas triploids have broader leaves, more or less drooping, carried on vigorous stems. In plantains and other interspecific hybrids, the leaves are yellow-green (or display various shades of greenness). Conversely, within the Lujugira-Mutika subgroup, there is a wide variability of colors, including the green but variegated leaves of 'Nasuuna,' the purplish leaves of 'Bitambi,' the glossy leaves of 'Namafura,' and various shades of dirty green of the lamina with or without the red midrib. In general, however, color variation is more pronounced in *M. acuminata*, while *M. balbisiana* tends to be monochromatic green.

Musa plants have a large leaf area, ranging from 1.27 to 2.80 m^2 in dessert bananas (Stover and Simmonds, 1987; Stover, 1988) and 0.68 to 0.92 m^2 in plantains (Anojulu, 1992). The total leaf area of Cavendish cultivars at flowering is approximately 16.9 to 25 m^2, providing a closed canopy that protects the soil from the impact of rain and oxidative insolation (Stover and Simmonds, 1987).

Leaf pigmentation and posture may affect photosynthetic efficiency and transpiration rates as described by Brun (1960, 1961a, 1961b, 1962, 1965) and Turner (1972), but there are surprisingly not many recent studies relating leaf morphology to photosynthetic efficiency or transpiration of the existing cultivars.

1.2.5 The Inflorescence

The apical meristem and flower initiation have been extensively discussed by Barker and Steward (1962), Ram and Steward (1962), Fahn et al. (1963), and Stover and Simmonds (1987). The time of flowering and fruit maturity are both dependent to a large extent on cultivar and environmental conditions. The following events characterize flowering: elongation of the corm internodes; suppression of leaf development, which is replaced by bract development; and, eventually, initiation and development of the inflorescence. Floral initiation occurs with the apical meristem ceasing to produce leaves (about 30–40 leaves in lowland tropical locations) and elongating through the center of the pseudostem, until the inflorescence bud emerges. Inflorescence emergence is described as shooting, at which stage the purplish green/greenish brown inflorescence appears vertical with a pointed tip and a broader base, but it soon bends over due to its weight. The inflorescence is a complex spike with a stout peduncle on which flowers are arranged in nodal clusters on transverse cushions (crown), subtended by large spathe-like bracts that are nearly ovate and usually purple-red in color (Purseglove, 1972). Each nodal cluster consists of two closely appressed rows of flowers, one above the other, and enclosed in a large subtending bract. The bracts and their axillary groups of flowers are arranged spirally around the axis and the bracts closely overlap each other, forming a tight conical inflorescence. In the cultivar 'Lwezinga' (rolled), however, the spiral is continuous and consequently the bracts form one long strip from the female to the male inflorescences. In other cultivars the spirally arranged bracts overlap in such a way that all bracts and flowers within are enclosed tightly and protected in a large bulbous structure. Each bract is a large red/purplish pointed structure that becomes completely reflexed as the flowers develop and that falls from the axis, leaving a large scar when the fruits start to mature. Some clones—for example, *Musa* (AAA-EA) 'Nakawere'—have yellowish-green bracts with streaks of brown, which makes the emerging inflorescence appear as yellow instead of purple. The flowers are bisexual but mostly unisexual in function and there are three types of flowers. The lower bracts of the inflorescence axis enclose the female flowers, the middle few bracts may enclose neutral or hermaphrodite flowers, whilst at the tip of the inflorescence are the male flowers. The female flower consists of an elongated inferior, trilocular ovary of three fused carpels, on top of which six tepals (five united and one free) are implanted, surrounding a thick style and five or six fleshy and nonfunctional stamens (staminodes). The ovary of female flowers constitutes two thirds of the length of the whole flower, and these are the flowers that eventually develop into individual fruits or fingers of the bunch. In the male flowers, the ovary constitutes one third of the length of the flower, usually deciduous, and contains one slender style, five stamens topped by long anthers that do not, however, produce functional pollen. Neutral or hermaphrodite flowers consist of an ovary that does not develop into fruits but forms short useless fingers. The ovary is about half the length of the flower. The stamens of neutral flowers do not produce pollen.

1.2.6 Bunch Morphology

Continued elongation of the main stalk of the inflorescence causes the bunch to hang over, the bracts open and fall, disclosing the female flower clusters or hands. The female flowers undergo further development without being pollinated or fertilized. Growth of the inflorescence stalk is rapid and the hands become separated by several centimeters of stalk. After the emission of a number of hands of female flowers (1–12 depending on the conditions and cultivar), the bud differentiates hands of hermaphrodite and then male flowers. The bracts and the male flowers open and fall, so that eventually a considerable length of stalk separates the male bud from the bunch proper. Further rapid growth causes the individual fruits to fill, and their increased weight bends the main stalk so that the bunch hangs down vertically, although some bananas have upright bunches. Although the physiology, anatomy, and development of musa cultivars are similar, the morphology tends to be variable, particularly for the Lujugira-Mutika subgroup. Karamura (1999) divided this subgroup of bananas into clone sets based on bunch and rachis characteristics (orientation and compactness), persistence of neutral flowers and bracts, and shape and apex of male bud. The lax pendulous banana bunched clone set is Musakala, whereas the most compacted and subhorizontal bunched clone set is Nakabululu. 'Nakitembe' and 'Nfuuka' clone sets have compact medium-sized oblique bunches but 'Nakitembe' has persistent floral parts on the rachis and fruits as well as imbricated male buds (Figure 1.3).

This bunch/inflorescence orientation is common among the widely grown Cavendish and Plantain subgroups. Within the diploids and other groups, including the highland AAA-EA, two more inflorescence orientations can be recognized. Firstly there are those that are carried obliquely (to the pseudostem). The cultivars 'Nfuuka' and 'Mbwazirume' are examples of obliquely oriented inflorescences (Figure 1.3). In the "small-bunched" cultivars such as 'Nabakululu,' 'Endirira,' and 'Katabunyonyi,' the orientation is clearly subhorizontal (Figure 1.3). Purseglove (1972) has implied that geotropism is a factor in determining inflorescence orientation, conferring desirable traits for the export trade as manifested by the AAA Cavendish and Gros Michel cultivars. The weight of the inflorescence too appears to play an important role. Among the Lujugira-Mutika, it can be generalized that the heaviest-bunch cultivars such as 'Rumenyamagali' tend to have pendulous orientation, while the small-bunched cultivars tend to have subhorizontal inflorescences. The medium-sized bunch cultivars are carried obliquely on the pseudostem. Another factor that appears to influence inflorescence orientation relates to the presence or absence of the male bud, the big purple terminal protuberance on the bunch. It has been observed that cutting off the male bud at shooting makes the inflorescence peduncle fail to elongate and make the U-downward turn, so that a normally pendulous inflorescence ends up subhorizontal. Among the Lujugira-Mutika, for example, 'Endiriira,' the only AAA cultivar without a male bud, always carries its inflorescence subhorizontally. It would appear that bananas have a "male-bud factor" that is responsible for inflorescence orientation. In breeding programs where bunch weight is a consideration, attention should be given to the male bud factor's effects on the fruit characters. Cultivars without male buds may be important sources of traits in breeding programs aimed at controlling Xanthomonas wilt disease of bananas (Biruma et al., 2007).

The male bud itself varies in many aspects, which include bract arrangement, color, shape and its apex, and presence and degree of waxiness. In the majority of cultivars, male bud bracts are purple and purplish green considering the external side of the bracts. In cultivars 'Nakawere' and 'Tereza' of the Lujugira-Mutika subgroup, however, the male bract is greenish-yellow with vertical brown streaks. The majority of *M. balbisiana* clones have a continuous uniform crimson color in the internal surface of the bracts (Figure 1.4), whereas in the *M. acuminata* cultivars the crimson color fades to yellow at the base of the bract (Figure 1.4). Male bracts in many cultivars reflex and roll back after opening and eventually fall off while still fleshy. In some cultivars such as 'Mbwazirume,' 'Namaliga,' and 'Nakitembe,' the bracts persist and form a "protective" cover over the persistent neutral and male flowers (Figure 1.3). Upon falling, the bracts leave prominent scars. Within the

(a) (b) (c) (d)

FIGURE 1.3 Bunch orientation in the Lujugira-Mutika subgroup. (a) Subhorizontal 'Nakabululu,' (b) oblique, with bare rachis 'Bitambi,' (c) oblique, rachis with persistent floral parts 'Mbwazirume,' (d) pendulous 'Musakala.'

Lujugira-Mutika subgroup, male bud shape and apex is variable. In general the cultivars with short and fat fingers, closely packed bunch, and oblique to subhorizontal bunch orientation have obtuse-angled male bud apices. Conversely, the long-fingered, loosely packed, and pendulous bunch cultivars have acute male bud apices.

The compound tepal of the male flower is white to cream but with a flush of pink among the plantains, 'Silks,' 'Mysore,' 'Kamaramasenge,' 'Bluggoes,' 'Pisang Awak,' and the 'Red/Green Red.' The lobes of the compound tepal as well as the stigma range from yellow to orange. The anthers are pink among the Lujugira-Mutika bananas and the 'Pisang Awak,' yellow in the 'Bluggoes,' orange in the Cavendish, and cream in most other groups. The ovule arrangement in the ovary does not vary much except being either two or four rowed in the different groups. This ovule arrangement can be clearly seen in the longitudinal sections of fruits.

FIGURE 1.4 Variation in the internal bract color: Varieties with a B genome have a homogeneous red color towards the base of the bract.

1.2.7 Fruit Morphology

Both Purseglove (1972) and Stover and Simmonds (1987) provide extensive coverage of the flowers and fruits of the musa plant, based largely on observations made on the Cavendish subgroup and French plantains (AAB), with little reference given to the Lujugira-Mutika subgroup. Seedless fruits develop parthenocarpically from female flowers. However, numerous aborted ovules, carried in an axile placentation in the ovary, can be seen as small brown objects in the center of the fruit. The fruit is elongated, curved, and more or less round in cross-section but with the triangular form of the ovary still visible. At the tip of the fruit, the perianth, androecium, and the style become withered but persist for a short time, separated from the fruit by a brown corky layer. The fruit is protected by an epidermis and an underlying parenchyma layer in which the vascular bundles and a series of latex tubes are found. Inside this "skin" lies the pulp, a tissue of large cells filled with starch that is partially converted to sugars during the ripening process. As the fruit develops, it slowly curves under a negative geotropic response.

There is variability in finger length, and consumers generally prefer long fingers to short ones for all uses. This is true for dessert as well as *matooke* (green cooking) bananas. The longer fingers are easily peeled during *matooke* preparation. Among the Lujugira-Mutika subgroup, the cultivars with long fingers—'Musakala,' 'Muvubo,' and 'Nakibizzi' (also called 'Mpologoma')—are increasingly becoming commercial cultivars in Uganda. Fruits are glabrous, less angular, sharply pointed or bottle-necked or almost blunt at the apex, without floral relics. Fruit length varies from 5 cm to more than 30 cm. Within the bunch, fruits vary in their arrangement, position, and number in a hand, shape of fruit, apex, waxiness, and rows and shapes of ovule they contain. The fruits may lack ovules like in 'Kattabunyonyi' (AAA-EA), or they may have two or four rows of ovules. There is also variation in the color of fruit skin, pulp (ranging from white, cream, ivory, and beige-pink), and absence or presence of stalk. Fruit characteristics are increasingly becoming important in the banana chips industry as well as in the table-fruit markets.

1.3 THE UNDERGROUND SYSTEM

The underground system of the musa plant consists of a subterranean stem or corm that bears the root system and developing suckers.

1.3.1 The Corm

The corm is the true stem of the plant, with a cultivar-dependent diameter ranging from 20–25 cm for a typical mature Cavendish and about 15–18 cm for most African plantains. It has short internodes that are completely covered by tightly appressed leaf scars. The longitudinal section of the corm looks like an inverted cone; it is differentiated internally into a central parenchymatous cylinder surrounded by a 1–3 cm thick cortex. The top of the cone is dome shaped, with the apical meristem at the crest of the dome (Price, 1995). During the vegetative growth stage of the musa plant, the apical meristem has the form of a flattened dome with little internal differentiation. At the transition from the vegetative to the floral stage, a profound change occurs in the growing point: What was more or less a quiescent mass of cells starts showing signs of intense and general mitotic activity so that the meristematic area becomes convex and rises above the surrounding leaf bases. The outermost layer of the central cylinder consists of a cambium-like meristematic tissue from which roots arise.

1.3.2 Root Systems

The *Musa* root system is a complex structure that supports multiple plant functions. For example, it ensures the optimal uptake of water and nutrients, provides anchorage to the plant, and produces plant growth regulators (De Langhe et al., 1983; Swennen et al., 1984; Stover and Simmonds, 1987; Price, 1995). Research on *Musa* roots started some 70 years ago (Skutch, 1932) predominantly on export dessert bananas, such as Gros Michel (Moreau and Le Bourdelles, 1963) and Cavendish cultivars (Beugnon and Champion, 1966; Lassoudière, 1978; Avilan et al., 1982).

Although extensive breeding efforts have been devoted to improve shoot traits of *Musa*, comparatively little has been done for roots, despite the interdependence of shoot growth and root development. For example, nematodes reduce root growth, which often results in yield decline in *Musa* (Swennen et al., 1988; Gowen and Quénéhervé, 1990). While the nematode pest has been considered an important priority in breeding programs at the International Institute of Tropical Agriculture (IITA) (IITA, 1997, 1998), the Centre Africain de Recherches sur Bananiers et Plantains (CARBAP) (Fogain et al., 1996, 1998), and the Fundación Hondureñea de Investigación Agricola (FHIA) (Rowe, 1991; FHIA, 1998), no systematic effort has been devoted to developing root systems that are less prone to nematode damage.

A comprehensive study of the root system of a wider gene pool is required to construct an ideotype target for the genetic improvement of plantains and bananas. Such studies were carried out at the International Institute of Tropical Agriculture (IITA) and by Bioversity International in Nigeria and Uganda, respectively. Experiments focused on elucidating relationships between root and shoot traits, assessing variability in root system size, assessing the biophysical effects on root development, and devising alternative methods for root evaluation (Blomme, 2000).

1.3.2.1 The Considerable Size of the *Musa* Root System

The *Musa* root system is adventitious. Cord roots arise, usually in groups of three or four, from a common primordium (Skutch, 1932; Riopel and Steeves, 1964; Turner, 1970). They grow through the cortex where cytolysis takes place to give space. Before the roots penetrate the soil, their tips broaden abruptly (Skutch, 1932). They are 5–10 mm thick (Riopel and Steeves, 1964), initially white and fleshy before turning somewhat corky with age. The mature roots have prominent lacunae in the cortex and have large vessels and phloem strands in the central portion of the stele. Xylem elements are formed at a level in the root at which elongation has ceased (Riopel and Steeves, 1964).

Champion and Olivier (1961) reported that, for the dessert banana 'Poyo,' the zone on the corm on which roots emerge (i.e., root-bearing zone) is negatively geotropic. Swennen et al. (1988) reported both a positive and a negative geotropic movement or widening of the root-bearing zone of sucker-derived plants. However, they found that the center of activity of the root-bearing zone becomes negatively geotropic with time. The number of cord roots varies considerably depending upon the health status of the plant. A healthy corm can bear 200 to 400 primary cord roots with a total length of 230 m (Summerville, 1944; Robin and Champion, 1962; Beugnon and Champion, 1966). Fawcett (1913) noted that growth rates of the tips may reach 60 cm per month, which is in agreement with later studies with 'Poyo' in Côte d'Ivoire, where growth rates of 2–3.5 cm a day were recorded (Lassoudière, 1978). Various factors, such as soil properties, climatic conditions, and pest and diseases, influence the root growth rate (Lassoudière, 1971).

Roots generally spread over 2–3 m and may extend up to 5 m from the plant, but most of the root system occurs within 60 cm of the stem (Avilán et al., 1982; Gousseland, 1983). Root distribution down the soil profile is strongly influenced by soil type (Irizarry et al., 1981) and drainage. Compact soils, impermeable soil layers, high clay content, and saturated soil conditions prevent or reduce root growth (Beugnon and Champion, 1966; Champion and Sioussaram, 1970; Godefroy, 1969; Lassoudière, 1971). Root systems are confined mostly to the upper 40 cm of soil because of unfavorable subsoil conditions. Araya et al. (1998) found that 65% of the total root weight was in the upper 30 cm of soil for the clone 'Valery' (AAA) growing on a sandy clay loam in Costa Rica. The same authors found not less than 79% and 88% of the roots in the first 45 and 60 cm of the soil profile, respectively. Besides the laterally spreading roots, there are a limited number of roots that grow vertically (Simmonds, 1966; Summerville, 1939).

Cord roots emerge in flushes for sucker-derived plants. Thus, Swennen et al. (1988) reported that, for the False Horn plantain 'Agbagba,' 40 cord roots were formed during the first 3 weeks after planting, from preformed roots that were present in the cortex of the sucker. The number of primary roots remained unchanged for the next 4 weeks, at which time another flush of root emergence was observed (Swennen et al., 1988). Similar emergence flushes were reported by Beugnon and Champion (1966) for the 'Poyo' dessert banana and Lavigne (1987) for a dessert banana grown in a rhizotron. New roots are formed continuously until flowering occurs (Champion and Olivier, 1961; Beugnon and Champion, 1966; Turner, 1970; Lavigne, 1987). However, roots may remain alive and functional beyond fruit maturity. For example, the roots of a harvested mother corm were still alive (and presumably functioning) during growth and at the harvest of the daughter stem (Walmsey and Twyford, 1968; Lassoudière, 1980).

The upper surface of corms in aging plantain fields can be seen above soil level (Stover, 1972; Swennen et al., 1988; Swennen, 1990), a phenomenon called "high mat," and causing newly formed roots to only penetrate the topsoil or die off before reaching the soil surface (Moreau and Le Bourdelles, 1963). The plants become weak and tip over easily because they are no longer firmly based in the soil (Swennen, 1990). Earthing up (adding soil around the plant) only slightly enhances root growth and plant vigor. However, mulch protects the roots (which would otherwise dry out) and improves the ramification and stability of the plants (Swennen, 1990).

1.3.2.2 The Effect of Soil and Climate on Root Systems

Studies on the effects of soil and climate on root growth conducted at IITA stations in Nigeria revealed significant influences of substrate type and climate on root growth (Blomme, 2000). Also, field-grown plants had a relatively larger root system (i.e., lower shoot–root ratio) under low-nutrient conditions. These plants might have invested more dry matter in the root system in order to explore a larger soil volume to produce a vigorous shoot. In addition, a reduction in soil bulk density (during the first months after plowing and harrowing) significantly enhanced root and shoot growth.

1.3.2.3 Root Branching: The Lateral Roots

The banana cord roots bear numerous lateral roots of much smaller diameter. These lateral roots, bearing root hairs, are believed to be primarily responsible for water and mineral uptake by the plant. Lateral root growth, however, is highly influenced by microenvironmental conditions (Blomme et al., 2003) and genotype (Swennen et al., 1986). As a rule, lateral roots are generally abundant on cord roots growing in organic layers in the topsoil and are absent from those growing deeper in the soil.

The first-order laterals usually emerge 12–15 cm from the tip of a cord root and may be as long as 15 cm (Laville, 1964). Second-order and third-order laterals may also be present (Riopel, 1966). When the apex of a primary cord root is damaged, due to biotic or abiotic factors, two or three first-order laterals may develop into secondary cord roots (Lassoudière, 1977). Lavigne (1987) reported that no secondary cord roots are formed if the damage occurs more than 20 cm from the apex. In a study of 10 varieties, Swennen et al. (1986) found lateral roots to account for up to 98% of total root length, with varietal differences for (1) total root length, (2) relative proportions of cord root and first-order and second-order laterals, (3) length of first-order laterals, and (4) proportion of cord roots covered by laterals. This work suggested the existence of genetic variability for these components of the root architecture.

In *Musa*, first-order lateral roots are initiated basipetally in the apical region of the primary axes. Lateral root primordia can be detected in the pericycle within 450–1200 μm of the apical meristem. Their circumferential position in the pericycle is adjacent to protoxylem poles and opposed to protophloem (Riopel and Steeves, 1964; Esau, 1965; Charlton, 1982). Therefore, all laterals arising in association with a given protoxylem form a longitudinal series or rank, the number of ranks per root being equivalent to that of protoxylem poles, which varies between 28 and 34 (Riopel and Steeves, 1964).

The distribution of lateral root primordia in the root tip has been analyzed on the basis of longitudinal and circumferential distances between successive primordia and between particular primordia and the apical meristem (Charlton, 1982). Lateral root primordia were rather regularly spaced in a longitudinal pattern within each rank, each lateral arising at a rather constant proportion of the distance between the root tip and the previous lateral. They were also less likely to occur close to each other than would be expected on a random basis. It was concluded that lateral roots are initiated in a nonrandom dispersed pattern (Riopel, 1966).

Root hairs appear 4–6 cm from the apex of the cord roots and reach their mature size at 8–12 cm distally from their place of origin. Their length may exceed 2 mm.

1.3.2.4 Allocation of Dry Matter to the Shoot and Root System

The percentage dry matter allocated to the shoot and the root portion of a *Musa* plant depends on the developmental stage and planting material (Figure 1.5) (Blomme, 2000). For example, juvenile *in vitro*–derived 'Calcutta 4' (AA) and 'FHIA3' (AAAB) plants allocated up to 45% of dry matter to the root system just after field establishment. Root dry-matter content declined to about 10–15% at flowering. In contrast, less than 20% of dry matter was allocated to the roots of juvenile sucker-derived 'Mbi Egome' (AAB) plants, while the corm contained up to 70% of the plant dry matter (Figure 1.5).

Percentage dry matter invested in the mat root system decreased gradually towards flower emergence (Figure 1.5). For example, there was a clear increase in the shoot–root ratio with increasing age for the *in vitro*-derived 'FHIA3' mat. The shoot–root ratio of juvenile sucker-derived 'Mbi Egome' plants was relatively high, compared to the *in vitro*–derived plants and was caused by the large corm size of the planting material. As for the *in vitro*–derived plants, the highest shoot–root ratio was observed during the late vegetative phase.

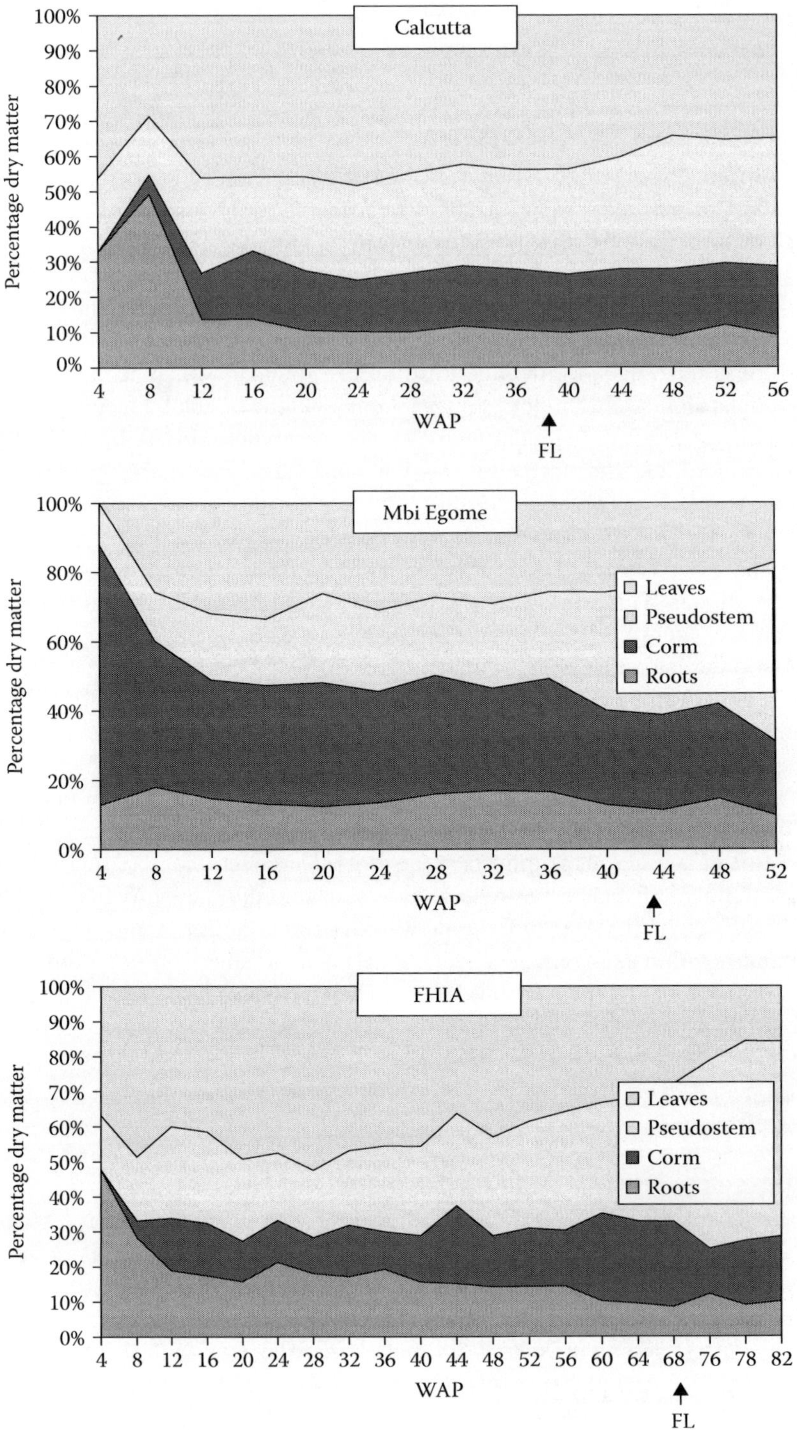

FIGURE 1.5 Dry matter partitioning between roots, corm, pseudostem (including bunch) and leaves in unthinned mats of 'Calcutta 4,' 'Mbi Egome,' and 'FHIA3' from planting until harvest (FL: flower emergence). (From Blomme, G., 2000, The interdependence of root and shoot development in banana (*Musa* spp.) under field conditions and the influence of different biophysical factors on this relationship, PhD diss., K.U. Leuven. Faculteit Landbouwkundige en Toegepaste Biologische Wetenschappen, Belgium.)

1.3.2.5 Genetic Variability in Root Growth and Relationships with Shoot Growth: Refining the Ideotype

Strong positive correlations between shoot and root development of a mat are observed during the vegetative and reproductive phases (Blomme, 2000). Thus, vigorous shoot growth in most plantains and cooking bananas is associated with a large root system. In contrast, the semi-dwarf dessert banana cultivar 'Valery' has a small root system. Despite the observed variability in root system size between genotypes, shoot–root dry weight ratios are similar formats of a wide variety of genotypes (Figure 1.6). These results indicate that breeding for an increased root system size, in addition to being cumbersome, goes along with breeding for a larger aerial part. The study of segregating populations should, however, be considered as different shoot–root relationships may be observed at the within-population level.

Strong positive relationships between shoot and root traits are also observed for the lateral shoots (i.e., suckers) (IITA, 1999; Blomme, 2000). Therefore, breeding for a regulated suckering (two to three well-developed lateral shoots present during the reproductive phase) may prove to be more beneficial for plant anchorage and productivity of the crop than breeding for a modified shoot–root ratio or root system size of the main plant. Under regulated suckering, lateral shoots will have a vigorous shoot as well as root development. In addition, their large corm size will give stability to the mat. For example, plantain hybrids with a regulated suckering will be less susceptible to toppling compared to plantain landraces, which predominantly exhibit an inhibited suckering.

Studies carried out in Uganda (Kawanda Agricultural Research Institute and Makerere University Agricultural Research Institute Kabanyolo) also found significant and positive relationships between root, corm, and aerial growth traits of complete mats in East African Highland bananas (AAA-EA) during the vegetative and early reproductive stage (Blomme, Ocan, et al., 2004; Bloome, Sebuwufu, et al., 2004; Sebuwufu et al., 2004a, 2004b). Additionally, on-farm and on-station studies focusing on the highland bananas 'Mpologoma,' 'Lwadungu,' 'Nakitembe,' 'Mbwazirume,' and 'Kibuzi,' the

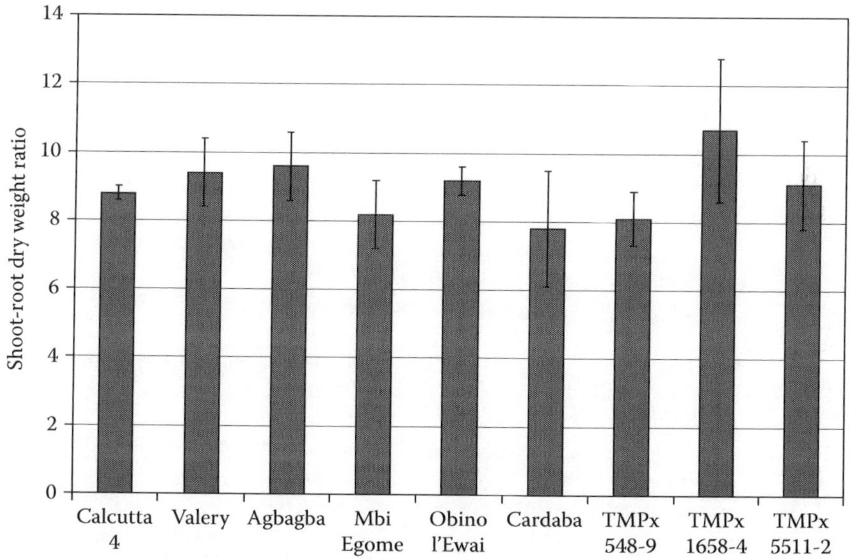

FIGURE 1.6 Shoot–root ratio for the mat of nine *Musa* spp. genotypes at flower emergence of the plant crop ['Calcutta 4' (AA); 'Valery' (AAA); 'Agbagba,' 'Mbi Egome,' and 'Obino l'Ewai' (AAB); 'Cardaba' (ABB); 'TMPx 548-9,' 'TMPx 1658-4,' and 'TMPx 5511-2' (tetraploid plantain hybrids)]. (From Blomme, G., 2000, The interdependence of root and shoot development in banana (*Musa* spp.) under field conditions and the influence of different biophysical factors on this relationship, PhD diss., K.U. Leuven. Faculteit Landbouwkundige en Toegepaste Biologische Wetenschappen, Belgium.)

dessert banana 'Sukali ndizi' (AAB), the plantain 'Gonja' (AAB), and the beer banana 'Kayinja' (ABB) revealed strong relationships between bunch weight, root, corm, and aerial growth traits. Hence, poor root development will adversely affect shoot and leaf canopy development and as a result reduce yield. Reciprocally, when leaves are affected by black leaf streak disease, the root system is reduced.

1.3.2.6 Alternative Methods for Root System Assessment

Two man-days are needed to excavate, wash, and assess the root system of one mature *Musa* plant. Also, attempts to correlate root growth of juvenile plants with that of adult plants were inconclusive, indicating that an early assessment of root system growth of juvenile plants may not give sufficient information about the root system size and development of mature plants (Blomme, 2000). Therefore, methodologies for fast and nondestructive root system assessment were developed (IITA, 1999; Blomme, 2000).

Due to the strong relationships between shoot and root traits, regression models were obtained to estimate root traits from easily measurable shoot traits (Table 1.1). More than 90% of the variability in root traits could be explained by the variability observed in shoot traits. However, as the shoot–root ratio is dependent on the developmental stage of a plant and on environmental conditions (Blomme, 2000), fine-tuning of these models will be needed when assessing plants grown under different environments.

Core root samples taken around a plant can also provide adequate information on the variability in mat root system size (Blomme, 2000). This method only takes about 5% of the time needed to excavate and assess the root system of a mature plant. These alternative methods may provide breeders and especially nematologists with adequate information on the roots of a *Musa* plant.

TABLE 1.1
Regression Models to Predict Root System Characteristics

	Trait#				
Trait#	**LA^**	**PC**	**HS**	**PL**	**R^2**
DR	0.001628***	0.596934**	–	–	0.93
NR	0.001459***	1.255633***	–	–	0.93
LR	0.066704***	23.476717**	–	–	0.94
AD	–	0.093835***	–	0.681434***	0.97
TL	0.099478***	–	14.69139***	–	0.92
TD	0.002066***	0.426590	0.17142*	–	0.93

Source: Blomme, G., The Interdependence of Root and Shoot Development in Banana (Musa spp.) under Field Conditions and the Influence of Different Biophysical Factors on This Relationship, PhD diss., K. U. Leuven, Faculteit Landbouwkundige en Toegepaste Biologische Wetenschappen, Belgium, 2000.

Note: Plants are 20 weeks old, using aerial growth characteristics and ploidy level as independent variables. #: LA: leaf area (cm^2); PC: pseudostem circumference (cm); HS: height of the tallest sucker (cm); PL: ploidy level; DR: root dry weight (g); NR: number of cord roots; LR: cord root length (cm); AD: average basal diameter of the cord roots (mm); TL: cord root length of the mat (mother plant and suckers) (cm); TD: root dry weight of the mat (g). *, **, *** Significant at $P < 0.05$, 0.01, and 0.001, respectively. ^: independent variables.

1.4 VALUE OF MORPHOLOGICAL VARIATION

1.4.1 Introduction

Musa breeding depends on variations existing within accessions, cultivars, landraces, or wild taxa related to crops. These genetic resources need to be investigated and evaluated to provide information on variation if they are to be utilized efficiently (Pickersgill, 1988). Crop-improvement strategies make use of information on variation with regard to yield, physiology, diseases, and pest responses as well as morphology.

Morphological variation, which is a prerequisite for breeding, constitutes one of the oldest indicators of diversity. Morphological variation is based on the scoring of numerous descriptors that can easily be observed and accessed without any special technical skill and at low cost. However, morphological variation is not a direct expression of the genome and can be influenced by environmental effects so that it expresses more of the phenotype than the genotype (Perrier, 1993).

1.4.2 Sources of Morphological Variation in Cultivated *Musa*

Variation is the basis of speciation, when acted upon by selection. In nature, variation is a result of basic evolutionary forces—mutation, recombination, genetic drift, and selection acting in different ecologies over time. In clonally propagated populations, however, variation may occur through mutations although other mechanisms can cause variation in banana (Nyine and Pillay, 2010). Selection is largely mediated through environmental factors in different areas or habitats.

1.4.2.1 Mutations

Bananas and plantains are clonally propagated and it is generally known that clonal propagation reduces variation within a population because relatively few genotypes are selected for conservation and use. However, clonal propagation can conserve heterozygotes as well as homozygotes, and the only source of variation in these musa plants has been mutations. One example of a group of bananas is the Lujugira-Mutika subgroup found in East Africa. These bananas were introduced in the region by Arab traders from India, through Madagascar on the eastern coast during the 15th century (Simmonds, 1962). Like all other cultivated bananas, they are vegetatively propagated and there is hardly any variation in the suckers being produced by the mother plant except through mutations and the chimerical nature of the plant, a condition that is not yet well understood. When a banana plant is propagated, it is considered to be genetically the same but in reality, among the AAA-EA cultivars, not all suckers produced on the same mat are always phenotypically identical (Karamura et al., 2004). The meristematic tissues of the mother plant may be giving rise to highly variable suckers. Plants composed of two or more genetically distinct tissues have been called chimera (Huxley, 1940; Crane and Laurence, 1956). Although chimerism has been observed in bananas for a long time, not only is its frequency of occurrence not well understood (De Langhe, 1964; Simmonds, 1966) but also not many in-depth studies have been attempted. However, the phenomenon is believed to be a major source of variation within the crop, a process that generates genetic and phenotypic variation that farmers have continued to exploit through selection breeding.

A number of morphological characters are prone to mutations, including plant stature, bunch and fruit shapes, and astringency in fruit pulp. The implications of chimerism are not usually obvious, depending on which traits of the plant have been affected. In the Lujugira-Mutika subgroup, the phenomenon varies, and its implication for genetic variability and conservation on farm has not been elucidated. It is not known how chimeric traits are selected by a given community and subsequently conserved on farm in banana-based farming systems (Karamura et al., 2008). Yet the farmers' role in initiating and adjusting the numbers and proportions of cultivars on farm based on different cultivar traits is fundamentally important to the way landraces should be evaluated and maintained in *ex situ* and *in situ* collections.

All Lujugira-Mutika bananas are resistant to Fusarium wilt races 1 and 2. Conversely, with regard to black Sigatoka, banana weevils, and nematodes, there is variation in resistance even within the same clone set. Similarly within the same subgroup, variation with regard to female pollen fertility exists, with the 'Nfuuka' clone set being the most fertile and the 'Musakala' clone set the least fertile; 'Nakitembe' and 'Nakabululu' clone sets fall in between. All these variability options can and have been exploited for both selection and crossbreeding as well as for commercial purposes.

1.4.2.2 Selection

The process of selection produces patterns of variation in crop populations as well as patterns of individual variation that would be useful to breeders. The former is of primary importance in evolution and the latter in identification. The example of the East African Highland bananas (EAHB) again illustrates this. Since their introduction in the East African region, the EAHB have diversified in this region, not only because of mutations but also due to continuous natural and farmers' selection. The morphological variation pattern existing in these bananas has been as a result of the action of selective forces on the genotypes, and such variation would presumably have an evolutionary significance to ensure survival of the selected types and lead to ultimate improvement and spread of these bananas. Some variation patterns might have little or no evolutionary significance. Bunch and fruit characters experience great selection pressure, particularly the compactness and shape of the bunch and fingers, since these are very important to farmers and other consumers for whom compactness and shape of the bunch in addition to fruit size and shape are known commercial traits. A compact, medium, shaped bunch can easily be packed, transported to market, and sold quite easily in that respect. Cylindrical or very lax bunches or variegated and astringent fruits are not preferred by traders, although they have also arisen through mutations (Rubaihayo and Gold, 1993). This farmers' selection breeding has been based largely on continuously changing morphological characteristics attributed to chimerism, and this has been the basis for plant identification and nomenclature since cultivar names are mainly based on the phenotypic characteristics of cultivars. If, for example, a new mat produces suckers whose characteristics differ from those of the mother mat, farmers can either discard the "different" suckers, or maintain them as mutants of the mother plant or as a different cultivar to which they give a name depending on whether the differences between mother and daughter are major or minor. The major differences with acceptable qualities will readily be selected and adopted widely by farmers and subsequently by traders and consumers. Hence a daughter plant may receive a name different from the mother if a major and consistent difference occurs. At the farm level, the major phenotypic and nomenclature changes are always associated with the floral parts, probably because these are most used by the farmers. Minor changes that do not normally alter nomenclature include changes in the color of the pseudostem, petioles, and midribs. In this case, the daughter plant retains the name of the mother plant but receives a second part of the name to indicate the minor change. The identities of these cultivars become crucially important tools in selection breeding and conservation efforts because the value, potential, and limitations of each clone will very much be influenced by their characters and correct identity. In summary the structure and morphological variation pattern in a crop varies according to its breeding system and the amount, intensity, and direction of selection to which they have been subjected (Pickersgill, 1994).

1.5 ECOLOGICAL ADAPTATION

The variability within a clonal population is more phenotypic than genotypic and many times is caused by slight environmental differences within the habitat. Clonal material is ideal for the study of morphological and physiological effects of diverse environmental factors on plants from different areas or habitats. When breeders deal with clones, particularly with those coming out as new materials from their breeding programs, it is important that they are able to recognize differences in one cultivar from another. Distinction of one cultivar to another has increasingly gained emphasis in connection with legislation about breeders' rights in different countries. The requirement is further accentuated by traders and buyers who must, in many instances, exercise care in selecting the correct cultivar for their needs. In addition, different growing conditions impose different characteristics to essentially the same clone. For example, 'Sukali Ndizi' (AAB) from dry regions is smaller in both finger and bunch sizes than that from wetter regions but the consumer markets prefer the former to the latter. Furthermore, the farmer wants to get the right price for the particular cultivar he grows. Thus, primarily for economic reasons, identification based on morphology is most important at the cultivar level and is only possible when morphological variation allows it in various types of environments. Since breeders tend to look much more for resistance and yields, it is not yet fully understood whether the observed phenotypic variability in their new materials imparts resistance or tolerance to biotic and edaphic constraints that threaten the very existence of the crop. Morphological variability seen in a new cultivar may be a reflection of genetic elasticity, and its understanding will advance efforts to exploit its potential. It therefore follows that breeders need to focus on the interrelationships between morphological variability on one hand and the plant's ability to survive environmental constraints on the other. The following examples briefly explain such interrelationships between morphological variability, food texture, and distribution incidences.

The Lujugira-Mutika bananas have been divided into four morphologically distinct clone sets, and the clone sets bring together clones that share characteristics important to farmers and consumers, though these characteristics were not used in grouping the clone sets (Pickersgill and Karamura, 1999). The clones in 'Nakabululu' and 'Nakitembe' clone sets sucker profusely, mature quickly, produce soft-textured/flavored food, and are widely distributed but not rich in diversity. Most clones in 'Nfuuka' are slow to produce suckers, take a long time to mature, and produce hard-textured food, not evenly distributed but rich in diversity. However, one subcluster of 'Nfuuka' shares with 'Nakabululu' and 'Nakitembe' the characteristics of rapid maturity and production of some textured food. 'Nfuuka' is more heterogeneous than other clone sets and overlaps with most of them. Its name means "I am changing" and reflects the farmers' perception that somatic mutations are particularly frequent in these clones. The final clone set, 'Musakala,' contains the higher-yielding clones (e.g., 'Lumenyamagali,' which means "I break bicycles"), grown on a commercial scale to supply urban markets. Clones in this set tend to be intermediate with regard to time to maturity and in quality of their food.

In conclusion, morphological variation in clonal crops such as bananas is mediated through mutation (though there is an increasing number of a crossbreeding program), thought to be somatic in nature, and through a phenomenon called chimerism. Both chimeric and somatic mutation variants are then subjected to natural and farmer selection in response to the changing environment and/or economic conditions to widen the variation and exploit its potential.

REFERENCES

Allen, R.N., E.B. Dettman, G.G. Johns, and D.W. Turner. 1988. Estimation of leaf emergence rates of bananas. *Aust. J. Agric. Res.* 39:53–62.

Anojulu, C.C. 1992. The role of plantain in environment (Monograph) IITA, Plantain Programme.

Araya, M., A. Vargas, and A. Cheves. 1998. Changes in distribution of roots of banana (*Musa* AAA cv. Valery) with plant height, distance from the pseudostem, and soil depth. *J. Hort. Sci. Biotechnol.* 73(4):437–440.

Avilán, R.L., R. Meneses, and R.E. Sucre. 1982. Distribución radical del banano bajo diferentes sistema de manejo de suelos. *Fruits* 37:103–109.

Barker, W.G. and F.C. Steward. 1962. Growth and development of the banana plant. *Ann. Bot.* 26, 389–423.

Beugnon, M. and J. Champion. 1966. Etude sur les racines du bananier. *Fruits* 21:309–327.

Biruma, M., M. Pillay, L. Tripathi, G. Blomme, S. Abele, M. Mwangi, et al. 2007. Banana Xanthomonas wilt: A review of the disease, management strategies and future research directions. *Afr. J. Biotechnol.* 6:953–962.

Blomme, G. 2000. The interdependence of root and shoot development in banana (*Musa* spp.) under field conditions and the influence of different biophysical factors on this relationship. PhD diss., no. 421. K.U. Leuven. Faculteit Landbouwkundige en Toegepaste Biologische Wetenschappen, Belgium.

Blomme, G., R. Swennen, and A. Tenkouano. 2003. Assessment of genotypic variation in the root architecture of *Musa* spp. under field conditions. *InfoMusa* 12(1):24–29.

Blomme, G., D. Ocan, H. Mukasa, G. Sebuwufu, P.R. Rubaihayo, A. Tenkouano, and R. Swennen. 2004. Review of research on relationships between root traits, shoot traits and bunch weight. Paper presented at the first International Congress on *Musa*: Harnessing Research to Improve Livelihoods, July 6–9, 2004, Penang, Malaysia.

Blomme, G., G. Sebuwufu, H. Mukasa, D. Ocan, A. Tenkouano, and R. Swennen. 2004. Review of on-station and on-farm research on the root system in Nigeria and Uganda. Paper presented at the first International Congress on *Musa*, Harnessing Research to Improve Livelihoods, July 6–9, 2004, Penang, Malaysia.

Brun, W.A. 1960. Simultaneous and continuous recordings of photosynthesis and transpiration from the upper and lower surfaces of intact banana leaves. *Plant Physiol.* 35(Suppl), vii–ix.

Brun, W.A. 1961a. Photosynthesis and transpiration of banana leaves as affected by severing of the vascular system. *Plant Physiol.* 36:577–580.

Brun, W.A. 1961b. Photosynthesis and transpiration from upper and lower leaf surfaces of intact banana leaves. *Plant Physiol.* 36(4):399–405.

Brun, W.A.1962. Rhythmic stomatal opening responses in banana leaves. *Plant Physio.* 37. (suppl).

Brun, W.A. 1965. Rapid changes in transpiration in banana leaves. *Plant Physiol.* 40, 797–802.

Champion, J. 1961. Indications preliminaries sur la croissance du bananaier 'Poyo.' *Fruits* 16:191–194.

Champion, J. 1963. *Le Bananier*. Maisonneuve and Larose. Paris.

Champion, J. and P. Olivier. 1961. Etudes préliminaires sur les racines de bananier. *Fruits* 16(7):371–374.

Champion, J. and D. Sioussaram. 1970. L'enracinement du bananier dans les conditions de la station de Neufchâteau (Guadeloupe). *Fruits* 25:847–859.

Charlton, W.A. 1982. Distribution of lateral root primordia in root tips of *Musa acuminata* Colla. *Ann. Bot.* 49:509–520.

Cheesman, E.E. 1947. The classification of the bananas. *Kew Bull.* 2:97–117.

Cheesman, E.E.1948a. Classification of the bananas II. The genus *Musa* L. *Kew Bull.* 2:106–117.

Cheesman, E.E. 1948b. Classification of the bananas III. Critical notes on species. *Kew Bull.* (3):11–28; (2):145–157; (3):323–328.

Cheesman, E.E. 1949. The classification of the bananas. *Kew Bull.* 4:23–28, 133–137, 265–272, 445–452.

Cheesman, E.E. 1950. The classification of the bananas. *Kew Bull.* 5:27–31, 151–155.

Compton, J.A. and T.A.J. Henderson. 1997. A morphometric analysis of the *Cimicifuga foetida* L. complex (Ranunculacceae). *Bot. J. Linn. Soc.* 123:1–123.

Crane, M.B. and W.J.C. Lawrence. 1956. *The genetics of garden plants.* 4th ed. London: Macmillan.

Davis, G. 1995. Banana and plantain in the East African highlands. In: *Bananas and Plantains*. S. Gowen, ed., 493–508. London: Chapman & Hall.

De Langhe, E. 1961. La taxonomie du bananier plantain en Afrique équatoriale. *J. d'Agriculture Tropicale et de Botanique Appliquée*, 8, 419–449.

De Langhe, E.A. 1964. The origin of variation in the plantain banana. *Mededelingen van de Landbouwhogeschool* (State Agricultural University of Ghent, Belgium) 39:45–80.

De Langhe, E., R. Swennen, and G.F. Wilson. 1983. Aspects hormonaux du rejetonnage des bananiers plantains. *Fruits* 38(4):318–325.

Esau, K. 1965. *Plant anatomy.* New York: Wiley and Sons.

Fahn, A., S. Stoler, and T. Fist. 1963. Vegetative shoot apex in banana and zonal changes as it becomes reproductive. *Bot. Gaz.* 124:246–250.

Fawcett, W. 1913. *The banana: Its cultivation, distribution and commercial uses.* London: Duckworth.

FHIA. 1998. Programa de banano y plátano. Informe técnico 1998. Banana and plantain programme. 1998 technical report. La Lima (HND): FHIA.

Fogain, R., R. Achard, M. Kwa, P. Ferrier, and J. Sarah. 1996. Nematode control on banana plantations in Cameroon: Summary of ten years of research on the efficacy of nematicide compounds. *Fruits* 51(3):151–161. In French.

Fogain, R., E. Fouré, and C. Abadie. 1998. Root disease complex of bananas and plantains in Cameroon Seminario Internacional sobre Producción de Plátano, Armenia (COL), 1998. Armenia (COL): Corpoica.

Godefroy, J. 1969. Le développement des racines du bananier dans divers sols: Relation avec la fertilité. *Fruits* 24:101–104.

Gousseland, J. 1983. Etude de l'enracinement et de l'émission racinaire du bananier 'Giant Cavendish' (*Musa acuminata* AAA, sous-groupe Cavendish) dans les andosols de la Guadeloupe. *Fruits* 38:611–623.

Gowen, S. and P. Quénéhervé. 1990. Nematode parasites of banana, plantains and abaca. In: *Plant parasitic nematodes in subtropical and tropical agriculture*, M. Luc, R.A. Sikora, and J. Bridge, eds., 431–460. Wallingford, UK: CAB International.

Huxley, J. 1940. *The new systematic*. Oxford: Clarendon Press.

IITA. 1997. Project 7: Improving plantain- and banana-based systems. *IITA annual report 1997*. Ibadan, Nigeria: IITA.

IITA. 1998. *IITA medium term plan 1999–2001: Towards sustainable development in Africa*. Ibadan, Nigeria: IITA.

IITA. 1999. Project 7: Improving plantain- and banana-based systems. *IITA annual report 1998*. Ibadan, Nigeria: IITA.

Irizarry, H., J. Vicente-Chandler, and S. Silva. 1981. Root distribution of plantains growing on five soil types. *J. Agric.* (University of Puerto Rico) 65(1):29–34.

Karamura, D. 1999. Numerical taxonomic studies of the East African highland bananas (*Musa* AAA-EA) in Uganda. INIBAP, France.

Karamura, D.A., B. Mgenzi, E.B. Karamura, and S. Sharrock. 2004. Exploiting indigenous knowledge for the management and maintenance of *Musa* biodiversity on farm. *Afr. Crop Sci. J.* 12(1):67–74.

Karamura, D., E. Karamura, W. Tushemereirwe, P.R. Rubaihayo, and R. Markham. 2008. Chimersim and its implications for conservation strategies of East African highland bananas: Proceedings of the International Conference on Banana and Plantain for Africa. *Acta Hort.* Leuven: ISHS

Karamura, E.B. and D.A. Karamura. 1995. Banana morphology. Part 2: The aerial shoot. In: *Bananas and plantains*, S.R. Gowen, ed., 190–205. London: Chapman and Hall.

Lassoudière, A. 1971. La croissance des racines du bananier. *Fruits* 26:501–512.

Lassoudière, A. 1977. Croissance et développement du bananier 'Poyo' en Côte d'Ivoire, thesis, L'Université Nationale de Côte d'Ivoire, Abidjan.

Lassoudière, A. 1978. Quelques aspects de la croissance et du développement du bananier 'Poyo' en Côte d'Ivoire: Le système radical. *Fruits* 33:314–338.

Lassoudière, A. 1980. Matière végétale élaborée par le bananier Poyo depuis la plantation jusqu'à la récolte du deuxième cycle. *Fruits* 35:405–446.

Lavigne, C. 1987. Contribution à l'étude du système racinaire du bananier. Mise au point de rhizotrons et premiers résultats. *Fruits* 42:265–271.

Laville, E. 1964. Etude de la mycoflore des racines du bananier 'Poyo.' *Fruits* 19:435–449.

Moreau, B. and J. Le Bourdelles. 1963. Etude du système racinaire du bananier 'Gros Michel' en Equateur. *Fruits* 18:71–74.

Nyine, M. and M. Pillay. 2010. The effect of banana breeding on the diversity of East African Highland banana. *Acta Hort.* (in press).

Perrier, X. 1993. Numerical analysis of genetic diversity in banana. In: *Breeding banana and plantain for resistance to diseases and pests*. J. Ganry, ed., 23–34. CIRAD/INIBAP, Montpellier, France.

Pickersgill, B. 1988. The needs of plant breeding, agriculture and horticulture for infraspecific classifications. In: *Prospects in systematics*, D.L. Hawksworth, ed., 363–376. Oxford: Clarendon Press.

Pickersgill, B. 1994, From descriptors to DNA: New tools and new tasks in the evaluation of genetic resources. In: *Evaluation and exploitation of genetic resources pre-breeding. Proceedings of the genetic resources section meeting of Eucarpia*. F. Balfourier and M.R. Perretant, eds., Clermont-Ferrand, France.

Pickersgill, B. and D. Karamura. 1999. Issues and options in the classification of cultivated bananas, with particular reference to the East African Highland bananas. In: *Taxonomy of cultivated plants: Third international symposium*, S. Andrews, A.C. Leslie and C. Alexander, eds., 159–167. Kew: Royal Botanic Gardens.

Price, N.S. 1995. Banana morphology. Part 1: Roots and rhizomes. In: *Bananas and plantains*, S.R. Gowen, ed., 179–205. London: Chapman & Hall.

Purseglove, J.W. 1972. *Tropical crops: Monocotyledons*. Vol. 2. London: Longmans.

Ram, Y M.M. and F.C. Steward. 1962. Growth and development of the banana plant. *Ann. Bot.* 25:657–673.

Riopel, J. 1966. The distribution of lateral roots in *Musa acuminata* 'Gros Michel.' *Am. J. Bot.* 53:403–407.

Riopel, J. and T.A. Steeves. 1964. Studies on the roots of *Musa acuminata* cv. Gros Michel 1: The anatomy and development of main roots. *Ann. Bot.* 28:475–494.

Robin, J. and J. Champion. 1962. Etudes des émissions des racines de la variété du bananier Poyo. *Fruits* 17:93–94.

Rowe, P.R. 1991. Genetic progress in banana and plantain. *Augura—Revista* (Colombia) 17(1):79–83.

Rubaihayo, P.R. and C.S. Gold. 1993. Rapid rural appraisal of highland banana production in Uganda. *InfoMusa* 2:15–16.

Sebuwufu, G., P.R. Rubaihayo, and G. Blomme. 2004a. Variability in the root system of East African banana genotypes. In: *Conservation through utilization of banana and plantain in the Great Lakes region of East Africa*, E. Karamura, D.A. Karamura, S. Sharrock and E. Frison, eds., 85–93. *Afr. Crop Sci. J.* (special issue)12(1):85–93.

Sebuwufu, G., P.R. Rubaihayo, and G. Blomme. 2004b. Genotypic variability in root traits and shoot–root ratio of *Musa* spp. Implications to the improvement of the root system. Paper presented at the first International Congress on *Musa*: Harnessing Research to Improve Livelihoods, July 6–9, 2004, Penang, Malaysia.

Simmonds, N.W. 1962. *The evolution of the bananas*. London: Longmans.

Simmonds, N.W.1966. *Bananas*. 2nd ed. London: Longmans.

Simmonds, N.W. and K. Shepherd. 1955. The taxonomy and origins of the cultivated bananas. *Linn. Soc. Bot.* 55:302–312.

Skutch, A.F. 1930. On the development and morphology of the leaf of the banana (*M. sapientum L.*). *Am. J. Botany*, 17, 252–271.

Stace, Clive A. 1980. *Plant taxonomy and biosystematics*. London: Edward Arnold.

Stover, R.H. 1972. *Banana, plantain and abaca diseases*. CMI, Kew.

Stover, R.H. 1988. Variation and cultivar nomenclature in the *Musa* AAA group, Cavendish subgroup. *Fruits* 43:353–357.

Stover, R.H. and N.W. Simmonds. 1987. *Bananas*. London: Longmans.

Summerville, W.A.T. 1939. Root distribution of the banana. *Queensland J. Agric. Sci.* 52:376–392.

Summerville, W.A.T. 1944. Studies on nutrition as qualified by development in *Musa cavendishii* Lambert. *Queensland J. Agric. Sci.* 1:1–127.

Swennen, R. 1990. *Plantain cultivation under West African conditions: A reference manual.* Ibadan, Nigeria: International Institute of Tropical Agriculture.

Swennen, R. and D. Vuylsteke. 1988. Bananas in Africa: Aspects of diversity and prospects for improvement. *Proceedings of IITA/CNR/IBPGR/UNEP workshop on crop genetic resources in Africa*, 17–20. Ibadan, Nigeria: IITA.

Swennen, R., G.F. Wilson, and E. De Langhe. 1984. Preliminary investigation of the effects of gibberellic acid (GA3) on sucker development in plantain (*Musa* cv. AAB) under field conditions. *Trop. Agric.* (Trinidad) 61(4):253–256.

Swennen, R., E. De Langhe, J. Janssen, and D. Decoene. 1986. Study of the root development of some *Musa* cultivars in hydroponics. *Fruits* 41(9):515–524.

Swennen, R., G.F. Wilson, and D. Decoene. 1988. Priorities for future research on the root system and corm in plantains and bananas in relation with nematodes and the banana weevil. In: *Nematodes and the borer weevil in bananas: Present status of research and outlook.* Proceedings of a workshop, Bujumbura, Burundi, 7–11 December 1987. Montpellier, France: INIBAP.

Taylor, S.E. 1969. Tattered banana leaves. *Mol. Bot. Garden Bull.* 57:14–17.

Taylor, S.E. and O.J. Sexton. 1972. Some implications of leaf tearing in Musaceae. *Ecology*, 53, 143–49

Turner, D.W. 1970. Banana roots. *Agric. Gaz. N.S.W.* 81:472–473.

Turner, D.W. 1972. Banana plant growth. 1. Gross morphology. 2. Dry matter production, leaf area and growth analysis. *Aust. J. Expt. Agric. Anim. Husb.*12:209–215 and 216–224.

Walmsley, D. and I.T. Twyford. 1968 The zone of nutrient uptake by the Robusta banana. Trop. Agric. Trinidad, 45, 113–119.

2 Evolution and Genetic Relationships in Banana and Plantains

Uma Subbaraya, Marimuthu S. Saraswathi, and Michael Pillay

CONTENTS

2.1 INTRODUCTION

Banana (*Musa* spp.) is the world's most important fresh fruit commodity in terms of volume of trade. The global banana industry started in the late 1800s as a result of technological advances like refrigerated shipping and is now more than a century old. Compared to any other agricultural product, banana exhibits colonial economic nationalism and contemporary neoliberal stages of growth and evolution in the world economy (Wiley, 2008). Banana has grown in its popularity for its versatility and adaptation to many agro-climatic conditions, resilience to climatic changes, nonspecific seasonal fruit production, year-round availability of fruits, high productivity per unit area (40–60 tons of fruits per hectare annually), and finally, if well managed, the plantation remains productive for 10–20 years. The spectrum of cultivar diversity with regards to maturity allows staggered but year-round availability of fruits. The popularity of the fruit led to its adoption and cultivation in more than 150 countries.

Banana is known by different names in the various countries of the world. These include: banana (Japanese, Italian, Portuguese, Serbo-Croat), *banane* (French, German), *banaan* (Dutch), *banan* (Danish), *banaani* (Finnish), *banan* (Russian/Polish), *banbán* (Hungarian), banana/*mpanána* (Greek), banana (Hebrew), *banema* (Guinea), *choui* (Vietnam), *Chiao* (Chinese), *futo* (New Caledonia), *futi* (West Polynesia), *hnget-pyaw* (Burmese), *ikindu/kitoke* (East African), *Kela* (India), *klue/klui* (Thai), *mauz* (Turkish), *maúz* (Arabic/Persian), *maso/ndizi* (Swahili), *pisang* (Malay/Indonesian), *plátano*/banana (Spanish), *saging* (Philippines), *usi* (New Guinea), *uch/ut* (Micronesian), and *vudi* (Fiji).

2.2 GROWTH OF GLOBAL BANANA INDUSTRY

The global area under banana and plantain cultivation is estimated to be 10.2 million hectares (ha) (FAO STAT, 2008) with the greatest contribution coming from Africa (5.8 million ha) followed by Asia and Pacific (2.04 million ha) and Latin America (1.31 million ha). During the last 5 years, the area under cultivation has increased by 6.87% and 0.68% for banana and plantain, respectively, with African continents (16.5%) followed by Asia (6.5%) contributing mainly to the increase. The global production is around 125.04 metric tons (mt), with bananas contributing 96.70 mt and plantains 34.30 mt (FAO STAT, 2009). During the last 5 years, banana has witnessed an increase in production by 45%, with Asia contributing 38% of the production. India is the largest banana producer followed by China, the Philippines, Brazil, Ecuador, and others. Although the global productivity of banana has increased from 15.6 to 18.8 tons/ha, that of plantain has increased marginally from 6.1 to 6.5 tons/ha. Asia, despite its minimum contribution to the increase in global area under banana and plantain cultivation, has witnessed a giant leap: a 42.3% increase in productivity. Costa Rica records the highest productivity of 52.54 tons/ha followed by India (35.87 tons/ha), Mexico, Colombia, and Ecuador. The major exporting countries are Ecuador, the Philippines, China, Costa Rica, Colombia, Panama, and Honduras, with the export trade worth about US$8.6 billion annually (FAO STAT, 2008).

2.3 CONSUMPTION AND UTILIZATION OF BANANA

Bananas are an important source of carbohydrate and dietary fiber. They contain high amounts of essential vitamins A_1, B_1, B_2, and C; minerals such as potassium (Chandler, 1995); and substantial quantities of starch and hemicelluloses (Ketiku, 1973; Mota et al., 2000). Plantains have a greater amount of vitamins A and C (20 mg/100 g) than bananas. Plantains are revered as a food equivalent across many African countries, Latin America, the Caribbean, and the Polynesian islands. Although the average global per capita consumption of banana and plantain is reported as 5.2 kg/person (Nayar, 2010), it is as high as 239 kg/person in Uganda, 223 in Burundi, 180 in Rwanda, 141 in Gabon, and 131 in Samoa (FAO STAT, 2008), increasing marginally over the past few years. It is worth noting that statistics provided by the FAO do not distinguish between plantain and banana.

2.3.1 Importance of Banana as an Edible Fruit

Bananas and plantains are the only group of fruits that also constitute a staple food for millions. They are consumed fresh, cooked, steamed, roasted, and brewed (Pillay et al., 2002). Besides the fruits, the flower buds and inner core of the pseudostem are also used as vegetables as well as for a wide range of therapeutic uses. Bananas are also processed into puree, juice, fig, jams, canned banana slices (Thompson, 1995), and beer and wine in Africa (Olaoye et al., 2006) (see Chapter 14).

Bananas are regarded as nature's secret of perpetual youth in the practice of traditional medicine in India, China, and ancient Persia (www.articlesbase.com, accessed 19 April 2010). The low lipid and high energy values make it ideal for obese and geriatric patients (Gasster, 1963). Bananas are useful for persons with peptic ulcer, infant diarrhea, celiac disease, and colitis (Seelig, 1969). Being low in salt and high in potassium chloride, it is a recommended dietary supplement to lower blood pressure (www.herbalextractsplus.com/banana.cfm, accessed 19 April 2010). Bananas can stimulate the production of hemoglobin in the blood and are useful for anemic patients. Bananas are rich in tryptophan, a type of protein that the body converts into serotonin, which keeps the mind in a relaxed state. Fruits are regarded as coolants, which lower both the physical and emotional temperature of expectant mothers. Bananas are rich in fiber and pectins, which absorb water and restore bowel movement. Bananas contain benign amino acids that are useful for the removal of stones in the kidney and gall bladder. Similar functions are also reported for the central core of the pseudostem, which is either consumed as a salad or cooked with pulses. Bananas are rich in lectins, which are sugar-binding proteins that can identify foreign invaders like viruses, attach themselves to pathogens and block their entry into the human body. The roots of banana are considered to have an antihelmintic effect.

2.3.2 Banana as an Alternative Source of Income

Banana leaves are popular in South India and Africa as hygienic dining plates and wrapping material. Leaf production alone is also an income source for small-scale farmers. The underground rhizome is exploited as animal feed in a composite mixture. The banana pseudostem either alone or in combination with rice straw has proved to be a good substrate for mushroom cultivation. The plant sap is also used as an indelible ink in industry.

Banana fiber is being used in the treatment of industrial and municipal wastes. Banana fibers also find use in the pulp industry because of their optimum burst, tear, and tensile indices (Uma et al., 2002). The fibers are used as a base material in cottage industries for making handicrafts and for making a wide range of goods, including bags, rope cordage, yarns, abrasive backing paper, tea bags, shoes, and so forth. Tear and tensile strength of banana fiber derived from plants like *M. textilis* make it amenable for printing of Japanese yens, and it is also blended with cotton in various ratios for use in the textile industry.

2.3.3 Indigenous Technical Knowledge (ITK)

Inhalation of the smoke from burning banana leaves is used to relieve wheezing in asthmatic patients. Tender leaves smeared with oil are used for dressing of wounds and blistered skin surfaces. Flower buds boiled with salt and oil are consumed by northeast Indian tribes to cure joint pains and promote blood circulation. When cooked with pulses or with coconut, flower buds are considered to be a remedy for heart ailments and for dissolving kidney stones. Banana pulp is added to cereal beer made of rice, sorghum, and so on to aid in fermentation and providing a fruity flavor. Dried and powdered pulp of wild cultivars such as 'Bhimkol' and 'Athiakol' are used as infant food as they are easily digestible. Both ripe and unripe fruits of *M. nagensium* are cooked with sorghum and other cereals and fed to pigs to improve their health. Rhizomes of all wild members of the sections *Eumusa* and *Rhodochlamys* are chopped into pieces and cooked with pulses as animal feed (Uma, 2006).

2.4 TAXONOMY

Botanically banana is classified in the genus *Musa*, which—together with two other genera, *Musella* and *Ensete*—are placed in the family Musaceae and order Zingiberales (Table 2.1). Members of this family are large herbs 2–9 m tall with an aerial trunk consisting of compacted leaf sheaths that grow directly from the top of the corm (Purseglove, 1976). The earliest description of dessert banana *Musa sapientum* and 'Silk' banana *M. paradisiaca* was provided by Linnaeus in 1753. Other monospecific nomenclature such as *M. sapientum* ssp. *paradisiaca* by Baker (1893), *M. paradisiaca* ssp. *sapientum* by Schumann (1900), and *M. sapidisiaca* by Jacob (1952) were not accepted by taxonomists. The earliest taxonomic references were made by Baker (1893), while Schumann (1900) mentioned 42 species in his monograph, of which 10 belonged to the genus *Physocaulis*, 20 to *Eumusa*, and 12 to *Rhodochlamys*. Sagot (1887) and Baker (1893) distinguished three subgenera in the genus *Musa*: *Physocaulis*, *Eumusa*, and *Rhodochlamys*. This classification was revised by Cheesman (1947), whose detailed taxonomic treatment was based on chromosome number, pseudostem stature, inflorescence characters, and seed morphology. Cheesman divided the genus *Musa* into four sections: *Eumusa* ($x = 11$), *Rhodochlamys* ($x = 11$), *Australimusa* ($x = 10$), and *Callimusa* ($x = 10$). This classification has been widely accepted, despite subsequent reports of basic chromosome numbers $n = 7$ for *M. ingens* and $n = 9$ for *M. beccari* (Shepherd, 1999). Recently the section *Ingentimusa* was added (Argent, 1976; Simmonds and Weatherup, 1990). In fact, several unresolved issues remain, including the evolutionary relationships among the four sections of the genus *Musa*, the separation of sections *Eumusa* and *Rhodochlamys*, which has apparently little taxonomic support, and the relationship between the wild progenitors and the cultivated clones (Simmonds, 1953, 1960; Shepherd, 1999; Ude et al., 2002a; Wong et al., 2003). To date, more than 36 species and subspecies have been identified and described (Daniells et al., 2001), and more recently 73 species were described by Häkkinen and Väre (2008), warranting a thorough revision of the genus *Musa* and the development of a classification key.

2.5 ORIGIN AND HISTORY

Banana with its versatile utilities has been recognized as one of the earliest crops to be domesticated by man. As a wild-seeded plant, banana must have been first recognized for its non-edible purposes like fiber, roofing, and ropes for navigation purposes. The earliest documentary evidence of banana is found in the Vedic period (approx. 1700 BCE), during which use of fruits like mangoes, gooseberries, and bananas was mentioned in the *Rig-Veda*, one of the four volumes of Veda written by the Indian sages. Further evidence is from the Indian epics, *Mahabharata* and *Ramayana*, which date to approximately 1400 BCE. Aranya Kanda and the forest trek of Valmiki's *Ramayana* state that members of royal heritage were draped in clothes woven of banana fiber.

Suggestions of banana fruits in the contemporary diet in *Ramayana* perhaps indicate the domesticated status of banana in the Indian subcontinent. Early epics of the Pali Buddhist monks in 500–600 BCE also describe bananas "as big as an elephant tusk." This could probably refer to plantains (Horn plantain) from the southern peninsula of greater India (including Sri Lanka), where plantain cultivation is in vogue today. It is taken for granted that banana and plantain originated in Southeast Asia and the Pacific and spread to the rest of the world. However, recent linguistic, anthropological, and archeological studies have contributed to an alternative migratory theory (Nayar, 2010).

Alexander's invasion in 327 BCE popularized this "holy fruit" in contemporary writings. This was followed by Megashthanes (350–290 BCE), the Greek traveler and geographer, in his travelogue, *Indica*, by Theophrastus (371–287 BCE), and by Pliny (23–79 CE), a Roman naturalist, in his book *Historia Naturalis*. Charaka (fourth century) from India has documented the various medicinal uses of bananas in his medical compilation, *Susrutha Samhita*. Reports on early cultivation of banana in human civilization were obtained from Buddhist sculptures of central India, *stupas* in Sanchi, carvings in Nalanda, and paintings in Ajantha and Ellora caves. Archeologists have also

TABLE 2.1
Conspectus of the Family Musaceae, Classification, Distribution, and Uses of *Musa* Species

Genus	Chromosome No. (2*n*)	Section	Distribution	Species	Uses and Utilization
Ensete	9	–	Asia (India, China, Thailand, Philippines), Africa	*E. superbum* (Roxb.) Cheesman, *E. glaucum* (Roxb.) Cheesman *E. ventricosum* (Welw.) Cheesman, *E. livingstoniana*, *E. proboscidea*, *E. gilletti* (De Wild) Cheesman, *E. buchanani*, *E. homblei* (Bequaert) Cheesman, *E. perrieri* (Claverie) Cheesman	Fiber, vegetable, ornamental
Musa	10	*Australimusa*	Queensland, Polynesia, Philippines, Australia	*M. lolodensis* Cheesman, *M. peekelii* Lauterb, *M. maclayi* von Muell. Ex Mikl-Maclay, *M. jackeyi* W. Hill, *M. bukensis* Argent, *M. textilis* Née, *M. Fehi*	Fiber, dessert fruit
		Callimusa	Indo-China, Malaysia, Philippines, Indonesia	*M. coccinea* Andrews, *M. violascens* Ridely, *M. gracilis* Holttum, *M. beccarii* Simmonds, *M. erecta*, *M. borneënsis* Beccari, *M. coccinea* Andrews, *M. exotica* R. Valmayor, sp. now *M. flavida*, *M. hotta*, *M. gracilis* Holttum, *M. salaccensis* Zoll	Ornamental
	11	*Eumusa*	All banana growing countries and continents	*M. acuminata* Colla, *M. balbisiana* Colla, *M. schizocarpa* Simmonds, *M. basjoo* Sieb, *M. flaviflora* Cheesman, *M. itinerans* Cheesman, *M. flavicarpa*, *M. sikkimensis* Kurz, *M. cheesmani* Simmonds, *M. nagensium* Prain, *M. halabanensis* Meijer, *M. ochracea* Shepherd	Food, fruit, vegetable, fiber, and medicinal applications
	11	*Rhodochlamys*	India, Malaysia, Indo-China, Myanmar, Cambodia, Vietnam, Philippines, Thailand	*M. ornata* Roxb., *M. rosea* Baker, *M. rubra* Wall. Ex Kurz, *M. mannii* H. Wendl. Ex Baker, *M. velutina* H. Wendl. and Drude, *M. laterita* Cheesman, *M. sanguinea* Hook. f., *M. aurantiaca* Mann ex Baker, *M. rosacea* Jacq.	Ornamental

found the banana depicted in ancient ruins such as the Buddhist temple of Bharhut, dating from the second century BCE, and the Javanese monument to Buddha erected in Borobodur in the year 850 CE. Yang Fu, a Chinese official in the T'ang dynasty (618–907), wrote an *Encyclopedia of Rare Things* wherein he describes the banana plant, which is possibly the first mention of the banana in Chinese texts.

Similarly plantains have a long history of domestication in Africa and their entry to the continent is reported to be about 1500–2000 years ago (Bruce, 1790; Purseglove, 1976; De Langhe et al., 1995; Doutrelepont et al.,1996; Mbida et al., 2000). Phytoliths of *Musa* and *Ensete* unearthed at Nkang in Cameroon are the first archeological indication of a cultivated crop, dating back 3000 years in Central Africa and providing a clear indication of an early food economy in Africa (Mbida et al., 2000).

Banana and plantain is a complex crop with vast diversity, utility, and spread, and addressing their origin as any other simple crop would be an underestimation. After their simultaneous and independent evolution across Asia, Polynesia, and Africa, metamorphosis of the earliest wild banana, a weedy, seeded, nonedible plant into a domesticated, parthenocarpic (nonseeded), edible tasty fruit occurred in a long evolutionary journey. Archeobotanical and paleoecological studies combined with radiocarbon dating and stratographic analysis clearly suggest the existence of banana as early as 6950 to 6440 BCE. Domestication and extensive cultivation was evidenced by increased phytolith counts in New Guinea. This dating was prior to the influence of Polynesians, suggesting that bananas and plantains indeed had their independent origin and evolution in the Pacific islands (Denham et al., 2003, 2004; Lentfer, 2003).

2.5.1 Origin of Diploid Bananas

Wild bananas are all diploids and the ancestry of cultivated bananas originates from two major species, *M. acuminata* and *M. balbisiana*, with occasional mention of the involvement of *M. textilis* and *M. schizocarpa*. Simmonds (1962a) proposed the early domestication of diploid banana—that is, *M. acuminata* in its primary center of origin, namely Malaysia, which has the greatest variability. A partial occurrence of parthenocarpy in *M. acuminata* ssp. *banksii* (Simmonds, 1962a, 1962b) was suggested, which contributed for the development of parthenocarpy. This was later confirmed by restriction fragment length polymorphism (RFLP) studies (Carreel, 1994). The development of parthenocarpy was also reported in *M. acuminata* ssp. *burmannica* distributed in Western Ghats of India and the cultivar 'Matti' using morpho-molecular characterization (Uma et al., 2005a). 'Matti' is considered to have evolved by continuous human selection for complete parthenocarpy similar to 'Pisang Rejang,' the possible seedy progenitor of 'Pisang Lilin' (Nasution, 1991).

Unlike *M. acuminata*, development of parthenocarpy in *M. balbisiana* has not been investigated in detail. Occurrence of parthenocarpic diploid and triploid *M. balbisiana* (BB and BBB) have been reported from the Philippines and Thailand by several workers, but the true genomic constitution of these plants still needs confirmation. Valmayor et al. (2002) reported the occurrence of parthenocarpy in *M. balbisiana* in the natural hybrids like 'Abuhon' of the Philippines. This is assumed to be a cross between seeded forms of *M. balbisiana* ('Pacol') with soft and scanty seeded 'Chuoi Hot.' Similarly, 'Saba,' 'Pisang Kepok,' 'Pisang Nipah,' 'Klui Lep Chung Kut,' 'Chuoi Ngu,' and 'Chuoi Mat' have been recognized as parthenocarpic *M. balbisiana* clones originating through chromosome restitution and repeated back-crossing with their parents (Brewbaker and Gorrez, 1956; Danh et al., 1998). Other triploid *M. balbisiana* (BBB) clones reported from the Philippines are 'Saba,' 'Cardaba,' 'Gubao,' 'Bigihan,' 'Pa-A Dlaga,' 'Saba Sa Hapon,' 'Pondol,' 'Turankong,' 'Sabang Puti,' 'Mundao,' 'Dali-an,' 'Kalimpos,' 'Binendito,' and so forth. During a visit by the senior author to the Field gene bank at Phakchong Banana Research station, Thailand, in 2002, some accessions reported to be BBB in the gene bank were found to exhibit the traits of *M. textilis*. A comprehensive study using morpho- and molecular approaches is warranted to decipher the actual genomic status of these plants. But edibility in terms of emptiness of seeds, softened seed coats, and pulpiness of fruits over its mucilaginous pulp has been observed in 'Bhimkol,' a widespread landrace of northeastern India. Human selection for pulpy fruits and fewer seeds is still practiced in BB landraces.

In general, parthenocarpy in banana is genetically controlled by three complementary genes with modifiers (Dodds, 1943a, 1945; Dodds and Simmonds, 1946a, 1946b, 1948). Sterility is also genetically controlled, supplemented by various forms of chromosomal aberrations and abnormal meiosis, all contributing to edibility in banana. Vegetative propagation favored the perpetuation of edibility through genetically controlled parthenocarpy and sterility in banana. Evolution for parthenocarpy could be expected through landraces like 'Bhimkol' over a period. 'Bhimkol' derived hybrids are available at the National Research Center for Banana (NRCB) and are being screened for parthenocarpy and hardiness.

Edibility in banana occurred over several years with acquisition of parthenocarpy complemented with the development of sterility. Human interventions for selection of edible diploids and nature's intervention to induce sterile mutants resulted in a vast diversity of AA diploids. Another major milestone was the development of edibility in terms of sweetness. *Musa acuminata* ssp. *banksii* derived diploids have starchier fruits, while sweet-pulped cultivars are suggested to be closer to *M. acuminata* ssp. *malaccensis* (Carreel et al., 1994). Similarly, clones from the different subspecies differed by translocation events in their chromosomes (Shepherd, 1999). Natural introgression between clones of the different subspecies and human selection for better edible diploids was a vital step in banana evolution.

2.5.2 Origin of Polyploid Bananas (Triploids and Tetraploids)

Development of polyploid bananas, more specifically triploids, followed the events leading to the origin of cultivated diploids. In nature, positive selection occurred in favor of heterozygotes that were nearly or completely seed-sterile (Simmonds, 1962, 1966). Vegetative cultivation and a non-preference for selection for seed-fertile clones over a long period must have led to accumulation of structural changes in chromosomes (Simmonds, 1962a). But Faure et al. (1993) indicated that sterility in diploid banana is not due to structural hybridity but is genetically controlled. Female restitution leading to the production of diploid (unreduced) gametes and hybridization within subspecies of *M. acuminata* ssp. or between different banana species led to the production of triploids. It is reported that production of unreduced diploid gametes is a frequent occurrence in diploid bananas in both the male and female sides (Dodds and Simmonds, 1946a).

By virtue of their superior traits like parthenocarpy, edible pulp, good bunch, and sturdiness, triploids were preferred for selection and clonal perpetuation by early man. Naturally occurring improved edible diploids of a particular subspecies, retaining fertility to some extent, hybridized with other subspecies, leading to the development of initial *M. acuminata* triploids. Raboin et al. (2005) used RFLP markers to trace $2n$ restitution and n gamete donors of Cavendish and Gros Michel. They were shown to have a common diploid ancestor from Madagascar and Comoro.

2.5.3 Development of Bispecific Polyploids

Musa balbisiana is the other wild *Eumusa* species that had broad distribution across South and Southeast Asia. Although no subspecies have been recognized within the species, recent studies have revealed sufficient variability within *M. balbisiana* (Sotto and Rabara, 2000; Ude et al., 2002b; Uma et al., 2005; Ge et al., 2005). *Musa balbisiana* has adopted to drier regions but exhibits copious growth in tropical and subtropical forests of the Philippines, Thailand, New Guinea, India, south China, Malaysia, Myanmar, Indonesia, and a few other regions (Valmayor et al., 2002; Uma et al., 2005a). Interspecific crosses between *M. acuminata* diploids and *M. balbisiana* occurred either in their primary origin or on their journey to new locations, resulting in different genomic combinations: AB, AAB, ABB, and ABBB, of which AAB and ABB encompassed a greater diversity.

The AAA genomic group forms the major allopolyploids valued for its commercial status in the global banana industry. It has various subgroups like Cavendish, Gros Michel, Red, Ibota, Lujugira-Mutika, and a few unique accessions (Uma and Sathiamoorthy, 2002).

The AB diploids evolved in a specific locality of Southeast Asia, especially southern India and Sri Lanka (Uma and Sathiamoorthy, 2002), from where it spread to eastern Africa in the recent past (Shepherd, 1999; De Langhe, 1996; De Langhe and Maret, 1999; Nayar, 2010). The main AB subgroups in India include 'Ney Poovan' and 'Kunnan,' while Daniells et al. (2001) recognized two AB groups as 'Ney Poovan' and 'Kamaramasenge.' No studies have been carried out to confirm the genetic relationships between 'Kamaramasenge' and 'Kunnan' subgroups. Molecular studies using SSR (simple sequence repeat) markers confirmed the occurrence of only two subgroups in the AB genome, 'Kunnan' and 'Ney Poovan' in India (Uma et al., 2010). The AB genotypes specific to South India—'Kunnan' and 'Safed Velchi'—did not show any evidence of having a *M. balbisiana* cytoplasmic genome (Carreel et al., 2002). It is possible that there are *M. balbisiana* types that are unique to India. India has a wide diversity of *M. balbisiana* clones (Uma et al., 2005b, 2006). Consideration of *M. balbisiana* from India together with samples from other regions is expected to provide a holistic picture on the evolution of this species.

2.5.4 AAB Genomic Group

Hybridization between *M. acuminata* and *M. balbisiana* led to a broader diversity of genotypes now classified as Plantain, Silk, Mysore, Pome, and other unique members with AAB genomes. The plantains are considered to have evolved from *M. acuminata* ssp. *banksii* as female parent and *M. balbisiana* as male parent (Carreel et al., 2002). Plantains seem to have evolved in New Guinea where a greater diversity for plantains and cultivated diploids is still present. There is some evidence to show that plantains also originated in the Philippines due to the natural availability of French plantains in Luzon (De Langhe and Valmayor, 1980; Valmayor et al., 2002). Although De Langhe (1996) suggested that southern India could be a possible place of origin for plantains, it is now accepted that the area is perhaps only a secondary center of diversification. Southern India has only French plantains and very little diversity has been observed in this group. Therefore, the epics suggesting the presence of bananas "as big as an elephant tusk" in 500–600 BCE mentioned at the beginning of this chapter raise doubts about the early occurrence of Horn plantains in India.

The Silk group of bananas is popular in South and South Asia, East Africa, Brazil, and other Latin American countries. Studies using RFLP (Carreel et al., 2002) reveal a paternal lineage from *M. acuminata* ssp. *malaccensis* for this group, unlike the other AAB subgroups. Subspecies *malaccensis* is naturally distributed in Malaysia, Indonesia, and Thailand. It appears that diploids evolved from this group hybridized with *M. balbisiana* to give rise to Silk clones (AAB).

The Mysore group (AAB) is a collection of hardy bananas that spread across almost the whole world. Chloroplast DNA studies shows that it paternally evolved from *Musa acuminata* ssp. *errans* and maternally from a *M. acuminata* ssp. with type II cpDNA (Carreel et al., 2002). It also has an AB diploid and diploid *M. balbisiana* in its ancestry. Subspecies *errans* is mainly found in the Philippines. But the natural occurrence of *M. swarnaphalya*, which morphologically resembles *M. acuminata* ssp. *banksii/errans* in northeast India, could suggest another center of evolution for *M. acuminata* ssp. *banksii* and ssp. *errans*–derived cultivated bananas. 'Pisang Kelat,' a unique AAB member reported to have a B genome of maternal origin and classified as BAA, is important from a pathological perspective since it is immune to most of the leaf spot diseases and Fusarium wilt (*Annual Report*, 1999).

The Maia Maoli/Popoulu (AAB) group is endemic to Oceania and specifically of Polynesian origin. The group is recognized for its unusually fat and squat fruits with blunt tips, with the Maoli group exhibiting longer fruits. Great diversity for fruit shape and size exists in Hawaii, west Polynesia, and Pohnpei. Popoulu fruits are very characteristic due to their tendency to split at ripeness (Ploetz et al., 2007).

The Iholena subgroup, distinguished by a coppery undersurface of leaves, salmon-colored flesh, and right-angled fruits, is traditionally grown in the highlands of Hawaii. They are consumed raw or cooked. Hawaiian diversity is represented by seven Iholena varieties. Iholena fruits are arranged loosely and at right angles to the bunch. Male flowers have characteristic lavender-colored stamens. The fruits remain pale yellowish-green throughout their development. The group is also sparsely represented in New Guinea, Samoa, French Polynesia, Tonga, and others (Ploetz et al., 2007). The uniqueness of the group and its diversity is attributed to a series of additional somatic mutations in basic hybrids and cultivars in niche locations of Polynesia (De Langhe and De Maret, 1999). Their genetic closeness to plantains is suggested by Carreel et al. (1994), with *M. acuminata* ssp. *banksii* contributing the AA genomes in both cases (Horry and Jay, 1988; Lebot et al., 1993; Carreel et al., 2002).

2.5.5 ABB Groups

All the ABB clones in various subgroups of 'Bluggoe,' 'Pelipita,' and 'Ney Mannan' have *M. balbisiana* in their lineage (Carreel et al., 2002) and may be referred to as true ABB while 'Pisang Awak' and 'Peyan,' the dessert cultivars of ABB, have a cytoplasmic DNA associated with the B genome and are referred to as BAB. It is debatable whether the subgroup 'Bontha,' which is starchy and primarily cooked and eaten, should be included in the 'Pisang Awak,' 'Peyan' (dessert bananas), 'Monthan,' 'Bluggoe,' and 'Vennutu Mannan' groups of bananas. Some 'Bluggoe' types are commercially used as ripe fruits in South India. Although it is suggested that cultivars with B-rich genomes are drought tolerant, 'Monthan' seems to be a proven exception and others need scientific confirmation.

2.5.6 *Australimusa*

The section *Australimusa* has edible varieties generally referred to as Fe'i banana. It has six recorded wild species: *M. lolodensis*, *M. peekelii*, *M. maclayi*, *M. jackeyi*, *M. bukensis*, *M. textilis*, and one cultivated species, *M. fehi* (Daniells et al., 2001). All are characterized by gigantic stature, red plant sap, and negatively geotropic bunching habit (erect bunches). Their distribution is restricted to New Guinea, eastern Indonesia, the Solomon Islands, and other Pacific islands (Argent, 1976; Nayar, 2010). But early Indian literature on Sunga dynasty (2 BCE) talks about a banana variety with red sap. Perhaps a banana with red sap did exist in India at some time in the past, which raises the question whether the *Australimusa* were present in Southern India. Archeological and archeobotanical studies may prove this point. Domestication of *Australimusa* also started with the occurrence of parthenocarpy and seed sterility. Mutations and human selections perpetuated the edible forms, especially Fe'i bananas. They were passively cultivated and protected in many of the west Pacific Islands (Mac Daniels, 1947). Morphotaxonomic studies (Cheesman, 1947), RFLP markers (Jarret et al., 1992), and Carreel (1994) suggested an interspecific origin for the Fe'i bananas within the *Australimusa*, with *M. maclayi* as the most probable ancestor. *Musa textilis*, a member of *Australimusa*, is endemic to Philippines. Several cultivated clones have evolved in nature as intersectional crosses between *M. textilis* and *M. schizocarpa*, and *M. textilis* and *M. balbisiana* (for example, 'Butuhan,' 'Karoina,' 'Umbubu,' and 'Mayalopa').

2.6 MOLECULAR CHARACTERIZATION

A classification system for banana and plantains was first developed by Simmonds and Shepherd (1955), who used a simple numerical system by scoring 15 morphological characters on a score from 1 to 5. A modified and revised scoring system was developed by Silayoi and Chomchalow (1987) and Singh and Uma (1996). The latter system incorporates the genetic variability available especially in India. A general key to identifying the various groups was based on 121 morphological characters using the high degree of variability for characters such as pseudostem blotching, pigmentation, and leaf length: breadth ratio, flowering and fruiting duration, number of hands and fingers, peduncle nature, pedicel length, fruit size, taste and nature of ripe fruit flesh, and so forth (IPGRI-INIBAP/CIRAD, 1996). User-friendly software was developed for classifying a banana genotype/clone by using a pair-wise discriminant function for the five genomes with 10 algorithms and 19 morphological characters.

Morphotaxonomic characterization and genomic classification (Simmonds and Shepherd, 1955) for identification of unique accessions, synonyms, and mutants in Indian germplasm identified 81 distinct clones and 23 mutants in the gene pool of 240 accessions at Tamil Nadu Agricultural University, Coimbatore. A data base has been created for 545 accessions that were characterized at the NRCB using the modified score card and detailed morphotaxonomic characterization using the "*Musa* descriptor." The morphotaxonomic classification system failed to distinguish 45 ABB accessions that appeared to be synonyms, and they were later confirmed with molecular characterization (Saraswathi et al., 2009b).

Four categories—that is, group, subgroup, clone set, and clone—were adopted in the classification of East African Highland bananas based on qualitative characters (Karamura, 1998). Complementing morphotaxonomic characterization with molecular characterization is more useful in drawing meaningful conclusions on banana diversity.

Molecular markers have been used widely for diversity analysis, characterization of germplasm, and fingerprinting of genotypes in *Musa*. Precise knowledge on the extent of diversity in a germplasm is a prerequisite for a strategic breeding program in a crop like *Musa* that is recalcitrant to breeding owing to parthenocarpy, sterility, and polyploidy.

2.6.1 Isozyme-Based Markers

A few studies were carried out to assess genetic diversity in *Musa* using isozyme markers (Bonner et al., 1974; Horry, 1993; Jarret and Litz, 1986a, 1986b). Jarret and Litz (1986a) identified polymorphisms in banana and plantains representing various ploidy levels for enzymes: malate dehydrogenase (MDH), phospho glucomutase (PGM), glutamate oxaloacetate transaminase (GOT), shikimate dehydrogenase (SKDH), and peroxidase. Bhat et al. (1992a, 1992b) observed a high degree of polymorphism for peroxidase, superoxide dismutase, esterases, and acid phosphatases in bananas and plantains. Espino and Pimentel (1990) used SKDH and MDH and found species-specific markers as well as markers for identification of interspecific hybrids.

While some enzymes were of little use in discriminating genomic groups (Espino and Pimentel, 1990; Bhat et al., 1992b), others like MDH were useful in differentiating the ABB and AAB from BB/BBB group of cultivars (Espino and Pimentel, 1990). Isozyme polymorphism was able to distinguish between the 'Saba' and 'Bluggoe' subgroups (Rivera, 1983) but failed to differentiate cultivars within the Cavendish subgroup (Jarret and Litz, 1986a).

The isozyme data of Lebot et al. (1993, 1994) suggested that the genes contributed by *M. acuminata* to the Pacific plantains are similar to those of the *M. acuminata* × *banksii* complex of Papua New Guinea. The Pacific plantains may have originated in Papua New Guinea rather than in Asia or the Malayan archipelago. Analysis of the anthocyanin composition of bracts in wild and cultivated forms by Horry and Jay (1988) suggested two independent centers of domestication for *M. acuminata*, one in Southeast Asia and the other in Papua New Guinea. They also suggested that the A genome of AAB plantains is more allied to Papuan *M. acuminata* ssp. *banksii* than Asian *M. acuminata*. Isozyme polymorphism among Indian and exotic germplasm showed that the exotic introductions were distinctly different from indigenous diploids. However, an indigenous red diploid ('Sannachenkadali') grouped with 'Pisang Berlin,' an exotic introduction, needs further investigation. Selvarajan et al. (2002) were able to differentiate the Sigatoka leaf-spot-resistant and susceptible varieties using SKDH.

2.6.2 DNA-Based Markers

2.6.2.1 Restriction Fragment Length Polymorphism (RFLP)

RFLP markers individually or in combination with other techniques have been used in the classification of *Musa* to amend the genome formula and subspecies/subgroup classification of some varieties. Delineation of the four sections of the genus *Musa* using RFLPs was reported by Gawal et al. (1992). The study grouped *Eumusa* and *Rhodochlamys* in one clade and *Australimusa* and *Callimusa* in the other. Phylogenetic analysis of species and subspecies of *Musa* showed that *M. schizocarpa* is very close to *M. acuminata (*Jarret et al., 1992). Fe'i bananas were distinct from the five species of *Australimusa*—including *M. maclayi*, its presumed ancestor—but showed that *M. lododensis* was closest to the Fe'i bananas. In a similar study conducted at the Centre de Coopération Internationale en Recherche Agronomique pour le Développement (CIRAD), nuclear probes were used to identify the alleles specific to *M. acuminata*, *M. balbisiana*, and *M. schizocarpa*. Starchy cultivars were found to be closely associated with *M. acuminata* ssp. *banksii* while sweet-pulped cultivars were closer to *M. acuminata* ssp. *malaccensis*. RFLP studies also confirmed the involvement of *M. acuminata* in the origin of parthenocarpy, as all the diploid parthenocarpic bananas contained *M. acuminata* alleles. Studies by Carreel et al. (2002) revealed that the parthenocarpic varieties are linked to *M. acuminata* ssp. *banksii* and *M. acuminata* ssp. *errans*. RFLP analysis provided evidence for a strong bias towards maternal transmission of chloroplast DNA (cpDNA) and paternal transmission of mitochondrial DNA in *M. acuminata*, suggesting two separate mechanisms of organelle transmission and selection. Knowledge of the organelle mode of inheritance constitutes an important point for phylogeny analyses in bananas and may offer a powerful tool to confirm the origin of hybrids.

RFLP of cpDNA among the *M. acuminata*–derived clones displayed a range of variation (Gawel and Jarret, 1991). The resultant cladogram showed clear clustering of the *M. acuminata* subspecies and differentiated *M. acuminata* and *M. balbisiana* cytoplasm. On the basis of cytosolic RFLP probes, the triploid Cavendish group (AAA) was found to be related to *M. acuminata* ssp. *errans* and *M. acuminata* ssp. *malaccensis*, whereas RFLP data from the nuclear genome did not show any association with either subspecies. Carreel et al. (2002) used RFLP in combination with heterologous mitochondrial and chloroplast probes to investigate the contribution of other *Musa* spp. besides *M. acuminata* and *M. balbisiana* to the origin of cultivated bananas. Generally, beer bananas could not be distinguished from similar AAA types used for cooking (Simmonds, 1966) but apparently their chloroplast genome was quite distinct for several cleavage sites when compared to dessert bananas.

2.6.2.2 Randomly Amplified Polymorphic DNA (RAPD)

RAPD studies have been widely used in *Musa* for varietal identification (Bhat and Jarret, 1995), species identification (Pillay et al., 2000), identification of genomic groups (Kaemmer et al., 1992; Howell et al., 1994; Pillay et al., 2000; Visser, 2000; Rekha et al., 2001), identification of genomes (Pillay et al., 2000), and genetic diversity (Bhat and Jarret, 1995; Pillay et al., 2001). Multivariate analysis of RAPD bands identified similar patterns of variation that matched the morphological variation in *Musa* (Howell et al., 1994). RAPD polymorphism was also able to differentiate *Musa* cultivars of Kenyan origin that were closely related (Kahangi et al., 2002; Onguso et al., 2004; Mizutani et al., 2004).

RAPDs was used to investigate the genetic variation and phylogenetic relationships among 'Silk' group representatives of both indigenous and exotic origin (Uma et al., 2004). RAPDs used to compare the genetic relationships in *M. balbisiana* accessions collected from the Indian mainland and Andaman and Nicobar Island (Uma et al., 2004) produced two distinct clusters based on geographical origin. Wild types from the Indian mainland were distinct from those collected from Andaman and Nicobar Islands. Further intraspecific relationships among *M. balbisiana* from different regions of India indicated the existence of considerable variation with a polymorphism of 74.6% (Uma et al., 2006). RAPDs were shown to detect differences between closely related organisms such as cultivars of plantain grown in Jamaica and Nigeria (Agoreyo et al., 2008).

Duplex RAPDs were used to identify clones of 'Thella Chakkarakeli' (AAA) at NRCB (unpublished).

2.6.2.3 Retrotransposon-Based Markers

The inter-retrotransposon amplified polymorphism (IRAP) markers were able to identify presence of the B genome in natural hybrids (Nair et al., 2005). Microarray analysis of long terminal repeat (LTR) retrotransposons showed that the copy number varies with the subgroup and genome composition (Teo et al., 2005). Therefore, it could used as potential marker in *Musa* for characterization of genome composition, identification of ancestral genomes (Teo et al., 2005), and classification of varieties. The IRAP primer designed based on the LTR sequence of banana Ty3-gypsy-like retroelement (*M. acuminata* Monkey retrotransposon, AF 1433332) was used by Nair et al. (2005) to identify the B genome in banana cultivars. A second primer pair designed from the B-specific bands of *M. balbisiana* 'Pisang Gala' was used to classify AAB and ABB cultivars. Among the 36 cultivars tested with this primer, the B-specific band was absent in the AA and AAA cultivars but present in all interspecific hybrids with the B genome, including AB, AAB, and ABB cultivars. In ABB genomes, the band intensity was high when compared to those observed in AAB genomes. B-genome dosage was also reported by Pillay et al. (2000) and Nwakanma et al., (2003a). IRAP has also been used by Asif and Othman (2005) to discriminate the Fusarium wilt-resistant and susceptible somaclones of cultivar 'Rasthali.'

Preliminary molecular characterization of *M. swarnaphalya* was carried out using RAPD and IRAP markers to confirm its uniqueness among wild *M. acuminata* and *M. balbisiana* that was observed from its morphological characterization. Three distinct groupings were observed. In one group, *M. swarnaphalya* grouped with *M. itinerans* and *M. nagensium* into a distinct cluster, suggesting its unique species status within section *Eumusa*. The three wild *acuminata* species formed one cluster, while all three *M. balbisiana* species (both wild and cultivated) grouped into the third cluster. Molecular characterization using retroelements (IRAP markers) also exhibited similar results, with *M. swarnaphalya* sharing 58% similarity with members of *Eumusa* (*M. acuminata* ssp. *burmannica* and *M. acuminata* ssp. *bumannicoides*) and was distinct from *Rhodochlamys* and *Ensete* members (Anon, 2005). IRAP was found to be more robust than RAPD in studying the intragroup diversity within Cavendish clones (Saraswathi et al., 2009a).

2.6.2.4 Amplified Fragment Length Polymorphism (AFLP)

AFLP has been used to determine the degree of intra- and intersectional and intra- and interspecific genetic variation in *Musa* (Ude et al., 2002a, 2003; Tugume et al., 2002; Wong et al., 2002). Studies of Ude et al. (2002b) indicated that there is much more genetic diversity within *M. balbisiana* than those suggested using morphological descriptors.

Bhat et al. (2004) showed that it was possible to identify differences between two accessions of the AAB genome cultivar 'Rasthali,' indicating the presence of intracultivar genetic variation. Significant genetic diversity was also present among AA, AB, and ABB Indian banana cultivars, supporting the notion that India, along with other neighboring Southeast Asian countries, is the center of diversity for cultivars of banana and plantain (Bhat et al., 2004). AFLP proved to be a useful tool in the identification of clones and related cultivars, as well as for distinguishing between banana cultivars from China (Wong et al., 2002). Molecular phylogeny using AFLP suggested that species of section *Rhodochlamys* had a close affinity with species of section *Eumusa*, and that section *Australimusa* could be merged with those of section *Callimusa* (Wong et al., 2002). Sequence related amplified polymorphism (SRAP) was found to be comparatively better than AFLP in analyzing the genetic diversity of banana and plantain (Youssef et al., 2010).

2.6.2.5 Minisatellites and Microsatellites

Kaemmer et al. (1992) used oligoprobes to differentiate various genomic groups in *Musa*. Subsequently, Bhat et al. (1995) used oligoprobes for cultivar identification and overall genome analysis to establish relatedness among *Musa* germplasm accessions. The frequency of microsatellites expected in bananas is one sequence-tagged microsatellite site (STMS) per 30 kp, as that of humans (Lagoda et al., 1998). Weising et al. (1996) were the first to use microsatellite-based markers to characterize banana and plantain. Later, the discrimination potential of STMS markers was explored by Grapin et al. (1998) to confirm subspecies organization and clarify some clonal relationships. Oriero et al. (2006) were able to distinguish diploid from triploid cultivars through hierarchical cluster analysis. Creste et al. (2003, 2005) used microsatellite markers to classify the genotypes according to their genomic group and subgroups. However, separation of wild diploid genotypes from the cultivated ones was difficult, indicating a common origin of these genotypes (Creste et al., 2004). Sotto and Volkaert (2004) reported that *M. acuminata* microsatellites are usually less polymorphic in *M. balbisiana* than in *M. acuminata*.

In a comparative analysis made by Crouch et al. (1999), all three assays—namely, RAPD, the variable number of tandem repeats (VNTR), and AFLP—detected a high level of polymorphism between parental genotypes and within progeny populations. However, AFLP had the highest multiplex ratio while VNTR detected the highest levels of polymorphism. Further VNTR and RAPD analyses suggested a high frequency of homologous recombination, which is quite useful in the introgression of useful traits from exotic germplasm.

De Langhe et al. (2005) tried to integrate the morphotaxonomic and molecular data while classifying the African plantains. The mismatch observed implies that either morphological characters are being influenced by the environment or the markers have not sampled critical regions of the genome.

Genetic diversity and phylogenetic studies were undertaken to identify possible progenitors of 'Silk' banana varieties (Uma et al., 2008). Microsatellite markers distinguished *acuminata* wild, *balbisiana* wild, and bispecific cultivated 'Silk' accessions. The results suggest that, irrespective of geographical locations of origin, the 'Silk' group of bananas has a narrow genetic base with more than 80% similarity and that its origin could be from a single genotype. The same study also clustered 'Khungsong' wild, a wild *M. balbisiana* with 'Silk' and *M. acuminata* instead of *M. balbisiana* as expected, suggesting that it may be involved in evolution of the 'Silk' group (Uma et al., 2008). Microsatellite markers were used to characterize 41 accessions from sections *Eumusa* and *Rhodochlamys* (Durai, 2008). The two groups were distinct, with *Rhodochlamys* showing more affinity with species of *M. acuminata* rather than those of *M. balbisiana*. Microsatellites were useful in identifying duplicates, synonyms, mutants, and unique accessions in the ABB group of Indian origin. This study proposed 'Bontha' as a new subgroup in ABB cooking bananas (Saraswathi et al., 2009b).

2.6.2.6 DArT Markers

Genetic diversity of carotenoid-rich bananas was evaluated by diversity arrays technology (DArT) (Amorim et al., 2009) towards the development of novel cultivars with functional properties.

2.6.2.7 Chloroplast and Mitochondrial DNA Polymorphism

Genetic information present in plant mitochondrial and chloroplast DNA is widely used in phylogenetic studies and population genetics because of its complementarities to nuclear Mendelian segregation.

Nwakanma et al. (2003a) were also able to define two lines of evolution in *Musa* using six chloroplast and two mitochondrial DNA probes. One lineage comprised of the sections *Australimusa* and *Callimusa* (with basic chromosome number $x = 10$) while species of section *Rhodochlamys* formed the other lineage. Species of *Eumusa* were distributed in both lineages. Section *Rhodochlamys* appeared as a sister group of section *Eumusa*, with *M. laterita* having a genome that was identical to some subspecies of the *M. acuminata* complex. The progenitors of the present-day bananas were evolutionarily distant from each other. *M. balbisiana* occupied a basal position in the cladogram, indicating its primitive status in the evolutionary pattern.

Nwakanma et al. (2003b) also amplified the intergenic spacer (ITS) region of seven *M. acuminata* and five *M. balbisiana* accessions using specific primers. A 700 bp fragment produced by all accessions was then digested with 10 restriction enzymes. Digestion with *Eco*RV produced markers for the A and B genomes of *Musa*. The non-coding cpDNA sequences were utilized by Swangpol et al. (2007) for assessing the lineage of *Musa* interspecific hybrids from Thailand.

The population structure of wild *M. balbisiana* in China was assessed with cpDNA polymerase chain reaction–restriction fragment length polymorphism (PCR-RFLP). A chloroplast DNA genealogy of 21 haplophytes identified two major clades that correspond to two geographical regions separated by the Beijiang and Xijiang rivers. Nuclear SSR data also revealed significant geographical structuring in banana populations (Ge et al., 2005).

2.7 CONCLUSION

Banana and plantain are ancient fruits domesticated by man to become an important commodity in terms of trade. Banana has been a versatile tropical crop with buoyancy to adapt to diverse climatic conditions, season-independent production, with nutritional opulence and high productivity, making it popular across 150 countries.

The history of banana cultivation is well documented. Its domestication has been proven through advanced archeobotanical and paleoecological studies. New studies are revealing information about its evolution. Banana taxonomy has undergone a number of revisions, but the development of classification keys to identify the greater diversity and variability of new accessions from previous inaccessible areas need revision. There is a need for more detailed studies on the parental contribution at the subspecies level. Further research on the diversity in *M. balbisiana*, its evolution, and the existence of subspecies is required. The existence of triploid *M. balbisiana* remains to be solved. The involvement of other sections, especially *Rhodochlamys*, which has a close genetic proximity to *Eumusa*, their coexistence, and the intersectional introgression, and natural occurrence of intersectional hybrids can provide a new dimension for the evolution of extant bananas. Compared to the genus *Musa*, *Ensete* is the least researched in spite of its importance as a food-fiber crop, especially in Ethiopia. Its possible role in the evolution of bananas needs investigation, especially with the identification of a new species *M. kuppiana* (Anon, 2005) in the forests of northeast states that exhibits a transitional status between *Musa* and *Ensete*.

Morphotaxonomic characterization has been an important tool in studying diversity and phylogeny and, complemented with molecular markers, has provided better resolution for the evolutionary studies. The application of molecular markers has more benefits in banana compared to seeded plants due to inherent barriers like parthenocarpy, sterility, polyploidy, and so forth. Molecular analysis of various genomes across global gene banks more specifically from the areas of *Musa* origin should lead to a better understanding and knowledge on *Musa* origin and evolution. But compared to other commercial crops, genomic research in banana and plantain is still gaining momentum.

REFERENCES

Agoreyo, B.O., K.D. Golden, and S.E. Brown. 2008. Analysis of genetic variability among plantain cultivars (*Musa paradisiaca L.*) using arbitrarily primed PCR technique. *Afr. J. Biotechnol.* 7(8):1041–1045.

Amorim, E.P., A.D. Vilarinhos, K.O. Cohen, V.B.O. Amorim, J.A.S. Serejo, S.O. Silva, et al. 2009. Genetic diversity of carotenoid rich bananas evaluated by DArT. *Genet. Mol. Biol.* 32:96–103.

Anon. 2005. Molecular characterization of *Musa swarnaphalya* using retro-elements. *Annual Report, NRCB* (ICAR), Triuchirapalli, India.

Argent, G. 1976. The wild bananas of Papua New Guinea. *Notes Roy. Bot. Garden Edinburgh* 35:77–114.

Asif, M.J. and R.Y. Othman. 2005. Characterization of Fusarium wilt-resistant and Fusarium wilt-susceptible somaclones of banana cultivar Rasthali (*Musa* AAB) by random amplified polymorphic DNA and retrotransposon markers. *Plant Mol. Biol. Reptr.* 23:241–249.

Baker, J.G. 1893. A synopsis of the genera and species of Musaceae. *Ann. Bot.* 7:189–229.

Bhat, K.V. and R.L. Jarret. 1995. RAPD and genetic diversity in Indian *Musa* germplasm. *Genet. Resour. Crop Evol.* 42:107–118.

Bhat, K.V., S.R. Bhat, and K.P.S. Chandel. 1992a. Survey of isozyme polymorphism for clonal identification in *Musa* I. Esterase, acid phosphatase and catalase. *J. Hort. Sci.* 67:501–507.

Bhat, K.V., S.R. Bhat, and K.P.S. Chandel. 1992b. Survey of isozyme polymorphism for clonal identification in *Musa*. Part II, Peroxidase, superoxide dismutase, shikimate dehydrogenase and malate dehydrogenase. *J. Hort. Sci.* 67:737–744.

Bhat, K.V., S.R. Bhat, K.P.S. Chandel, S. Lakhanpaul, and S. Ali. 1995. DNA fingerprinting of *Musa* cultivars with oligo deoxyribonucleotide probes specific for simple sequence repeats motifs. *Genet. Anal. Biomed. Eng.* 12:45–51.

Bhat, K.V., Y. Amaravathi, P.L. Gautam, and C.V. Karayil. 2004. AFLP characterization and genetic diversity analysis of Indian banana and plantain cultivars (*Musa* spp.). *Plant Genet. Resour.* (GBR) 2:121–130.

Bonner, J.W., R.M. Warner, and J.L. Brewbaker. 1974. A chemosystematic study of *Musa* cultivars. *HortScience* 9:325–328.

Brewbaker, J.L. and D.D. Gorrez. 1956. Classification of Philippine *Musa*. III.(A). Saguing maching (*M. banksii* Fv. M) (B). Alisanay, a putative hybrid of *M. textilis* and *M. banksii. Philipp. Agric.* 40:258–268.

Bruce, J. 1790. *Travels to discover the source of Nile in years 1768, 1769, 1779, 1771, 1772, and 1773*. 5 vols. Edinburgh: Edinburgh University Press.

Carreel, F. 1994. Etude de la diversite des bananiers (genre *Musa*) al aide des marquers RFLP. Thesis, Institut National Agronomiques, Paris–Grignon.

Carreel, F., S. Fauré, D. González de Léon, P.J.L. Lagoda, X. Perrier, F. Bakry, et al. 1994. Evaluation de la diversité génétique chez les bananiers diploïdes à l'IRFA-CIRAD. *Fruits* (special issue):25–40.

Carreel, F., D. Gonzalez de Leon, P. Lagoda, C. Lanaud, C. Jenny, J.P. Horry., and H. Tezenas du Montcel. 2002. Ascertaining maternal and paternal lineage within *Musa* by chloroplast and mitochondrial DNA RFLP analysis. *Genome* 45:679–692.

Chandler, S. 1995. The nutritional value of bananas. In: *Bananas and plantains*, S. Gowan, ed., 468–480. London: Chapman & Hall.

Cheesman, E.E. 1947. Classification of the bananas. II. The genus *Musa* L. *Kew Bull.* 2:106–117.

Creste, S., A.T. Neto, R. Vencovsky, S. De O. Silva, and A. Figueira. 2003. Genetic characterization of banana cultivars from Brazil using microsatellite markers. *Euphytica* 132:259–268.

Creste, S., A.T. Neto, R. Vencovsky, S. De O. Silva, and A. Figueira. 2004. Genetic diversity of *Musa* diploid and triploid accessions from the Brazilian banana breeding program estimated by microsatellite markers. *Genet. Resour. Crop Evol.* 51:723–733.

Creste, S., T. R. Benatti, M. R. Orsi, A. M. Risterucci, and A. Figueira. 2005. Isolation and characterization of microsatellite loci from a commercial cultivar of *Musa acuminata*. *Molecular Ecology Notes* 6:303–306.

Crouch, J.H., H.K. Crouch, H. Constandt, A. Van Gysel, P. Breyne, M. Van Montagu, et al. 1999. Comparison of PCR-based molecular marker analyses of *Musa* breeding populations. *Mol. Breed.* 5:233–244.

Danh, L.D., H.H. Nhi, and R. Valmayor. 1998. Banana collection, characterization and conservation in Vietnam. *InfoMusa* 7:10–13.

Daniells, J., C. Jenny, D. Karamura, and K. Tomekpe. 2001. *Musalogue: A catalogue of musa germplasm: Diversity in the genus Musa.* Montpellier, France: INIBAP.

De Langhe E.A.L. 1996. Banana and plantain: The earliest fruit crop? Focus Paper No. 1. In: *INIBAP annual report 1995*. Montpelier, France: INIBAP.

De Langhe, E. and P. De Maret. 1999. Tracking the banana: Its significance in early agriculture. In: *Prehistory of food: Appetites for change*, C. Gosden and J. Hather, eds., 377–396. New York: Routledge.

De Langhe, E. and R. Valmayor. 1980. French plantains in South East Asia. *IBPGR/ RECSEA Newsl.* 4:3.

De Langhe, E., R. Swennen, and D. Vuylsteke. 1995. Plantain in the early Bantu world. *Azania* 29–30:147–160.

De Langhe, E., M.Pillay, A. Tenkouano, and R. Swennen. 2005. Integrating morphological and molecular taxonomy in *Musa* L. The case of the African plantains (*Musa* spp. AAB group). *Plant Syst. Evol.* 255:225–236

Denham, T.P., S.G. Haberle, C. Lentfer, R. Fullagar, J. Field, M. Therin, et al. 2003. Origins of agriculture at Kuk Swamp in the highlands of New Guinea. *Science* 301:189–193.

Denham, T. P., S. Haberle, and C. Lentfer. 2004. New evidence and revised interpretations of early agriculture in highland New Guinea. *Antiquity* 78:839–857.

Dodds, K.S. 1943a. The genetic system of banana varieties in relation to banana breeding. *Emp. J. Expt. Agric.* 11:89–98.

Dodds, K.S. 1943b. Genetical and cytological studies of *Musa*. Part V, Certain edible diploids. *J. Genet.* 45:113–138.

Dodds, K.S. 1945. Genetical and cytological studies of *Musa*. Part VI, The development of female cells of certain edible diploids. *J. Genet.* 46:161–179.

Dodds, K.S. and N.W. Simmonds. 1946a. Genetical and cytological studies of *Musa*. Part IX, The origin of an edible diploid and the significance of interspecific hybridization in the banana complex. *J. Genet.* 48:285–293.

Dodds, K.S. and N.W. Simmonds. 1946b. Genetical and cytological studies of *Musa*. Part VIII, The formation of polyploid spores. *J. Genet.* 47:223–241.

Dodds, K.S. and N.W. Simmonds. 1948. Sterility and parthenocarpy in diploid hybrids of *Musa*. *Heredity* 2:101–117.

Doutrelepont, H., L. Vrydaghs, E. De Langhe, R. Swennen, C. Mbida, B. Janssens, et al. 1996. Banana in Africa. In: *First European meeting on phytolith research*, A. Pinilla, M.J. Machado, and J.J. Tresserras, eds., 23–26. September 1996, Madrid, Spain.

Durai, P. 2008. Genetic diversity and phylogenetic analysis of the genus *Musa* sect. *Rhodochlamys* (Musaceae) through morphotaxonomic traits and microsatellite markers. PhD diss., Bharathidasan University.

Espino, R.R.C. and R.B. Pimentel. 1990. Electrophoretic analysis of selected isozymes in BB cultivars of Philippines bananas. In: *Proceedings of the international workshop on identification of genetic diversity in the genus Musa*, R.L. Jarret, ed., 36–40. Montpellier, France: INIBAP.

FAO STAT. Various years. http://www.faostat.fao.org

Faure, S., J.L. Noyer, J.P. Horry, F. Bakry, C. Lanaud, and D. Gonzalez de Leon. 1993. A molecular based linkage map of diploid bananas. *Theor. Appl. Genet.* 87:517–526.

Gasster, M. 1963. The banana in geriatric and low-calorie diets. *Geriatics* 18:782–786

Gawel, N.J. and R.L. Jarret. 1991. Chloroplast DNA restriction fragment length polymorphisms (RFLPs) in *Musa* species. *Theor. Appl. Genet.* 81:783–786.

Gawel, N.J., R.L. Jarret, and A.P. Whittemore. 1992. RFLP-based phylogenetic analysis of *Musa*. *Theor. Appl. Genet.* 84:286–290.

Ge, X.J., M.H. Liu, W.K. Wang, B.A. Schaal, and T.Y. Ching. 2005. Population structure of wild bananas *Musa balbisiana* in China determined by SSR finger printing and cpDNA PCR-EFLP. *Mol. Breed.* 14:933–944.

Grapin, A., J.L. Noyer, F. Carreel, D. Dambier, F.C. Baurens, C. Lanaud, and P.J.L. Lagoda. 1998. Diploid *Musa acuminata* genetic diversity assayed with STMS. *Electrophoresis* 19:1374–1380.

Häkkinen, M. and H. Väre. 2008. Typification and check-list of *Musa* names (Musaceae) with nomenclatural notes. *Adansonia* 30(1):63–112.

Horry, J.P. 1993. Taxonomy and genetic diversity of diploid bananas. In: *Breeding banana and plantain for resistance to diseases and pests*, J. Ganry, ed., 35–42. Montpellier, France: CIRAD.

Horry, J.P. and M. Jay. 1988. Distribution of anthocyanin in wild and cultivated banana varieties. *Phytochemistry* 27:2667–2672.

Howell, E.C., H.J. Newburry, R. Swennen, L.A. Withers, and B.V. Ford Lloyd. 1994. The use of RAPD for identifying and classifying *Musa* germplasm. *Genome* 37:328–332.

IPGRI-INIBAP/CIRAD. 1996. *Descriptor for banana (Musa spp.)*. Rome: IPGRI.

Jacob, K.C. 1952. *Madras Bananas, a Monograph*. Madras: Government Printer.

Jarret, R.L. and R.E. Litz. 1986a. Isozymes as genetic markers in banana and plantain. *Euphytica* 35:539–549.

Jarret, R.L. and R.E. Litz. 1986b. Enzyme polymorphisms in *Musa acuminata* Colla. *J. Heredity* 77:183–188.

Jarret, R. L., N. Gawel, A. P. Whittemore, and S. Sharrock. 1992. RFLP based phylogeny of *Musa* species in Papua New Guinea. *Theor. Appl. Genet.* 84:579–584.

Kaemmer, D., R. Afza, K. Weising, G. Kahl, and F.J. Novak. 1992. Oligonucleotide and amplification fingerprinting of wild species and cultivars of banana. *Biotechnology* 10:1030–1035.

Kahangi, E.M., M.A. Lawton, and C.Y. Kumar. 2002. RAPD profiling of some banana varieties selected by small-scale farmers in Kenya. *J. Hort. Sci. Biotechnol.* 77:393–398.

Karamura, D. 1998. *Numerical taxonomic studies of the East African Highland bananas (Musa AAA-EA) in Uganda*. Montpellier, France: INIBAP.

Ketiku, A.O. 1973. Chemical composition of unripe (green) and ripe plantains (*Musa paradisiaca). J. Sci. Food Agric*. 24:703–707.

Lagoda, P.J.L., J.L. Noyer, D. Dambier, F.C. Baurens, A. Grapin, and C. Lanaud. 1998. STMS in the Musaceae. *Molec. Ecol.* 7:657–666.

Lebot, V., K.M. Aradhya, R. Manshardt, and B. Meilleur. 1993. Genetic relationships among cultivated bananas and plantains from Asia and the Pacific. *Euphytica* 67:163–175.

Lebot, V., A.B. Meilleur, and R.M. Manshardt. 1994. Genetic diversity in eastern Polynesian *Eumusa* bananas. *Pac. Sci.* 48:16–31.

Lentfer, C.J. 2003. *Tracing antiquity of banana cultivation in Papua New Guinea*. Sydney: Pacific Biol. Foundation.

Linnaeus, C. 1753. Species plantarum, exhibentes plantas rite cognitas, ad genera relatas, cum differentiis specificis, nominibus trivialibus, synonymis selectis, locis natalibus, secundum systema sexuale digestas. Holmiae, Impensis Laurentii Salvii. [L. Salvius, Stockholm.]. [1 May 1753] [Starting point for Spermatophyta, Pteridophyta, Sphagnaceae, Hepaticae, Fungi (incl. slime moulds and lichen-forming fungi) and Algae (pro parte).].

Mac Daniels, L.H. 1947. A study of the Fe'i bananas and its distribution with reference to Polynesian migrations. *Bernice P. Bishop Mus. Bull.* No. 190.

Mbida, C.A., W. Van Neer, T.H. Doutrelepont, and L. Vrydaghs. 2000. Evidence for banana cultivation and animal husbandry during the first millennium BC in the forest of South Cameroon. *J. Arch. Sci.* 27:151–162.

Mizutani, F., J.M. Onguso, E.M. Kahangi, and D.W. Ndivitre. 2004. Genetic characterization of banana and plantains in Kenya by RAPD markers. *Sci. Hort.* 99:9–20.

Mota, R.V., F.M. Lajolo, C. Ciacco, and B.R. Cordenunsi. 2000. Composition and functional properties of banana flour from different varieties. *Starch-Stärke* 52:63–68.

Nair, A.S., C.H. Teo, T. Schwarzacher, and P.H. Harrison. 2005. Genome classification of banana cultivar from South India using IRAP markers. *Euphytica* 144:285–290.

Nasution, R.E. 1991. A taxonomic study of *Musa acuminata* Colla with its intraspecific taxa in Indonesia. *Memoirs of the Tokyo University of Agriculture*, vol. 32. Tokyo.

Nayar, N.M. 2010. The bananas: Botany, origin, dispersal. *Hort. Rev.* 36:117–164.

Negash, A., A. Tsegave, R. Van Treuren, and B. Visser. 2002. AFLP analysis of *Ensete* clonal diversity in south and southwestern Ethiopia for conservation. *Crop Sci.* 42:1105–1111.

Nwakanma, D.C., M. Pillay, B.E. Okoli, and A. Tenkouano. 2003a. PCR-RFLP of the ribosomal DNA internal transcribed spacers (ITS) provides markers for the A and B genomes in *Musa* L. *Theor. Appl. Genet.* 108:154–159.

Nwakanma, D.C., M. Pillay, B.E. Okoli, and A. Tenkouano. 2003b. Sectional relationships in the genus *Musa* L. inferred from PCR-RFLP of organelle DNA sequences. *Theor. Appl. Genet.* 107:850–856.

Olaoye, O.A., A.A. Onilude, and O.A. Odowu. 2006. Quality characteristics of bread produced from composite flours of wheat, plantain and soybeans. *Afr. J. Biotechnol.* 5:1102–1106.

Onguso, J.M., E.M. Kahangi, D.W. Ndiritu, and F. Mizutani. 2004. Genetic characterization of cultivated bananas and plantains in Kenya by RAPD markers. *Sci. Hort.* 99:9–20.

Oriero, C.E., O.A. Odunola, Y. Loco, and I. Ingelbrecht. 2006. Analysis of B-genome derived SSR markers in *Musa spp. Afr. J. Bot.* 5:126–128.

Pillay, M., D.C. Nwakanma, and A. Tenkouano. 2000. Identification of RAPD markers linked to A and B genomes in *Musa. Genome* 43:763–767.

Pillay M., E. Ogundiwin, D.C. Nwakanma, G. Ude, and A. Tenkouano. 2001. Analysis of genetic diversity and relationships in East African banana germplasm. *Theor. Appl. Genet.* 102:965–970.

Pillay, M., A. Tenkouano, and J. Hartman. 2002. Bananas and plantains: Future challenges in *Musa* breeding. In: *Crop improvement, challenges in the twenty-first century*, M.S. Kang, ed., 223–252. New York: Food Products Press.

Ploetz, C.P., A.K. Kepler, J. Daniells, and S.C. Nelson. 2007. Banana and plantain—an overview with emphasis on Pacific island cultivars—Musaceae. www.traditionaltree.org .pdf, 1–26 (accessed 23 April 2010).

Purseglove, J.W. 1972. *Tropical crops: Monocotyledons*, vol. 2. London: Longmans.

Purseglove, J.W. 1976. The origins and migrations of crops in tropical Africa. In: *Origins of African plants domestication*, J.R. Harlan, J.M.J. de Wet., and A.B.L. Slemler, eds., 291–309. The Hague: Mouton.

Raboin, L.M., F. Carmel, J.L. Noyer, F.C. Baurens, J.P. Horry, F. Bakry, et al. 2005. Diploid ancestors of triploid export banana cultivars: Molecular identification of 2*n* restitution gamete donors and *n* gamete donors. *Mol. Breed.* 16:333–341.

Rekha, A., K.V. Ravishankar, L. Anand, and S.C. Hiremath. 2001. Genetic and genomic diversity in banana based on D2 analysis and RAPD markers. *InfoMusa* 10:29–34.

Rivera, R.N. 1983. Protein and isozyme banding patterns among Philippines cooking bananas and their wild parents. *Paradisiaca* 6:7–12.

Sagot, P. 1887. Sur le genre bananier. *Bull. de la Société Botanique de France* 34:328–330.

Saraswathi, M.S., S. Uma, K. Prasanya Selvam, S. Ramaraj, P. Durai and M.M. Mustaffa. 2009a. Robustness of IRAP and RAPD marker systems in studying the intra-group diversity of Musa Cavendish (AAA) clones. Paper presented at the ISHS/Promusa Banana symposium, September 14–18, 2009, Phoenix City, Guangzhou, China.

Saraswathi, M.S., S. Uma, E. Vadivel, P. Durai, S.A. Siva, G. Rajagopal, and S. Sathiamoorthy. 2009b. Diversity analysis in Indian cooking bananas (ABB) through morpho-taxonomic and molecular characterization. Paper presented at the ISHS/Promusa Banana symposium, September 14–18, 2009, Phoenix City, Guangzhou, China.

Schumann, N.K. 1900. Musaceae. In *Das pilanzenreion*, A. Enler, ed. 4:1–45.

Seelig, R.A. 1969. Fruit and vegetables facts and pointers—Bananas. Washington, DC: United Fresh Fruit and Vegetable Association.

Selvarajan, R., S. Anitha, S. Uma, S. Saroja, and S. Sathiamoorthy. 2002. Isozyme analysis: A tool to differentiate resistant/susceptible source of *Musa* accessions to Sifatoka leaf spot disease. In: *Abstracts of the Global Conference on Banana and Plantain*, 28–31 October 2002, Bangalore, India, p. 141.

Shepherd, K. 1999. *Cytogenetics of the Genus Musa*. Montpellier, France: INIBAP.

Silayoi, B. and N. Chomchalow. 1987. Cytotaxonomic and morphological studies of Thai banana cultivars. In *Banana and plantain breeding strategies*, G.J. Persley and E.A. De Langhe, eds., 157–160. Canberra: ACIAR.

Simmonds, N.W. 1953. The classification of the bananas. *Kew Bul. III* 8:571–572; III 8:573–574.

Simmonds, N.W.1954. A survey of Cavendish group of bananas. *Trop. Agric.* 31:126–130.

Simmonds, N.W. 1960. Notes on banana taxonomy. *Kew Bull.* 14:198–212.

Simmonds, N.W. 1962a. *The evolution of the bananas*. London: Longman.

Simmonds, N.W. 1962b. The classification and nomenclature of the bananas and potatoes: Some implications. *Proc. Linn. Soc. Lond.* 173:111–113.

Simmonds, N.W. 1966. *Bananas*, 2nd ed. London: Longmans.

Simmonds, N.W. and K. Shepherd. 1955. The taxonomy and origins of the cultivated bananas. *J. Linn. Soc. Bot.* 55:302–312.

Simmonds, N.W. and S.T.C. Weatherup. 1990. Numerical taxonomy of the cultivated bananas. *Trop. Agric.* (Trinidad), 67:90–92.

Singh, H.P. and S. Uma. 1996. *Banana—Cultivation in India*. Directorate of Extension, Ministry of Agriculture, Govt. of India.

Sotto, R.C. and R.C. Rabara. 2000. Morphological diversity of *Musa balbisiana* Colla in the Philippines. *InfoMusa* 9:28–30.

Sotto, R.C. and H. Volkaert. 2004. The application of *Musa acuminata* microsatellites in *Musa balbisiana*. Paper presented at the 4th international symposium on Molecular and Cellular Biology of Banana, July 6–9, 2004, Penang, Malaysia.

Swangpol, S., H. Volkaert, R.C. Sotto, and T. Seelanan. 2007. Utility of selected non-coding chloroplast DNA sequences for lineage assessment of *Musa* interspecific hybrids. *J. Biochem. Molec. Biol.* 40:577–587.

Teo, C.H., S.H. Tan, C.L. Ho, Q.Z. Faidah, Y.R. Othman, J.S. Heslop Harrison, et al. 2005. Genome constitution and classification using retrotransposon based markers in the orphan crop banana. *J. Plant Biol.* 48:96–105.

Thompson, A.K. 1995. Banana processing. In: *Bananas and plantains*, S. Gowen, ed., 481–499. London: Chapman & Hall.

Tugume, A.K., G.W. Lubega, and P.R. Rubaihayo. 2002. Genetic diversity of East African Highland bananas using AFLP. *InfoMusa* 11:28–32.

Ude, G., M. Pillay, D. Nwakanma, and A. Tenkouano. 2002a. Analysis of genetic diversity and sectional relationships in *Musa* using AFLP markers. *Theor. Appl. Genet.* 104:1239–1245.

Ude, G., M. Pillay, D. Nwakanma, and A. Tenkouano. 2002b. Genetic diversity in *Musa acuminata* and *Musa balbisiana* and some of their natural hybrids using AFLP markers. *Theor. Appl. Genet.* 104:1246–1252.

Ude, G., M. Pillay, E. Ogundiwin, and A. Tenkouano. 2003. Genetic diversity in African plantain core collection using AFLP and RAPD markers. *Theor. Appl. Genet* 107:248–255.

Uma, S. 2006. *Farmer's knowledge of wild Musa in India*. Rome: FAO.

Uma, S. and S. Sathiamoorthy. 2002. *Names and synonyms of bananas and plantains of India*. Tiruchirapalli, India: National Research Center for Banana (ICAR).

Uma, S., S. Kalpana, and S. Sathiamoorthy. 2002. *Annual report of the ICAR-funded project "Physico-chemical and structural characteristics of banana pseudostem fiber."* Tiruchirapalli, India: National Research Centre for Banana.

Uma, S., S. Sudha, M.S. Saraswathi, M. Manickavasagam, R. Selvarajan, P. Durai, S. Sathiamoorthy, and S.A. Siva. 2004. Analysis of genetic diversity and phylogenetic relationships among indigenous and exotic Silk group of banana using RAPD markers. *J. Hort. Sci. Biotech.* 79:523–527.

Uma, S., S. Sathiamoorthy, and P. Durai. 2005a. *Banana—Indian genetic resources and catalogue*. Tiruchirapalli, India: NRCB (ICAR).

Uma, S., S.A. Siva, M.S. Saraswathi, P. Durai, T.V.R.S. Sharma, D.B. Singh, et al. 2005b. Studies on the origin and diversification of Indian wild banana *(Musa balbisiana)* using arbitrarily amplified DNA markers. *J. Hort. Sci. Biotech.* 80:575–580.

Uma, S., S.A. Siva, M.S. Saraswathi, M. Manickavasagam, P. Durai, R. Selvarajan, et al. 2006. Variation and intraspecific relationships in the Indian wild *Musa balbisiana* (BB) population as evidenced by RAPD. *Genet. Res. Crop Evol.* 55:349–355.

Uma, S., M.S. Saraswathi, S.A. Siva, M.S. Dhivya Vadhana, M. Manickavasagam, P. Durai, et al. 2008. Diversity and phylogenetic relationships among wild and cultivated bananas (Silk-AAB) revealed by SSR markers. *J. Hort. Sci. Biotechnol.* 83:239–245.

Uma, S., M.S. Saraswathi, K. Udhayaanjali, S.A. Siva, and P. Durai. 2010. *Diversity analysis in bispecific diploid (AB) banana clones using microsatellites* (in press).

Valmayor, R.V., R.R.C. Espino, and O.C. Pascua. 2002. *The wild and cultivated bananas of the Philippines*. Los Banos, Laguna: PARRFI and BAR.

Venkataramani, K.S. 1955. The banana complex. *J. Indian Bot. Soc.* 34:79–84.

Visser, A.A. 2000. Characterization of banana and plantain using RAPD. *Acta Hort.* 540:113–123

Weising, K, R.W.M. Fung, D.J. Keeling, R.G. Atkinson, and R.C. Gardner. 1996. Characterization of microsatellite from *Actinidia chinensis. Mol. Breed.* 18:7–10.

Wiley, J. 2008. *The banana: Empires, trade wars and globalization*. Lincoln: University of Nebraska Press.

Wong, C., R. Kiew, G. Argent, O. Set, S.K. Lee, and Y.Y. Gan. 2002. Assessment of the validity of the sections in *Musa* using AFLP. *Ann. Bot.* 90:231–238.

Wong, C., G. Argent. R. Kiew, O. Set, and Y. Y. Gan. 2003. The genetic relations of *Musa* species from Mount Jaya, New Guinea and a reappraisal of the sections of *Musa* (Musaceae). *Gard. Bull.* 55:97–111.

Youssef, M.A., R. Rivera, A.C. James, R. Ortiz, and R.M. Escobedo. 2010. Analysis of the genetic diversity of banana and plantain using SRAP and AFLP. Paper presented at the Plant & Animal Genomes XVIII Conference, January 6–9, 2010, San Diego, California.

3 Genetic Resources for Banana Improvement

Markku Häkkinen and Richard Wallace

CONTENTS

3.1 INTRODUCTION

Bananas (inclusive of plantains) represent a major source of carbohydrates in the diets of a large percentage of the world's population. In terms of worldwide production of food crops, bananas and plantains rank fourth behind rice, wheat, and corn. The majority of the fruit produced is consumed locally with only a small percentage (~15%) of the worldwide production entering the export market (Panis et al., 2007). The development of new edible banana cultivars is required as a result of the existing parthenocarpic varieties' susceptibility. Diseases and pests that attack bananas are discussed in detail in other chapters of this book. Although progress has been made in the development of new varieties of edible bananas by (employing both traditional and non-traditional methods of breeding) urgency associated with the development of additional improved cultivars.

Most of the edible bananas known today are derived from two wild-seeded species, *Musa acuminata* Colla and *M. balbisiana* Colla (Colla, 1820). The fruits of these plants are barely edible and contain numerous seeds with only a small amount of edible pulp (Simmonds, 1962). The edible bananas grown today belong to one of three categories. One of the groups displays the characteristics of *M. acuminata*, another displays the characteristics of *M. balbisiana*, and a third group displays characteristics of both (Simmonds, 1962, 1966). Diploids of *M. acuminata* gave rise to seedless fruit when they became parthenocarpic and sterile (Simmonds, 1962). These parthenocarpic fruit-producing plants were vegetatively propagated via divisions known as "pups" or "suckers." As the production scale was increased and large numbers of these sterile cultivars were grown in close proximity, pathogens began to attack the plants (Simmonds, 1966). Employing these parthenocarpic plants has resulted in them becoming susceptible to diseases and pests due to a lack of genetic diversity. This chapter focuses on the wild-seeded bananas in the genus *Musa* that represent some of the best sources of genetic diversity that can be used in the breeding of new edible bananas.

3.2 WILD SEEDED *MUSA* SPECIES

Over the last 150 years, various botanists have categorized wild bananas into sections or subgenera. Linnaeus was the first to assign scientific nomenclature to bananas in his book *Species Plantarum*, published in 1753. He then established the modern botanical nomenclature that is still in wide use today. Sagot (1887) and Baker (1893) distinguished three subgenera for the genus *Musa*: *M.* subgen. *Physocaulis*, *M.* subgen. *Eumusa*, and *M.* subgen. *Rhodochlamys*. Cheesman (1947, publ. 1948) developed a clear and coherent classification system for the genus *Musa*. His original grouping of the species in the genus *Musa* into four sections proved to be very useful and it has, therefore, been widely accepted. However, he indicated that "the groups have deliberately been called sections rather than subgenera in an attempt to avoid the implication that they are of equal rank." He further pointed out that this work "may stimulate investigation of a genus that is difficult to collect and study, but sufficiently interesting and important in both economic and its more strictly botanical aspects to repay the investigators." In Cheesman's classification, the genus was divided into four sections: *Australimusa* $2n = 20$, *Callimusa* $2n = 20$, *Musa* (*Eumusa*) $2n = 22$, and *Rhodochlamys* $2n = 22$ (Cheesman, 1947, 1948a, 1948b, 1949a, 1949b, 1950; Shepherd, 1959, 1999; Champion, 1967; Simmonds and Weatherup, 1990; Häkkinen and Sharrock, 2002; Häkkinen, 2004a, 2007, 2009a, 2009b; Häkkinen et al., 2007; Häkkinen and Väre, 2008c). Argent (1976) added one more section, *Ingentimusa* $2n = 14$, comprised of a single species, *Musa ingens* Simmonds.

As evidenced from the above discussion, the genus *Musa* contains a number of members with various characteristics. Some of the members of the different sections are compatible in a conventional breeding sense and some intersectional hybrids have been reported (*vide infra*). The incorporation of genetic traits (disease and pest resistance, drought and cold tolerance, etc.) from the sections listed above and described in the subsequent paragraphs will play an important role in the development of new, improved hybrid bananas for use by future generations.

3.3 CLASSIFICATION OF THE WILD SPECIES

3.3.1 *Musa* Section *Australimusa*

Musa section *Australimusa* (originally reported in *Kew Bulletin* 2:11 in 1947 by Cheesman; type: *Musa textilis* and by Née, published in *Anales de Ciencias Naturales* 4:123 in 1801) originated from the Philippines through the Indonesian archipelago to Irian Jaya, Papua New Guinea, and northern Australia. The section *Australimusa* consists of nine described species:

M. boman Argent (Argent, 1976)
M. bukensis Argent (Argent, 1976)
M. fitzalanii M. Muell. (Mueller, 1875)
M. jackeyi W. Hill (Hill, 1874)
M. johnsii Argent (Argent, 2001)
M. lolodensis Cheesman (Chessman, 1950)
M. maclayi M. Muell (Mueller, 1885; Argent, 1976)
M. peekelii Lauterb (Lauterbach 1914; Häkkinen and Väre, 2009)
M. textilis Née (Nee, 1801)

There is also a group of parthenocarpic edible types of *Australimusa* that are known as Fe'i bananas (*Musa fehi* Bertero, in Viellard, 1862; MacDaniels, 1947). These cultivars are distinguished by their erect bunches and red sap and are found almost exclusively in the Pacific region. The red sap from this group of plants is unusually stable to light exposure and has been employed as dye and ink. The fruit from this group of plants must be cooked prior to eating as it is very astringent when raw. As is the case with other members of the genus *Musa*, various parts (leaves, fiber, pseudostems, etc.)

FIGURE 3.1 *Musa textilis* Née. (Courtesy of Markku Häkkinen.)

of plants from the *Australimusa* section are used in a variety of applications in the region where they occur. Hybrids between the members of *Australimusa* and *Musa* (formerly known as *Eumusa*; Simmonds, 1956, publ. 1957) sections have also been identified (Carreel, 1994).

3.3.2 *Musa* Section *Callimusa* Cheesman

Musa section *Callimusa* Cheesman (originally published by Cheesman in 1947 in *Kew Bulletin* 2:112; type: *Musa Coccinea*, and by Andrews in *Botanist's Repository* 1, plate 47, in 1797) originated on the Asian continent from southern China to northern Vietnam, peninsular Malaysia, Sumatra, and Borneo. The section *Callimusa* consists of 22 described species:

- *M. azizii* Häkkinen (Häkkinen, 2005a)
- *M. barioensis* Häkkinen (Häkkinen, 2006a)
- *M. bauensis* Häkkinen and Meekiong (Häkkinen and Meekiong, 2005a)
- *M. beccarii* N. W. Simmonds (Simmonds, 1960; Häkkinen et al., 2005)
- *M. borneensis* Becc. (Beccari, 1902; Häkkinen and Meekiong, 2005b)
- *M. campestris* Becc. (Beccari, 1902; Häkkinen, 2003b, 2004a, 2004b)
- *M. coccinea* Andrews (Andrews, 1797)
- *M. exotica* R. V. Valmayor (Valmayor, 2001)
- *M. gracilis* Holttum (Cheesman, 1950)
- *M. hirta* Becc. (Beccari 1902, Häkkinen 2004b)
- *M. lawitiensis* Nasution and Supard (Nasution and Supardiyono, 1998; Häkkinen, 2006b)
- *M. lokok* Geri and Ng (Geri and Ng, 2005)
- *M. lutea* R. V. Valmayor (Valmayor, Danh, and Häkkinen, 2004)
- *M. monticola* Argent (Argent, 2000)
- *M. muluensis* M. Hotta (Hotta, 1967)
- *M. paracoccinea* A. Z. Liu and D. Z. Li (Liu and Li, 2002)
- *M. sakaiana* Meekiong, Ipor, and Tawan (Meekiong, Ipor, and Tawan, 2005)
- *M. salaccensis* Zoll. ex Backer 1924 (Backer, 1924; Häkkinen and Väre, 2009)
- *M. tuberculata* M. Hotta (Hotta, 1967)
- *M. violascens* Ridl. (Ridley, 1893)
- *M. viridis* R. V. Valmayor, L. D. Danh and Häkkinen (Valmayor, Danh and Häkkinen, 2004)
- *M. voonii* Häkkinen (Häkkinen, 2004a)

FIGURE 3.2 *Musa campestris* Becc. (Courtesy of Markku Häkkinen.)

Ornamental bananas of the section *Callimusa* (Cheesman, 1947; Häkkinen, 2004a; Häkkinen and Väre, 2008c) are very attractive plants of the genus *Musa* L. that are largely grown for their brightly colored inflorescences. Some of the colors found in the bracts of the members of this section include pink, red, orange, yellow, purple, and white. In some cases, the fruit itself is colored, which further increases the ornamental potential of the plants. The fruits from plants in this section are unsuitable for eating due to the large number of seeds and small amount of edible pulp. The members of the *Callimusa* section are smaller plants (in both height and diameter) than the members of the section *Musa*.

3.3.3 *Musa* Section *Musa* L.

Musa section *Musa* (originally published in Species Plantarom 1043 in 1753; Type: *Musa paradisiaca* L. Autonym *Musa* [*Eumusa*] and by Cheesman in *Kew Bulletin* 2:108 in 1947) originated in southern India through the Asian continent to the Philippines, Sumatra, Borneo Indonesia, Papua New Guinea, and northern Australia. The section *Musa* consists of 15 described species:

- *M. acuminata* Colla (Colla, 1820; Cheesman, 1948; Simmonds, 1956; Häkkinen and De Langhe, 2001)
- *M. balbisiana* Colla (Colla, 1820; Cheesman, 1948; Simmonds, 1956; Argent, 1976)
- *M. banksii* F. Muell. (Mueller, 1864)
- *M. basjoo* Iinuma (Iinuma, 1874)
- *M. celebica* Warb. (Warburg, 1900)

FIGURE 3.3 *Musa balbisiana* Colla. (Courtesy of Markku Häkkinen.)

M. cheesmanii N. W. Simmonds (Simmonds, 1956)
M. griersonii Noltie (Noltie, 1994)
M. itinerans Cheesman (Cheesman, 1949; Häkkinen et al., 2008)
M. lanceolata Warb. (Warburg, 1900)
M. nagensium Prain (Prain, 1904; Häkkinen, 2008)
M. ochracea K. Sheph. (Shepherd, 1964)
M. schizocarpa N. W. Simmonds (Simmonds, 1956)
M. sikkimensis Kurz (Kurz, 1877)
M. thomsonii (King) Baker (Baker, 1893; Cheesman, 1948; Cowan and Cowan, 1929)
M. yunnanensis Häkkinen and H. Wang (Häkkinen and Wang, 2007, 2008b)

The vast majority of the banana and plantain cultivars grown worldwide for fruit production are contained in this section of the genus *Musa*.

Although the majority of the members of this section are well known, the most recently discovered member of this section, *M. yunnanensis*, has interesting characteristics that may warrant its further exploration in banana breeding research. The plant was originally discovered in the Mekong River watershed at elevations of 500–1800 m (Häkkinen and Wang, 2007). It tolerates seasonal frost damage (and snow), which occurs during January and February in its habitat. The plant is also grown at higher elevations (up to 2100 m) where the stems and leaves are used as animal fodder. Three additional varieties of *M. yunnanensis* have also recently been reported (Häkkinen and Wang, 2008b). In addition to the resistance to cold displayed by *M. yunnanensis* and its varieties, it also appears to be disease resistant, as no common banana diseases have been observed in the different populations studied to date (M. Häkkinen, personal observation).

3.3.4 *Musa* Section *Rhodochlamys* (Baker) Cheesman

Musa section *Rhodochlamys* (originally reported by Cheesman in *Kew Bulletin* 2:110 in 1948; type: *Musa ornata* Roxb Fl Ind 2:488 in 1824, Basionym: *Musa* spp. *Rhodochlamys* Baker in *Annals of*

Botany [Oxford] 7:204 in 1893) originated on the Asian continent from Tibet to Cambodia, including western Yunnan. The members of this section are unique within the genus *Musa* in that they are tolerant to drought. In their native environments, plants in this section go dormant during the dry season but grow very quickly when the rainy season arrives. They then reach maturity and produce fruit and seeds prior to the return of the dry season when they again go dormant. The section *Rhodochlamys* consists of 12 described species from continental Asia ranging from Tibet to Cambodia:

M. aurantiaca Baker (Baker, 1893; Häkkinen and Väre, 2008a)
M. chunii Häkkinen (Häkkinen, 2009a)
M. dasycarpa Kurz (Kurz, 1867; Häkkinen and Väre, 2008b)
M. laterita Cheesman (Cheesman, 1949b; Häkkinen, 2001)
M. mannii Baker (Baker, 1892; Hooker, 1893; Häkkinen, 2007)
M. ornata Roxb. (Roxburgh, 1824; Häkkinen and Väre, 2008c)
M. rosea Baker (Baker, 1893; Häkkinen, 2006c)
M. rubinea Häkkinen and Teo (Häkkinen and Teo, 2008)
M. rubra Kurz (Kurz, 1867; Hooker, 1895; Häkkinen, 2003a)
M. sanguinea Hook. f. (Hooker, 1872; Häkkinen, 2007)
M. siamensis Häkkinen and Rich. H. Wallace (Häkkinen and Wallace, 2007)
M. zaifui Häkkinen and H. Wang (Häkkinen and Wang, 2008a)

The majority of the members of this section possess erect inflorescences with brightly colored bracts. Like the members of the section *Callimusa*, most of the members of this section are grown largely for their ornamental value (Hakkinen, 2005b). Some intersectional hybrids between the

FIGURE 3.4 *Musa siamensis* Häkkinen and Rich. H. Wallace. (Courtesy of Richard Wallace.)

FIGURE 3.5 *Musa x georgiana* Rich. H. Wallace. (Courtesy of Richard Wallace.)

members of the sections *Musa* and *Rhodochlamys* have been reported in the past (Simmonds, 1962; Shepherd 1999; Uma et al., 2006), but these hybrids were not described or characterized in detail by the authors. A new intersectional hybrid, *Musa x georgiana* Rich. H. Wallace (Figure 3.5) has been recently described and illustrated (Wallace and Häkkinen, 2009). The hybrid has *M. balbisiana* (one of the seeded members of the section *Musa* and an ancestor of edible bananas) as its female parent and *M. velutina* (a brightly colored member of the section *Rhodochlamys*) as its male parent. Due to the ever-increasing pest and disease pressure edible bananas are experiencing, incorporating valuable genetic characteristics (drought tolerance, disease and/or pest resistance, etc.) from other members of the *Musa* genus may represent an important component to the development of new hybrid edible bananas. In this respect, intersectional hybrids such as *M. x georgiana* may provide useful information for banana breeding programs.

3.3.5 *Musa* Section *Ingentimusa* Argent

The sap from this plant (originally reported in *Notes of the Royal Botanical Gardens of Edinburgh* 35(1):111 in 1976; type: *Musa ingens*, and by N. W. Simmons in *Kew Bulletin* 14:198 in 1960) is white-milky in young suckers and watery in older parts. The bracts are dull cream becoming gray-brown on lifting, with the fruit ripening yellow to coppery brown. Simmonds (1960) notes after describing *M. ingens*: "It will probably prove to be best placed in a new section of the genus." This seems to be the only logical course of action since it shows no obvious morphological affinity with any other banana species, and the chromosome number is still unique in the genus, confirming its isolated position.

In Papua New Guinea, *M. ingens* occurs in the Western Highlands District: Kubor range, Kamang, Minj valley, above the Tsau River north of Banz (observed by Argent); Madang District: Bundi Kara near Bundi Patrol Post (obs.); Eastern Highlands District: Kassam Pass (obs.); eastern slopes of Mt. Piora (obs.); Morobe District: eastern slopes of Mt Shungol (obs.); Sarawaket Mountains near Kasonombe (obs.); above Mindik (obs.); and the Northern District: northeast slopes of Mt. Victoria above Kokoda [(obs.), Argent 1976]. This species is common in several parts of the

FIGURE 3.6 *Musa ingens* N. W. Simmonds. (Courtesy of Jeff Daniells.)

highlands and undoubtedly will be found in many more localities. Variation in male bud shape may have geographical basis but too little information is at present available. The altitudinal range of the species is from 1000 to 2100 m, although fruiting specimens appear to be restricted to the lower 600 m of the altitudinal range. The seed germinated at sea level in Lae (6°44'0" S, 147°0'0" E), Papua New Guinea, but did not establish; suckers brought down to the lowlands became progressively more emaciated and died unless placed in air conditioning at least during the night. The species thus seems intolerant of continuous high temperatures.

3.4 CONCLUSION

The genus *Musa* is composed of five sections—*Australimusa*, *Callimusa*, *Musa*, *Rhodochlamys*, and *Ingentimusa*—that possess a rich diversity of genetic traits and morphological characteristics. Contained within the section *Musa* (*Eumusa*) are the edible bananas and plantains, which are of great importance both as an export crop and as a staple carbohydrate to a large portion of the world's population. Due to the ever-increasing pest and disease pressure edible bananas and plantains are experiencing, incorporating valuable genetic characteristics (drought tolerance, disease and/or pest resistance, etc.) from other members of the genus *Musa* will continue to represent an important component in the development of new hybrid edible bananas for use by future generations.

REFERENCES

Andrews, H.C. 1797. Musaceae. *Bot. Repos.* 1:343, plate 47.

Argent, G.C.G. 1976. The wild bananas of Papua New Guinea. *Notes Royal Bot. Gard. Edinb.* 35(1):77–114.

Argent, G.C.G. 2000. Two interesting wild Musa species (Musaceae) from Sabah, Malaysia. *Gard. Bull. Sing.* 52:77–114.

Argent, G.C.G. 2001. Contributions to the flora of Mount Jaya. 6, A new banana, *Musa johnsii* (Musaceae) from New Guinea. *Gard. Bull. Sing.* 53:1–7.

Backer, C.A. 1924. Musaceae. *Flora van Java Alf.* 3:13–141.

Baker, J.G. 1892. Scitamineae. In: *Flora of British India*, J.D. Hooker, ed., chap 6, 225–264. London: L. Reeve.

Baker, J.G. 1893. A synopsis of the genera and species of *Museae. Ann. Bot.* (Oxford) 7:189–229.

Bakry, F., F. Carreel, C. Jenny, and J.P. Horry. 2009. Breeding plantation tree crops. In: *Tropical species*, S. Mohan Jain and P.M. Priyadarshan, eds., 3–50. New York: Springer.

Beccari, O. 1902. Nota sui banani selvatici di Borneo. *Nelle foreste di Borneo*. Tipografia di Salvadore Landi, Firenze, 611–624.

Carreel, F. 1994. Etude de la diversité des bananiers (genre *Musa*) à l'aide des marqueurs RFLP. Thesis, INA, Paris–Grignon.

Champion, J. 1967. Notes et documents sur les bananiers et leur culture. Vol. 1, *Botanique et génétique des bananiers*. Institut Française de Recherches Fruitières Outre-Mer (I.F.A.C.). Paris: Setco.

Cheesman, E.E. 1947 (publ. 1948). Classification of the bananas. II, The genus *Musa* L. *Kew Bull.* 2:106–117.

Cheesman, E.E. 1948a. Classification of the bananas. III, Critical notes on species. *Kew Bull.* 3:11–28, 145–157.

Cheesman, E.E. 1948b. Classification of the bananas. III, Critical notes on species. *Kew Bull.* 3:323–328.

Cheesman, E.E. 1949a. Classification of the bananas. III, Critical notes on species. *Kew Bull.* 4:23–28, 133–137, 265–272.

Cheesman, E.E. 1949b. Classification of the bananas. III, Critical notes on species. *Kew Bull.* 4:445–452.

Cheesman, E.E. 1950. Classification of the bananas. III, Critical notes on species. *Kew Bull.* 5:27–31, 151–155.

Colla, L.A. 1820. Memoria sul genere *Musa*. Mem. Reale Accad. *Sci. Torino* 25:384–394.

Cowan, A.M. and J.M. Cowan. 1929. *The trees of northern Bengal, including shrubs, woody climbers, bamboos, palms and tree ferns*. Calcutta: Bengal Secretariat Book Depot.

Geri, C. and F.S.P. Ng. 2005. *Musa lokok* (Musaceae), a new species of banana from Bario, Borneo. *Gard. Bull. Sing.* 57:279–283.

Häkkinen, M. 2001. *Musa laterita*: An ornamental banana. *Fruit Gard. 33*(4):6–7.

Häkkinen, M. 2003a. Taxonomic history and identity of *Musa rubra* Wall. ex Kurz. *Philipp. Agric. Sci.* 86:92–98.

Häkkinen, M. 2003b. *Musa campestris* Beccari varieties in northern Borneo. *Philipp. Agric. Sci.* 86(4):424–435.

Häkkinen, M. 2004a. *Musa voonii*, a new *Musa* species from northern Borneo and discussion of the section *Callimusa* in Borneo. *Acta Phytotax. Geobot.* 55(2):79–88.

Häkkinen, M. 2004b. *Musa campestris* Becc. (Musaceae) varieties in northern Borneo. *Folia Malaysiana* 5(2):81–100.

Häkkinen, M. 2005a. *Musa azizii*, a new *Musa* species (Musaceae) from northern Borneo. *Acta Phytotax. Geobot.* 56(1):27–31.

Häkkinen M. 2005b. Ornamental bananas: Notes on the section *Rhodochlamys* (Musaceae). *Folia Malaysiana* 6(1–2):49–72.

Häkkinen, M. 2006a. *Musa barioensis*, a new *Musa* species (Musaceae) from northern Borneo. *Acta Phytotax. Geobot.* 57(1):55–60.

Häkkinen, M. 2006b. *Musa lawitiensis* Nasution & Supard. (Musaceae) and its intraspecific taxa in Borneo. *Adansonia*. 28(1):55–65.

Häkkinen, M. 2006c. A taxonomic revision of *Musa rosea* (Musaceae) in Southeast Asia. *Novon* 16:492–496.

Häkkinen, M. 2007. Ornamental bananas: Focus on *Rhodochlamys. Chronica Horticulturae* 47(2):7–12.

Häkkinen, M. 2008. Taxonomic identity of *Musa nagensium* (Musaceae) in Southeast Asia. *Novon* 18(3):336–339.

Häkkinen, M. 2009a. *Musa chunii*, a new *Musa* species (Musaceae) from Yunnan China and taxonomic identity of *Musa rubra*. *J. Syst. Evol. 47* (1):87–91. 33 (4) 6–7.

Häkkinen, M. 2009b. Lectotypification of two *Musa* sections (Musaceae). *Nordic J. Bot.* 27:207–209.

Häkkinen, M. and E. De Langhe. 2001. *Musa acuminata* in Northern Borneo. Montpellier, France: INIBAP.

Häkkinen, M. and K. Meekiong. 2005a. A new species of *Musa* from Borneo. *Syst. Biodivers*. 2(2):169–173.

Häkkinen, M. and K. Meekiong. 2005b. *Musa borneensis* Becc. (Musaceae) and its intraspecific taxa in Borneo. *Acta Phytotax. Geobot*. 56(3):213–230.

Häkkinen, M. and S. Sharrock. 2002. Diversity in the genus *Musa*—Focus on *Rhodochlamys*. *Institute for the Improvement of Banana and Plantain annual report 2001*. Montpellier, France: INIBAP, 16–23

Häkkinen, M. and C.H. Teo. 2008. *Musa rubinea*, a new *Musa* species (Musaceae) from Yunnan, China. *Folia Malaysiana* 9(1):23–33.

Häkkinen, M. and H. Väre. 2008a. A taxonomic revision of *Musa aurantiaca* (Musaceae) in Southeast Asia. *J. Syst. Evol.* 46(1):89–92.

Häkkinen, M. and H. Väre. 2008b. Taxonomic history and identity of *Musa dasycarpa*, *M. velutina* and *M. assamica* (Musaceae). *J. Syst. Evol.* 46(2):230–235.

Häkkinen, M. and H. Väre. 2008c. Typification and check-list of *Musa* names (Musaceae) with nomenclatural notes. *Adansonia* 30(1):63–112.

Häkkinen, M. and H. Väre. 2009. Typification of *Musa salaccensis* and nomenclatural notes on *Musa* (Musaceae). *Adansonia* 31(1):41–46.

Häkkinen M. and R. Wallace. 2007. *Musa siamensis*, a new *Musa* species (Musaceae) from SE Asia. *Folia Malaysiana* 8(2):61–70.

Häkkinen, M. and H. Wang. 2007. New species and variety of *Musa* (Musaceae) from Yunnan, China. *Novon* 17(4):440–446.

Häkkinen, M. and H. Wang. 2008a. *Musa zaifuii*, a new species (Musaceae) from Yunnan, China. *Nordic J. Bot*. 26 (1–2):42–46.

Häkkinen, M. and H. Wang. 2008b. *Musa yunnanensis* (Musaceae) and its intraspecific taxa in China. *Nordic J. Bot*. 26:317–324.

Häkkinen, M., M. Suleiman, and J. Gisil. 2005. *Musa beccarii* Simmonds (Musaceae) varieties in Sabah, northern Borneo. *Acta Phytotax. Geobot.* 56(2):137–142.

Häkkinen, M., G. Hu, H. Chen, and Q. Wang. 2007. The detection and analysis of genetic variation and paternity in *Musa* section *Rhodochlamys* (Musaceae). *Folia Malaysiana* 8(2):71–86.

Häkkinen, M., H. Wang, and X.J. Ge. 2008. *Musa itinerans* (Musaceae) and its intraspecific taxa in China. *Novon* 18(1):50–60.

Heslop-Harrison, J.S. and T. Schwarzacher. 2007. Domestication, genomics and the future for the banana. *Ann. Bot*. 100:1073–1084.

Hill, W. 1874. *Musa jackeyi*. *Rep. Brisbane Bot. Gard*. 7.

Hooker, J.D. 1872. *Musa sanguinea*. *Bot. Mag*. 98: table 5975.

Hooker, J.D. 1893. *Musa mannii*. *Bot. Mag*. 119: table 7311.

Hooker, J.D. 1895. *Musa rubra*. *Bot. Mag*. 121: table 7451.

Hotta, M. 1967. Notes on the wild bananas of Borneo. *J. Jap. Bot*. 42:344–352.

Iinuma, Y. 1874. *Musa basjoo*. Sintei Somoku Dzusetsu [Illustrations and descriptions of plants], 2nd ed. 3: pl. 1.

Kurz, S. 1867. Note on the plantains of the Indian archipelago. *J. Agric. Soc. Ind.* 14:295–301.

Kurz S. 1877. The banana: Pomological contribution. *J. Agric. Soc. Ind.*, N.S.V:112–168.

Lauterbach, C.A.G. 1914. Eine neue Musaceae Papuasiens. *Botanische Jahrbücher für Systematik* 50(suppl):306–307.

Linnaeus, C. 1753. *Sp. Pl. Laurentii Salvii*. Stockholmiae.

Liu, A.Z., D.Z. Li, and X.W. Li. 2002. Taxonomic notes on wild bananas (*Musa*) from China. *Bot. Bull. Acad. Sin.* 43:77–81.

MacDaniels, L.H. 1947. A study of the Fe'i banana and its distribution with reference to Polynesian migration. *Bernice P. Bishop Museum Bull.* 190:1–66.

Meekiong, K., I.B. Ipor, and C.S. Tawan. 2005. A new banana: *Musa sakaiana* (Musaceae) from Sarawak, Malaysia. *Folia Malaysiana* 6(3–4):131–138.

Mueller, F.J.H. von. 1864. *Musa banksii*. *Fragmenta Phytographiae Australiae* 4:132.

Mueller, F.J.H. von. 1875 *Musa fitzalanii*. *Fragmenta Phytographiae Australiae* 9:188.

Mueller, F.J.H. von. 1885. *Musa maclayi* in Miklouho Maclay. *Proc. Linn. Soc. New South Wales* 10:348–355.

Nasution, R.E. and A. Supardiyono. 1998. New species: *Musa lawitiensis. Bull. Kebun Raya Ind.* 8:128–130.
Née, L. 1801. *Musa textilis. Anales Ci. Nat.* 4:123.
Noltie, H.J. 1994. *Musa griersonii. Edinburgh J. Bot.* 51:171.
Panis, B., I. Van den Houwe, B. Piette, and R. Swennen. 2007. Cryopreservation of the banana germplasm collection at the International Transit Centre-Biodiversity International. *Adv. Hort. Sci.* 21(4):235–238.
Prain, J. 1904. *Musa nagensium.* Pt. 2, Natural History. *Asiat. Soc. Bengal*, 73:21.
Ridley, H.N. 1893. *Musa violascens. Trans. Linn. Soc.* Ser. 2, 3:384.
Roxburgh, W. 1824. *Musa ornata Fl., Ind.* W. Carey and N. Wallich, eds., 2–488. Calcutta: Mission.
Sagot, P. 1887. Sur le genre bananier. *Bull. Soc. Bot.* (France) 34:328–330.
Shepherd, K. 1959. Two new basic chromosome numbers in Musaceae. *Nature* 183:1539.
Shepherd, K. 1964. *Musa ochracea. Kew Bull.* 17:461.
Shepherd, K. 1999. Cytogenetics of the genus *Musa.* Montpellier, France: International Network for the Improvement of Banana and Plantain.
Simmonds, N.W. 1956 (publ. 1957). Botanical results of the banana collecting expedition 1954–5. *Kew Bull.* 3:463–489.
Simmonds, N.W. 1960. Notes on banana taxonomy. *Kew Bull.* 14(2):198–212.
Simmonds, N.W. 1962. *The evolution of bananas*. London: Longmans.
Simmonds, N.W. 1966. *Bananas*, 2nd ed. London: Longmans.
Simmonds, N.W. and S.T.C. Weatherup. 1990. Numerical taxonomy of the wild bananas. *New Phytol.* (London) 115:567–571.
Uma, S., M.S. Saraswathi, P. Durai, and S. Sathiamoorthy. 2006. Diversity and distribution of section *Rhodochlamys* (Genus *Musa*, Musaceae) in India and breeding potential for banana improvement programmes. *Pl. Genet. Resources Newslett.* 146:17–23.
Valmayor, R.V. 2001. Classification and characterization of *Musa exotica*, *M. alinsanya* and *M. acuminata* ssp. errans. *Philipp. Agric. Sci.* 84(3):325–331.
Valmayor, R.V., L. Danh, and M. Häkkinen. 2004. Rediscovery of *Musa splendida* A. Chevalier and description of two new species (*Musa viridis* and *Musa lutea*). *Philipp. Agric. Sci.* 87(1):110–118.
Viellard, E. 1862. Plantes utiles de la Nouvelle Caledonie. *Annales des Sciences Naturelles* 4(16):28–76.
Wallace, R. and M. Häkkinen. 2009. *Musa* x *georgiana*, a new intersectional hybrid banana with edible banana breeding relevance and ornamental potential. *Nordic J. Bot.* 27:182–185.
Warburg, O. 1900. *Musa celebica.* In: H.G.A.Engler, ed., *Pflanzenr.*, 4:45:22.

4 Genomes, Cytogenetics, and Flow Cytometry of *Musa*

Michael Pillay and Abdou Tenkouano

CONTENTS

4.1 INTRODUCTION

Banana and plantain (*Musa* spp.) are vegetatively propagated crops that are both essential components of the diet and important sources of income for about 400 million people in over 120 countries in the tropical and subtropical zones (Jones, 2000). In this chapter the term *banana* is used collectively for both bananas and plantains.

Banana production in many countries is seriously threatened by a complex of fungal and bacterial diseases, nematodes, viruses, and insect pests (Pillay et al., 2002). Banana breeding programs use diverse techniques, from conventional (Pillay et al., 2002; Pillay and Tripathi, 2006, 2007) to mutation breeding (http://mvgs.iaea.org/, accessed 23 April 2010) and asexual biotechnological approaches such as somaclonal variation, protoplast engineering, and somatic hybridization (Hwang, 2002; Hwang and Ko, 1988, 1989, 2002, 2004; Assani et al., 1996, 2005; Matsumoto et al., 2002) to create improved varieties. Plant breeding is multidisciplinary in nature, making use of knowledge from several scientific disciplines, including genetics, agronomy, horticulture, crop protection, cytogenetics, and many others (Fehr, 1987). Chromosomes have a special importance in plant breeding since they harbor the genes that constitute genetic linkage groups. Therefore, changes in chromosome structure and function are important determinants of trait inheritance and expression. Thus, segregation distortions in the construction of linkage maps in banana were explained by chromosomal structural rearrangements (Fauré et al., 1993). Meiotic analysis of *Musa* revealed chromosome inversions and translocations (Wilson, 1946) and other meiotic abnormalities (Adeleke et al., 2004). Therefore, better knowledge of chromosome structure and evolution in *Musa* is required.

Over the last 20 years, chromosome studies have shifted from a purely cytogenetic level and dying science to the molecular level ushering in an era of molecular cytogenetics. The development of new techniques for genome analysis has made it possible to examine in greater detail the structure of plant genomes. These techniques include restriction fragment length polymorphism (RFLP), the various methods using polymerase chain reaction (PCR) such as random amplified polymorphic DNA (RAPD), microsatellites or simple sequence repeats (SSR) or variable number of tandem repeats (VNTR), amplified fragment length polymorphism (AFLP), fluorescent and genomic *in situ* hybridization (FISH and GISH), diversity array techniques (DArT; Jaccoud et al., 2001), and high-resolution DNA melting (high resolution DNA melting [HRM]) (de Koeyer et al., 2009). High-throughput technologies such as single nucleotide polymorphisms (SNPs) or small-scale insertions/deletions (indels) allow easy and unambiguous identification of alleles or haplotypes (Bhattramakki and Rafalski, 2001).

The exploitation of plant genomes and genes for plant breeding is now becoming better recognized with the advent of molecular markers, marker-assisted selection, and molecular and cytogenetic maps. Although the genus *Musa* has benefited immensely from such studies, there is scope for greater in-depth research into the cytogenetics and molecular genetics of the genus. Some progress has been made since our last treatment of this topic (Pillay et al., 2004). This chapter provides a summary of our current understanding of cytogenetics, genomes, and flow cytometry in *Musa* and their applications in genome analysis and breeding. This chapter first considers the physical characteristics of the genomes in *Musa* and the methods used for their identification, then examines the current status of conventional and molecular cytogenetics, and finally discusses the role of flow cytometry in genome analysis.

4.2 POLYPLOIDY, ORIGIN, AND VARIATION IN CULTIVATED *MUSA*

Banana is a polyploid crop with over 50 species, not all of which are edible. Wild species are diploid ($2n = 2x = 22$), while the cultivated varieties are primarily triploid ($2n = 3x = 33$) with a few tetraploids ($2n = 4x = 44$) that are mainly derived either from natural or artificial hybridization (Robinson, 1996). *Musa* is divided into five sections (*Australimusa*, *Callimusa*, *Rhodochlamys*, *Eumusa*, and *Ingentimusa*) that vary in the basic number of chromosomes (Stover and Simmonds, 1987; Purseglove, 1988). The *Callimusa* and *Australimusa* have a basic chromosome number of $x = 10$, while *Eumusa* and *Rhodochlamys* have a basic chromosome number of $x = 11$. *Ingentimusa* with a single species, *M. ingens*, has a chromosome number of $2n = 14$. The *Callimusa* and *Rhodochlamys* consist of nonparthenocarpic (seed-bearing) species that have no nutritional value but are important as ornamentals. *Australimusa* consists of parthenocarpic edible types, collectively known as Fe'i bananas, with erect fruit bunches and a red sap that is diagnostic for the section. The section is important for food and fiber, and a valuable dark red dye is obtained from the pseudostems. The *Australimusa* has the highest vitamin A content among all bananas (Englberger et al., 2003a, 2003b, 2003c). *Eumusa* is the largest, most widely distributed and diversified, and most important section containing all edible bananas (Horry et al., 1997). *Musa acuminata* (A genome) is the most widespread of the *Eumusa* species and is found throughout the range of the section, with Malaysia (Simmonds, 1962) or Indonesia (Nasution, 1991; Horry et al., 1997) as the center of diversity. New molecular data show that the first center of domestication for *Musa* was in the Philippines–New Guinea area (Carreel et al., 2002). *Musa balbisiana* is the other wild *Eumusa* species that had broad distribution across South and Southeast Asia. *Musa balbisiana* has adopted to drier regions but exhibits copious growth in tropical and subtropical forests of the Philippines, Thailand, New Guinea, India, South China, Malaysia, Myanmar, Indonesia, and a few other regions (Valmayor et al., 2002; Uma et al., 2005).

Intraspecific hybridization between and among the various subspecies of *M. acuminata* produced a range of diploid cultivars with AA genomes. Diploid AAs produced triploid AAA types by mechanisms such as chromosome restitution (Shepherd, 1999). Interspecific hybridization between AA diploids and *M. balbisiana* (BB) gave rise to the many AAB and ABB types (Robinson, 1996). Other genomic groups including AB, ABBB, AAAB, and AABB also exist (Simmonds and Shepherd, 1955).

It is possible that a range of diverse *M. balbisiana* genotypes were involved in the hybridizations, creating the variability that is observed in extant B genome containing banana clones. Isoenzyme analysis (Lebot et al., 1993), sequence tagged microsatellite analyses (Kaemmer et al., 1997), amplified fragment length polymorphisms (Ude et al., 2002a; Wang et al., 2005), and chloroplast DNA analysis (Ge et al., 2005) coupled with morphological variation (Shepherd, 1988; Hari, 1989; Sotto and Rabara, 2000) revealed that *M. balbisiana* is a polymorphic species. Although variability in *Musa* cultivars is mainly attributed to natural mutations (de Langhe, 1964), other mechanisms such as somaclonal variation, transposable element activity, new genome combinations, polyploidy, genome duplications, mitotic recombination, and recombination of novel alleles could have played a role in this process (Nyine and Pillay, 2010). Data on the rate of natural mutations in crop plants are scarce. Pillay et al. (2003a) assessed the rate of natural mutation in the banana cultivar 'Sukali Ndizi' to be $F = 0.0003$. At this mutation rate, it is expected that greater variation should be present in extant banana cultivars. Although somaclonal variation is defined as the genetic variation in plants regenerated from cultured somatic cells (Larkin and Scowcroft, 1981), Cullis (2005) indicated that somaclonal variation may not be due only to *in vitro* propagation but may occur naturally in plant somatic and reproductive tissues. Much of the natural morphological variability that was observed in plantains was also observed in somaclonal variants in plantains (Vuylsteke et al., 1991). Somaclonal variation was cited as the reason for the change in bunch morphology observed in tissue-cultured 'Agbagba' variants that also showed increased fertility (Vuylsteke et al., 1991). Somaclonal variation has been associated with genomic changes in *Musa*. Thus, Oh et al. (2007) found at least one labile portion of the banana genome that is highly reactive to stress induced during tissue culture, suggesting that other mechanisms may also be responsible for variation in banana. Transposable elements have been identified in *Musa* (Balint-Kurti et al., 2000) although no direct links have been established, as yet, between variation and the transposable elements. It is known that transposable elements bring about variation in plants and animals (Kidwell and Lisch, 1997), and retrotransposons of rice were involved in mutations induced by tissue culture (Hirochika et al., 1996).

4.3 GENOMES IDENTIFIED IN *MUSA*

Genome composition has played an important role in the classification of bananas (Simmonds and Shepherd, 1955). Four genomes—A, B, S, and T—are presently known in cultivated bananas (Pillay et al., 2004). The A, B, and S genomes are present in section *Eumusa*, with the S genome characteristic of *M. schizocarpa*, and the T genome confined to the section *Australimusa*. Most modern cultivars of banana have one or more copies of the A and B genomes (Simmonds and Shepherd, 1955), whereas the S and T genomes occur in only a few (du Montcel, 1989; Sharrock, 1989; Carreel, 1995). Classification of edible bananas into genomic groups is based on a system created by Simmonds and Shepherd (1955). This system assigns a score of 1 to 5 for 15 selected morphological features that differentiate *M. acuminata* from *M. balbisiana*. To determine the genome composition of a cultivar, the 15 morphological characters are scored, with the total score determining the relative contribution of the A and B genomes to the constitution of a clone. The system is complicated by polyploidy and requires chromosome counts of a clone before its genome constitution can be determined. The classification system described above was revised by Silayoi and Chomchalow (1987) who introduced the new BBB clones in the system. The existence of pure BBB clones in *Musa* remains questionable and is not supported by molecular methods (Pillay et al., 2000) or field observations (Buddenhagen, personal communication). Thus, Valmayor et al. (1981) endorsed the original scheme proposed by Simmonds and Shepherd (1955).

Although Simmonds and Shepherd's (1955) classification system is useful and reliable and agrees with recent molecular methods, it has some limitations. It can only be applied to diploid, triploid, and tetraploid clones with the A and B genomes and does not cater for cultivars with the S and T genome characteristics. It is also difficult to study the evolution and taxonomy of the genus only by means of morphological markers because of the wide range of genetic variation existing in *Musa* (Robinson, 1996; Ortiz, 1997).

Banana clones that share similar characteristics and are considered to have arisen from a single clone by mutation are classified as subgroups. A list of the subgroups and some important clones that constitute each genomic group in *Musa* can be found in Robinson (1996) and Jones (2000). It is worth noting that banana clones with similar genomes may share similar characteristics but can be very different. For example, the AAA group contains the sweet dessert bananas as well as the cooking bananas of the East African highlands. The latter is starchy even in the ripe stage and is generally cooked before eating. This may suggest the presence of different A genomes in *Musa* (Ude et al., 2002a). Various combinations of the different A genomes produce different characteristics. The identification of different A genomes remains an interesting research topic. Sequencing of the *Musa* genome may be able to provide some answers to this question. Similarly, the AAB group includes the plantains that are cooked before becoming palatable and the 'Pome' subgroup (AAB) that is widely used as dessert bananas in some countries. Although the clones of the ABB group are rather homogeneous because all clones are cooked, some are also used as dessert banana, when overripe, in Malawi.

Cultivars with the S and T genomes have not spread throughout the world as did those with the A and B genomes. They have been identified only in Papua New Guinea (Shepherd and Ferreira, 1984; Arnaud and Horry, 1997). Genomic groups with the S genome include diploid AS, triploid AAS, and tetraploid ABBS, while T genome groups include AAT, AAAT, and ABBT. Molecular marker methods were used to show that *M. schizocarpa* and one or more species of section *Australimusa* played a role in the origin of some New Guinea cultivars (Carreel, 1995). Further confirmation for the involvement of the S and T genomes in some of these cultivars was elucidated by genomic *in situ* hybridization (D'Hont et al., 2000).

Several different methods have been used either directly or indirectly to identify the genomes in *Musa*.

4.3.1 *In Situ* Hybridization

Osuji et al. (1997a) were the first to apply molecular cytogenetic techniques for genome identification in *Musa*. This was followed by physical mapping of the 18S-25S and 5S ribosomal RNA genes of banana (Dolezelova et al., 1998). These studies used GISH and FISH, respectively, to differentiate the chromosomes present in different genome combinations. Samples used in the above studies were comprised of only the A and B genomes and/or their combinations. The study by D'Hont et al. (2000) included accessions with the A, B, S, and T genomes and some of their combinations. Genomic DNA from species representing the different genomes (*M. acuminata*, AA; *M. balbisiana*, BB; *M. schizocarpa*, SS; and *M. augustigemma*, TT) were used as probes to identify chromosomes from the A, B, S, and T genomes, respectively. Together, these studies made it possible to differentiate the four genomes in banana cultivars and hybrids using different fluorochromes. For example, plantains (AAB) were shown to have 22 A genome chromosomes and 11 chromosomes of B genome origin. Similarly, 'Wompa' (AS) displayed 11 chromosomes each from the A and S genomes, respectively (D'Hont et al., 2000). More interesting and for practical breeding reasons was the identification of genomes in synthetic hybrids of *Musa*. In a diploid hybrid ('TMP2x') from the cross 'Obino l'Ewai' (AAB) × 'Calcutta 4' (AA), GISH experiments showed that there were 22 A genome chromosomes, strongly suggesting that each parent contributed one set (11 chromosomes) of the A genome. Perhaps more interesting was the observation of 33 A genome and 11 B genome chromosomes in the tetraploid hybrid 'TMPx 4698-1' from the same cross. This result arose from a normal gamete (A) from 'Calcutta 4' and a $2n = 3x$ egg (AAB) from 'Obino l'Ewai.' GISH results were also useful in settling cases of disputed genomic constitution. For example, the cultivar 'Karoina' considered to be either ATT or AAT from morphological descriptors and molecular markers was found to be AAT. However, exceptions were also identified. For example, 'Pelipita' (ABB) was shown to have 8 A chromosomes and 25 B chromosomes instead of the expected 11 A and 22 B. Similarly, 21 A and 12 B chromosomes were identified in two plantains from Cameroon, 'M. bouroukou' and 'Nyombe' (D'Hont et al., 2000).

FISH studies showed a high degree of cross-hybridization between the A and B genomes of banana, suggesting that the two genomes are incompletely differentiated and share common DNA sequences (Osuji et al., 1997a). D'Hont et al. (2000) reported greater cross hybridization between the A and B genomes than between the A and S genomes, while the least cross hybridization was observed between the T and the A or B genomes. The intensity of cross hybridization is, perhaps, a reflection of the sequence homologies and affinities between the genomes. It also reflects the genetic distances between cultivars representing the different genomes and corresponds with morphological (Simmonds and Weatherup, 1990) and molecular markers analyses (Carreel, 1995; Ude et al., 2002a, 2002b).

4.3.2 Flow Cytometry

The sizes of the A, B, S, and T genomes were shown to be significantly different from each other by flow cytometric analysis of DNA content (D'Hont et al., 1999). The S and T genomes are larger than the A and B genomes, while the T genome has the highest DNA content. This information was used as a diagnostic tool to identify the different genomes in *Musa*. Genome size variation was also used to determine the species involved in interspecific diploid clones as well as the ploidy of banana clones. However, caution has to be exercised when using DNA content *per se* to identify and cluster similar genomes in *Musa* since the study by Kamaté et al. (2001) showed that DNA content is quite variable in cultivars with similar genomes. For example, the DNA content for AAB clones ranged from 1.61 pg to 1.79 pg, while those for ABB clones ranged from 1.70 pg to 2.23 pg.

4.3.3 Isozymes and Retrotransposons

Isozyme patterns were used to distinguish the A and B genomes in *Musa* (Espino and Pimentel, 1989). Hybridization of the *monkey* retrotransposon to a *Hin*dIII digest of genomic DNA of *Musa* produced bands that were specific to the A and B genomes (Balint-Kurti et al., 2000). The only exception was the cultivar 'Chakrakeli' (AAB), which did not show the B-specific band. It was suggested that 'Chakrakeli' is probably of AAA constitution. This assertion is probably true since 'Chakkarakeli' is classified as AAA in Stover and Simmonds (1987). Due to differences in spelling of the names in the two literature sources, proper identification of the cultivar is necessary to confirm these results. Mandal et al. (2001) used isoenzymes to identify banana varieties in India. Amongst four isozymes, esterase was found to be the most efficient in identifying eight cultivars out of ten. Teo et al. (2005) used retrotransposons to determine the genome constitution of hybrid bananas. One of the criticisms that can be leveled against the above studies is the small sample size. Consequently, a more comprehensive study with a variety of accessions representing different genomic combinations may be needed to determine the ultimate value of isozymes and retrotransposons as genome markers.

4.3.4 Molecular Markers for A and B Genome Sequences

Identification of PCR markers for detection of A and B genome sequences in *Musa* was reported by Pillay et al. (2000). Unique banding profiles for the differentiation of *M. acuminata* (A genome) and *M. balbisiana* (B genome) were obtained with three RAPD primers (A17, A18, D10) from OPERON Technologies (Alameda, Calif., USA). An interesting aspect of the B genome markers was that one of the RAPD fragments was diagnostic for clones with two B genomes. These primers were tested on a sample of 40 accessions representing landraces and hybrids of different ploidy and genome combinations. PCR assays made it possible to elucidate the genome composition of all the plants and were generally in agreement with genomic designations obtained from phenotypic descriptors. Some exceptions were observed. The clones 'Monthan Saba' and 'Bluggoe' were previously classified as BBB by Vakili (1967) and BBB and ABB, respectively, by Valmayor et al. (1981). Both 'Monthan Saba' and 'Bluggoe' were shown to be ABB with the RAPD marker system. Using morphological markers, 'Klue Tiparot' was considered to be a tetraploid but was reclassified as a triploid (Jenny et al. 1997; Horry et al., 1998). The RAPD marker system showed that 'Klue Tiparot' was a triploid ABB and resolved the conflicting classifications derived from morphological markers. In later experiments we found that 'Lep Chung Kut' usually classified as BBB in Jones (2000) and Stover and Simmonds (1987) is an ABB clone. This result puts into question the existence of naturally occurring BBB clones. As shown in Lysak et al. (1999), 'Red Dacca' (AAA) clustered with the AAB clones on the basis of nuclear DNA content. The same study also reported that the DNA content of 'Red Dacca' was slightly lower than expected for other AAA clones. Our RAPD marker system showed that 'Red Dacca' has an AAB genome constitution. It is known that RAPD can produce varying results in different laboratories. Researchers planning to use the RAPD marker system are encouraged to use the 'Red Hot' Taq polymerase from AB Gene (UK) to obtain comparable results.

The internal transcribed spacer (ITS) regions of the ribosomal RNA genes also provided markers for the A and B genomes in *Musa* (Nwakanma et al., 2003). A 530 bp fragment unique to the A genome and two fragments of 350 bp and 180 bp specific for the B genome were identified. Interspecific cultivars with both A and B genomes possessed all three fragments. A dosage effect was observed for the B genomes since the staining intensity of accessions with two B genomes was approximately two times that of accessions with a single B genome.

Nair et al. (2005) used the IRAP primer to identify the B genome in banana cultivars. The study was able to identify misidentification of some samples.

4.4 CYTOGENETICS OF *MUSA*

4.4.1 Mitotic versus Meiotic Chromosomes

Banana chromosomes are very small, with size in the range of 1–2 μm, and stain poorly with conventional stains (Dolezel et al., 1998). Even with improved staining techniques (Pillay and Adeleke, 2001; Adeleke et al., 2002) the small chromosomes are difficult to characterize because they condense tightly, making measurements of the long and short arms very difficult, if not impossible. The mitotic metaphase chromosomes of bananas appear as small dotted structures without any cytological markers in the fully contracted state. The chromosomes are also homogeneous, making it difficult to identify individual chromosomes and impeding the development of karyotypes. A karyotype of *Musa* was reported by Dantas et al. (1993). The only chromosomes that are easily identifiable are the pair of satellite chromosomes that are also the longest chromosomes. Wang et al. (1993) also reported a single pair of chromosomes with satellites in wild AA bananas. The rest of the chromosomes appear to be metacentric to submetacentric.

Since the 1960s chromosome banding techniques were used to provide cytological markers along the length of chromosomes, enabling easy identification of homologous pairs and individual chromosomes of a karyotype. These techniques exploited the variation of staining intensity along chromosomes based on chromatin condensation in euchromatic versus heterochromatic regions (Danilova and Birchler, 2008).

Many different banding techniques are available, each giving a different banding pattern based on the specific biochemical consequences of each method (Stace, 1980). The common types applied to plants include C-, G-, Q-, R-, and Hy-banding. Banding techniques of banana chromosomes were not successful in the authors' laboratory and in other leading cytogenetics laboratories (Dolezelova et al., 1998). Perhaps a refinement of current techniques and more extensive studies may be required to determine the potential of banding techniques in *Musa* and its role in *Musa* breeding.

Most cytological studies in *Musa* have been restricted to mitotic metaphase chromosomes. It appears that either mitotic prometaphase or pachytene chromosomes may be more promising for producing karyotypes in *Musa* (Adeleke et al., 2002). Our unpublished work showed that mitotic prometaphase chromosomes are generally about three to five times longer than those at the fully condensed metaphase stage. In such chromosomes the primary and secondary constrictions are more easily distinguished than in the overcondensed metaphase stage. In most plants with very small chromosomes, researchers are shifting to the pachytene stage of meiosis for studying chromosome morphology. Compared with metaphase chromosomes, pachytene chromosomes present higher axial resolution because the chromosomes are much longer (Wang et al., 2009). Axial resolution is defined as the smallest physical distance that can be detected by microscopy (De Jong et al., 1999). The extent of compaction of mitotic and meiotic chromosomes varies in plant species: 25 times in *Arabidopsis* (Fransz et al., 2000), 6.2 times in rye (Zoller et al., 2004), and 15.9 times in maize (Wang et al., 2006). Further advantages of pachytene analysis are useful landmarks for identifying individual chromosomes, including the visibility of centromeres, chromomeres, telomeres, knobs, nucleoli, and the possibility to distinguish eu- and heterochromatin (Schulz-Schaeffer, 1980). A technique to analyze pachytene chromosome in *Musa* has been reported by Adeleke et al. (2002). With further studies of pachytene chromosomes, it may be possible to identify chromosomal structural changes that occur during evolution and breeding. New techniques such as FISH using chromosomes' specific DNA sequences may also be useful in identifying banana chromosomes (Hribova et al., 2008).

4.4.2 Molecular Cytogenetics

Over the last two decades molecular cytogenetics has considerably contributed to a better understanding of plant genome structure and evolution (Valarik et al., 2004). FISH and GISH research has facilitated identification of parental genomes in hybrids and individual chromosomes in some plants (Taketa et al., 2000), physical mapping of genes on chromosomes (Dolezelova et al., 1998; Fuchs et al., 1998), integration of genetic and physical maps with marker-tagged bacterial artificial chromosomes (BAC) clones (Fahrenkrug et al., 2001; Yuan et al., 2000), analysis of genomic distribution of mobile genetic elements (Balint-Kurti et al., 2000), and other repetitive DNA sequences (Gortner et al., 1998; Heslop-Harrison, 1996).

In addition to genome identification, FISH studies in banana were useful in providing some degree of chromosome identification (Osuji et al., 1997a; Dolezelova et al., 1998). The very strong FISH signals in the centromeric regions of some banana chromosomes (Osuji et al., 1997a) could be useful in identifying individual *Musa* chromosomes since stronger hybridization signals were observed only in some chromosomes. The differential hybridization signals along the length of the chromosome may be due to the organization of the different classes of repetitive DNA. Our unpublished research has shown that the chromosomes of *Musa* contract differentially during prometaphase and metaphase. The centromeric regions appear to be more highly condensed than the distal regions, suggesting that condensation of chromosomes in *Musa* begins in the centromeric region. The distal regions lag behind in this process and often appear as lightly stained "tails." This may provide another reason for the stronger hybridization signals observed in the centromeric regions. Uneven staining patterns are considered characteristic of prometaphase chromosomes, especially in plants with small chromosomes (Fukui, 1996). They are caused by differential condensation of the chromatin fiber and are called condensation patterns. Condensed regions at the proximal regions are called primary condensations, whereas small condensations at the interstitial or terminal regions are termed faint, unstable, or small condensations. Centromeric regions in other species with small chromosomes such as *Arabidopsis* are also known to stain more brightly (Maluszynska and Heslop-Harrison, 1993).

FISH studies identified a single pair of sites for the 18S-25S rDNA in *Musa* (Dolezelova et al., 1998; Osuji et al., 1998) whereas the 5S rDNA genes were localized on two to four different chromosomes (Bartos et al., 2005). All known banana repetitive DNA sequences were not useful in chromosome identification since they were dispersed on all chromosomes of *M. acuminata*. The *monkey* transposons were concentrated in the nucleolar organizer regions and appeared dispersed throughout the genome (Balint-Kurti et al., 2000). The only exception was a part of the retrotransposon *monkey* and the repetitive DNA clone *Radka*14 that localized to the secondary constriction (Balint-Kurti et al., 2000; Valarik et al., 2002). Chromosomes with secondary constrictions are even identifiable without FISH. Other useful cytogenetic markers are DNA sequences from large insert libraries such as BAC (Hribova et al., 2008). In order to position BACs and other DNA sequences along the chromosomes, molecular cytogenetic maps are needed. Cytogenetic maps can be highly informative to support the construction of physical maps and map-based cloning projects and to position genes in large heterochromatic regions where linkage distances are inaccurate due to low levels of meiotic recombination (Lambie and Roeder, 1986; Zhong et al., 1999).

Several BAC libraries are available for banana, including a BAC library from *M. acuminata* cv Calcutta 4 (Vilarinhos et al., 2003, 2006), binary BAC (BIBAC) library of *M. acuminata* cv Tuu Gia (Ortiz-Vanquez et al., 2005), and a BAC library of *M. balbisiana* cv Pisang Klutuk Wulung (Safar et al., 2004). Although BAC-FISH was used successfully to localize chromosome specific BAC clones in other plants, such as potato, rice, and cotton, localization of BAC clones on mitotic chromosomes of *M. acuminata* using FISH was not very successful (Hribova et al., 2008). Only one of eight BAC clones tested produced a single locus signal on the chromosomes of *M. acuminata*. The clone signal corresponded with the location of the 5S rDNA genes. The study by Hribova et al. (2008) showed that BAC libraries appear to have low efficiency for developing cytogenetic markers in banana.

Similar problems were encountered in cotton (Wang et al., 2009). Cotton contains high levels of secondary compounds, including polysaccharides, phenolics, and other organic constituents that could have prevented the availability of target DNA sequences to the hybridization probe during BAC-FISH analysis (Wang et al., 2009). Similar reasons may be provided for the poor BAC-FISH resolution of banana chromosomes.

Highly condensed mitotic metaphase chromosomes are not useful for FISH since they limit the optical resolution of adjacent FISH targets (Kulikova et al., 2001).

The optical axial resolution of FISH targets in metaphase chromosomes is limited to 1–3 Mbp (Lichter et al., 1990; Cheng et al., 2002). For higher resolution, pachytene chromosomes are more appropriate because their complements measure 10–40 times longer and display a differentiated pattern of euchromatic and heterochromatic regions (Zhong et al., 1996; Cheng et al., 2002). Although the production of high-quality pachytene spreads is technically more demanding than mitotic metaphase spreads (De Jong et al., 1999), it appears that this may be the only way to map BACs on banana chromosomes. High-resolution pachytene chromosome mapping of BACs has been achieved in many crops, such as in rice (Cheng et al., 2001) and cotton (Wang et al., 2009). FISH-based mapping will be valuable to align the physical and cytogenetic maps in banana.

4.5 FLOW CYTOMETRY IN *MUSA*

Flow cytometry is extensively used in basic and applied research. Flow cytometry has been used in ploidy analysis, nuclear genome composition, and nuclear DNA content in *Musa*.

4.5.1 Ploidy Analysis in *Musa*

Since ploidy manipulation is critical to banana breeding, ploidy determination is an essential component of any *Musa* improvement program. For example, crossing the triploid plantain with diploid accessions generates diploid, triploid, tetraploid, aneuploid, and hyperploid progeny (Vuylsteke et al., 1993; Osuji et al., 1997b). Oselebe et al. (2006a) also found that progenies of $2x$-$2x$ crosses were predominantly diploid (99.7%), those of $2x$-$4x$ crosses were mainly diploid (96.2%), while the $4x$-$2x$ crosses produced predominantly triploid progenies (94.1%). Ploidy polymorphism has been attributed to abnormal meiosis (Asker, 1980), selective affinity of chromosomes of the A and B genomes, and complex microsporogenesis leading to the formation of gametes with varying chromosome constitutions. For selection of a hybrid of a desired ploidy level, there is need for a rapid and precise method for ploidy screening at an early stage of plant development (Dolezel et al., 1997). Ploidy determination in *Musa* was done primarily by examining plants at the morphological level. The characters of the plant were scored and compared with its resemblance to the progenitor species, *M. acuminata* and *M. balbisiana* (Stover and Simmonds, 1987). Although this system was quite accurate, it had some difficulties. For example, only mature plants could be used and banana takes between 12 and 24 months to reach full maturity when the characters are fully expressed and can be measured. In addition, morphological characters can be greatly influenced by the environment and this method can lead to inconsistent results. Conventional root-tip methods for ploidy determination are labor intensive and difficult due to the small size of *Musa* chromosomes and are not practical when thousands of plants are involved. Alternative methods for rapid ploidy determination in *Musa* including gametophytic and sporophytic characters such as pollen size, stomatal size, and density have been reported (Vandenhout et al., 1995; Tenkouano et al., 1998). These methods also have their limitations. Determination of pollen size cannot be used for early screening since plants require 9 to 12 months before flowering (Tenkouano et al., 1998). Tenkouano et al. (1998) found that the average number of chloroplasts in stomatal guard cells of triploid and tetraploids was, respectively, 1.30 and 1.53 times higher than in diploids. Consequently, the number of chloroplasts in stomatal guard cells can be used to determine ploidy levels in *Musa* germplasm. Flow cytometry is still a quicker and easier method for ploidy determination in banana. Many studies have used flow cytometry to measure ploidy in banana (Dolezel et al., 1997; Pillay et al., 2000, 2006; Oselebe et al. 2006a, 2006b). These studies have tested the validity of some of the ploidy levels determined entirely from morphological characters. For example, 'Klue Tiparot' classified as a natural tetraploid (ABBB) is now known to be a triploid ABB with $2n = 3x = 33$ chromosomes both by classical microscopy and flow cytometry (Jenny et al., 1997; Horry et al., 1998; Pillay et al., 2000). Similarly, Horry et al. (1998) found that 'Pisang Jambe,' classified as a tetraploid AAAA, is actually a triploid AAA ($2n = 3x = 33$), while the triploid '(Kluai) Ngoen' (AAB) from Malaysia is a tetraploid AAAB ($2n = 4x = 44$).

Nsabimana and van Staden (2006) were the first to use frozen banana tissue to determine ploidy of the banana. Their study revealed that the clones 'Pomme,' 'Kamaramasenge,' and 'Gisubi Kayinja' that were previously thought to be diploid were actually triploid. Likewise, 'Gisubi Kagongo' and 'Dibis' that were considered to be triploid and tetraploid, respectively, were found to be diploid and triploid, respectively. There appears to be some confusion on the ploidy of 'Kamaramasenge,' which was reported to be an AB diploid (Pillay et al., 2001) and as a triploid in the above study and by Dolezelova et al. (2005). This anomaly can best be solved by confirming the identity of the plant and counting its chromosomes using conventional root-tip techniques. One has to be cautious when assigning new ploidy levels to banana clones and ensure that no transcription errors were committed when plants are transferred from one germplasm collection to another with regards to the names of plants and the ploidy. Records of the germplasm collection at the Onne Station of the International Institute of Tropical Agriculture in Nigeria indicated a clone named 'Diby 2' that was recorded as a triploid. Whether the clone called 'Dibis' in the study by Nsabimana and van Staden (2006) is the same as 'Diby 2' requires confirmation. Similarly, the word "kayinja" is typically used for beer banana in Uganda and it is well known that it is a triploid. In all likelihood, 'Gisubi Kayinja' is a triploid and was perhaps mislabeled as a diploid.

Flow cytometry was used to confirm the ploidy level of the dessert banana cultivar 'Sukali Ndizi' that was known to be a diploid with an AB genome composition since its introduction into Uganda in 1903 (Pillay et al., 2003a). Flow cytometry and conventional chromosome analysis together with molecular marker technology showed that 'Sukali Ndizi' is a triploid with AAB genomes. Likewise, 'Pisang Awak' was known to be triploid (Robinson, 1996). However, Pillay et al. (2006) used flow cytometry and conventional root-tip chromosome counts to determine the tetraploid status of 'Pisang Awak' and two other cultivars ('Foulah 4' and 'Nzizi') that had been previously classified as triploids. Perhaps different cytotypes of 'Pisang Awak' exist. Similarly, cultivars that were previously classified as diploids, including 'Too' and 'Toowoolee,' were found to be triploids. Careful screening of *Musa* genotypes with flow cytometric and molecular techniques will undoubtedly help to resolve similar problems and provide a more consistent classification system. Flow cytometric methods for determining ploidy should allow breeders to screen in the early stages of development for euploids in segregating populations of banana hybrids (Dolezel et al., 1994), thus saving valuable space, time, and resources currently being used to grow out aneuploids and hyperpolyploids.

It appears that a detailed morphological description of field-grown plants, conventional chromosome counting, and flow cytometry is essential to confirm the status of plants making up the AB and ABB genome groups in *Musa*.

Breeding schemes that depend on mutation, somaclonal variation, protoplast fusion, cell suspension cultures, anther and microspore culture, protoplast engineering, and colchicine doubling of chromosomes have used flow cytometry as an indispensable tool (Novak et al., 1989; Megia et al., 1993; Panis et al., 1993; Cote et al., 1996; Matsumoto and Oka, 1998; Gimenez et al., 2001; Assani et al., 2003, 2005; Xiao et al., 2008, 2009).

4.5.2 DNA Content, Genome Size, and Base Composition

Flow cytometric analysis of the nuclear DNA content is based on the analysis of the relative fluorescence intensity of nuclei stained with a DNA fluorochrome (Dolezel, 1991; Dolezel et al., 1997). The nuclear DNA content of many plant species, including *Musa*, has been estimated by flow cytometry. However, relatively few studies have been conducted in *Musa* to estimate the DNA content and genome size (Arumuganathan and Earle, 1991; Lysak et al., 1999; D'Hont et al., 1999; Kamate et al., 2001). One of the major shortcomings of these studies is the small sample size representing the different genomic groups. While two of these studies (Lysak et al., 1999; Kamate et al., 2001) involved accessions with only the A and B genomes, only one study included plants with the S and T genomes in its analysis (D'Hont et al., 1999). However, there is a large difference in the DNA content estimates obtained by D'Hont et al. (1999) when compared with the studies by Lysak et al. (1999) and Kamate et al. (2001). The first report of the total DNA content and genome size in *Musa* was provided by Arumuganathan and Earle (1991). A 2C DNA content of 1.81 pg with a genome size of 873 Mbp was reported. The term "2C" represents the DNA content of a diploid nucleus and is represented by the G_1 peak in flow cytometric analysis. Afza et al. (1993) reported a 2C value of 1.23 pg for *M. acuminata* subsp. *banksii*; 2C = 1.26 pg for *M. acuminata* subsp. *Errans*, and 2C = 1.26 for the AA diploid cultivar 'Pisang Mas.' They estimated that 2C = 1.14pg in *M. balbisiana*. The DNA content for AA genome accessions ranges from 1.22 to 1.27 pg (Lysak et al., 1999) and 1.20 to 1.33 pg (Kamate et al., 2001). This corresponds to the estimate of Afza et al. (1993). Similarly, genome size estimates of *Musa* ranges from 534 to 615 Mbp (Lysak et al., 1999) and 560 to 610 Mbp (Kamate et al., 2001). The estimate of DNA content by Arumuganathan and Earle (1991) is similar to those reported for triploid accessions in other studies. It is likely that Arumuganathan and Earle (1991) measured a triploid rather than a diploid accession.

With regards to individual genomes in *Musa*, the B genome is considered to have the smallest nuclear DNA content, with estimates ranging from 1.03 pg/2C to 1.16 pg/2C in different studies

(Lysak et al., 1999; D'Hont et al., 1999; Kamate et al., 2001). Estimates for the average size of the A genome have been placed at 1.25 pg/2C and 1.27 pg/2C (Kamate et al., 2001) while D'Hont et al. (1999) reported that the A genome is 1.11 pg/2C. The S and T genomes are reported to contain 1.18 pg/2C and 1.27 pg/2C DNA, respectively (D'Hont et al., 1999). As indicated earlier, the DNA estimates of D'Hont et al. (1999) are significantly lower, and a meaningful comparison of DNA content in the A, B, S, and T genomes is not possible. However, the DNA content of the S and T genomes is reported to be greater than those of the A and B genomes (D'Hont et al., 1999). Nevertheless, it is clear that there are significant differences of 12–15% in DNA content between the A and B genomes. The size difference is probably due to the presence of repetitive sequences that were found to be in greater abundance (10.5 Mbp) in *M. acuminata* (Valarik et al., 2002). The expectation is that a triploid accession with two B genomes (ABB) will invariably have less DNA than a triploid accession with one B genome (AAB). However, this is not the case, as shown by the results of Lysak et al. (1999) and Kamate et al. (2001). The bigger sample size used in Kamate et al. (2001) showed a greater number of outliers with respect to genome designation and DNA content. For example, the genome size of 'Simili Radjah' (ABB) was typical of that of a tetraploid. Further, an 11% difference in DNA content was noted within subspecies of the *M. acuminata* complex. This contrasts with the 3.9% variation in DNA content among the AA accessions reported by Lysak et al. (1999). Differences in the estimation of DNA content have been reported for other crops such as *Pisum sativum* (Cavallini and Natali, 1990; Cavallini et al., 1993) and *Glycine max* (Graham et al., 1994). However, both the *P. sativum* and *G. max* results were not supported by the studies of Baranyi and Greilhuber (1996) and Greilhuber and Obermayer (1997), respectively. Consequently, interpreting DNA content data in *Musa* must be approached with great caution, until more accurate methods for DNA determination become available.

Lysak et al. (1999) used DNA content to cluster the clones with different genome combinations. With the exception of one clone, 'Red Dacca,' all the clones with similar genomes clustered together. The clustering corresponded with accepted taxonomic classification of *Musa* based on other characters.

4.5.3 Gene Content and Density

Gene content and density have not been widely studied in banana. The gene content and density of two BAC clones from *M. acuminata* was revealed by sequencing the BAC clones (Aert et al., 2004). One of the clones (MuH9) had a gene density of one gene per 6.9 kb while the other (MuG9) had a gene density of one gene per 10.5 kb. The genomic base composition of *Musa* is estimated to have a median value of 40.8% GC.

4.6 CYTOGENETIC AND FERTILITY RELATIONSHIPS IN *MUSA*

Most of the widely grown banana cultivars are characterized by either low male and female fertility or complete male and female sterility (Pillay et al., 2002). The high level of sterility is due to production of gametes with uneven chromosome numbers. The development of silver staining for *Musa* chromosomes made it possible to visualize the meiotic chromosomes in *Musa* (Pillay and Adeleke, 2002). Cytological studies of microsporogenesis in *Musa* highlighted a number of meiotic anomalies that showed direct relationships between meiotic behavior and fertility (Adeleke et al., 2002). Sporad configurations for a number of accessions were shown to exceed the usual tetrad in both diploids and triploids. These unusual sporad configurations were related to abnormal chromosome movements during sporogenesis. Lagging chromosomes and univalents were observed in anaphase I of many cultivars. These chromosomes were excluded from the nuclear material of the daughter cells as they proceed into the resting phase. There was a strong relationship between tetrad formation and pollen fertility. For example, the AA diploids 'Calcutta 4' and 'Long Tavoy' showed very high pollen fertility levels that corresponded strongly with the high percentage of tetrads observed in these accessions. Interestingly, the diploids 'Pitu' and 'SF265' had lower pollen fertility although they did not manifest any irregular sporad formation. In general, pollen fertility decreased in the triploids, with the plantains (AAB) showing a very marked decline in pollen fertility. Univalents were observed frequently in *Musa*, indicating the erratic chromosome movements observed generally in this genus at anaphase. Further studies are required to show the full extent in which chromosome behavior and sterility in *Musa* are correlated.

4.7 CONCLUSION

New techniques in cytogenetics are constantly evolving. While serious efforts have been made in developing genetic maps for many crops species, much less effort is being devoted to physical mapping. The majority of linkage maps developed in plant species is not integrated with any type of physical map (Cheng et al., 2001). In plants with both large and small chromosomes, recombination is mainly distributed along the distal half of the chromosomes. Many studies have shown that recombination in the centromeric regions, which may account for almost 50% of the length of the chromosomes, is suppressed (Werner et al., 1992; Gill et al., 1993; Delaney et al., 1995, 1995b). Therefore there is need to align linkage and physical maps. Physical maps are being developed for many crops mainly by direct visualization of DNA sequences on chromosomes by FISH. In plants like banana with small chromosomes, it seems that FISH on pachytene chromosomes may be the route to follow. Application of BAC-FISH on pachytene complements can contribute significantly to the construction of physical maps and to map base cloning by confirming the physical location of markers on the linkage groups (Tang et al., 2009).

REFERENCES

Adeleke, M.T.V., M. Pillay, and B.E. Okoli. 2002. An improved procedure for examining meiotic chromosomes in *Musa*. *HortScience* 37:959–961.

Adeleke, M.T.V., M. Pillay, and B.E. Okoli. 2004. Meiotic irregularities and fertility relationships in diploid and triploid *Musa* L. *Cytologia* 69:387–393.

Aert, R., L. Sági, and G. Volckaert. 2004. Gene content and density in banana (*Musa acuminata*) as revealed by genomic sequencing of BAC clones. *Theor. Appl. Genet.* 109:129–139.

Afza, R., D. Kammer, J. Dolezel, J. Konig, M. Van Duren, G. Kahl, and F.J. Novak. 1993. The potential of nuclear DNA flow-cytometry and DNA fingerprinting for Musa improvement programmes. In: *Proc. Int. Symp. Breeding Banana and Plantain for Resistance to Disease and Pests*, J. Ganry, ed., 65–75. Montpellier, France: CIRAD.

Arnaud, E. and J-P. Horry. 1997. *Musalogue, a catalogue of Musa germplasm, Papua New Guinea collecting missions, 1988–1989*. Montpellier, France: INIBAP.

Arumuganathan, K. and E.D. Earle. 1991. Nuclear DNA content of some important species. *Plant Mol. Biol. Rep.* 9:208–218.

Asker, S. 1980. Gametophytic apomixis: Elements and genetic regulation. *Hereditas 93*:277–293.

Assani, A., F. Bakry, F. Kerbellec, R. Haicour, G. Wenzel, and B. Foroughi-Wehr. 2003. Production of haploids from anther culture of banana [(*Musa balbisiana* (BB)]. *Plant Cell Rep.* 21:511–516.

Assani, A., D. Chabane, R. Haicour, F. Bakry, G. Wenzel, and B. Foroughi-Wehr. 2005. Protoplast fusion in banana (*Musa* spp.): Comparison of chemical (PEG: polyethylene glycol) and electrical procedure. *Plant Cell Tissue Org. Cult.* 83:145–151.

Balint-Kurti, P.J., S.K. Clendennen, M. Dolezelova, M. Valarik, J. Dolezel, P.R. Beetham, and G.D. May. 2000. Identification and chromosomal localization of the *monkey* retrotransposon in *Musa* sp. *Mol. Gen. Genet.* 263:908–915.

Baranyi, M. and J. Greilhuber. 1996. Flow cytometric and Feulgen densitometric analysis of genome size in *Pisum*. *Theor. Appl. Genet.* 92:297–307.

Bartos, J., O. Alkhimova, M. Doleželová, E. De Langhe, and J. Doležel. 2005. Nuclear genome size and genomic distribution of ribosomal DNA in *Musa* and *Ensete* (Musaceae): Taxonomic implications. *Cytogenet. Genome Res.* 109:50–57.

Bhattramakki, D. and A. Rafalski. 2001. Discovery and application of single nucleotide polymorphism markers in plants. In: *Plant genotyping: The DNA fingerprinting of plants*, R.J. Henry, ed., 179–191. Wallingford: CAB International.

Carreel, F. 1995. Etude de la diversite genetique des bananiers (genre Musa) a l'aide des marqueurs RFLP. PhD diss., Institut National Agronomique Paris–Grignon.

Carreel, F., D. Gonzalez de Leon, P. Lagoda, C. Lanaud, C. Jenny, J-P. Horry, and H. Tezenas du Montcel. 2002. Ascertaining maternal and paternal lineage within *Musa* by chloroplast and mitochondrial DNA RFLP analysis. *Genome* 45:679–692.

Cavallini, A. and L. Natali. 1990. Nuclear DNA variability within *Pisum sativum* (*Leguminosae*): Cytophotometric analysis. *Plant Syst. Evol.* 173:179–183.

Cavallini, A., L. Natali, G. Cionini, and D. Gennai. 1993. Nuclear DNA variability within *Pisum sativum* (*Leguminosae*): Nucleotypic effects on plant growth. *Heredity* 70:561–565.

Cheng, Z, G.G. Presting, C.R. Buell, R.A. Wing, and J. Jiang. 2001. High resolution chromosome mapping of bacterial artificial chromosomes by genetic markers reveals the centromere location and the distribution of genetic recombination along chromosome 10 of rice. *Genetics* 157:1749–1757.

Cheng, Z., C.R. Buell, R.A. Wing, and J. Jiang. 2002. Resolution of fluorescence in-situ hybridization mapping on rice mitotic prometaphase chromosomes, meiotic pachytene chromosomes and extended DNA fibers. *Chromosome Res.* 10:379–387.

Côte, F.X., R. Domergue, S. Monmarson, T. Schwendiman, C. Teisson, and J.V. Escalant. 1996. Embryogenic cell suspensions from the male flower of *Musa* AAA cv. Grand Nain. *Physiol. Plant.* 97:285–290.

Cullis, C.A. 2005. Mechanisms and control of rapid genomic changes in flax. *Ann. Bot.* 95:201–208.

Danilova, T.V. and J.A. Birchler. 2008. Integrated cytogenetic map of mitotic metaphase chromosome 9 of maize: Resolution, sensitivity, and banding paint development. *Chromosoma* 117:345–356.

Dantas, J.L.L., K. Shephard, W. dos S. Soares Filho, Z.J.M. Cordeiro, S. de Oliveira Silva, and A. Silva Souza. 1993. Citogenetica e melhoramento genetico da bananeira (*Musa* spp.) *Documentos EMBRAPA-CNMPF* 48.

De Jong, J.H., P. Fransz, and P. Zabel. 1999. High resolution FISH in plants: Techniques and applications. *Trends Plant Sci.* 4:258–263.

de Koeyer, D., K. Douglass, A. Murphy, S. Whitney, L. Nolan, Y. Song, et al. 2009. Application of high-resolution DNA melting for genotyping and variant scanning of diploid and autotetraploid potato. *Mol. Breed.* 25(1):67–90.

Delaney, D.E., S. Nasuda, T.R. Endo, B.S. Gill, and S.H. Hulbert. 1995. Cytologically based physical maps of the group-2 chromosomes of wheat. *Theor. Appl. Genet.* 91:568–573.

De Langhe, E.A. 1964. The origin of variation in the plantain banana. *Mededelingen van de Landbouwhogeschool* 39:45–80. Ghent: State Agricultural University of Ghent, Belgium.

D'Hont A., A. Paget Goy, C. Jenny, J. Escoute, M. Layssac, J.C. Glaszmann, and F. Carreel. 1999. Paper presented at the Plant and Animal Genomes Conference VII, January 17–21, 1999, San Diego, CA, USA.

D'Hont, A., A. Paget-Goy, J. Escoute, and F. Carreel. 2000. The interspecific genome structure of cultivated banana, *Musa* spp, revealed by genomic DNA *in situ* hybridization. *Theor. Appl. Genet.* 100:177–183.

Dolezel, J. 1991. Flow cytometric analysis of nuclear DNA content in higher plants. *Phytochem. Anal.* 2:143–154.

Dolezel, J., M. Dolezelova, and F.J. Novák. 1994. Flow cytometric estimation of nuclear DNA amount in diploid bananas (*Musa acuminata* and *M. balbisiana*). *Biol. Plant.* 36:351–357.

Dolezel, J., M.A. Lysak, I. Van den Houwe, M. Dolezelova, and N. Roux. 1997. Use of flow cytometry for rapid ploidy determination in *Musa*. *InfoMusa* 6:6–9.

Dolezel, J., M. Dolezelova, N. Roux, and I. Van den Houwe. 1998. A novel method to prepare slides for high resolution chromosome studies in *Musa* spp. *InfoMusa* 7:3–4.

Dolezelova, M., M. Valárik, R. Swennen, J.P. Horry, and J. Dolezel. 1998. Physical mapping of the 18S-25S and 5S ribosomal RNA genes in diploid bananas. *Biol. Plant.* 41:497–505.

Dolezelova, M., J. Dolezel, I. Van Den Houve, N. Roux, and R. Swennen. 2005. Ploidy levels revealed. *InfoMusa* 14:34–36.

Ducreux, M., E. Aguillar, and A. Grapin. 2001. Plant regeneration from protoplasts of dessert banana cv. Grande Naine (*Musa* spp., Cavendish sub-group AAA) via somatic embryogenesis. *Plant Cell Rep.* 20:482–488.

du Montcel, T. H. 1989. *M. acuminata* subspecies *banksii*. In: *Identification of genetic diversity in the genus Musa*, R. Jarret, ed., 211–218. Los Banos, Philippines: INIBAP.

Englberger, L., W. Aalbersberg, P. Ravi, E. Bonnin, G.C. Marks, M.H. Fitzgerald, and J. Elymore. 2003a. Further analyses on Micronesian banana, taro, breadfruit and other foods for provitamin A carotenoids and minerals. *J. Food Compos. Anal.* 16:219–236.

Englberger, L., I. Darnton-Hill, T. Coyne, M.H. Fitzgerald, and G.C. Marks. 2003b. Carotenoid-rich bananas: A potential food source for alleviating vitamin A deficiency. *Food Nutr. Bull.* 24:303–318.

Englberger, L., J. Schierle, G.C. Marks, and M.H. Fitzgerald. 2003c. Micronesian banana, taro, and other foods: Newly recognized sources of provitamin A and other carotenoids. *J. Food Compos. Anal.* 16:3–19.

Espino, R.R.C. and R.B. Pimentel. 1989. Electrophoretic analysis of selected isozymes in BB cultivars of Philippine bananas. In: *Identification of genetic diversity in the genus Musa*, R. Jarret, ed., 36–40. Los Banos, Philippines: INIBAP.

Fahrenkrug, S.C., G.A. Rohrer, B.A. Freking, T.P. Smith, K. Osoegawa, C.L. Shu, et al. 2001. A porcine BAC library with tenfold genome coverage: A resource for physical and genetic map integration. *Mamm. Genome* 12:472–474.

Fauré, S., J.L. Noyer, J.P. Horry, F. Bakry, C. Lanaud, and D. González de León. 1993. A molecular marker-based linkage map of diploid bananas (*Musa acuminata*). *Theor. Appl. Genet.* 87:517–526.

Fehr, W.R. 1987. *Principles of cultivar development.* Vol. 1. New York: Macmillan.

Fransz, P.F., S. Armstrong, J.H. de Jong, L.D. Parnell, G.van Drunen, C. Dean, et al. 2000. Integrated cytogenetic map of chromosome arm 4S of *A. thaliana*: Structural organization of heterochromatic knob and centromere region. *Cell* 100:367–376.

Fuchs, J., M. Kuhne, and I. Schubert.1998. Assignment of linkage groups to pea chromosomes after karyotyping and gene mapping by fluorescent *in situ* hybridization. *Chromosoma* 107:272–276.

Fukui, K. 1996. Plant chromosomes at Mitosis. In: *Plant chromosomes: Laboratory methods*, K. Fukui and S. Nakayama, eds., 1–18. Boca Raton, FL: CRC Press.

Ge, X-J., M.H. Liu, W.K. Wang, B.A. Schaal, and T.Y. Chiang. 2005. Contrasting population structuring of wild banana, *Musa balbisiana*, in China based on evidence between microsatellite fingerprinting and cpDNA PCR-RFLP. *Mol. Ecol.* 14:933–944

Gill, K.S., B.S. Gill, and T.R. Endo. 1993. A chromosome region specific mapping strategy reveals gene-rich telomeric ends in wheat. *Chromosoma* 102:374–381.

Gimenez, C., E. De Garcia, N.X. De Enrech, and I. Blanca. 2001. Somaclonal variation in banana: Cytogenetic and molecular characterization of the somaclonal variant CIEN BTA-03. *In Vitro Cell. Dev. Biol. Plant* 37:217–222.

Gortner, G., M. Nenno, K. Weising, D. Zink, W. Nagl, and G. Kahl. 1998. Chromosomal localization and distribution of simple sequence repeats and the *Arabidopsis*-type telomere sequence in the genome of *Cicer arietinum* L. *Chrom. Res.* 6:97–104.

Graham, M.J., C.D. Nickell, and A.L. Rayburn. 1994. Relationship between genome size and maturity group in soybean. *Theor. Appl. Genet.* 88:429–432.

Greilhuber, J. and R. Obermayer. 1997. Genome size and maturity group in *Glycine max* (soybean). *Heredity* 78:547–551.

Hari, P.C. 1989. Morphological study of *Musa acuminata* ssp. *malaccensis* (Ridley) Simmonds to determine taxonomic characteristics. *Saussurea* 19:187–214.

Heslop-Harrison, J.S. 1996. Electron microscopy in plant chromosomes. In: *Plant chromosomes: Laboratory methods*, K. Fukui and S. Nakayama, eds., 225–245. Boca Raton, FL: CRC Press.

Hirochika, H., K. Sugimoto, H. Otsuki, H. Tsugawa, M. Kanda. 1996. Retrotransposons of rice involved in mutations induced by tissue culture. *Proc. Natl. Acad. Sci. (USA)* 93:7783–7788.

Horry, J-P., R. Ortiz, E. Arnaud, J.H. Crouch, R.S.B. Ferris, D.R. Jones, et al. 1997. Banana and plantain. In: *Biodiversity in trust: Conservation and use of plant genetic resources in CGIAR centres*, D. Fuccillo, L. Sears, and P. Stapleton, eds., 67–81. Cambridge: Cambridge University Press.

Horry, J-P., J. Dolezel, M. Dolezelova, and M.A. Lysak. 1998. Do natural A*x*B tetraploid bananas exist? *InfoMusa* 7:5–6.

Hribova, M., M. Dolezelova, and J. Dolezel. 2008. Localization of BAC clones on mitotic chromosomes of *Musa acuminate* using fluorescence *in situ* hybridization. *Biologia Plantarum* 52(3):445–452.

Hwang, S.C. 2002. Somaclonal variational approach to breeding Cavendish banana for resistance to Fusarium wilt race 4. Paper presented at global conference on banana and plantain, October 28–31, 2002, Bangalore, India.

Hwang, S.C. and W.H. Ko. 1988. Mutants of Cavendish banana resistant to race 4 of *Fusarium oxysporum* f. sp. *cubense. Plant Prot. Bull.* (Taiwan) 30:386–392.

Hwang, S.C. and W.H. Ko. 1989. Improvement of fruit quality of Cavendish banana mutants resistant to race 4 of *Fusarium oxysporum* f. sp. *cubense. Plant Prot. Bull.* (Taiwan) 31:131–138.

Hwang, S.C. and W.H. Ko. 2004. Cavendish banana cultivars resistant to Fusarium wilt acquired through somaclonal variation in Taiwan. *Plant Dis.* 88:580–588.

Jaccoud, D., K. Peng, D. Feinstein, and A. Kilian. 2001. Diversity arrays: A solid state technology for sequence information independent genotyping. *Nucleic Acids Res.* 29:e25.

Jenny, C., F. Carreel, and F. Bakry. 1997. Revision on banana taxonomy: Klue Tiparot (*Musa* sp.) reclassified as a triploid. *Fruits* 52:83–91.

Jones, D.R. 2000. Introduction to banana, abaca and enset. In: *Diseases of banana, abaca and enset*, D.R. Jones, ed., 1–36. Wallingford, UK: CABI Publishing, CAB International.

Kaemmer, D., D. Fischer, R.L. Jarret, F-C. Baurens, A. Grapin, D. Dambier, et al. 1997. Molecular breeding in the genus *Musa*: A strong case for STMS marker technology. *Euphytica* 96:49–63.

Kamaté, K., S. Brown, P. Durand, J.M. Bureau, D. De Nay, and T.H. Trinh. 2001. Nuclear DNA content and base composition in 28 taxa of *Musa*. *Genome* 44:622–627.

Kidwell, M.G. and D. Lisch. 1997. Transposable elements as sources of variation in animals and plants. *Proc. Natl. Acad. Sci. (USA)* 9:7704–7711.

Kulikova, O., G. Gualtieri, R. Geurts, D.J. Kim, D. Cook, T. Huguet, et al. 2001. Integration of the FISH pachytene and genetic maps of *Medicago truncatula*. *Plant J*. 27:49–58.

Lambie, E.I. and G.S. Roeder. 1986. Repression of meiotic crossings over by a centromere (Cen30) in *Saccharomyces cerevisiae*. *Genetics* 114:768–769.

Larkin, P.J. and W.R. Scowcroft. 1981. Somaclonal variation—A novel source of variability from cell cultures for plant improvement. *Theor. Appl. Genet*. 60:197–214.

Lebot, V., K.M. Aradhya, R. Manshardt, and B. Meilleur. 1993. Genetic relationships among cultivated bananas and plantains from Asia and the Pacific. *Euphytica* 67:163–175.

Lichter, P., C.J. Tang, K. Call, G. Hermanson, G.A. Evans, D. Housman, and D.C. Ward. 1990. High-resolution mapping of human chromosome 11 by in situ hybridization with cosmid clones. *Science* 247:64–69.

Lysak, M.A., M. Dolezelova, J.P. Horry, R. Swennen, and J. Dolezel. 1999. Flow cytometric analysis of nuclear DNA content in *Musa*. *Theor. Appl. Genet*. 98:344–1350

Maluszynska, J. and J.S. Heslop-Harrison. 1993. Molecular cytogenetics of the genus *Arabidopsis, in situ* hybridization of rDNA sites, chromosome numbers and diversity in centromeric heterochromatin. *Ann. Bot*. 71:479–484.

Mandal, A.B., A. Maiti, B. Chowdhury, and R. Elanchezhian. 2001. Isoenzyme markers in variety identification of banana. *In Vitro Cell. Dev. Biol. Plant* 37:599–604.

Matsumoto, K. and S. Oka. 1998. Plant regeneration from protoplasts of a Brazilian dessert *banana* (*Musa* spp., AAB group). *Acta Hort*. 490:455–462.

Matsumoto, K., A.D. Vilarinhos, and S. Oka. 2002. Somatic hybridization by electrofusion of banana protoplasts. *Euphytica* 125:317–324.

Megia, R., R. Haicour, S. Tizroutine, V.B. Trang, L. Rossignol, D. Sihachakr, et al. 1993. Plant regeneration from cultured protoplasts of the cooking banana cv. Bluggoe (*Musa* spp., ABB group). *Plant Cell Rep*. 13:41–44.

Nair, A.S., C.H. Teo, T. Schwarzacher, and J.S. Heslop-Harrison. 2005. Genome classification of banana cultivars from South India using IRAP markers. *Euphytica* 144:285–290.

Nasution, R.E. 1991. A taxonomic study of the species *Musa acuminata* Colla with its infraspecific taxa in Indonesia. *Mem. Tokyo Univ. Agric* 32:1–122.

Novak, F.J., R. Afza, M. van Duren, M. Perea-Dallos, B.V. Conger, and T. Xiaolang. 1989. Somatic embryogenesis and plant regeneration in suspension cultures of dessert (AA and AAA) and cooking (ABB) banana (*Musa* spp.). *Biotechnology* 7:154–159.

Nsabimana, A. and J. Van Staden. 2006. Assessment of genetic diversity of Highland bananas from the National Germplasm Collection at Rubona, Rwanda, using RAPD markers. *Sci. Hort.* 113:293–299.

Nyine, M. and M. Pillay. 2010. The effect of banana breeding on the diversity of East African Highland banana. *Acta Hort.* (in press).

Nwakanma, D.C., M. Pillay, B.E. Okoli, and A.Tenkouano. 2003. Sectional relationships in the genus *Musa* L. inferred from PCR-RFLP of organelle DNA sequences. *Theor. Appl. Genet*. 107:850–856.

Oh, T.J, M.A. Cullis, K. Kunert, I. Engelborghs, R. Swennen, and C.A. Cullis. 2007. Genomic changes associated with somaclonal variation in banana (*Musa* spp). *Physiologia Plantarum* 129(4):766–774.

Ortiz, R. 1997. Secondary polyploids, heterosis and evolutionary crop breeding for further improvement of the plantain and banana (*Musa* spp. L.) genome. *Theor. Appl. Genet*. 94:1113–1120.

Ortiz-Vazquez, E., D. Kaemmer, H.B. Zhang, J. Muth, M. Rodriguez-Mendiola, C. Arias-Castro, et al. 2005. Construction and characterization of a plant transformation-competent BIBAC library of the black Sigatoka-resistant banana *Musa acuminata* cv. Tuu Gia (AA). *Theor. Appl. Genet.* 110:706–713.

Oselebe, H.O., A. Tenkouano, and M. Pillay. 2006a. Ploidy variation of *Musa* hybrids from crosses. *Afr. J. Biotechnol.* 5:1048–1053.

Oselebe, H.O., A. Tenkouano, M. Pillay, I.U. Obi, and M.I. Uguru. 2006b. Ploidy and genome segregation in *Musa* breeding populations assessed by flow cytometry and randomly amplified polymorphic markers. *J. Hort. Sci.* 131:780–786.

Osuji, J.O., J. Crouch, G. Harrison, and J.S. Heslop-Harrison. 1997a. Identification of the genomic constitution of *Musa* L. genotypes (bananas, plantains and hybrids) using molecular cytogenetics. *Ann. Bot.* 80:787–793.

Osuji, J.O., D. Vuylsteke, and R. Ortiz. 1997b. Ploidy variation in hybrids from interploid $3x \times 2x$ crosses in *Musa*. *Tropicultura* 15:37–39.

Osuji, J.O., J. Crouch, G. Harrison, and J.S. Heslop-Harrison. 1998. Molecular cytogenetics of *Musa* species, cultivars and hybrids: Location of 18S-5.8S-25S and 5S rDNA and telomere-like sequences. *Ann. Bot.* 82:243–248.

Panis, B., A.W. Wauwe, and R. Swennen. 1993. Plant regeneration through direct somatic embryogenesis from protoplasts of banana (*Musa* spp.). *Plant Cell Rep.* 12:403–407.

Pillay, M. and M.T.V. Adeleke. 2001. Silver staining of *Musa* L. chromosomes. *Cytologia* 66:33–37.

Pillay, M. and L. Tripathi. 2006. Banana: An overview of breeding and genomics research in *Musa*. In: *Genome mapping and molecular breeding in plants*, Vol. 4, *Fruits and nuts*, C. Kole, ed., 281–301. Heidelberg: Springer-Verlag.

Pillay, M. and L. Tripathi. 2007. Banana breeding. In: *Breeding major food staples*, M.S. Kang and P.M. Priyadarshan, eds., 393–428. New York: Blackwell Science.

Pillay, M., D.C. Nwakanma, and A. Tenkouano. 2000. Identification of RAPD markers linked to A and B genome sequences in *Musa*. *Genome* 43:763–767.

Pillay, M., E. Ogundiwin, D.C. Nwakanma, G. Ude, and A. Tenkouano. 2001 Analysis of genetic diversity and relationships in East African banana germplasm. *Theor. Appl. Genet.* 102:965–970.

Pillay, M., A. Tenkouano, and J. Hartman. 2002. Bananas and plantains: Future challenges in *Musa* breeding. In: *Crop improvement, challenges in the twenty-first century*, M.S. Kang, ed., 223–252. New York: Food Products Press.

Pillay, M., J. Hartman, C. Dimkpa, and D. Makumbi. 2003a. Establishing the genome of 'Sukali Ndizi.' *African Crop Sci. J.* 11:119–124.

Pillay, M., C. Dimkpa, G. Ude, D. Makumbi, and W. Tushmereirwe. 2003b. Genetic variation in the banana cultivar 'Sukali Ndizi' grown in different regions of Uganda based on RAPD and AFLP markers. *Afr. Crop Sci. J.* 11:87–95.

Pillay, M., A.Tenkouano, G. Ude, and R. Ortiz. 2004. Molecular characterization of genomes in *Musa*. In: *Banana improvement: Cellular, molecular biology, and induced mutations*, M.S. Jain and R. Swennen, eds., 271–286. Plymouth, UK: Science Publishers.

Pillay, M., E. Ogundiwin, A. Tenkouano, and J. Dolezel. 2006. Ploidy and genome composition of *Musa* germplasm at the International Institute of Tropical Agriculture. *Afr. J. Biotechnol.* 5:1224–1232.

Purseglove, J.W. 1988. *Tropical crops: Monocotyledons*. London: Longman Scientific and Technical.

Robinson, J.C. 1996. *Bananas and plantains*. Crop Production Science in Horticulture 5. Wallingford, UK: CAB International.

Safar, J., J.C. Noa-Carrazana, J. Vrana, J. Bartos, O. Alkhimova, X. Sabau, et al. 2004. Creation of a BAC resource to study the structure and evolution of the banana (*Musa balbisiana*) genome. *Genome* 47:1182–1191.

Schulz-Schaeffer, J. 1980. *Cytogenetics, plants and humans*. Berlin: Springer-Verlag.

Sharrock, S. 1989. Collecting *Musa* in Papua New Guinea. In: *Identification of genetic diversity in the genus Musa*, R. Jarret, ed., 140–157. Los Banos, Philippines: INIBAP.

Shepherd, K. 1988. Observations on *Musa* taxonomy. In: *Identification of genetic diversity in the genus Musa*, R. Jarrett, ed., 158–165. Los Banos, Philippines: INIBAP.

Shepherd, K. 1999. *Cytogenetics of the genus Musa.* Montpellier, France: IPGRI; Rome: INIBAP.

Shepherd, K. and F.R. Ferreira. 1984. The PNG Biological Foundation banana collection at Laloki, Port Moresby, Papua New Guinea. *South East Asia Newslett.* 8:28–34.

Silayoi, B. and N. Chomchalow. 1987. Cytotaxonomic and morphological studies of Thai banana cultivars. *ACIAR Proceedings* 21:157–160.

Simmonds, N.W. 1962. *The Evolution of the Bananas*. London: Longmans Green and Co.

Simmonds, N.W., 1966. *Bananas*, 2nd ed. Tropical Agriculture Series. London: Longman.

Simmonds, N.W. and K. Shepherd. 1955. The taxonomy and origins of the cultivated bananas. *J. Linn. Soc. Bot.* 55:302–312.

Simmonds, N.W. and S.T.C. Weatherup. 1990. Numerical taxonomy of the wild bananas (*Musa*). *New Phytol.* 115:567–571.

Sotto, R.C. and R C. Rabara. 2000. Morphological diversity of *Musa balbisiana* Colla in the Philippines. *InfoMusa* 9(2):28–30.

Stace, C. 1980. *Plant taxonomy and biosystematics*. London: Edward Arnold.

Stover, R.H. and N.W. Simmonds. 1987. *Bananas*, 3rd ed. Essex, England: Longman Scientific and Technical.

Taketa, S., H. Ando, K.Takeda, G.E. Harrison, and J.S. Heslop-Harrison. 2000. The distribution, organization and evolution of two abundant and widespread repetitive DNA sequences in the genus *Hordeum*. *Theor. Appl. Genet.* 100:16–176.

Tang, X., J. de Boer, H. J. van Eck, C. Bachem, R.G. Visser, and H. de Jong. 2009. Assignment of genetic linkage maps to diploid *Solanum tuberosum* pachytene chromosomes by BAC-FISH. *Chromosome Res.* 17:899–915.

Tenkouano, A., J.H. Crouch, H.K. Crouch, and D. Vuylsteke. 1998. Ploidy determination in *Musa* germplasm using pollen and chloroplast characteristics. *HortScience* 33(5):889–890.

Teo, C.H., S.H. Tan, C.H. Ho, Q.Z. Faridah, Y.R. Othman, J.S. Heslop-Harrison, et al. 2005. Genome constitution and classification using retrotransposon based markers in the orphan crop banana. *J. Plant Biol.* 48:96–105.

Ude, G., M. Pillay, D. Nwakanma, and A. Tenkouano. 2002a. Genetic diversity in *Musa acuminata* Colla and *M. balbisiana* Colla and some of their natural hybrids using AFLP markers. *Theor. Appl. Genet.*104:1246–1252.

Ude, G., M. Pillay, D. Nwakanma, and A. Tenkouano. 2002b. Analysis of genetic diversity and sectional relationships in *Musa* using AFLP markers. *Theor. Appl. Genet.* 104:1239–1245.

Uma, S., S. Sathiamoorthy, and P. Durai. 2005. *Banana—Indian genetic resources and catalogue*. Tiruchirapalli, India: NRCB (ICAR).

Vakili, N.G. 1967. The experimental formation of polyploidy and its effect in the genus *Musa*. *Am. J. Bot.* 54:24–36.

Valárik, M., H. Simkova, M. Hribova, J. Safar, M. Dolezelova, and J. Dolezel. 2002. Isolation, characterization and chromosome localization of repetitive DNA sequences in bananas (*Musa* ssp.). *Chromosome Res.* 10:89–100.

Valarik, M., J. Bartos, P. Kovarova, M. Kubalakova, J.H. De Jong, and J. Dolezel. 2004. High-resolution FISH on super-stretched flow-sorted plant chromosomes. *Plant J.* 37:940–950.

Valmayor, R.V., F.N. Rivera, and F.M. Lomoljo. 1981. *Philippine banana cultivar names and synonyms*. NPGRL, Institute of Plant Breeding Bulletin No. 3. Los Banos: University of the Philippines.

Valmayor, R.V., R.R.C. Espino, and O.C. Pascua. 2002. *The wild and cultivated bananas of the Philippines*. Los Banos, Laguna: PARRFI and BAR.

Vandenhout, H.R., R. Ortiz, D. Vuylsteke, R. Swennen, and K.V. Bai. 1995. Effect of ploidy on stomatal and other quantitative traits in plantain and banana hybrids. *Euphytica* 83:117–122.

Vilarinhos, A.D., P. Piffanelli, P. Lagoda, S. Thibivilliers, X. Sabau, F. Carreel, et al. 2003. Construction and characterization of a bacterial artificial chromosome library of banana (*Musa acuminata* Colla). *Theor. Appl. Genet.* 106:1102–1106.

Vilarinhos, A., F. Carreel, M. Rodier, I. Hippolyte, A. Benabdelmouna, D. Triaire, et al. 2006. Characterization of translocations in banana by FISH of BAC clones anchored to a genetic map. Paper presented at the International Conference Plant and Animal Genome XIV, January 14–18, 2006, San Diego, California.

Vuylsteke, D., R. Swennen, and E. De Langhe. 1991. Somaclonal variation in plantains (*Musa* spp. AAB group) derived from shoot-tip culture. *Fruits* 46:429–439.

Vuylsteke, D., R. Swennen, and R. Ortiz. 1993. Development and performance of black Sigatoka-resistant tetraploid hybrids of plantain (*Musa* spp., AAB group). *Euphytica* 65:33–42.

Wang, Z., Z. Lin, and K. Pan. 1993. Cytogenetical studies in *Musa* (*Eumusa*). In: *Current banana research and development in China*, 29–43. Guangzhou: South China Agricultural University.

Wang, X-L., T-Y. Chiang, N. Roux, G. Hao, and X-J. Ge. 2005. Genetic diversity of wild banana (*Musa balbisiana* Colla) in China as revealed by AFLP markers. *Genet. Resour. Crop Evol.* 54:1125–1132.

Wang, C-J.R., L. Harper, and W. Zacheus Cande. 2006. High-resolution single-copy gene fluorescence in situ hybridization and its use in the construction of a cytogenetic map of maize chromosome 9. *Plant Cell* 18:529–544.

Wang, K., Y. Zaijie, C. Shu, J. Hu, Q. Lin, W. Zhang, et al. 2009. Higher axial-resolution and sensitivity pachytene fluorescence in situ hybridization protocol in tetraploid cotton. *Chromosome Res.* 17:1041–1050.

Werner, J.E., T.R. Endo, and B.S. Gill. 1992. Towards a cytogenetically based physical map of the wheat genome. *PNAS* 89:11307–11311.

Wilson, G.B. 1946. Cytological studies in the Musae I. Meiosis in some triploid clones. *J. Genet.* 31:341–358.

Xiao, W., X. Huang, Y.R. Wei, J.T. Zhao, X.M. Dai, and X.L. Huang. 2008. Plant regeneration from protoplast culture of *Musa* AAB Silk cv. Guoshanxiang. *Act. Hort. Sin.* 35(6):873–878, in Chinese with English abstract.

Xiao, W., X. Huang, Q. Gong, X-M. Dai, J-T. Zhao, Y-R. Wei, et al. 2009. Somatic hybrids obtained by asymmetric protoplast fusion between *Musa* Silk cv. Guoshanxiang (AAB) and *Musa acuminate* cv. Mas (AA). *Plant Cell Tissue Organ Cult.* 97:313–321.

Yuan, Q., F. Liang, J. Hsiao, V. Zismann, and M-I. Benito. 2000. Anchoring of rice BAC clones to the rice genetic map in silico. *Nucl. Acids Res.* 28:3636–3641.

Zhong, X.B., P. Fransz, J. Wennekes-van Eden, P. Zabel, A. van Kammen, and J.H. de Jong. 1996. High resolution mapping on pachytene chromosomes and extended DNA fibers by fluorescence *in-situ* hybridisation. *Plant Mol. Biol. Report.* 14:232–242.

Zhong, X.-B., J. Bodeau, P.F. Fransz, V.M. Williamson, A. van Kammen, J.H. de Jong, et al. 1999. FISH to meiotic pachytene chromosomes of tomato locates the root-knot nematode resistance gene *Mi-1* and the acid phosphatase gene *Aps-1* near the junction of euchromatin and pericentromeric heterochromatin of chromosome arms 6S and 6L, respectively. *Theor. Appl. Genet.* 98:365–370.

Zoller, J.F., R.G. Herrmann, and G. Wanner. 2004. Chromosome condensation in mitosis and meiosis of rye (*Secale cereale* L.). *Cytogenet. Genome Res.*105:134–144.

5 Genetics of Important Traits in *Musa*

Eli Khayat and Rodomiro Ortiz

CONTENTS

5.1 INTRODUCTION

Despite parthenocarpy and female sterility, the genus *Musa* portrays a wealth of alleles that govern important agronomic as well as consumer traits. Initially, cultivars were selected by plant vigor, yield, and hardiness—traits closely linked to higher ploidy levels (Simmonds, 1962). However, while most triploid cultivars are useless to breeders due to the high occurrence of female sterility, tetraploids are undesirable due to early senescence, fruit drop, short shelf life, and a weak pseudostem (Pillay et al., 2002). Nearly all domesticated bananas and plantains are inter- or intraspecific triploid hybrids that constitute all possible combinations of the A and B genomes that are represented by *Musa acuminata* Colla and *M. balbisiana* Colla, respectively. In addition to the growth characteristics, triploids were preferred by farmers due to their seedless phenotype. In this regard, cultivated bananas differ from their wild diploid ancestors (AA, BB genomes), which bear large hard seeds.

Hybridization of diploid *Musa* species, followed by recombination events and mutations, generated enough genetic diversity to allow domestication of diploid cultivars bearing parthenocarpic fruit (Simmonds, 1962). Sexual polyploidization ($2n \times n$) resulted in almost-sterile triploid cultivars, which multiplied through sucker propagules in the centers of origin. During the long history of cultivation of triploid cultivars, new mutants originated and farmers selected among them for specific genetic combinations that showed desirable phenotypes. Domestication occurred after vegetative propagation of a few selected high-yielding phenotypes and linkages containing favorable combinations. Given that parthenocarpy was a fundamental precondition, the gene pool of cultivated clones remained stagnant throughout the history of banana domestication. Ironically, the abundance of alleles for each trait in the genus *Musa* is not resourceful enough to alter specific characteristics in commercial cultivars.

In the 80 years of banana and plantain breeding, very few qualitative or quantitative traits have been improved in commercially important cultivars when compared to other crop species. However, recent advancements in banana structural and functional genomics have opened new horizons for banana breeding through genetic engineering (Crouch et al., 1998; Khayat et al., 2004a). The technology's impressive contributions to insights in population and molecular genetics have been fueled by advances in computational methodology (Khayat et al., 2004a), and indeed these insights and methodologies have spurred developments in the understanding of the banana genome and to some extent the mapping of genes that control specific traits on virtual chromosomes. Still, the complexity of positional cloning in banana, combined with the lack of saturated linkage maps, hinders gene-targeted hybridization breeding. The inability of researchers to use map-based cloning is a serious limitation to identifying strategic loci in segregating populations. However, the recent progress made in banana genomics (Cheung and Town, 2007) will undoubtedly assist breeders to identify interesting loci and genes that govern important traits. To live up to such expectations will be a tall order due to considerable obstacles remaining to be surmounted. The main barriers are a long-cycling time span from seed to seed in the triploid lines, and the low fertility of most polyploid cultivars.

Through domestication, several morphological and physiological traits were emphasized to distinguish domesticated crops from their wild ancestors. These characteristics are referred to as the "domestication syndrome" (Poncet et al., 1998). Normally, syndrome characteristics include a more compact growth habit, increased earliness, reduction/loss of seed dispersal and dormancy, gigantism, and increased morphological diversity in the consumed portion of the plant (Frary and Doganlar, 2003). Studies on the domestication syndrome and domestication process have revealed that numerous traits that distinguish crop plants from their wild relatives are often governed by a relatively small number of loci having effects of unequal magnitude (Frary and Doganlar, 2003).

The differences between domesticated bananas and their wild relatives involve parameters that govern yield and fruit characteristics. Above all, farmers tended to select hybrids exhibiting parthenocarpic fruit.

5.2 COMPONENTS OF YIELD

The components of yield are well defined by banana and plantain producers and breeders. These include bunch weight, number of marketable hands in a bunch, number of fingers in a hand, finger size, cycling time, and number of stems per hectare. In the export industry (mainly Cavendish), the most commonly used yield parameter is box per stem ratio. This appraisal combines weight and fruit quality. These parameters are widely used for selection of hybrids (Osuji et al., 1997; Krikorian et al., 1993; Crouch et al., 2000) and somaclones (Khayat et al., 2004b).

The merit of selecting yield components is limited due to masking of the traits by diseases and pests. It was therefore argued that subtle quantitative traits are best selected for in subtropical regions that are devoid of diseases and pests (Khayat et al., 2004). To this end, somaclonal selections of Cavendish were highly productive in one of the most marginal climates for banana growth: northern Israel. The yield performance of these selections was proven superior not only in subtropical areas such as in Israel and South Africa (Eckstein et al., 1998) but also for tropical climates such as the Davao region of the Philippines (Khayat et al., 2004), Africa, and Latin America (unpublished results).

The shift from diploid to polyploid karyotypes greatly increased the bunch weight and fruit size of both banana and plantain hybrids (Ortiz and Vuylsteke, 1995a). The combination of A and B genomes contributes to productivity.

Banana and plantain breeders should quantify additive and nonadditive components of genetic variance to choose appropriate selection methods to improve quantitative traits. Furthermore, narrow sense heritability, which is the ratio between the additive genetic variance and the total phenotypic variance, is a scale-independent quantity that plays an important role in the theory of selection methods because the response to selection can be predicted if the breeder knows the target trait heritability. Triploid parent-offspring regression analysis suggests that bunch weight, number of fruits, and fruit length had intermediate to low heritability in tetraploid plantain-banana hybrids from $3x$–$2x$ crosses (Ortiz, 2000). This indicates that they could not be predicted using the phenotype of the triploid parent and that progeny testing will be required to assess and select the best parents for population improvement of such traits. Progeny testing in secondary triploid hybrids (from $4x$–$2x$ crosses) shows, however, that general combining ability was much greater than specific combining ability for yield and associated traits (Tenkouano et al., 1998). This result suggests therefore that the $4x$–$2x$ breeding scheme should aim to accumulate favorable alleles in potential tetraploid and diploid parents through recurrent selection.

5.3 PLANT ARCHITECTURE

Due to cultural impacts, banana breeders emphasized structural characteristics of the pseudostem, the bunch, and roots. Plant size, from rhizome to the petiole, greatly fluctuates in many cultivars. Cavendish populations (*M. acuminata* Cavendish, AAA), particularly plants produced by tissue culture, are notoriously known to mutate to either "dwarf" or "tall" phenotypes. In general, the dwarf mutants develop a small deformed bunch (Reuveni et al., 1996; Khayat et al., 2004). For practical reasons, lower stature plants are considered advantageous. However, in most cultivars, taller plants are positively correlated with a larger bunch. For instance, short Cavendish cultivars 'Buena Vista' and 'Dwarf Cavendish' showed a yield reduction of approximately 11% in comparison to 'Grand Nain' (Eckstein et al., 1998). An unusual exception to this rule was recently selected in Western Galilee (Israel). A somaclonal mutant of 'Zelig' (a Cavendish selection) registered under the name 'Adi' was selected for a combined low stature and large bunch (Figure 5.1). The average plant height of the selected clone is approximately 230 cm, compared to 315 cm of its originator 'Zelig.' The average bunch weight and finger size of 'Adi' are both larger than 'Zelig' (Khayat, 2009).

The "dwarf" and "tall" mutants in Cavendish have been characterized as under- or oversensitive to the hormone gibberellic acid, respectively. An *in vitro* bioassay was developed where a low concentration of gibberelic acid 3 (GA3) was included in the growth medium during the elongation stage of the tissue culture. Wild-type plants in the GA3 containing medium elongated above the plants that were placed on a substrate devoid of the hormone (control treatment). "Dwarf" mutants on the hormone medium grew below the height of the control while oversensitive mutants grew above (Khayat et al., 2004). A mutant that was depicted as "Long and Narrow Leaves" (LNL) was created by exhaustive cycling of tissue culture, leading to induction of retrotransposon expression and transposition. These experiments (Khayat et al., 2004b) led to the conclusion that the long-internode mutation is caused by insertions of retro-elements in a chromosomal locus having an effect on plant height. An extensive number of cycles in tissue culture enhances the activity of retrotransposable elements in the genome. Similar results were reported in rice where

FIGURE 5.1 'Adi,' a new Cavendish selection. Comparison of height between the selections 'Adi,' 'Zelig,' and 'Jaffa.'

specific genes were mutated by transposition of Tos 17, a rice retrotransposon activated by tissue culture (Hirochika et al., 1996). Interestingly, in rice at least 50% of the Tos 17 landing sites were detected as coding regions. The "Landing Pad" hypothesis of retrotransposable elements may explain the high frequency of "dwarf" mutants in *Musa*. The conserved *Ty1-copia* group of retrotransposons, highly expressed in banana (Khayat et al., 2004), is ubiquitous to various dicotyledonous and monocotyledonous plant species, including potatoes, onions, fava beans, barley, and rye (Amar et al., 1997; Kumar et al., 1997).

The process of tissue culture also induces mutations mediated by DNA methylation (Peraza-Echeverria et al., 2001). It was pointed out that the source of explants had an effect on the degree of polymorphism. *In vitro* plants produced from floral meristems showed greater frequencies of methylation in comparison to sucker meristems.

Other mechanisms of point mutations, such as deletions or nucleotide replacements, could not account for the high incidence of this mutation. "Semi-dwarf" mutants were isolated by Tang (2005) in Cavendish. The author describes the Cavendish selection 'TC1–229' as having a stature of 70 cm lower than its originator, 'Tai-Chiao No 3.' The dwarf French plantain 'Njock Kon' was postulated by Swennen and Vuylsteke (1987) to be a mutation of a giant French cultivar. This finding was later confirmed by Osuji et al. (1997). The height of the pseudostem—that is, the distance between the soil and the petioles of the highest leaves—is used for subgrouping plantain cultivars into giant, medium, and small. *Musa* clones with short internodes are called "dwarf" cultivars (Ortiz and Vuylsteke, 1998a). Such a dwarfism is controlled by the single recessive gene *dw* (Ortiz and Vuylsteke, 1995b), which appears to be close to the centromere and shows a dosage effect at the tetraploid level.

The inhibition of lateral bud growth due to substances released by the terminal bud is a limiting factor for plantain perennial productivity. The inheritance of apical dominance is controlled by the major recessive gene *ad* (Ortiz and Vuylsteke, 1994a). The dominant *Ad* allele improves the suckering of the crop as measured by the height of the tallest sucker in plantain-banana hybrids. The *Ad* gene may regulate gibberellic acid production, but this allele shows incomplete penetrance and variable expressivity.

Another important morphological characteristic is the form of the fruit bunch. For the export industry, a large cylindrical bunch is highly valued for its characteristic to reduce abrasion of the fingers with each other (Robinson et al., 1993). A formula was proposed to characterize the bunch conformation. To assess conicalness of the bunch, Champion (1967) proposed the formula $(D_1 - D_2)/L$, where D_1 is the diameter of the largest (basal hand), D_2 is the diameter of the smallest (apical hand), and L is the length of the bunch. Ortiz and Vuylsteke (1998b) indicate that bunch orientation might be an oligogenic trait regulated by the epistatic effects of at least three dominant loci.

Completion of the vegetative state and flower initiation and development are critical components of yield. Stover and Simmonds (1987) proposed that bunch size is affected by the timing of the transition from vegetative to reproductive growth as well as by the timing of transition from male to female organs in the developing inflorescence (Stover and Simmonds, 1987).

5.4 FRUIT PARTHENOCARPY

Parthenocarpy entails the development of ovaries into fruit without fertilization. The mechanism of the parthenocarpic trait was extensively studied in various fruit crops. Genes for parthenocarpy are thought to express changes in the pattern of hormone production, transport, and/or metabolism. In addition, parthenocarpy is also implemented as a strategy to overcome a growth substance concentration threshold during the critical period of anthesis and to promote ovary growth in such a way that pollination and fertilization are no longer needed or possible (Nitsch, 1970).

By far, most cultivated banana and plantain cultivars are parthenocarpic and sterile, and consequently the trait is well documented in the literature (Ortiz and Vuylsteke, 1992; Ortiz and Vuylsteke, 1995a, 1995b). Over 50 years ago it was discovered by Simmonds that the parthenocarpy trait in banana is governed by three complementary genes, and that the diploid variety 'Calcutta 4' is deficient in one of the three, while the other two are homozygous and dominant (Simmonds, 1953). Ortiz and Vuylsteke (1995a) reported a wide variation of yield components in euploid segregants of F_1 hybrids between the French plantain 'Obino Ewai' (AAB) × 'Bobby Tannap' and the diploid 'Calcutta 4' (AA). The segregating population for one of the three elements (P_1) is correlated to bunch weight, hands per bunch, fruit weight, fruit length, and fruit circumference but not to fruit filling. Unfortunately, as of today the loci harboring the genes for fruit parthenocarpy have not yet been mapped. Moreover, the biochemical mechanism in banana remains obscure.

5.5 FRUIT RIPENING AND SENESCENCE

The genetic control of fruit ripening and senescence is relatively well understood due to the resemblance of physiological and biochemical processes in climacteric fruits of other better-studied species. Tomatoes serve as a model species in which both the genetic and biochemical changes during ripening were studied in great detail (Barry and Giovannoni, 2006; Giovannoni, 2007). Long shelf life is among the most important marketing parameters of fruit quality. In fact, the choice of Cavendish as the main export cultivar was largely based on its extended shelf life compared to other triploid and diploid cultivars (Serano et al., 2003).

The majority of ripening-related enzymes have been identified and cloned either by comparing the protein sequences to orthologs from other plant species or by using a differential display of gene expression between ripe and unripe fruit (Clendennen and May, 1997; Khayat et al., 2004). Among these are enzymes in the pathways leading to ethylene biosynthesis, starch hydrolysis, sucrose biosynthesis, tissue softening, chlorophyll degradation, and lipid degradation.

The significant variation in fruit ripening time among plantain and banana hybrids and landraces suggests that this characteristic could be improved through crossbreeding (Ferris et al., 1999). Banana and plantain differ largely in traits that relate to the conversion of starch to sugar in the pulp. The conversion of starch to soluble sugars—sucrose, glucose, and fructose—entails the catalysis of a series of chemical reactions. It is well established that plantain and cooking bananas (especially cultivars containing at least one copy of the B genome) degrade starch slower and less completely in comparison to dessert bananas that are composed of the A genome (Cordenunsi and Lajolo, 1995; Robinson, 1996). However, there are several exceptions to the rule. The cultivar 'Populou CMR,' which is an AAB hybrid, contains a significant amount of sugars; the same holds for the cultivars 'French Clair' and 'Batard' (Ngalani et al., 1999).

Starch formation in the fruit is largely mediated by the enzyme starch synthase. The gene encoding this enzyme was isolated from banana (Clendennen and May, 1997; Medina-Suarez et al., 1997) and its expression was analyzed in different stages of fruit ripening. Very little information can be extracted from the literature about the catalytic characteristics of starch synthase in banana fruit. The question of whether the difference in starch accumulation between A and B genomes in the preclimacteric stage is due to differences in the catalytic activity of starch synthase remains unresolved. Degradation of starch in the fruit is mediated by at least three classes of enzymes: namely, starch phosphorylase responsible for the phospholitic pathway and α and β amylases that mediate the hydrolytic pathway. Da Mota et al. (2002) reported two forms of starch phosphorylases in banana fruit: type I with high affinity to branched glucan chains and type II with low affinity to glucan chains. Clearly bananas and plantains express both the hydrolytic as well as the phosphorolytic pathway (Seymour, 1993). Another unresolved issue is related to phosphorylation versus hydrolysis of starch breakdown in the different banana and plantain cultivars. At the cell level, type I phosphorylase is highly correlated to starch degradation and phosphorylation of glucose, suggesting a coarse control of the rate of starch degradation.

Following starch degradation, bananas and plantains synthesize and store sucrose as an end product. Sucrose biosynthesis in higher plants can be mediated either by sucrose phosphate synthase (SPS), which uses uridinated glucose and fructose 6-phosphate as substrates, or by sucrose synthase (SuS) that uses uridinated glucose and fructose as substrates. While SPS is unidirectional, SuS is bidirectional, and in various tissues it competes with invertases to catabolize sucrose. SPS is considered a rate-limiting enzyme in many plant tissues, including banana fruit (Nascimento et al., 1997). In photosynthetic and storage tissues of various crop species, the reaction of SPS is regulated allosterically by sucrose, and by post-translational modification of serine phosphorylation (Huber and Huber, 1996). Surprisingly, in banana fruit, the enzyme was shown to be transcriptionally regulated. SPS transcripts are up-regulated in concert with the physiological events associated with the climacteric rise (Nascimento et al., 1997). There are no indications that the transcription of banana fruit SPS is activated by ethylene. Given that sucrose is the main constituent that contributes to fruit sweetness, it is highly conceivable that the enzyme plays a significant role in the trait of fruit sweetness.

Sink activity and mobilization of carbohydrates to the developing fruit were not extensively studied in bananas. Sugar transporters across the cell membranes, hydrolysis of sucrose, and mobilization of carbohydrates to the plastids are envisioned to have a strong impact on yield. These metabolic processes were shown to institute the activity of the sink, and consequently have a strong impact on yield. In tomatoes, it was realized that sink activity was mapped to a quantitative trait loci (QTL) harboring cell-wall-bound acid invertase. This gene was correlated with high yield (Fridman et al., 2004).

Ethylene gas is considered to be the trigger of transition of all climacteric fruit from the preclimacteric stage to the climacteric rise (Stover and Simmonds, 1987). Consequently, inhibition of ethylene synthesis is expected to delay ripening and extend the shelf life of the green fruit. This was confirmed by using tools of genetic engineering in tomatoes (Hamilton et al., 1990). The rate-limiting step in the pathway leading to ethylene synthesis in plants is the formation of its precursor molecule, the amino acid ACC from AdoMet. The reaction is catalyzed by the enzyme ACC synthase (ACCsyn). The final step is mediated by ACC oxidase (ACCox) that catalyzes the reaction in which ACC is converted to ethylene (Yang and Hoffman, 1984). Hamilton et al. (1990) inhibited ethylene biosynthesis in transgenic tomatoes by silencing the gene encoding ACC oxidase. The transgenic plants with an antisense construct to ACC oxidase produced approximately 3% of the ethylene produced in the wild type. Similar results were achieved in tomatoes by antisense silencing of ACC synthase (Oeller et al., 1991). Both constructs improved the shelf life of the transgenic tomato fruit. However, since ethylene is a regulator of other ripening processes, total abolishment of ethylene synthesis resulted in fruit with poor aroma and deficient in pigments. These traits could be reversed by exposure of the harvested fruit to exogenous ethylene.

Genes encoding ACC synthase and ACC oxidase were isolated from banana (Lopez-Gomez et al., 1997; Huang et al., 2006). While ACC synthase is a multigene family represented by at least eight members in the banana genome, ACC appears in a single locus suggesting a single copy. The pattern of expression of the ripening-related transcript isoform of ACCsyn as well as ACCox correlates well with the pattern of ethylene synthesis. In banana, unlike most climacteric fruit, soon after the climacteric rise there is a sharp decline in ethylene evolution, which is correlated to the *in vivo* activity of ACC oxidase (Liu et al., 1999). Mutants in ACCsyn1 or ACC oxidase are expected to have an extended shelf life along with matching phenotypes that are related to fruit ripening—that is, delayed chlorophyll degradation, inhibition of synthesis of aroma compounds, delayed conversion of starch to soluble sugars, and inhibition of fruit softening. These biochemical processes could be induced by application of exogenous ethylene. Exploring tools of biotechnology, gene silencing, using RNA interference (RNAi) or antisense technologies is expected to prolong the green life of the fruit. Given that ACCox is a single-copy gene, silencing it may hamper biological processes not associated with fruit ripening. On the other hand, given that ACCsyn1 is specifically associated with fruit ripening, it would presumably be a better candidate for post-transcriptional silencing of fruit ripening.

Transcription factors that act upstream of the ethylene as well as non-ethylene-mediated regulation of fruit ripening were identified and cloned in tomatoes by positional cloning of a mutated locus mapped to chromosome 5. The gene product was identified as two MADS-box genes residing side by side in the same locus (Vrebalov et al., 2002). These factors cosegregate with the well-studied *Rin* mutant in tomatoes. The *Rin* mutation is widely used in hybrid tomatoes for extension of the shelf life of tomato fruit. The heterozygous mutant for the *Rin* allele exhibits firm fruit with normal pigmentation. A MADS-box gene (MADS:*MuMADS1*) was recently isolated from banana (Liu et al., 2009). Homology data suggest that the transcript encoding *MuMADS1* is related to the AGAMOUS subfamily. Expression was shown in the stamen and pistil, as well as in the rhizome. The expression levels were enhanced in fruit that were exposed to external ethylene. The analysis performed by Liu et al. (2009) suggests that *MuMADS1* is involved in fruit ripening.

5.6 NEMATODE RESISTANCE

Nematodes cause massive damage to cultivated banana and plantains. The damages to commercial plantations can reach up to 50% of yield loss in untreated soils. The susceptibility of banana and plantains to nematodes is facilitated by the roots' soft texture and shallow penetration in the soil. The problem is accentuated in large-scale banana plantations where monocropping is a common practice. Banning of nematicides like methyl bromide in various parts of the world exacerbated the problem and left farmers with inappropriate and unreliable alternatives (for more details, see Chapter 7 in this book). Although the *Musa* genus contains a wide range of resistant alleles for both horizontal (nonspecific to a certain nematode) as well as vertical resistance (specific to a certain nematode), for the most part natural hybrids are considered susceptible. However, several natural diploid *Musa* AA lines were reported to have resistance to the most virulent banana nematode in the tropics, *Radopholus similis* (Pinochet and Rowe, 1978). Screening methods were instituted to determine resistance levels of banana germplasm to nematodes, in particular for resistance to *R. similis* and *Pratylenchus coffeae* (Collingborn and Gowen, 1997). Several cultivars of diploid and triploid bananas were demonstrated to exhibit various degrees of resistance to nematodes. For instance, the mixed (AB) diploid 'Kunanis' as well as the triploid 'Yangambi Km5' (AAA) were found to be highly resistant to *R. similis* (Sarah et al., 1992; Collingborn and Gowen, 1997; Fogain and Gowen, 1997). The resistance in 'Yangambi Km5' is incomplete as the roots of this cultivar develop lesions, but the number of nematodes is rather low. The phenomenon may stem from exhibiting the defense mechanism of the plant to the nematodes.

Very limited information exists regarding the inheritance of resistance in commercial cultivars. For the most part, "race" specific (pathotype) resistance of the host is governed by a single dominant gene, while resistance to a wider range of nematodes (horizontal resistance) is due to more than a single allele (oligo or polygenic mode of resistance). Dochez et al. (2009) reported on crosses between the two diploid genotypes: 'TMB2x 6142-1' susceptible to *R. similis* and 'TMB2x 8075-7,' which is highly resistant. Approximately 40% of the segregants expressed a resistant phenotype, 34% exhibited susceptibility, and 14% exhibited partial resistance, while the remainder was inconclusive. The heterozygous nature of both parental lines contributed to the complexity of interpretation of the results. The authors concluded that resistance in the segregants was controlled by two separate alleles. According to the model developed by Dochez et al. (2009), resistance is conferred by both genes in concert. Double-recessive alleles of one of the genes suppress the other gene, resulting in susceptibility. Interestingly, resistance of the 'TMB2x 8075-7' is inherited from its parental line 'Calcutta 4' ('C-4'). However the resistance of 'C-4' to *R. similis* is unequivocal (Moens et al., 2002). This parental line is considered highly resistant to black leaf streak disease. It would be interesting to determine the genetic source of the diverse resistance and the mode of inheritance to hybrid progenies of which 'C-4' was used as a parental line.

A novel biotechnological approach was used by Michaeli et al. (2004) to confer resistance to a Cavendish (AAA) selection by expressing a silencing double-stranded RNA (for RNA interference) transcript encoding part of the nematode Collagen 5 gene (Col 5). When the nematode feeds on the plant, the double-stranded RNA is absorbed by the nematode and the RNAi silencing mechanism is activated in the reproductive organs of the nematode, resulting in defective embryos. Plants were tested in a potted experiment as well as in a field trial for resistance to nematodes. Various transgenic lines proved to be resistant to a range of nematodes sharing a consensus sequence in the Col 5 gene. RNase protection assays revealed that the double-stranded RNA molecules remained active despite their mobilization through the digestive tract of the nematodes.

5.7 RESISTANCE TO BLACK LEAF STREAK DISEASE

Black leaf streak disease (BLSD) is a fungal disease causing the most serious economic, environmental, and public health issues in banana-growing areas (Hernandez and Witter, 1996). BLSD is caused by the ascomycete fungus *Mycosphaerella fijiensis* Morelet. The fungus was first reported in Fiji in 1964, but according to Stover (1978) the center of origin of the pathogen is Papua New Guinea and the Solomon Islands. Transfer of germplasm from Southeast Asia to Latin America and Africa spread the disease to these regions. The disease became an epidemic in most of the world's tropical banana-growing countries. Gradually, *M. fijiensis* replaced a former epidemic in *Musa* caused by the related fungus *M. musicola*, which was first identified in Java in 1902 (Jones, 1990). The differences between *M. fijiensis* and *M. musicola* were observed in the rate of lesion development on the surface of the leaves of infected plants (Moulion-Pefura et al., 1996). In addition to the smaller lesions in *M. musicola* compared with *M. fijiensis*, the spread of tissue oxidation in the leaf lamina is limited to the vicinity of the site of infection. Advanced strains of *M. fijiensis* developed an evolutionary capacity to rapidly adapt to a necrotrophic phase transition. The dead tissue can offer no effective defense against the pathogen.

Effective activation of defense mechanisms against fungal pathogens largely relies on the plant's rapid recognition that the pathogen is present. When a plant contains genetic resistance to a particular disease, the pathogen will elicit a cascade of responses in the plant that induce the appropriate mechanism of defense in the plant. Some defense mechanisms initiated by *R* (resistance) genes, to name a few, including hypersensitive responses that lead to a programmed death of plant cells (Goodman and Novacky, 1994; Jones and Dangl, 1996), induced synthesis of antimicrobial compounds (for example, phytoalexins), synthesis of antimicrobial proteins (pathogenesis-related [PR] proteins, defensins, and so forth), cell-wall reinforcement in the infected areas, and blockage of vessels (reviewed by Dixon et al., 1994). Importantly, resistant and susceptible plants both contain this basic defense machinery. What is lacking in the susceptible plants is recognition of the disease mediated by specific *R* genes and other elicitor receptor proteins. The vast majority of *R* genes isolated to date contain nucleotide-binding sites (NBS) and leucine-rich repeats (LRR). The latter motif probably mediates the aspect of disease specificity (reviewed by Baker et al., 1997; Hammond-Kosack and Jones, 1997).

In recent years, knowledge of genes regulating plant disease resistance has increased considerably (Chisholm et al., 2006). However, identification of genes conferring susceptibility is narrow (Van Damme et al., 2005; Yang et al., 2006; Consonni et al., 2006). The information gap between the nature of resistance and that of susceptibility is likely due to differences in their genetic tractability. These findings demonstrate that NBS-LRR genes can condition disease susceptibility and resistance and may have implications for *R* gene deployment (Lorang et al., 2007). This postulation may shed a new light on susceptibility as a trait, particularly in parthenocarpic species like bananas and plantains that ceased to evolve early in evolutionary history, due to lack of ability to cross hybridize.

"Gene-for-gene" type resistance (Flor, 1946) is commonly triggered by enabling of a genetically dominant resistance *R* gene product by a dominant, pathogen-derived, avirulence (*Avr*) gene product. Given that recognition of the *Avr* product is crucial, a single allele of the receptor product is sufficient for activation of the cascade reaction of the downstream defense mechanism. Recognition of pathogen signaling plays a key role in the activation. In 1995, Fullerton and Olsen reported 63 distinct virulent strains of *M. fijiensis*. Consequently, resistance through introgression of a single receptor protein that activates a downstream defense mechanism is unlikely. In fact, a segregation analysis for the resistance to BLSD in a cross between two triploid plantains and 'C-4' revealed that the trait is governed by no fewer than three different loci (Ortiz and Vuylsteke, 1994b). As a consequence of the oligogenic nature of the trait, availability of fully resistant material is scarce. Analyzing the genetic inheritance of resistance, Ortiz and Vuylsteke (1994b) suggested a major recessive gene (bs_1) and two minor additive modifier genes (bsr_i). These authors distinguish between severity of symptoms on the leaves and the delay in the infection as the youngest leaf exhibiting spots expresses it. It was suggested that the youngest spotted leaf is an indication of a gene for gene response. Craenen and Ortiz (1997) showed further that the intralocus interaction in the bs_1 locus apparently regulates the appearance of symptoms on the leaf surface, whereas the additive effect and the intralocus interaction of the bs_i locus affect disease development in the host plant. An extensive search of banana *R* genes belonging to the subfamily containing NBS-LRR motifs revealed 52 contiguous sequences in the 'C-4' genome (Miller et al., 2008).

Another class of *R* genes belongs to the kinase superfamily. To this end, 13 different serine/threonin kinase proteins, classified as *Pto* type, were isolated from bananas and characterized (Peraza-Echeverria et al., 2007). Seven of these were classified as *Pto*-type genes based on their similarity to tomato *Pto*-type resistance genes. This class of *R*-gene candidates is unusual in the sense that it recognizes more than one *Avr* protein and as such may confer resistance against more than one pathogen (Kim et al., 2002).

It has been noted that the climatic conditions of high temperature and humidity enhance the disease and delay the defense mechanism employed by the plant. Under these climatic conditions banana cultivars rarely express complete resistance to *M. fijiensis*. In highly resistant cultivars, the disease development is delayed to a point where the impact on yield is negligent, but nevertheless the defense is inadequate (Carlier et al., 2000). This resistance has also been described as nonspecific with no interactions between signals from the pathogen and its host (Fullerton, 2002). These modes of resistance are defined as "innate immunity traits." Examples of innate traits that confer banana resistance to BLSD were suggested by Craenen and Ortiz (1997) and include stomata density and pseudostem waxiness.

Finally, true and complete resistance of banana germplasm to *M. fijiensis* has not been shown to exist, although a level of tolerance/resistance occurs in the resistant diploids and their progenies. The precise denotation of the terms *resistance* and *tolerance* as used in reports is not always well defined. The term *resistance* is appropriately used in less-severe diseases due to any cause, tissue susceptibility included. Tolerance, in the sense of a relatively small yield reduction arising from relatively severe infections, has been reported in various hybrids containing introgressions from 'C-4' or other "resistant" diploid material. Tetraploid cultivars that exhibit high tolerance to the disease under experimental conditions, including field trials, are often disappointing when exposed to a wide range of isolates. In some cases the susceptibility is amplified in older plants under the stress of harsh environmental conditions.

5.8 CONCLUDING REMARKS

The discipline of banana genetics is expected to undertake major changes in the next few years. The prospects of elucidating the biochemical and genetic mechanisms of genes that govern important traits in the *Musa* genome will undoubtedly accelerate breeding in the postgenomic era. Elucidation of the complete *Musa* genomes (A and B) will open opportunities that did not exist for breeders until now. Novel bioinformatic tools for inter- and intragenome comparisons will assist *Musa* breeders to overcome the lack of high-resolution physical and genetic maps. Modern computation tools for synteny projection of chromosome profiles will circumvent the need of positional cloning to identify strategic QTL. Finally, if accepted by the public, genetic engineering may be very resourceful for the creation of new and improved traits in the *Musa* genome.

REFERENCES

Amar, K., S.R. Pearce, K. McLean, G. Harrison, J.S. Heslop-Harrison, R. Waugh, et al. 1997. The Ty1-copia group of retrotransposons in plants: Genomic organisation, evolution, and use as molecular markers. *Genetica* 100:205–217.

Baker, B., P. Zambryski, B. Staskawicz, and S.P. Dinesh-Kumar. 1997. Signaling in plant-microbe interactions. *Science* 276:726–733.

Barry, C.S. and J.J. Giovannoni. 2006. Ripening in the tomato green-ripe mutant is inhibited by ectopic expression of a protein that disrupts ethylene signaling. *Proc. Natl. Acad. Sci. (USA)* 103(20):7923–7928

Carlier, J., E. Fouré, F. Gauhl, D.R. Jones, P. Lepoivre, X. Mourichon, et al. 2000. Black leaf streak. In: *Diseases of banana, abaca and enset*, D.R. Jones, ed., 37–79, Wallingford: CABI Publishing.

Champion, J. 1967. *Les bananiers et leur culture*, Vol. 1. Paris: Setco.

Cheung, F. and C.D. Town. 2007. A BAC end view of the *Musa acuminata* genome. *BMC Plant Biol.* 7:29.

Chisholm S.T., G. Coaker, B. Day, and B.J. Staskawicz. 2006. Plant interactions with bacterial pathogens. *Cell* 124:803–814.

Clendennen, S.K. and G.D. May. 1997. Differential gene expression in ripening banana fruit. *Plant Physiol.* 115463–115469.

Collingborn, F.M.B. and S.R. Gowen. 1997. Evaluation de cultivars de bananiers pour la resistance a *Radopholus similis* et *Pratylenchus coffeae*. *InfoMusa* 6,3.

Consonni, C., M.E. Humphry, H.A Hartmann, M. Livaja, J. Durner, L. Westphal, et al. 2006. The protective role of silicon in the Arabidopsis–powdery mildew pathosystem. *Nat. Genet.* 38:716–720.

Cordenunsi, B.R. and F.M. Lajolo. 1995. Starch breakdown during banana ripening: Sucrose synthase and sucrose phosphate synthase. *J. Sci. Food Agric.* 43:347–351.

Craenen, K. and R. Ortiz. 1997. Effect of the bs_1 gene in plantain-banana hybrids on response to black Sigatoka. *Theor. Appl. Gent.* 95:497–505.

Crouch, H.K., J.H. Crouch, S. Madsen, D.R. Vuylsteke, and R. Ortiz. 2000. Comparative analysis of phenotypic and genotypic diversity among plantain landraces (*Musa* spp., AAB group). *Theor. Appl. Genet.* 101:1056–1065.

Crouch, J.H., D. Vuylsteke, and R. Ortiz. 1998. Perspectives on the application of biotechnology to assist the genetic enhancement of plantain and banana (*Musa* spp.) *Electron. J. Biotechnol.* 1(1):11–22.

Da Mota, R.V., B.R. Cordenunsi Joa, R.O. Nascimento, E. Purgatto, R.M. Maria, and R.F.M. Lajolo. 2002. Activity and expression of banana starch phosphorylases during fruit development and ripening. *Planta* 216:325–333.

Dixon, R.A., M.J. Harrison, and C. Lamb. 1994. Early events in the activation of plant defense response. *Annu. Rev. Phytopath.* 32:479–501.

Dochez, C., A. Tenkouano, R. Ortiz, J. Whyte, and D. De Waele. 2009. Host plant resistance to *Radopholus similis* in a diploid banana hybrid population. *Nematology* 11:329–335.

Eckstein, K., C. Fraser, J. Husselmann, and N. Temple Murray. 1998. The evaluation of promising new banana cultivars. *Acta Hort.* 490:57–69.

Ferris, R.S.B., R. Ortiz, and D. Vuyslteke. 1999. Fruit quality evaluation of plantains, plantain hybrids, and cooking bananas. *Postharvest Biol. Tech.* 15:73–81.

Flor, H.H. 1946. Genetics of pathogenicity in *Melampsora lini*. *J. Agric. Res.* 73:335–357.

Fogain, R. and S.R. Gowen. 1997. "Yangambi Km5" (*Musa* AAA, Ibota subgroup): A possible source of resistance to *Radopholus similis* and *Pratylenchus goodeyi*. *Fundam. Appl. Nematol.* 20:1–6.

Frary, A. and S. Doganlar. 2003. Comparative genetics of crop plant domestication and evolution. *Turk. J. Agric. For.* 27:59–69.

Fridman, E., F. Carrari, Y.S. Liu, A.R. Fernie, and D. Zamir. 2004. Zooming in on a quantitative trait for tomato yield using interspecific introgressions. *Science* 305:1786–1789.

Fullerton, R.A. 2002. Pathogenic variability in *Mycosphaerella fijiensis* and its implications for disease resistance bananas and plantains. In: *Acorbat memorias XV reunión*, Cartagena de Indias, Colombia, 27 October–2 November 2002, 167–172, Medellin: Asociación de Bananeros de Colombia.

Fullerton, R.A. and T.L. Olsen. 1995. Pathogenic variability in *Mycosphaerella fijiensis* Morelet, cause of black Sigatoka in banana and plantain. *New Zealand J. Crop Hort. Sci.* 23:39–48.

Giovannoni, J. 2007. Fruit ripening mutants yield insights into ripening control. *Curr. Opin. Plant Biol.* 10:283–289.

Goodman, R.N. and A.J. Novacky. 1994. *The hypersensitivity response of plants to pathogens: A resistance phenomenon*. St. Paul, MN: APS Press.

Hamilton, J.A., G.W. Lycette, and D. Grieson. 1990. Antisense gene that inhibits ethylene in transgenic plants. *Nature* 346:284–287.

Hammond-Kosack, K.E. and J.D.G. Jones. 1997. Plant disease resistance genes. *Ann. Rev. Plant Phys. Plant Mol. Biol.* 48:575–607.

Hernandez, C.E. and S.G. Witter. 1996. Evaluating and managing the environmental impact of banana production in Costa Rica: A systems approach. *AMBIO* 25:171–178.

Hirochika, H., K. Sugimoto, Y. Otsuki, H. Tsugawa, and M. Kanda. 1996. Retrotransposons of rice involved in mutations induced by tissue culture. *Proc. Natl. Acad. Sci. (USA)* 93:7783–7788.

Huang F.C., Y.Y. Do, and P.L. Huang. 2006. Genomic organization of a diverse ACC synthase gene family in banana and expression characteristics of the gene member involved in ripening of banana fruit. *J. Agric. Food Chem.* 54:3859–3868.

Huber, S.C. and J. Huber. 1996. Role and regulation of sucrose-phosphate synthase in higher plants. *Annu. Rev. Plant Physiol. Plant Mol. Biol.* 47:431–44.

Jones, A.M. and J.L. Dangl. 1996. Logjam at the Styx: Programmed cell death in plants. *Trends Plant Sci.* 1:114–119.

Jones, D.R. 1990. Black Sigatoka in the Southeast Asian-Pacific region. *Musarama* 3:2–5.

Khayat, E. 2009. Adi, a novel Cavendish banana selection expressing lower stature, higher fruit yield and quality than existing Cavendish lines including its ancestral line 'Zelig.' US patent no. 0222959P.

Khayat E., T. Gontmakher, E.Goren, M. Kipershlag, M. Amira, S. Regev, et al. 2004a. Discovery of functional genes in the *Musa* genome. In: *Banana improvement: Cellular and molecular biology and induced mutations*, M. Jain and R. Swennen, eds., 321–330. Enfield, NH: Science Publishers.

Khayat E., A. Duvdevani, E. Lahav, and B.A. Ballesteros. 2004b. Somaclonal variation in banana (*Musa* acuminata cv. Grande Naine). Genetic mechanism, frequency and application as a tool for clonal selection. In: M.S. Jain and R. Swennen (Eds.), *Banana improvement: Cellular, molecular biology, and induced mutations*, Enfield, NH: Science Publishers, pp. 97–110.

Kim, Y.J., N.C. Lin, and G.B. Marti. 2002. Two distinct *Pseudomonas* effector proteins interact with the *Pto* kinase and activate plant immunity. *Cell* 109:589–598.

Krikorian, A.D., H. Irizarry, S.S. Caronauer-Mitra, and E. Rivera. 1993. Clonal fidelity and variation in plantain (*Musa* AAB) regenerated from vegetative stem and floral axis tips in vitro. *Ann. Bot.* 71:519–535.

Kumar, A., S.R. Pearce, K. McLean, G. Harrison, J.S. Heslop-Harrison, R. Waugh, et al. 1997. The *Ty1-copia* group of retrotransposons in plants: Genomic organisation, evolution, and use as molecular markers. *Genetica* 100:205–217

Liu, J., B. Xu, L. Hu, M. Li, W. Su, J. Yang, et al, 2009. Involvement of banana MADS-box transcription factor gene in ethylene-induced fruit ripening. *Plant Cell Rep.* 28:103–111.

Liu, X., S. Shiomi, A. Nakatsuka, Y. Kubo, R. Nakamura, and A. Inaba. 1999. Characterization of ethylene biosynthesis associated with ripening in banana fruit. *Plant Physiol.* 121:1257–1265.

Lopez-Gomez, R., A. Campbell, G.Y. Dong, S.F. Yang, and M.A. Gomez-Lin. 1997. Ethylene biosynthesis in banana fruit: Isolation of genomic clone to ACC oxidase and expression studies. *Plant Sci.* 123:123–131.

Lorang, J., T. Sweat, and J. Wolpert. 2007. Plant disease susceptibility conferred by a "resistance" gene. *Proc. Natl. Acad. Sci. (USA)* 107:14861–14866.

Medina-Suarez, R., K. Manning, J. Fletcher, J. Aked, C.R. Bird, and G.B. Seymour. 1997. Gene expression in the pulp of ripening bananas. *Plant Physiol.* 115:453–461.

Michaeli, S., D. Kenigsbuch, O. Livneh, D. Levy, and E. Khayat. 2004. Plants resistant to cytoplasm-feeding parasites. U.S. Patent, Appl. No: 10/568,914.

Miller, R.N.G., D. Bertioli, F.C. Baurens, C.M.R. Santos, P.C. Alves, N. Martins, et al. 2008. Analysis of non-TIR NBS-LRR resistance gene analogs in *Musa acuminata* Colla: Isolation, RFLP marker development and physical mapping. *BMC Plant Biol.* www.biomedcentral.com/1471–2229/8/15

Moens, T., J.A. Sandoval, J.V. Escalant, and D. De Waele. 2002. Evaluation of progeny from a cross between Pisang Berlin and *M. acuminata* spp *burmannicoides* Calcutta 4 for evidence of segregation with respect to resistance to black leaf streak disease and nematodes. *InfoMusa* 11:20–22.

Moulion-Pefura, A., A Lassoudiere, J. Foko, and D.A. Fontem. 1996. Comparison of development of *Mycosphaerella fijiensis* and *Mycosphaerella musicola* on banana and plantain in the various ecological zones in Cameroon. *Plant Dis.* 80:950–954.

Nascimento, J.R.O., B.R. Cordenunci, F.M. Lajolo, and M.J.C. Alcocer. 1997. Banana sucrose-phosphate synthase gene expression during fruit ripening. *Planta* 203:283–288.

Ngalani, J.A., J. Tchango Tchango, and M. Reynes. 1999. Starch and sugar transformation during the ripening of banana and plantain cultivars grown in Cameroon. *Trop. Sci.* 39:115–119.

Nitsch, J.P. 1970. Hormonal factors in growth and development. In: *Food science and technology*, A.C. Hulme, ed., Vol. 1, 427–472. New York: Academic Press.

Oeller, P.W., L.M. Wong, L.P. Taylor, D.A. Pike, and A. Theologis. 1991. Reversible inhibition of tomato fruit senescence by antisense RNA. *Science* 254:437–439.

Ortiz, R. 2000. Understanding the *Musa* genome: An update. *Acta Hort.* 54:157–168.

Ortiz, R. and D. Vuylsteke. 1992. Inheritance of black Sigatoka resistance and fruit parthenocarpy in the triploid AAB plantain. In: *Agronomy abstracts*. Madison, WI: American Society of Agronomy, Crop Science Society of America, and Soil Science Society of America.

Ortiz, R. and D. Vuylsteke. 1994a. Genetic analysis of apical dominance and improvement of suckering behavior in plantain. *J. Am. Soc. Hort. Sci.* 119:1050–1053.

Ortiz, R. and D. Vuylsteke. 1994b. Inheritance of black Sigatoka disease resistance in plantain-banana (*Musa* spp.) hybrids. *Theor. Appl. Gent.* 89:146–152.

Ortiz, R. and D. Vuylsteke. 1995a. Effect of the parthenocarpy gene P_1 and ploidy on fruit and bunch traits of plantain-banana hybrids. *Heredity* 75:460–465.

Ortiz, R. and D. Vuylsteke. 1995b. Inheritance of dwarfism in AAB plantains. *Plant Breed.* 114:466–468.

Ortiz, R. and D. Vuylsteke. 1998a. Quantitative variation and phenotypic correlations in banana and plantain. *Sci. Hort.* 72:239–253.

Ortiz, R. and D. Vuylsteke. 1998b. Segregation of bunch orientation in plantain and banana hybrids. *Euphytica* 101:79–82.

Osuji, J.O., E.O. Bosa, D. Vuylsteke, and R. Ortiz. 1997. Multivariate pattern of quantitative trait variation in triploid banana and plantain cultivars. *Sci. Hort.* 71:197–202.

Peraza-Echeverria, S., V.A. Herrera-Valencia and K.A. James. 2001. Detection of DNA methylation in micropropagated banana plants using methylation-sensitive amplification polymorphism. *Plant Sci.* 161:359–367.

Peraza-Echeverria, S., A. James Kay, B. Canto-Canche, and E. Castillo-Castro. 2007. Structural and phylogenic analysis of *Pto*-type disease resistance gene candidates in banana. *Mol. Genet. Genomics* 278:443–453.

Pillay, M., A.Tenkouano, and J. Hartman. 2002. Banana and plantains: Future challenges in Musa breeding. In: *Crop improvement in the 21st century*, M. Kang, ed., 223–251. New York: Food Product Press.

Pinochet, T. and P. Rowe. 1978. Reaction of two banana cultivars to three different nematodes. *Plant Dis. Rep.* 62:727–729.

Poncet, V., F. Lamy, J.J. Enjalbert, A. Sarr, and T. Robert. 1998. Genetic analysis of the domestication syndrome in pearl millet (*Pennisetum glaucum* L., Poaceae): Inheritance of the major characters. *Heredity* 81:648–658.

Reuveni, O., Y. Israeli, and E. Lahav. 1996. Somaclonal variation in banana and plantain (*Musa* species). In: *Biotechnology in agriculture and forestry*, Vol. 36, *Somaclonal variation in crop improvement II*, Y.P.S. Bajaj, ed., 174–196. Berlin: Springer Verlag.

Robinson, J.C. 1996. *Bananas and plantains*. Wallingford: CAB International.

Robinson, J.C., D.J. Nel, and K. Eckstein. 1993. A field comparison of ten Cavendish sub-group banana cultivars and selections (*Musa* AAA) over four crop cycles in the subtropics. *J. Hort. Sci.* 68:511–521.

Sarah, J.L., F. Blavignac, C. Sabatini, and M. Boisseau. 1992. Une methode de laboratoire pour le criblage varietal des bananiers vis-a-vis de la resistance aux nematodes. *Fruits* 47:559–564.

Serano, M., M.V. Saenz, and A. Vargas. 2003. Algunas caracteristicas poscosecha del fruto de cinco diploids (*Musa* AA) y cuatro triploids (*Musa* AAB, AAA) de banano. *Corbana* 29:1–15.

Seymour, G.B. 1993. Banana. In: *Biochemistry of fruit ripening*. G.B. Seymour, J.E. Taylor, and G.A. Tucker, eds., 83–106. London: Chapman and Hall.

Simmonds, N.W. 1953. Segregation in some diploid bananas. *J. Genet.* 51:458–469.

Simmonds, N.W. 1962. *The evolution of the bananas*. London: Longman Green and Co.

Stover, R.H., 1978. Distribution and probable origin of *Mycosphaerella fijiensis* in Southeast Asia. *Trop. Agric.* 55:65–68.

Stover, R.H. and N.W. Simmonds. 1987. *Bananas*. New York: Longman Scientific and Technical, John Wiley and Sons.

Swennen, R and D. Vuylsteke 1987. Morphological taxonomy of plantain (*Musa* cultivars AAB) in West Africa. In: *Banana and plantain breeding strategies*, G.J. Persley and E.A. De Langhe, eds., 165–171. Proc. Workshop, Cairns, Australia, October 13–17, 1986. ACIAR Proc. No. 21.

Tang, C.Y. 2005. Somaclonal variation: A tool for improvement of Cavendish banana cultivars. *Acta Hort.* 692:61–65.

Tenkouano, A., R. Ortiz, and D. Vuylsteke. 1998. Combining ability for yield and plant phenology in plantain-derived populations. *Euphytica* 104:151–158.

Van Damme, M., A. Andel, R.P. Huibers, R. Panstruga, P.J. Weisbeek, and G. Van den Ackerveken. Identification of Arabidopsis loci required for susceptibility to the downy mildew pathogen *Hyaloperonospora parasitica*. 2005. *Mol. Plant-Microbe Interact.* 18:583–592.

Vrebalov, J., D. Ruezinsky, V. Padmanabhan, R. White, D. Medrano, R. Drake, et al. 2002. A MADS-box gene necessary for fruit ripening at the tomato ripening-inhibitor (*Rin*) locus. *Science* 296:343–346.

Yang, B., A. Sugio, and F.F. White. 2006. The virulence factor AvrXa7 of *Xanthomonas oryzae* pv. *oryzae* is a type III secretion pathway-dependent nuclear-localized double-stranded DNA-binding protein. *Proc. Natl. Acad. Sci. (USA)* 103:10503–10508.

Yang, S.F. and N.E Hoffman. 1984. Ethylene biosynthesis and its regulation in higher plants. *Annu. Rev. Plant Physiol.* 35:155–189.

6 Major Diseases of Banana

Guy Blomme, Simon Eden-Green, Mohammed Mustaffa, Bartholemew Nwauzoma, and Raman Thangavelu

CONTENTS

6.1 GENERAL INTRODUCTION

Banana is produced in over 100 countries and used in various ways. While large plantations owned by multinational companies are grown mainly for export, small-scale farmers rely on banana for food and income. Although the former cultivation systems are monocultures of a single cultivar, banana cultivation for local consumption relies on a multitude of cultivars suited to different farming systems and uses (Aurore et al., 2008). The tropical and subtropical environments that are ideal for growing bananas and plantains are at the same time regions with high disease and pest pressure. The creation of disease- and pest-resistant varieties is the priority for genetic improvement programs. Bananas are affected by a number of fungal, bacterial, and viral diseases. The most important fungal diseases are the Sigatoka diseases (yellow and black) and Fusarium wilt. Banana Xanthomonas wilt, a relatively new bacterial disease, is becoming a serious threat to bananas in East Africa in much the same way that blood disease posed in Indonesia and Moko in Central America. A large number of banana viruses affect banana in different regions, with some becoming more predominant in some areas. This chapter discusses the major fungal, bacterial, and viral diseases of banana. The interested reader in diseases of banana should also consult the treatment of this topic by Jones (2000a).

6.2 SIGATOKA DISEASES OF BANANA

6.2.1 INTRODUCTION

A wide range of leaf spot diseases affect banana (*Musa* spp.) in different parts of the globe, many of them of minor economic importance, but some cause severe reduction in production and quality, leading to important economic loss. Among the economically important leaf spot diseases are the fungal diseases black leaf streak (black Sigatoka), caused by *Mycosphaerella fijiensis* Morelet; yellow Sigatoka, also known as Sigatoka, caused by *M. musicola* Leach; and Mycosphaerella speckle, caused by *M. musae* (Spreg.) Syd. Symptoms of black and yellow Sigatoka are sometimes difficult to differentiate. In general, the first symptom is the appearance on the upper leaf surface of pale yellow streaks (yellow Sigatoka) or dark brown streaks on the lower surface (black Sigatoka), each 1–2 mm long, that enlarge to form necrotic lesions with yellow halos and light gray centers (Mourichon et al., 1997). The initial symptoms of Mycosphaerella speckle are light brown or tan-colored irregular blotches on the lower leaf surface that may show as smoky patches on the upper surface. The three diseases can be recognized by the primary lesions. This stage of the disease is the most recognizable and can be used to distinguish between the three *Mycosphaerella* leaf spot diseases. As the disease progresses, spots become gray in the center but keep a brown border (Crous and Mourichon, 2002).

6.2.2 Origin and Geographical Distribution of Sigatoka Diseases

Southeast Asia is not only the place of origin of *Musa*, but it is also the place of origin of the Sigatoka disease complex. Thus, yellow Sigatoka was first identified in the Java island of Indonesia in 1902 (Zimmerman, 1902). However, the disease became an epidemic in the Sigatoka valley of Fiji in 1903, from where the disease took its name. It is in the Fiji islands that black Sigatoka was first identified in 1963 (Rhodes, 1964), but the disease was probably not widespread in Southeast Asia and the Southern Pacific region at that time (Ploetz, 2001). Both diseases have since become pandemic. In Asia, yellow Sigatoka has been recorded in China, Malaya, Taiwan, the Philippines, Indonesia, and India, where it is prevalent in the following states: Assam, Bihar, West Bengal, Kerala, Tamil Nadu, Karnataka, Maharashtra, and Gujarat (Rawal, 2000). The first record of *M. musicola* in Africa was in Uganda in 1932, but the disease was not widely distributed until the 1950s (Simmonds, 1962). The disease has not been recorded in Egypt, Israel, and the Canary Islands (Rawal, 2000). Black Sigatoka has become the most important foliar disease of banana and plantain in Latin America, Asia, Africa, and the Pacific Islands (Meredith and Firman, 1970; Stover, 1978; Graham, 1969). In Latin America, black Sigatoka was observed for the first time in Honduras in 1972 (Stover, 1983). It now occurs in south central Mexico and Bolivia, as well as northwestern Brazil. In the Caribbean, it is found in Cuba, Jamaica, the Dominican Republic, and southern Florida (Ploetz, 2001). Black Sigatoka was accidentally introduced into Africa and was first recorded in Gabon in 1978 (Frossard, 1980). A suspected occurrence in Zambia in the early 1970s (Raemaekers, 1975) was shown not to be associated with *M. fijiensis* (Tushemereirwe and Waller, 1993). From Gabon, black Sigatoka rapidly spread into Cameroon in 1980 (Fouré, 1985, 1987) and Congo in 1985 (Mourichon, 1986). The disease reached Nigeria in 1986 (Wilson and Buddenhagen, 1986). In East Africa, black Sigatoka was first reported from the island of Pemba (Tanzania) in 1987 (Dabek and Waller, 1990), Kenya (Kung'u et al., 1992), and Malawi (Ploetz et al., 1992). Mobambo and Naku (1993) reported the presence of the disease in Eastern Zaire (Democratic Republic of Congo) and other Central African countries. Mourichon and Fullerton (1990) gave a comprehensive account on the global distribution of both diseases.

Mycosphaerella eumusae was discovered in the mid 1990s as a member of the Sigatoka disease complex of banana (Beveraggi et al., 1995; Burt et al., 1999). The disease originated from Southeast Asia and parts of Africa, where it affects varieties that are resistant to both *M. fijiensis* and *M. musicola* (Mourichon et al., 1997). So far, there has been no official confirmation of *M. fijiensis* in India, but *M. musicola* remains a major production constraint (Nwauzoma et al., 2008). All commercial cultivars such as 'Rasthali' (AAB), synonym 'Silk,' 'Karpuravalli' (ABB), synonym 'Pisang Awak,' 'Monthan' (ABB), and those of the AAA Cavendish subgroup exhibit different levels of susceptibility. The actual distribution and epidemiology of Sigatoka leaf spot diseases remain ambiguous due to similarity in symptoms and life cycle of the pathogens. However, rapid and robust species-specific molecular-based diagnostic tools have been developed to detect and quantify the pathogens (Arzanlou et al., 2007).

Mycosphaerella fijiensis is considered a quarantine pathogen in *Musa*-producing areas and continues to occupy new ecological niches. *Mycosphaerella musicola* exists in colder environments compared to *M. fijiensis*, and warmer temperatures favor faster development of ascospores and germ tubes in *M. fijiensis*. Moreover, *M. fijiensis* is more vigorous and aggressive than *M. musicola*, producing four times as many ascospores in the same period (Peterson et al., 2005), causing premature fruit ripening. Yield losses of between 20 and 90% have been reported in Nigeria (Mobambo et al., 1993) and elsewhere (Jones, 2002a, 2000b; Marin et al., 2003; Fouré, 1985).

6.2.3 Symptom Description

Symptoms of black and yellow Sigatoka diseases are very similar and often difficult to distinguish with the naked eye, which hampers disease management. Tiny, chlorotic spots at the lower (abaxial) surface of third and fourth unrolled or opened leaves are the first symptoms of black Sigatoka disease, about 2–3 weeks after infection. These spots grow into tiny brown streaks and later change into streaks darker in color and sometimes with a purple tinge that appear on the upper (adaxial) surface of the leaf. The lesions later coalesce, becoming fusiform or elliptical and darker, to give the characteristic black streaking of the leaves. Fouré (1982, 1987) divided the disease development into six stages as seen in AAB plantain cultivars:

1. Stage 1: Presence of minute specks yellowish in color and less than 1 mm in length only present on the lower or abaxial surface of the leaf. They are often more abundant near the left side of the leaf, particularly towards the tip but not visible in transmitted light.
2. Stage 2: There is the transition from specks into streaks, 3–4 mm long and 1 mm wide. They are reddish-brown in color and more visible on the lower surface of the leaf than on the upper part.
3. Stage 3: The streaks coalesce (2–3 cm in length) and the color changes from reddish-brown to dark brown. When the streaks are many and more or less evenly distributed, the entire leaf surface blackens.
4. Stage 4: The transition from streak to spot is characterized by the development of light brown, water-soaked, elliptical spots, very much visible early in the morning with the presence of dew or rainfall.
5. Stage 5: The round, elliptical spots become completely black and are surrounded by water-soaked border.
6. Stage 6: The center of the spot dries, turns light gray, and becomes surrounded by a narrow, well-defined black ring that in turn is encircled by a bright yellow halo. These spots remain visible even after the leaf has dried out completely.

There is no sharp distinction between one symptom stage and the next. Fouré's (1982, 1987) system of classification is convenient for disease assessment procedures. The rate at which initial specks change into streaks depends on climatic conditions, plant age, vigor, and cultivar as well as disease development and severity (Meredith and Lawrence, 1969). On resistant varieties, symptom development is very slow and may not progress to stage 4 (appearance of spots) until leaf senescence (Jacome and Schuh, 1992; Fullerton, 1994). Resistance of *Musa* species to *M. fijiensis* seems to be more related to postinfection production and activation of pathogen-related proteins (Lepoivre et al., 1993) and phytoalexins (Luis et al., 1993) than to small changes in structure and preformed substances. According to Beveraggi et al. (1995), partial resistance in the cultivar 'Fougamou' was linked to a preexisting antifungal plant phenolic compound, while in the case of the highly resistant cultivar 'Yagambi Km 5,' an active mechanism was induced following penetration of stomata by the pathogen.

Lesions caused by yellow Sigatoka may look similar to those caused by black Sigatoka, but one can distinguish yellow from black Sigatoka by examining the fungal hyphae producing asexual spores. For example, the conidiophores of *M. fijiensis* occur in clumps (sporodochia) with basal scars at the points of attachment while *M. musicola* has no scars. Conidia (asexual spores) and the structures producing male sexual spores (spermagonia) occur on the underside of the leaf in infections caused by *M. fijiensis*, whereas *M. musicola* produces conidia mostly on the upper surface of the leaf. Symptom development is slower in *M. musicola* than in *M. fijiensis*. Recently, polymerase chain reaction (PCR)–based diagnostics have become available for the diagnosis and detection of the three major causal agents of Sigatoka diseases (Arzanlou et al., 2007).

6.2.4 Epidemiology

Epidemiology involves studying the factors that influence disease development such as host–pathogen interactions and the environmental and edaphic factors. A thorough understanding of disease epidemiology is a prerequisite for the development of management strategies, including durable host resistance (Craenen, 1998). The main factors that influence the production, release, and spread of black and yellow Sigatoka inoculum are rainfall, temperature, dew, and wind. Well-defined seasonal and day-to-day trends in inoculum production can be related to availability of free moisture on the surface of the leaf and to minimum temperature. Maximum temperatures are really limiting factors, provided free moisture is present (Stover, 1972). However, the principal factor in the release of ascospores is rain. Short and violent rains, separated by sunny periods, are the most favorable periods for the release of ascospores. Gauhl and Pasberg-Gauhl (1994a, 1994b) and Gauhl (1994) described seasonal variation in production of ascospores in Nigeria and conidia in Costa Rica. Both black and yellow Sigatoka are spread by ascospores, with conidia being an additional source of inoculums for yellow Sigatoka. Conidia are only produced in black Sigakota in the absence of rainy weather. Conidia are produced on young leaves in the early streak stage in black Sigatoka and in the mature spot stage in yellow Sigatoka. About 30,000 and 1,200 conidia per spot are produced by *M. musicola* and *M. fijiensis*, respectively. The conidia are separated from the conidiophores by water or wind. Mature ascospores are produced 4 weeks after the appearance of streaks in *M. musicola* and 2 weeks in *M. fijiensis*, and are actively released to young, actively growing leaves mainly by wind and water. Germination of spores occurs on the leaf when a film of water is present, after about 2–6 hours, depending on the temperature. Following germination, stomatal penetration occurs after about 48–72 hours favored by temperature above 20°C (Stover, 1980) and fluctuations in humidity (Goods and Tsirch, 1963). Once infection is established, hyphae emerge from the stomata and either develop into conidiophores or grow across the surface to infect adjacent stomata. Marin et al. (2003) stated that *M. fijiensis* has greater penetration ability into the stomata and greater spotting than *M. musicola*.

6.2.5 Disease Management

Although plant diseases can hardly be completely eradicated, their impact can be reduced to acceptable economic levels. A combination of cultural and chemical practices is recommended for the management of Sigatoka disease complex: field sanitation, host nutrition, and sound cultural practices; fungicides; detrashing (deleafing); pruning; ensuring good drainage and canopy aeration; resistant cultivars. Chemical control is achieved by alternating protectant fungicides like mancozeb or chlorothalonil with systemic fungicides such as benzimidazole or triazole. Systemic fungicides are more suitable than protectant fungicides even though they pose the risk of potential resistance in pathogen population (Marin et al., 2003). This, in addition to concern for the environment, has resulted in the development of biological control methods using epiphytic bacteria isolated from banana leaves against *M. fijiensis* (Jiménez et al., 1987). Some of the bacterial species used in the biological control of Sigatoka disease complex are *Pseudomonas* spp., *Bacillus* spp., and *Serratia marcescens*.

The simplest way for the home gardener to control the disease is to destroy the severely diseased leaves or remove them and place them top-side down on the ground to reduce the chance of spore dispersal into the banana canopy. However, breeding for resistant cultivars has been credited as the most appropriate long-term control method. One of the strategies to overcome leaf spot diseases in *Musa* is through the hybridization of resistant AA diploids with triploids to obtain tetraploids that combine disease resistance with good agronomic traits (Ortiz et al., 1995). Several tetraploid hybrids have been developed by the International Institute for Tropical Agriculture (IITA), Nigeria, and the Fundación Hondureña de Investigación Agrícola (FHIA) in Honduras (Rowe and Rosales, 2000), as well as other international breeding programs in Brazil, Cameroon, and Guadeloupe (Fullerton, 1994; Carlier et al., 2000). Genetic transformation and somaclonal variation have also been used to incorporate or identify durable disease resistance (Trujillo and Garcia, 1996; Sági et al., 1997; Crouch et al., 1998; Nwauzoma et al., 2002). Future disease management programs will continue to be directed on an integrated pest management (IPM) program that embraces early detection of the pathogens that cause the Sigatoka diseases and encompasses eradication as well as cultural, chemical, and improved quarantine management strategies.

6.3 FUSARIUM WILT DISEASE IN BANANA

6.3.1 Introduction

Fusarium wilt, commonly referred to as Panama disease, is regarded as one of the most devastating diseases of banana and ranked as one of the top six important plant diseases in the world. Although Fusarium wilt probably originated in Southeast Asia (Ploetz and Pegg, 1997), the disease was first discovered in Eagle Farm, Brisbane, Queensland, Australia, in 1876 in banana var. Sugar (Silk AAB) (Bancroft, 1876). In 1940, Snyder and Hansen proposed the name *Fusarium oxysporum* Schlecht. f. sp. *cubense* (E. F. Smith). Presently, Fusarium wilt has been reported in all the banana-growing regions of the world (Asia, Africa, Australia, and tropical America) except some islands in the South Pacific, the Mediterranean, Melanesia, and Somalia (Ploetz and Pegg, 2000).

Race 1 of this disease completely wiped out the commercially important 'Gros Michel' cultivar in Central America in the 1950s (Stover, 1962). The Cavendish clones, which are resistant to race 1, were identified as a substitute for 'Gros Michel' for the export industry. Of late, these Cavendish cultivars also succumb to Fusarium wilt in many banana-producing countries such as Taiwan (Stover and Malo, 1972), Vietnam, Malaysia, Indonesia, Cambodia, China, the Philippines, South Africa, and Australia (Ploetz, 1990). The disease poses a serious threat to the multibillion-dollar export industry and also to the livelihoods of millions of small-scale banana growers (Ploetz, 2005).

6.3.2 General Symptoms

The fungus infects the roots of banana plants, colonizing the vascular system of the rhizome and pseudostem, and inducing characteristic wilting symptoms usually 5–6 months after planting, and the symptoms are expressed both externally and internally (Stover, 1962). The external symptom initially appears as yellowing of the leaf margins of older leaves, which gradually spreads across the entire leaf surface. Generally yellowing progresses from the oldest to the youngest leaves, but in some cases, the yellowing may be seen in the middle position (third or fourth leaves). The leaf petiole turns brown and leaves gradually collapse at the petiole and hang down to form a "skirt" of dead leaves around the pseudostem (Figure 6.1). The youngest leaves often stand erect, giving the plant a "spiky" appearance. Newly emerging leaves are paler in appearance, are reduced in size, and exhibit wrinkling and a distorted appearance (Moore et al., 1995). In the pseudostem, longitudinal splits may develop near the soil level. The plant also produces many suckers before it dies.

The internal symptoms are characterized by vascular discoloration beginning with yellowing of the vascular tissues in the roots and corm/rhizome (Figure 6.2), which progress to form yellow, red, or brown discolored vascular strands in the pseudostem and sometimes in the stalk. The disease also spreads to suckers and the internal symptoms can be seen even after 1 or 2 months of emergence of suckers (Moore et al., 1995).

FIGURE 6.1 Fusarium wilt-infected plant with yellowing of leaves in banana cv 'Cavendish.'

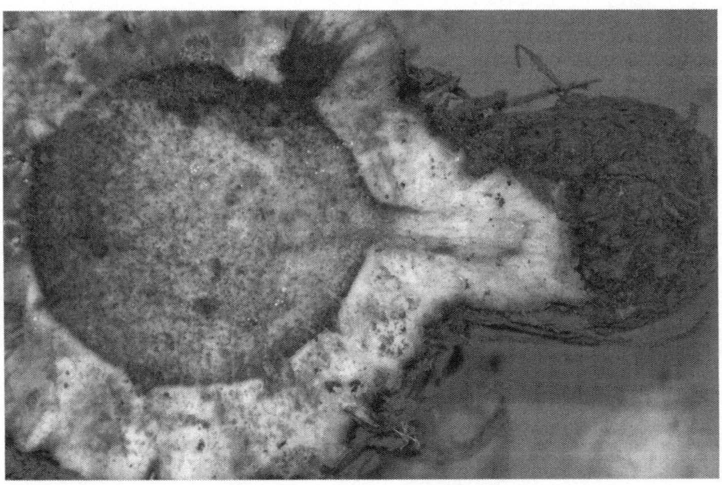

FIGURE 6.2 Vascular discoloration due to Fusarium wilt in banana.

6.3.3 Causal Organism

The disease is caused by the soil-borne hyphomycete *Fusarium oxysporum* Schlecht. f. sp. *cubense* (*Foc*). On potato dextrose agar (PDA), fungal colonies produce white aerial mycelia that may turn purple in the center. Sporodochia are cream to orange on carnation leaves on carnation leaf agar (CLA), and sclerotia are blue and submerged. Three types of asexual spores—macroconidia, microconidia, and chlamydospores—are formed. Micro- and macroconidia are produced on branched and unbranched monophialides, and the size of these spores range from 27 to 55 μm × 3.3 to 5.5 μm, and from 5 to 16 μm × 2.4 to 3.5 μm, respectively. Microconidia are produced on false heads, are one- or two-celled, and oval- to kidney-shaped. Macroconidia are abundant, four to eight celled, thin walled and delicate, slightly sickle shaped with an attenuated apical cell and a foot-shaped basal cell. Chlamydospores are thick walled, terminal, and intercalary, usually globose, 7–11 μm in diameter, and are formed singly or in pairs in hyphae or conidia but may also be found in clusters or short chains. These spores are formed during the latter stages of disease in dead host-plant tissue and in the soil. They are capable of surviving in the soil for several years. No perfect stage (teleomorph) of *F. oxysporum* is known (Su et al., 1986; Jones, 2000a; Ploetz, 2000).

6.3.4 Infection Processes of *Foc*

The fungus survives as chlamydospores in soil and on plant debris and invades the host through the root hairs, root tips, and natural wounds along the lateral root base and also through the wounds caused by farm implements, insect pests, and parasitic nematodes (Rishbeth and Naylor, 1957; Lee et al., 2009). Once *Foc* enters the root, it colonizes the cortex and enters the xylem in the vascular system, rhizome, and pseudostem (Rishbeth, 1955). After colonization, the pathogen blocks the plant's vascular system, which will lead to wilting and finally to plant mortality (Ploetz and Pegg, 2000). After plant death, the fungus grows out of the xylem into the surrounding tissues, forming many chlamydospores, which are returned to the soil after the plant decays.

6.3.5 Survival and Spread

The fungus survives primarily in the soil and on plant debris as chlamydospores for more than 30 years (Moore et al., 1995; Ploetz, 2005). The pathogen is also able to infect and persist in the roots of alternative hosts such as *Paspalum*, *Panicum*, *Ixophorus*, *Commelina*, and *Chloris inflata*, which have been identified as nonsymptomatic hosts (Waite and Dunlap, 1953; Sun, 1977). The pathogen populations tend to be higher and survive longer in light-textured soils than in heavy alkaline soils. In the suppressive soil, which contains more microbial population, the pathogen development is suppressed and this soil type has been reported in Central America, the Canary Islands, Australia, and South Africa (Moore et al., 1995).

Spread of the pathogen locally, nationally, and internationally is through infected rhizomes or suckers, and also through soil attached to infested planting material or implements/vehicles (Ploetz, 1994). Surface water is also responsible for the short- and long-distance dispersal of the disease. The spread between the plants within the field is also affected by roots (Moore et al., 1995). Wind that carries infested dust and trash is also involved in the spread of the fungus. There is no evidence that the disease is disseminated in true seed (Ploetz and Pegg, 1997).

6.3.6 Classification

Foc can be classified into many races based on their ability to infect certain banana cultivars and also vegetative compatibility groups (VCG) (Ploetz and Pegg, 1997). Four major races have been reported (Moore et al., 1995) and their characteristics are:

1. Race 1: It occurs throughout the world and cultivars like 'Gros Michel' (AAA), 'Silk' (AAB), 'Pome' (AAB), abacá, 'Maqueño' (AAB), 'Pisang Awak' (ABB), and I.C.2 (AAAA) (Ploetz et al., 1990; Bentley et al., 1995) are affected. Recent studies conducted by National Research for Banana (NRCB) in India has indicated that some VCGs of race 1 also attack race 2 suscepts and vice versa.

2. Race 2: This race attacks mainly 'Bluggoe' and 'Monthan' (ABB) and other closely related cooking bananas. It also affects tetraploid 'Bodles Altafort' (hybrid between 'Gros Michel' and 'Pisang lilin') and enset (*Ensete ventricosum*—an important food crop in Ethiopia).
3. Race 3: Waite (1963) first identified the wilt disease in several species of *Heliconia* in Central and South America in the mid 1900s and named the causal strain as race 3 of *F. oxysporum* f. sp. *cubense*. This was recorded in Australia, Honduras, and Costa Rica. This is nonpathogenic or weakly pathogenic on *Musa* spp. (Ploetz, 1990).
4. Race 4: It occurs in most of the banana-growing regions and is most destructive since it affects race 1 and race 2 susceptible clones as well as the Cavendish cultivars and 'Pisang Mas' (AA). This race is separated into two groups: subtropical race 4 (SR4) (VCGs, 0120, 0121, 0129, 01211 found in Australia, 0120 in Canaries and South Africa, and 0122 in the Philippines) and tropical race 4 (TR4) (VCG-01213–01216 complex). SR4 affects Cavendish plants that have been predisposed to disease by cold temperatures in the subtropics, whereas TR4 attacks Cavendish more aggressively under tropical conditions in the absence of any predisposing factors. There are no reports on occurrence of *Foc* TR4 in the major banana-growing countries in Asia, including India (Thangavelu et al., 2001).

6.3.7 Disease Management

Effective management of this disease is possible only with integration of different strategies available, since no single method is able to contain the disease effectively. This approach not only reduces the initial inoculum but also increases the plant resistance, delays the disease development, and reduces the effectiveness of initial inoculum (Agrios, 1997). Quarantine and sanitation mainly involve restricting the movement of corm, suckers, plant or plant parts, soil, and so forth from infected areas to uninfected/clean areas. Trenches may be dug to separate the diseased areas to prevent the movement of fungal spores through surface runoff water (Moore et al., 1999).

Creation of anaerobic condition by flooding is successful in managing the disease at least for the first cycle of the crop (Wardlaw, 1961; Stover, 1962). In India, crop rotation with paddy and flooding for 3–4 months before planting banana is effective (Sun and Huang, 1985; Thangavelu et al., 2001); interplanting with cassava lowered the inoculum (Buddenhagen, 2009a). In Central America, certain fungicides such as Vapam, mylone, allyl alcohol, or formaldehyde controlled the disease very effectively for 2–3 years (Stover, 1962).

Soil application of calcium compounds and phosphate salts, such as $Ca(OH)_2$, $Ca(NO_3)_2\ 4H_2O$, $CaCO_3$, $CaSO_4$, K_2HPO_4, and $NaH_2PO_4\ 2H_2O$, strongly inhibited chlamydospore germination and promoted lysis of germ tubes of *Foc* in soil (Sun and Huang 1985; Huang et al., 1989). However, application of ammonia nitrogen increased the disease severity, whereas the nitrate nitrogen forms decreased it (Huber and Watson, 1974).

Of late, biological control of Fusarium wilt disease is becoming more popular because it is environmentally friendly and also offers a potential alternative to the use of resistant banana varieties (Weller et al., 2002; Fravel et al., 2003). Biological control agents (BCAs) like *Trichoderma*, *Pseudomonas*, *Streptomyces*, and nonpathogenic *Fusarium* (np*Fo*) of both rhizospheric and endophytic origin are effective against Fusarium wilt diseases (Lemanceau and Alabouvette, 1991; Alabouvette et al., 1993; Larkin and Fravel, 1998; Weller et al., 2002; Thangavelu et al., 2003). The suggested modes of action involved in the control of the diseases are: mycoparasitism, antibiotic production, production of cell-wall-degrading enzymes such as chitinases and glucanases, competition for space and nutrients, and induced resistance. *Pseudomonas fluorescens* strain WCS 417 (Nel et al., 2006) and endophytic isolates of nonpathogenic *F. oxysporum* (np*Fo*) (Gerlach et al., 1999; Nowak, 1998; Lian, 2009) are effective.

The use of fungicides is part of an integrated approach for Fusarium wilt management (Stover, 1962). Davis et al. (1994) demonstrated that the phosphonate fungicides were effective in controlling Fusarium wilt disease. Several studies have indicated that chemical R&H–3888 (nitrile) was most effective, followed by EP-161 (methyl isothiocyanate), Vapam (sodium *n*-methyl dithiocarbamate), allyl alcohol, and mylone (3, 5-dimethyl tetrahydro-1, 3, 5 2H-thiadiazine-2-thione) (Corden and Young, 1959, 1960). In India, stem injection with carbendazim decreased the disease incidence in cultivar 'Rasthali' (Silk), but this treatment proved to be ineffective in South Africa (Lakshmanan et al., 1987; Herbert and Marx, 1990). The fungicide benomyl and demethylation inhibitors (DMI) fungicides prochloraz, propiconazole, and cyproconazole/propiconazole significantly reduced the incidence of *Foc* when applied as root-dip and soil-drench treatments 1 week after planting (Nel et al., 2007). Root dipping of young plants with propiconazole or the biological Sonata also gave some effect in protecting the young seedlings (Buddenhagen, 2009b). In South Africa, the spread of early detected *Foc* was stopped by treatment of the infested sites with methyl bromide, which was effective only for 3 years. In the Philippines, heat treatment of soil was used to control the spread of the pathogen, but effective only for a few years (Herbert and Marx, 1990; Ploetz, 2000).

6.3.8 Host Resistance

This is the most economical and practical long-term option and must be part of an integrated disease management program for the small-scale farmers in many developing countries (Daniells, 2009). The Taiwan Banana Research Institute (TBRI) had released race 4 *Foc*-resistant Cavendish clones viz. 'Tai Chiao no. 1,' 'Tai Chiao no. 3,' 'TC3-1035,' '5,' 'GCTCV-218' (Farmosana), 'GCTCV-119,' and 'GCTCV-215-1,' which were resistant to *Foc* TR4 (Hwang and Ko, 2004). The FHIA (Fundación Hondureña de Investigación Agrícola) breeding program has produced several resistant hybrids to both *Foc* race 1 and race 4, such as 'FHIA-01' (Goldfinger, AAAB) (Moore et al., 1995; Jones, 2000a) and 'SH-3640/10' (High Noon) (De Beer, 1997). EMBRAPA (Empresa Brasileira de Pesquisa Agropecuaria) in Brazil has developed 'Prata,' 'Maca,' and 'Prata Ana' tetraploids, which showed resistance against *Foc* (de Matos et al., 1999). Recently, seven diploid (AA) and 11 tetraploid (AAAB) banana hybrids resistant to VCGs 0124 and 0125 have been developed (de Matos et al. 2009). In India, nine hybrids resistant to *Foc* race 1 have been developed. Among these, three were diploids ('H-02-09,' 'H-02-10,' and 'H-02-15'), two were triploids ('H-02-08' and 'NPH-02-01'), and four were tetraploids ('H-02-19,' 'H-02-23,' 'H-02-26,' and 'H-02-4') (Kumar et al., 2009). The Tropical Research and Education Center (TREC) of the University of Florida has identified four dessert clones ('Pisang Ceylon,' 'FHIA-01' [AAAB], 'FIAH-02' [AAAA], and 'FHIA-17' [AAAA]) and three cooking clones ('Kumunamba' [AAB], 'Kandrian' [ABB], and 'Saba' [ABB]) resistant to VCG 01210, 0124, and 1210 of *Foc* (Ploetz et al., 1990). Gamma irradiation of 'Dwarf Parfitt,' an extra-dwarf Cavendish banana cultivar with poor agronomic characters and resistant to subtropical race 4, resulted in the selection of 'DPM 25,' with improved productivity and also tolerance to subtropical race 4 of *Foc* (Smith et al., 2006).

6.3.9 Conclusion

Fusarium wilt disease caused by *Fusarium oxsporum* f. sp. *cubense* is considered as a highly devastating disease of banana. The disease affects almost all the important cultivars grown in different parts of the world. The pathogen of the disease can survive as chlamydospores for more than 30 years, which is very dangerous for the perennial system of cultivation. As the fungus *Fusarium* is highly variable in nature, wide diversity among the Fusarium wilt pathogen has been reported, which is again making difficulties in identifying the strains as well as development of suitable effective management practices including quarantine. Among the four races of *Foc*, tropical race 4, which is infecting commercially important Cavendish cultivars, is spreading very fast and posing a dangerous situation for the sustainable banana production throughout the world. Although several management practices such as cultural, chemical, and biocontrol methods have been evolved, no single method has been found effective so far. Hence all the available management strategies have to be integrated to contain the disease effectively.

6.4 BANANA BACTERIAL WILT DISEASES

6.4.1 Introduction

Such was the speed with which banana Xanthomonas wilt (BXW) has come to prominence as a major disease of banana in Africa that it received little or no attention in textbooks or monographs produced before the turn of the last century (Jeger et al., 1995; Thwaites et al., 1999; Ploetz et al., 2003). Following spread of the disease to central Uganda and the Democratic Republic of Congo (DRC), there has been a dramatic increase in investigations, which are documented in several comprehensive reviews (Biruma et al., 2007; Smith et al., 2008; Tripathi et al., 2009). These studies have brought into focus some remarkable similarities and also some important differences between diseases caused by a taxonomically diverse group of bacterial vascular pathogens of banana. Implications for disease management are considered here.

6.4.2 The Diseases and Their Distribution

The main groups of vascular pathogens of banana are summarized in Table 6.1. *Xanthomonas campestris* pv. *musacearum* (*Xcm*) was originally reported from Ethiopia (Yirgou and Bradbury, 1968) as causing an endemic disease of enset (*Ensete ventricosum*) and was subsequently observed in cultivated bananas (Yirgou and Bradbury, 1974). Although a potential threat to the region was noted, the disease remained confined to Ethiopia until around 2000, when isolated outbreaks were identified in central Uganda (Tushemereirwe et al., 2003) and eastern DRC (Ndungo et al., 2004). It has since spread rapidly and now affects at least six countries in the Great Lakes region of East Africa. Although enset is generally cultivated only in southern and central Ethiopia, it grows in a wild state in northeastern DRC, where natural infections have been noted (Ndungo et al., 2008).

TABLE 6.1
Bacterial Wilts of Banana

Common Name	Causal Agent	Distribution and Natural Hosts
Banana Xanthomonas wilt, BXW (enset wilt, banana bacterial wilt BBW). Also known as *kiwatoka* (Uganda) and *Unyanjano wa migomba* (Tanzania)	*Xanthomonas campestris* pathovar *musacearum (Xcm)*	Ethiopia, Uganda, D. R. Congo, Rwanda, Tanzania, Kenya. Reports from Burundi are as yet unconfirmed (Enset and all cultivated banana types; *M. balbisiana* less susceptible). Cultivars with persistent bracts and flowers may escape the disease
Moko diseases (including Bugtok)	*Ralstonia solanacearum* biovar 1, race 2	Americas: (Mexico, Central and South America, southern Caribbean, Jamaica): all cultivated types and wild *Heliconia* Philippines: recognized as Moko on AAA types; inflorescence infection of ABB types incompletely systemic and recognized as Bugtok/Tibaglon/Tapurok disease
Blood disease	"Blood disease bacterium" (*Ralstonia* sp.)	Indonesia: all cultivated and some wild *Musa* species
Bacterial wilt of *solanaceous* crops (brown rot, Granville wilt)	*Ralstonia solanacearum* biovar 1, race 1	Central/South America, may be introduced elsewhere: occasionally or experimentally on AA diploids, only

Recent genomic studies (Aritua et al., 2008) support a probable eastern African origin of *Xcm* and indicate that it may share common ancestry with pathogens of graminaceous hosts (sorghum, sugarcane, maize) from Ethiopia, now classified as *X. vasicola*. However, a change of name awaits formal taxonomic revision.

The origins and recent history of BXW in eastern Africa share striking similarities to those of banana blood disease in Indonesia. Blood disease was originally reported when intensive banana cultivation was introduced to two small offshore islands in the Salayar archipelago, to the southeast of the Indonesian island of Sulawesi (formerly Celebes; Rijks, 1916). In subsequent investigations, Gäumann (1921, 1923) found the disease to be widespread in cultivated bananas and also in wild (forest) *Musa* spp. in southern Sulawesi. At some locations, farmers reported historical epidemics ("waves" of disease) going back for many years, and had coined the local name "penyakit darah" on account of the reddish "blood-like" bacterial ooze that emerges from cut vascular tissues. The disease apparently remained confined to Sulawesi until the late 1980s when an outbreak was identified in west Java (Eden-Green and Sastraatmadja, 1990). Since then, it has spread across the Indonesian archipelago, from Aceh in the west to Irian Jaya in the east, a distance of over 4000 km. Gäumann (1921) showed the causal agent was a Gram-negative bacterium that he named *Pseudomonas celebensis*. Confusingly, this species was briefly reclassified as *Xanthomonas*, along with other nonfluorescent pseudomonads (Dowson, 1943). The epithet "celebensis" was then used for another organism and the specific binomial is no longer valid. On the basis of DNA homologies and fatty acid profiles, recent isolates have been shown to be a homogeneous group belonging to phylotype IV of the *Ralstonia solanacearum* complex, together with isolates from other hosts in Indonesia (Fegan and Prior, 2005, 2006). However, phenotypic properties, notably pathogenicity for banana but not for any solanaceous host, differ from those generally accepted for *R. solanacearum* and, pending formal taxonomic description, the causal agent is referred to as the blood disease bacterium (BDB).

Epidemics of Moko were first reported over 100 years ago in the 'Moko' (= 'Bluggoe') cultivar and came to prominence in the 1950s as a threat to commercial plantations of dessert bananas in Central America (Sequeira, 1998). Strains of *Ralstonia* (formerly *Pseudomonas*) *solanacearum* pathogenic to banana, collectively referred to as race 2, are thought to have evolved into specialized niches in Central and South America. Banana (B) strains, persisting in soils and spread mainly through cultivation and planting practices, originated in plantations developed from land previously growing wild *Heliconia* species, from which isolates pathogenic to banana were obtained (Buddenhagen, 1960). Some isolates from *Heliconia* were indistinguishable from B strains on Kelman's (1954) tetrazolium (TZC) medium but caused stunting, distortion, and slow wilting of young banana plants (D strains), whereas others produced colonies with dense red formazan pigment from TZC and were pathogenic to *Heliconia* and plantain (AAB) but not dessert bananas (H strains) or mildly pathogenic only to *Heliconia* (R strains; French, 1986). The pathogenicity of D, but not R, strains increased after serial passage through banana (Sequeira and Averre, 1961; French and Sequeira, 1970), supporting the view that pathogenicity to banana evolved from naturally infected *Heliconia*.

Isolates showing a small fluidal round (SFR) colony morphology on TZC were first reported from an unusual outbreak of Moko disease on 'Bluggoe' bananas in Honduras in the early 1960s (Buddenhagen and Elsasser, 1962), apparently originating in planting materials introduced from Venezuela (Sequeira, 1998). In contrast to B strains, these produced copious amounts of bacterial exudates (ooze) from infected inflorescences, a feature considered an adaptation to insect transmission. SFR and other colony variants (SFR-C and SFR-A strains) were associated with other insect-transmitted epidemics originating in Colombia and spreading in 'Chato' ('Bluggoe') bananas in the Amazon basin into Peru and Brazil (French and Sequeira, 1970). Strains with SFR characteristics are now widely distributed in South and Central America and the southern Caribbean (Lehmann-Danziger, 1987; Phelps, 1987).

The polyphyletic origins of banana pathogens within the *R. solanacearum* group are supported by studies on the molecular phylogeny and taxonomy. From analyses of 16s-23s rDNA interspatial region and endoglucanase sequences, Prior and Fegan (2005) and Fegan and Prior (2006) showed that Moko and Bugtok strains clustered into four sequevar subgroups within a single division (Phylotype II). These findings are consistent with earlier multilocation genotype (MLG) clusters proposed by Cook et al. (1991) from whole-genome restriction fragment length polymorphism (RFLP) analyses. Banana (B) and *Heliconia* (H) strains clustered with Bugtok isolates in sequevar 3 (MLG24), whereas insect-transmitted SFR and Amazon (A) strains grouped into two different and more distantly related clusters (sequevar 4/MLG25 and sequevar 6/MLG28), to which two new isolates from Brazil were related. In contrast, BDB isolates, previously classified as a novel MLG (Cook et al., 1991; Cook and Sequeira, 1994), clustered in a more distantly related division (Phylotype IV), together with other isolates of *R. solanacearum* and *R. syzygii* from Indonesia. These results suggest that capabilities for insect transmission in banana pathogens have evolved from several independent origins.

6.4.3 Symptoms and Disease Cycle

Symptoms of banana bacterial wilts are very much related to the mode of infection, the host cultivar, and, in the case of Moko, the strain of the pathogen. Where infection takes place via mature leaves, roots, or rhizome, systemic invasion proceeds via the corm into the pseudostem of the affected plant, and frequently into daughter suckers. In this "bottom-up" mode of infection, early symptoms are a progressive transient yellowing and flaccidity starting with the oldest leaves, which become necrotic and collapse at the base of the petiole. Pale green or whitish/yellowish panels develop in one or more of the youngest leaves, which become flaccid and necrotic. The development of fruit bunches is arrested and some of the fingers may ripen prematurely or split. Young suckers or followers may show a generalized wilting with little foliar discoloration. Internally, discolored vascular bundles, initially cream or yellow but later becoming brown or black, may be seen throughout the plant, eventually extending into the daughter suckers. In fruit-bearing plants, discoloration tends to be concentrated in the fruit stem and the younger, central leaf bases. Within a few minutes of cutting, vascular tissues of plants affected by Moko or blood disease exude viscous bacterial ooze ranging in color from cream to reddish brown or black. A distinguishing feature of BXW is that the ooze tends to be yellow and extremely copious, often filling airspaces within the leaf bases as well as vascular tissues.

"Top-down" infection via the inflorescence, whether by insect transmission to male flower buds or via contaminated cutting knives used in debudding, typically gives rise to a sequence of symptoms that are very specific to bacterial wilts and serve as useful distinguishing features from Fusarium wilt. The male bud shows blackening, shriveling, or distortion of one or more bracts and associated male flowers, progressing to the whole bud and into the peduncle. Within a few days of infection, droplets of cream-colored or pale yellow bacterial ooze (Figure 6.3) may emerge on recently exposed scars of bracts and peduncle cushions. The ooze is said to be attractive to stingless bees (*Trigona* spp.) and other generalist feeders (Buddenhagen and Elsasser, 1962). Depending on the cultivar and stage of fruit development at the time of infection, necrosis may extend into the younger fruit bunches. Often fruit clusters at first appear outwardly normal but subsequently exhibit premature ripening. By this stage, all fruits typically show internal discoloration and rotting, which can vary in color (red, brown, or black) and texture (firm or gelatinous). As infection progresses, internal vascular discoloration can be traced from the male bud into the peduncle, down the fruit stem, and eventually into the corm and other parts of the plant. In this case, the youngest leaves, which are subtended from the flower stem rather than the corm, may be the first to show symptoms, and experienced farmers or field staff may recognize this as the first tell-tale sign of infection.

Following inflorescence infection in ABB cultivars, and occasionally other types, bacterial invasion can be incompletely systemic and may remain confined to the affected pseudostem or fail to spread into all of the daughter suckers. This is typified by Bugtok disease of 'Saba,' 'Cardaba,' and similar cultivars in the Philippines (Soguilon et al., 1994) but has also been described in 'Bluggoe'

FIGURE 6.3 Bacterial ooze from banana infected with Xanthomonas wilt.

infected by SFR strains in Central America (Black and Delbeke, 1991). Suckers that become infected will rapidly die, but those that do not will survive to produce flower buds that can be infected by insects. In this way, infected mats may persist indefinitely and act as sources of inoculum for further spread of the disease.

6.4.4 Disease Cycle and Implications for Control

Opportunities for control interventions relate to the ecology and epidemiology of different strains of the pathogen, to the host cultivar, and to cultivation practices. In commercial plantations of dessert (AAA) cultivars developed in Central America in the 1950s, pruning, desuckering, and debudding were routinely practiced for agronomic reasons. There were thus ample opportunities for spread of B strains on cultivation tools and contaminated planting materials and, although occasional insect transmission of B strains probably occurred, routine elimination of male flower buds apparently prevented the development of insect-transmitted epidemics. Once these factors were understood, Moko was successfully brought under control by the early 1960s. It generally remains of minor concern under plantation conditions although continued vigilance is necessary.

In contrast, ABB cultivars such as 'Bluggoe' are confined largely to smallholder and domestic plantings, where they are rarely debudded in time to prevent inflorescence infection (if at all) or with precautions to prevent spread on cutting tools. Most ABB types have dehiscent male flowers and bracts, and probably other characteristics that render them particularly susceptible to inflorescence infection—features that strains such as SFR appear to have evolved to exploit. These factors have contributed to epidemics of Moko that, like those of BXW and blood disease, have proven much more difficult to control. The two pathosystems reach their apotheosis in the Philippines, where Bugtok in smallholder plantings of ABB/BBB cooking bananas such as 'Saba' and 'Cardaba,' and Moko in 'Dwarf Cavendish' plantations, behave as two different diseases but are caused by the same strain of *R. solanacearum* (Fegan, 2005; Fegan and Prior, 2006).

6.4.4.1 Infection via Soil and Roots

During the development of commercial plantations on newly cleared land in Central America, the incidence of Moko was closely related to the distribution of infected wild species of *Heliconia* (Sequeira and Averre, 1961). Although passive spread via the soil was slow, areas abandoned to bacterial wilt slowly encroached on healthy ones (Buddenhagen, 1960). Fallowing or rotation regimes were successfully developed to reduce the risk of infection (Sequeira, 1962). "Buffer zones" around infected plants, free of banana and other potential hosts of the pathogen, were also advocated to reduce the risk of reinfection (Stover, 1972), although there are conflicting reports on the role of alternative weed hosts and specific recommendations have varied (see discussion in Jeger et al., 1995). The persistence of bacteria in soil following previous disease infection was also recognized for blood disease (Gäumann, 1923) but experimental data are still lacking.

These findings contributed to early recommendations for control of BXW by uprooting and burying diseased plants. However, there is an increasing body of data to suggest that *Xcm* survives poorly in soil or plant residues (Mwebaze et al., 2006) and that, following rigorous uprooting of infected mats/removal of diseased plants, replanting can safely take place after a 6-month fallow period (Turyagyenda et al., 2007). Planting practices such as avoiding topsoil and allowing the cut surfaces of corms to heal for a few days before planting have been advocated to further reduce the risks of infection during replanting (Mwangi et al., 2007). However, a study carried out in Ethiopia by Shehabu et al. (2010) found that paring and air-drying of banana suckers before planting increased soil-mediated Xanthomonas wilt infections. Likewise, the practice of using discarded leaves, flower stalks, buds, or other infected plant residues as mulch presents risks for reintroducing soil-borne inoculum from external sources (Mwangi and Nakato, 2009).

Race 2 of *R. solanacearum* can reportedly infect roots without mechanical injury (Kelman and Sequeira, 1965), but this has not been systematically investigated and it is likely that for all diseases most soil-borne infection probably occurs through mechanical injury of roots and corms during planting or cultivation (Tumushabe et al., 2006).

6.4.4.2 Spread on Cutting and Cultivation Tools

The early recognition that the majority of Moko infections occurred via cutting tools contaminated during pruning and harvesting (Buddenhagen and Sequeira, 1958; Sequeira, 1958) led to standard adoption and rigid enforcement of field sanitation practices that remain successful elements of Moko control under plantation conditions (Stover, 1972). Similar practices have been recommended for BXW but are difficult for smallholders to adopt. Field observations on the spread of BXW in East Africa provide strong circumstantial evidence of the importance of mechanical transmission both within and between plantings, especially in areas where cultivation practices such as leaf pruning and debudding are practiced intensively (Mwangi and Nakato, 2009). In Ethiopia, Addis et al. (2008) confirmed the high susceptibility of leaves, floral rachis, and pseudostems but to a much lesser extent of roots to infection via contaminated cutting knives. Buregyeya et al. (2008) showed that cutting knives transmitted disease up to 19 days (stainless knife) or 6 days (steel panga) after artificial contamination with sap from plants infected with *Xcm*—quite long enough for the disease to be spread by farmers, or by itinerant traders who frequently undertake harvesting operations themselves, using knives that may have been contaminated from distant fields. Similar considerations apply to blood disease in Indonesia, where interisland spread has been attributed to traders using their own harvesting knives (Supriadi, 2005). However, in practice the cutting of leaves for packing or padding, rather than harvesting mature fruit bunches, may be more important for spread of infection. Pseudostems bearing mature fruit bunches are no longer in active growth and observations on Moko (SFR strain) reported by Hunt (1992) indicated that harvesting mature bunches with contaminated knives did not result in disease transmission to the mat.

The standard treatment originally recommended for disinfecting knives against Moko was formaldehyde (Stover, 1972), but now that its toxicity is better appreciated other treatments need to be found. Quaternary ammonium compounds (quats), which are now widely used in human hygiene, food, and agriculture, showed promise in early trials with Moko and one compound was considered cheaper and only marginally less effective than formaldehyde (Buddenhagen and Sequeira, 1958). In East Africa, domestic bleach (sodium hypochlorite, widely known by the trade name of Jik) at dilutions of up to 20% is currently recommended for cleaning tools in fields affected by BXW. However, hypochlorite is not very stable and neither this, nor the alternative recommendation of heating over a fire, is very practical for smallholders. Other strategies include avoidance of pruning where there are risks that BXW is present, strict avoidance of harvesting fruits or leaves using knives provided by traders, and of course use of forked sticks rather than knives to twist and break off male buds (see Section 6.4.4.3).

6.4.4.3 Inflorescence Infection

The probable role of inflorescence infection in blood disease epidemics was postulated by Gäumann (1923) who observed natural infections and showed that bacteria applied to female flowers could infect the fruits. However, these infections were not attributed to insect vectors at the time, and the significance of these observations was overlooked until investigation by Buddenhagen and Elsasser (1962) of an unusual outbreak of Moko disease on 'Bluggoe' bananas in Honduras. The authors observed up to 100 bees and wasps per hour frequenting a single male inflorescence, and recovered bacteria exhibiting SFR characteristics from 5% of stingless bees (*Trigona corvina*) and wasps collected from a patch of diseased plants over a 20-day period. Inflorescence symptoms developed rapidly: bacteria exuded from pedicels and bract scars after 15–25 days of infection and attracted insects including *Trigona* spp., wasps (*Polyoba* spp.), and fruit flies (*Drosophila* spp.; Buddenhagen and Kelman, 1964). Healthy plants were infected when bacteria were carried by insects to moist, newly exposed pedicels ("cushions") within about 2 days of abscission of the male flowers. Bagging whole inflorescences to exclude insects and breaking off the male buds before the first male flower cushions became exposed prevented infections, which otherwise spread rapidly to neighboring plants. Transmissions between 'Bluggoe' plantings more than 1 mile (1.6 km) apart were apparently rare, but in isolated cases plants up to 5 miles (8 km) apart became infected and acted as new foci of infection.

Similar groups of insects have been implicated as vectors of blood disease in Indonesia (Leiwkabessy, 1999, cited by Supriadi, 2005; Subandiya et al., 2005) and BXW in Uganda (Tinzaara et al., 2006), DRC (Fiaboe, Beed, et al., 2008), and Ethiopia (Shimelash et al., 2008).

However, although the causal bacteria have frequently been recovered from insects, very few experimental transmissions have been demonstrated, and the possible role of stingless bees in the spread of BXW has not been confirmed (Namu, 2008). Symptoms of inflorescence infection appear to be uncommon at altitudes exceeding 1700 m above sea level in the Great Lakes region of East Africa, presumably reflecting reduced vector activity at cooler temperatures (Addis et al., 2004; Mwangi et al., 2007; Ndungo et al., 2008). Bats and birds have also been implicated in spreading BXW over longer distances. Buregyeya et al. (2008) recovered *Xcm* from certain birds and bats for up to 5 days after they had visited diseased banana flowers, but further work is needed to confirm their epidemiological significance.

Removal of male flower buds after fruit clusters have formed has long been standard agronomic practice in commercial dessert banana plantations and also under intensive East African Highland (EAH) production in East Africa. This undoubtedly reduces the risk of transmission by insects but may make matters worse unless strict precautions are taken to avoid spread on contaminated knives. For dwarf cultivars such as 'Dwarf Cavendish,' buds can be broken off by twisting the peduncle by hand, and a forked stick can be used for taller plants. Experiments under on-farm conditions in Uganda have shown that removal of male flower buds in this manner is highly effective in preventing BXW infection (Blomme et al., 2009), provided this is done immediately after formation of the last female flower cluster. This practice has been particularly successful in preventing infection of 'Pisang Awak' (ABB, 'Kayinja') and farmers who have adopted it have come back into full production. The effectiveness of debudding suggests that, in practice, infection takes place only via the male flower parts: in dehiscent cultivars, male flowers and bracts are shed approximately daily as new ones emerge, so there is continuous availability of infection courts over a period of several weeks. However, a preliminary report suggests that infection can also occur via the female part of the inflorescence, and at a high frequency, when bananas are grown in isolation and there are high levels of inoculum in adjacent plants (Fiaboe, Kubiriba, et al., 2009). It is not yet clear whether these infections took place via female bract scars or the flowers themselves. In other studies (Tinzaara et al., 2006), *Xcm* has been recovered from nectar of male flowers and while male flowers may be too short lived to allow time for bacteria to multiply and invade the peduncle, this would not apply in female flowers or in cultivars with persistent male flowers.

6.4.4.4 Planting Materials

Daughter suckers are readily invaded if bacteria reach the corm, but this may not always occur depending on cultivar and route of infection. Thus contaminated planting materials present significant risk of infection, even if the remainder of the mat appears healthy. Spread of Moko between plantation operations in Central America is well documented (Buddenhagen, 1961) and is thought to be the means that the disease reached the Philippines as early as the 1940s (Rillo, 1979; Hayward, 2006). Similar mechanisms are likely to have contributed to the spread of blood disease in Indonesia and BXW in East Africa. The use of tissue-cultured plantlets eliminates the risks for commercial producers, but access to disease-free planting materials remains a challenge for smallholders (www.c3project.iita.org, accessed 18 May 2010).

6.4.4.5 Other Modes of Infection

Some strains of *R. solanacearum* are known to be spread in rivers, and flooding or irrigation has been suggested as means of dispersal of Moko (Sequeira, 1998). Circumstantial observations have suggested that BXW may be dispersed via wind-blown rain, on the feet or hooves of man and animals, or by browsing animals, but experimental evidence is lacking. Trade may play an important role in the long-distance spread of BXW, both from cutting by contaminated knives and from movement and subsequent discard of infected plant parts. An outbreak in Muleba district, Tanzania, is thought to have originated from infected male buds used as stoppers for containers of banana beer imported from Uganda (Mwangi and Nakato, 2009).

6.4.5 Screening and Breeding for Resistance

Several authors have reported differential responses of cultivated and noncultivated *Musa* to Moko (Stover, 1972), blood disease (Gäumann, 1921; Supriadi, 2005), and BXW (Ssekiwoko et al., 2006; Welde Michael et al., 2006; Tripathi et al., 2008) following either natural infection or experimental inoculation. Although no cultivated type has been reported to be truly resistant to any of these diseases, differences in speed and severity of symptom development have been noted following mechanical inoculation, and some clones have shown partial recovery with production of new suckers. However, such plants are likely to be tolerant rather than resistant, and symptomless plants can serve as hidden sources of inoculum. Stover (1972) reported that 34 of 345 accessions tested in Honduras showed some resistance to a Moko SFR strain injected into the pseudostem. These included the ABB clone 'Pelipita,' the 'Manang' accession of *M. acuminata* spp. *banksii*, and *M. balbisiana*. The latter has been reported to survive mechanical inoculation with B strains of Moko and also *Xcm* (Ssekiwoko et al., 2006; Tripathi et al., 2008).

In the absence of cell-mediated resistance, a more promising breeding strategy may be to exploit differences in susceptibility to inflorescence infection. The ABB clone 'Pelipita' has been suggested as a replacement for 'Bluggoe' because it has indehiscent male bracts and flowers, and is much less susceptible to insect transmission (Stover and Richardson, 1968). Cavendish types, which tend to retain male flowers and bracts, are less susceptible to inflorescence infection to Moko than the 'Gros Michel' plants affected by Moko in Central America. 'Dwarf Cavendish' is not widely grown in East Africa but observations in Ethiopia suggest that it is also less susceptible to BXW (Addis et al., 2004). Some EAH (AAA) and plantain (AAB) cultivars do not shed male flowers and bracts and hence exhibit apparent resistance, or escape, from BXW (Mwangi and Nakato, 2009). Others have dehiscent flower parts but appear to be less susceptible to infection either because the flowers are less attractive to insects or the flower scars are less conducive to penetration or survival of *Xcm* (Mwangi et al., 2006). Recently, Buddenhagen (2009b) has drawn attention to clones from Indonesia that do not produce male buds at all. One of these, known locally as 'Pisang Puju,' has fruit characteristics very similar to 'Pisang Kepok' (Saba) and is proposed as a replacement. However, although able to escape insect transmission, it is likely that these clones will be susceptible to infection of roots or leaves damaged by contaminated cultivation tools.

These features underline the importance of considering different routes of infection when screening for resistance that may prove useful in the field. Mechanical inoculation by injection of bacteria into pseudostem or leaf bases is a convenient technique for screening large numbers of plants (Tripathi, Odipio, et al., 2008) but may be unnaturally severe, especially if high doses of inoculum are used. Welde Michael et al. (2006) found all of 40 banana cultivars were susceptible following inoculation of a high dose of *Xcm* into leaf bases, but acknowledged that the inoculation method could have masked differences in susceptibility to infection via inflorescences.

Given the lack of resistance and difficulties in conventional breeding of bananas, transgenic approaches merit attention. Tripathi et al. (2008) report promising progress in generating several East African Highland banana (EAHB) and 'Pisang Awak' lines with enhanced defense mechanisms induced by introduction of ferredoxin-like amphiphatic protein (*pflp*) from sweet pepper. These are currently being evaluated for resistance to *Xcm* under containment conditions.

6.4.6 Impact and Future Outlook

Once main avenues of infection were recognized and appropriate phytosanitary measures devised, Moko was quickly brought under control in commercial plantations of dessert bananas, although at some considerable costs (Stover, 1972; Sequeira, 1998). It has remained a serious and intractable problem for smaller farmers since epidemics were first described in South and Central America in the 19th and early 20th centuries. Within the past 50 years, well-documented pandemics driven by insect transmission in highly susceptible ABB cultivars have continued to limit or destroy banana cultivations in both the Americas and the Philippines (French and Sequeira, 1970; Lehmann-Danziger, 1987; Sequeira, 1998; Molina, 2006). A similar pandemic has occurred in Indonesia, following spread of blood disease from Sulawesi to west Java in the late 1980s (Davis et al., 2001; Supriadi, 2005; Buddenhagen, 2009b) and now threatens the Southeast Asian mainland (Promed, 2009). In East and Central Africa, high densities of 'Kayinja' ('Pisang Awak') and similar susceptible cultivars have driven the insect-transmitted spread of BXW in some regions (Eden-Green, 2004), but the situation is more complex. Under intensive cultivation of AAA-EAHB types, where banana tends to have greatest value for income and food security, insect transmission is probably rare, but cultivation practices favor mechanical transmission.

Much has been written about the social, economic, cultural, and environmental impacts of banana bacterial wilts on affected smallholder communities but these aspects are difficult to quantify. Kalyebara et al. (2006) estimated that, in purely economic terms, cumulative losses to BXW in Uganda could reach US$5.6 billion over 10–15 years, representing an annual loss of about US$200 of food and income per farming household. Adoption of control measures could reduce losses by 40% in *matooke* but less in brewing types. Similar conclusions were reached independently from surveys carried out in 2005 by Karamura (2006) and co-workers, who predicted that, unchecked, the disease would spread throughout Uganda, resulting in a cumulative loss of over $4 billion by 2010, significant price increases, and reduction in production. However, in the short term, moderate production losses could produce shortages and higher prices that would benefit producers (Abele and Pillay, 2007). Increased market prices have acted as incentives to farmers prepared to adopt debudding to control BXW in 'Kayinja' juice bananas (Blomme and Karamura, 2006). Perhaps the greatest concerns arise in areas destabilized through conflict and civil unrest, where annual crops become disrupted or neglected and bananas assume increased importance for food security (Abele et al., 2007).

In purely technical terms, banana bacterial wilts are not difficult to control. Avoidance and sanitation measures are known and easy to implement, although complete eradication is more difficult to achieve. The problem for smallholder systems lies both in communicating knowledge and in gaining ownership and adoption of control measures that, to be fully effective, need to be applied on a community scale. Considerable efforts have been made in Uganda to gain community adoption through carefully structured interactions such as working groups and task forces, involving technical staff, local government, community leaders, and farmers, a process that has been termed participatory development communication (PDC; Nankinga and Okaasai, 2006; Odoi, 2006). These measures have met with considerable success in reducing the incidence and spread of BXW in smallholder communities (Kubiriba, 2009), but dealing with resurgence of infection remains a problem.

6.5 BANANA VIRUSES

6.5.1 Introduction

Viral diseases of banana constitute a very large group of important infective agents that reduce banana and plantain production in different parts of the world. Since banana is a vegetatively propagated crop, the viruses simply multiply in them and are spread during planting. Tissue culture can also help in spreading banana viruses if infected plants are multiplied in culture. For this reason, proper and careful indexing of planting materials for viral pathogens using molecular tools is necessary. The following viral pathogens have been reported from different banana growing areas of the world and they include: banana streak virus (BSV), banana bunchy top virus (BBTV), banana bract mosaic virus (BBMV), and banana mosaic or infectious chlorosis (also called cucumber mosaic virus, CMV).

6.5.2 Banana Streak Virus

6.5.2.1 Distribution

Viral leaf streak was first reported on *Musa* (AAA group Cavendish subgroup) 'Poyo' from Côte d'Ivoire (Lassoudiére, 1974). The causal agent was identified in Morocco in 1986 and called banana streak badnavirus (BSV) (Lockhart, 1986) and the disease was called banana streak virus. BSV is a para-retrovirus (Bouhida et al., 1993). BSV has been identified in different banana-growing areas of the world, including Benin, Côte d'Ivoire, Ghana, Malawi, Madagascar, Mauritius, Nigeria, Rwanda, South Africa, Tanzania, Australia, Brazil, the Canary Islands, and Trinidad (Lockhart and Olszewski, 1993). The disease has also been found in India where it is most severe on 'Mysore,' an important commercial AAB group, and in Latin America and Australia (Jones and Lockhart, 1993; Diekmann and Putter, 1996; Riechel et al., 1996). Although BSV disease occurs in all banana-producing areas, the amount of field infection may greatly vary.

6.5.2.2 Structure and Composition

Badnaviruses are a group of plant viruses with non-enveloped virions and average particle size of 120–150 nm × 30 nm and double-stranded DNA (dsDNA) genome of 7.4–7.8 kbp (Lockhart, 1990). Particle sizes of 60–900 nm in length have been observed in purified preparations. Badnaviruses differ from other plant viruses in particle morphology, genome size, vector relationships, and histopathology. Under the microscope, particles of CMV and BSV are distinctly different; whereas CMV is isometric (28–30 nm), BSV has a bacilliform shape (30 × 130–150 nm). Badnavirus particles contain a single species of dsDNA. Under nondenaturing conditions, the DNA migrates in agarose as three or more bands, corresponding to linear, circular, and twisted configurations. The virus codes for three proteins from three well-defined open reading frames (ORFs). ORF 111 polyprotein is the most characterized part of the virus.

6.5.2.3 Host Range and Viral Transmission

Most badnaviruses are present in clonally propagated tropical crops and spread in nature by vegetative propagation. The individual viruses typically have a restricted natural host range, frequently limited to a few species within a single plant genus. BSV is serologically related to the sugarcane bacilliform *virus* (ScBV; Lockhart and Autrey, 1988), which also produces streak symptoms in experimentally inoculated plants (Lockhart and Autrey, 1991). However, preliminary host-range studies with BSV isolates indicate differences in the ability to infect species within the Musaceae as well as sugarcane (Lockhart and Autrey, 1991). ScBV is transmitted from sugarcane to banana by the pink sugarcane mealy bug *Saccharicaccus sacharri* and from banana to banana by *Planococcus citri* in a semipersistent manner. Spread within short distances is accomplished when the vector (mealy bug) makes contact with foliage within the crop canopy from infected to healthy plant. The spread of badnaviruses occurs primarily within a single cultivated species. There is no evidence that mechanical contact, cutting tools, or cultural operations can spread BSV.

In most tropics, sugarcane is grown in close proximity to banana, and transmission from the former to the latter undoubtedly occurs. In India where sugarcane is grown in rotation with rice, BSV is very common, especially in 'Mysore' (AAB). Almost all the clones of this cultivar carry BSV, suggesting that the disease symptoms were of genetic origin. It is now known that BSV can be transmitted to new hybrids through the activation of integrants present in the parent B genomes, and tissue culture triggers symptom expression. The mode of replication in badnaviruses encourages a high level of mutation, and a number of serologically distinct, naturally occurring isolates of BSV have been identified. Hence, it would be useful to know if all strains are vector transmitted, and whether there is any vector specificity, to explain why BSV spreads in some areas and not in others.

6.5.2.4 Symptom Expression

Jones and Lockhart (1993) have reviewed symptoms associated with BSV and found that they are similar to cucumber mosaic virus (CMV), especially in early stages. However, necrotic streaks and periodicity of symptom expression are characteristic features of BSV. Plants may not show streak symptoms on all leaves, and for several months, emerging leaves may not exhibit streak symptoms at all or show only light symptoms. This explains why plants in quarantine need to be grown for some time and examined regularly for symptoms. For this reason, *Musa* pathologists may have mistaken BSV for CMV. A few photographs reputed to show symptoms of CMV in some publications look identical to those of BSV (Wardlaw, 1961; Stover, 1972). Stover (1972) and Simmonds (1987) showed that incidence of CMV in Horn plantain (AAB) in Honduras was extremely high, with greater symptoms being expressed in cooler conditions. In Morocco, BSV-infected banana (*Musa spp.* [AAA group] cv 'Dwarf Cavendish') had broken or continuous streaks and/or sparse or concentrated spindle-shaped lesions on the leaf lamina that often turned necrotic upon aging (Lockhart, 1986).

The following symptoms are associated with BSV-infected *Musa* germplasm in Nigeria: discrete whitish/yellow short streak on the leaf lamina turning darker yellow or orange with age, distortion of leaf lamina, internal necrosis of the pseudostem, length-wise cracking of outer leaf sheaths, and cigar leaf death (die-back) followed by plant death and distorted bunch or fingers (Gauhl and Pasberg-Gauhl, 1995). Others are abnormal emergence of flowers and bunches, shortened and distorted peduncle, and bunch bursting out from the side of the pseudostem (Gauhl and Pasberg-Gauhl, 1994b). Diseased plants have reduced growth and vigor, resulting in poor yield due to small or absence of bunches (Ortiz, 1996). These symptoms are more severe in the rainy season than in the dry season (Dahal et al., 1998). Pronounced streak symptoms occur sporadically throughout the year when inflorescence initiation coincides with an episode of increased replication.

6.5.2.5 Diagnosis and Detection

The successful control of most diseases depends on accurate diagnosis. It is known that BSV can persist in the host for a long time without producing any symptoms. The integration of more than a full-length genome of the virus in the host cell is a challenge to quarantine authorities and has serious implications on the exchange of planting materials across international boundaries. Previously, diagnosis of BSV was based on the recognition of symptoms. Unfortunately, the inherent difficulties in the use of symptoms as the only diagnostic technique resulted in confusion in differentiating BSV from CMV. In addition, the detection and diagnosis of badnaviruses is hindered by their low concentrations in host plants and the presence of secondary products such as tannins, mucilage, and other compounds (PBIP, 1997). Detection by visual observations is not reliable because of periodicity of symptoms, and the use of indicator plants is not applicable because BSV has a restricted host range. Detection by electrophoresis of virus-specific dsRNAs in plants has been found to be a highly sensitive method (Dodds et al., 1984). However, this technique is only applicable to plant viruses whose genomes consist of single-stranded RNA (ssRNA). But BSV and other badnaviruses that have dsDNA genomes do not have dsRNA replicative intermediates and therefore cannot be detected using this method. Serological methods, using immunoprecipitation, enzyme immunoassay (EIA), and immunoelectron microscopy (IEM) have been advocated (Lockhart, 1986). Both EIA and IEM provide rapid, sensitive, and reliable methods for the detection of badnaviruses when a viral antigen occurs in plant tissue. The application of these techniques is, however, restricted by the occurrence of serologically distinct badnavirus populations within a given crop. This serological heterogeneity among isolates has been illustrated in BSV using EIA and IEM (Lockhart, 1990).

To overcome these difficulties, an alternative approach based on the polymerase chain reaction (PCR) amplification of the conserved badnavirus genomic sequences was introduced (Sidi et al., 1988). This method has the potential to detect any variant of BSV. Thus, an isolate of BSV can be purified and the genome of this isolate is cloned and sequenced, allowing for the synthesis of BSV-specific primers. These primers are used to screen a diverse range of *Musa* germplasm to confirm their ability to detect BSV isolate from different sources. Medberry et al. (1990) employed this method to identify and analyze the nucleotide sequences of three badnaviruses (CoYMV, RTBV, and KTSV) and identified three conserved regions in the ORF 111, which include the $tRNA^{Met}$ binding domain, ribonuclease H (*RNAse* H), and reverse transcriptase (RT) regions. The PCR-based amplification method therefore appears to represent a workable solution to the problem of serological and genomic heterogeneity among BSV and other badnavirus triple-antibody sandwich enzyme-linked immunosorbent assay (TAS-ELISA) has been used to detect the virus (Thottapilly et al., 1997). Immunocapture PCR (IC-PCR) has also been used to differentiate the episomal and retroviral sequences in banana. Real-time PCR has also been used for quick detection of virus genomes, but the technique is expensive and may not be affordable by most tissue-culture laboratories.

6.5.2.6 Economic Importance

BSV has been reported to cause losses of 7–90% in Côte d'Ivoire (Lassoudière, 1979). BSV infection can reduce bunch size and distort fruit shape in banana. It has been reported that yield loss varies from cultivar to cultivar and with climatic conditions. In India, BSV severely infects the Mysore group of banana, namely 'Poovan,' 'Red Banana,' 'Robusta,' 'Nendran,' 'Cheni Champa,' and also the Cavendish group ('Basrai' and 'Grand Nain'). It is likely that BSV disease occurs in all banana-producing areas, although the amount of field infection may vary greatly.

6.5.2.7 Control and Elimination

The use of virus-free planting materials remains the best way to control badnaviruses. A combination of meristem culture and thermotherapy is another reliable method (Quak, 1977). However, this does not guarantee "virus-free" materials (George and Sherrington, 1984; Drew et al., 1989; Nwauzoma et al., 2004). This was further confirmed in thermotherapy studies with the closely related sugarcane bacilliform virus. As the ambient temperature was increased, the replication rate of the sugarcane virus also increased. Virions of badnaviruses are thought to be stable at room temperature for several weeks but infectivity is lost on exposure to 53–55°C for 10 minutes.

It has been demonstrated that the BSV genome is integrated into the *Musa* genome (Ndowora et al., 1999; INIBAP, 1997). Ortiz and Vuylsteke (1998) suggested the use of virus-tolerant genotypes as the most appropriate short-term control measure. *In vitro* culture in the presence of "anti-viral" compounds can be used to eliminate viruses from germplasms. One of the most effective and commonly used compounds is Ribavirin (1-D-ribofuranosil-1-H-1, 2,4-triazole-3-carboxamide). The actual mode of action of Ribavirin is still unclear; it appears that it has RNA-specific activity but it has been demonstrated that Ribavirin does not inhibit RNA polymerases (Lerch, 1987; Senula et al., 2000). Since there are no reports of its activity against dsDNA viruses of plants, it cannot be regarded as a potential therapeutic agent for BSV elimination.

6.5.3 Banana Bunchy Top Virus (BBTV)

6.5.3.1 Distribution

Bunchy top virus disease (BBTV) is a major production constraint of banana in most tropical countries, especially in Asia. The disease was first reported in Fiji and subsequently in Taiwan at about 1900 and Egypt in 1901. It is believed that the disease spread to Australia and Sri Lanka in 1913 through infected suckers from Fiji. It was introduced to India from Sri Lanka in 1943 and is now predominant in the following states: Andhra Pradesh, Tamil Nadu, Orissa, Maharastra, Bihar, Karnataka, West Bengal, Assam, and Uttar Pradesh (Selvarajan et al., 2007). Some important local cultivars grown by small-scale growers in several Asian countries are susceptible. Bunchy top is present in some parts of Africa and the South Pacific, but has not been recorded in South and Central America or the Caribbean.

6.5.3.2 Structure and Composition

BBTV was formally classified as a luteovirus due to the following characteristics: non-sap-transmissible nature but persistently transmitted by aphids, and presence of yellowish symptoms restricted to the phloem tissues. An analysis of infected plants (Dale et al., 1986) and electrophoretic patterns of dsRNA from infected plants (Iskra et al., 1989) supported the earlier classification as a luteovirus. However, Harding et al. (1991) and Thomas and Dietzgen (1991) used a modified method of Wu and Su (1990) to purify 18–20 nm isometric virus-like particles (VLPs) from infected plants. These particles had ssDNA of 1 kb with a molecular weight of about 20.000. Furthermore, Harding et al. (1993) reported that the sequence of the first component had circular ssDNAs of 1.111 kb, with other structures that resembled the geminiviruses. Burns et al. (1995) showed that BBTV has a multicomponent genome with at least six ssDNA components, namely components 1–6, and the circular genomic DNA of each component varies in size from 1.018 kb to 1.111 kb. All the components had a major common region (CR-M) and the stem loop common region (CR-S). Five out of the six components had single large ORFs in the virion sense. Xie and Hu (1995) showed that the sequence of three BBTV genomic components from a Hawaiian BBTV isolate had components 1, 3, and 4 that were almost identical to components 1, 2, and 5 of Harding et al. (1993) and Burns et al. (1995). Wu et al. (1994) sequenced two more BBTV components from Taiwan and found that both isolates encoded for replicase-associated proteins. Certainly, BBTV is not a luteovirus but is more similar to the geminivirus, but differs from the latter in that its irons are isometric. Probably, BBTV is a member of a new plant virus group, very much likely a babuvirus of the family *Nanoviridae*. It is isometric and measures 20 kDa and has a multicomponent genome. The virus is highly concentrated in the phloem of infected plants.

6.5.3.3 Transmission and Host Range

The disease is transmitted by an aphid, *Pentalonia nigronervosa*, especially at the nymphal stage through suckers and divided corms of infected plants. It can also be transmitted through infected micropropagated plantlets obtained from infected plants (Drew et al., 1989). It spreads very rapidly in summer when the vector population is very high and active. Drew et al. (1989) also showed that tissue-culture plants did not show characteristic symptoms until they were transferred to the greenhouse, where a majority expressed symptoms. The virus affects all the species of the genus *Musa* (*M. acuminata*, *M. balbisiana*, *M. sinensis*, and *M. paradisica* with their hybrids and Fe'i bananas). All commercial and widely grown cultivars of banana and plantain are susceptible. There are no reports of infection on wild diploid *Musa* species.

6.5.3.4 Disease Symptoms

Symptoms of the presence of BBTW include intermittent, dark green dots and streaks of different length on the leaf sheath, petiole, veins, and midrib. Typical streaks may not be present in some cultivars. Emerging leaves are shorter, brittle in texture, and narrow (Figure 6.4). Infected plants do not produce flowers, and when infection occurs late, bunches are formed but the fingers do not develop to maturity. In some cultivars, there is marginal chlorosis of the leaf lamina and vein flecking. Sometimes, the tip of the bract of the male bud turns green and when infection occurs very late in the season, the plant shows dark green streaks on the tip of the bracts.

6.5.3.5 Detection and Control

BBTV can be detected through symptoms, the presence of the aphid vector, serology, and nucleic acid assay. Advances in research in tissue culture and immunoassay diagnostic tools have enabled the production and use of healthy planting materials as a component to manage banana viruses. Transgenic virus resistance (TVR) has been widely used as an effective strategy to control plant viruses. This strategy could be of great relevance to bananas since there has been no immunity to BBTV, and conventional breeding programs are yet to prove effective. Genetic transformation of the cultivars 'Grand Nain' using *Agrobacterium* and 'Bluggoe' using microprojectile bombardment for BBTV resistance has been reported (Selvarajan et al., 2007).

FIGURE 6.4 Symptoms of banana bunchy top disease.

6.5.4 Banana Bract Mosaic Virus (BBrMV)

Banana bract mosaic virus is a potyvirus of the family *Potyviridae*. BBrMV was prominently reported first from the island of Mindanao, Philippines (Magnaye and Espino, 1990). Thomas and Magnaye (1996) confirmed the presence of the causal agent of the disease in the widely cultivated local cultivar 'Embul' (AAB group, Mysore) in Sri Lanka in 1996. BBrMV is also present in Ghana, Nigeria, and Malawi (Selvarajan et al., 2007). In India, BBrMV is found in the southern states of Tamil Nadu and Kerala, where it is referred to as "Kookan."

6.5.4.1 Structure and Composition

The genome of the virus is a positive single-stranded RNA (ssRNA) of approximately 10 kb particle size. The particles are flexuous rod shaped and measure 750 × 15 nm in size. Under the electron microscope, flexuous rod-shaped particles from the leaf sheath and bracts of infected 'Nendran' and 'Robusta' cultivars have been observed (Selvarajan et al., 2007). The virus titer is greater in bracts and the midrib than in the leaf sheath. Cytopathological observations reveal the presence of pinwheel inclusions and flexuous rod-shaped particles, with typical features of potyvirus. The coat protein appears as one major polypeptide with a molecular weight of 38 kDa and three minor polypeptides of 63, 53, and 22 kDa (Selvarajan et al., 2007). The open reading frame (ORF) of the virus amino acid sequence was found to be similar to the C-terminal half of the maize dwarf mosaic potyvirus coat protein.

6.5.4.2 Disease Symptoms

Characteristic symptoms include the presence of spindle-shaped, pinkish to reddish streaks on the pseudostem, midrib, and peduncle. Typically spindle shaped, mild mosaic streaks on bracts, peduncle, and fingers have also been observed. In the cultivar 'Nendran,' the leaf orientation changes, giving it the appearance of "Traveler's Palm" (*Ravenala madagascariensis*), with spindle-shaped mosaic patterns on both upper and lower leaf surfaces. Infected plants show a very long or extremely short peduncle, production of immature bunches, and raised corky growth on the peduncle. In 'Red Banana' and 'Robusta,' presence of strips on leaves and shriveled, pencil-like fingers have been observed.

6.5.4.3 Transmission and Detection

BBrMV is transmitted in a nonpersistent manner by three banana aphids (*Rhopalosihum maidi*, *Aphis gossypii*, and *Pentalonia nigronervosa*) and by the cowpea aphid *Aphis craccivora*. According to Sreenivasalu et al. (2006) quoted by Selvarajan et al. (2007), these vectors occur in banana and plantain in the semiarid tropics. Reverse transcriptase PCR (RT-PCR) was found to be more sensitive than ELISA and immunocapture PCR (IC-PCR) to detect the virus (Rodoni et al., 1999). Selvarajan et al. (2006) also found RT-PCR more sensitive than DAC-ELISA and dot immunobinding assay (DIBA) to detect viral isolates. RT-PCR amplification of viral RNA from crude plant sap has been reported (Selvarajan et al., 2007). Other detection techniques are nucleic acid probe specific to BBrMV (Thomas et al., 1997) and digoxigenin-labeled virus-specific probes.

6.5.4.4 Economic Importance and Control

In India, yield losses of 52.5% and 70% have been reported in the cultivar 'Nendran' and 'Robusta,' respectively (Cherian et al., 2002). Use of virus-free planting materials is the best way to control the spread of the virus.

6.5.5 Banana Mosaic or Cucumber Mosaic Virus (CMV)

Banana mosaic or infectious chlorosis caused by cucumber mosaic virus is a disease of relatively minor importance to banana, infecting more than 1,000 species of plants as hosts in the tropics, subtropics, and temperate regions of the world, including Australia and parts of Asia. CMV is distributed worldwide and has perhaps the widest host range of any plant pathogenic virus. First described in New South Wales (NSW) Australia in 1930, one of the isolates, IB, is restricted to Asia, while IA and II subgroups are worldwide (Selvarajan et al., 2007). In India, the disease occurred in the early 1940s in the state of Maharashtra and later reported in Tamil Nadu and other parts of the country. CMV isolates can be divided into two major serological subgroups defined as DTL and ToRS, and this provides valuable information on the geographical location of and symptoms produced by CMV isolates (Haase et al., 1989).

6.5.5.1 Structure and Composition

CMV consists of a spherical particle of 28–30 nm in diameter containing ssRNA and is naturally transmitted by aphid vectors or by seed (Lockhart and Jones, 2000), with viral particles of about 29 nm and composed of 180 subunits. The particles have a sediment (S) value of 98. The virions are made of 18% RNA, are highly labile, and rely on RNA-protein interactions for integrity. The genomic RNAs are SS in addition to sense RNAs with 5′ cap structures and 3′ conserved regions that can be folded into tRNA-like structures. The virus encodes five proteins, distributed on three genomic RNAs. RNA 1 is 3.3 kb and is the only monocistronic RNA that encodes the 1a protein that is required for viral replication. RNA 2 with 3 kb encodes the 2a protein, the viral polymerase, and the 2b protein, expressed from a low-abundance subgenomic RNA and RNA 4A. The 2b ORF is overlapped on the carbolic acid (COOH) terminal portion of the 2a ORF, which inhibits host post-transcriptional gene silencing (PTGS). RNA 3 has a particle size of 2.2 kb and is packaged in individual particles as a subgenomic RNA. RNA 4 (1 kb) encodes the movement protein (MP) from the 5 ORF and the coat protein (CP). Both 3a protein and the CP are necessary for intercellular movement of the virus through plasmodesmata, while 2b ORF is important in long-distance movement of CMV. Also, Hu et al. (1995) identified banana isolates that belong to the subgroup I, DTL serotype.

6.5.5.2 Disease Symptoms and Host Range

CMV has an incredible host range, infecting over 1,000 hosts comprising of 85 plant families and approximately 365 genera in both monocots and dicots. Symptoms, which depend on the strain of the virus pathogen and growth temperature, include yellow streaks or flecks on leaves in a mosaic pattern, leaf yellowing, and leaf mosaic that appears on all the leaves of infected plants. Rotting of cigar leaf, internal necrosis of pseudostem, and death of the entire plant are associated with CMV infection. Chlorotic banding and a rosette appearance of leaf arrangement are also symptoms of CMV (Selvarajan et al., 2007).

6.5.5.3 Viral Transmission

The disease is acquired through a wide range of host plants that grow in proximity with banana from where it is transmitted by aphid vectors in a nonpersistent manner: cotton aphid, *Aphis gossypii*, and corn aphid, *Rhopalosipum maidis*. Transmission has also been recorded by *Myzus persicae*, *Macrosiphum pisi*, and *Rhopalosipum prunifolae*. It can also be transmitted by the parasitic plant dodder (*Cuscuta* sp.), in which it replicates, and through seeds. The virus is most frequently transmitted mechanically and through sap of purified virions and viral RNA.

6.5.5.4 Disease Management

One of the control measures is the use of pathogen-free planting material and control of alternate hosts (weeds, legumes, cucurbits, and members of the *Solanaceae*, such as tomato). Meristem culture has also been developed to eradicate viruses of various vegetative propagating plants, including CMV. It was found that CMV particles could invade meristems and that a gradient of increasing virus concentration from the dome to the successive primordial exists (Walkey and Webb, 1968). This means that the probability of obtaining virus-free plants is inversely related to the size of the meristem excised. Thermo- and chemotherapy coupled with meristem culture can be used when meristem culture alone fails (Kartha, 1986). Berg and Bustamante (1974) observed that heat treatment (35–43°C) for 100 days performed on rhizomes with meristem culture was inefficient for cleaning commercial banana cultivars.

REFERENCES

Abele, S. and M. Pillay. 2007. Bacterial wilt and drought stresses in banana production and their impact on economic welfare in Uganda: Implications for banana research in east African highlands. *J. Crop Improvement* 19(1/2):173–191.

Abele, S., E. Twine, and C. Legg. 2007. *Food security in Eastern Africa and the Great Lakes*. Crop crisis control final report. Ibadan, Nigeria: IITA.

Addis, T., F. Handoro, and G. Blomme. 2004. Bacterial wilt (*Xanthomonas campestris* pv. *musacearum*) on enset and banana in Ethiopia. *InfoMusa* 13(2):44–45.

Addis, T., L.F. Turyagyenda, T. Alemu, E. Karamura, and G. Blomme. 2008. Garden tool transmission of *Xanthomonas vasicola* pv. *musacearum* on banana and enset in Ethiopia. Paper presented at the conference on Banana and Plantain in Africa: Harnessing International Partnerships to Increase Research Impact, October 5–9, 2008, Mombasa, Kenya.

Agrios, G.N.1997. *Plant pathology*. New York: Academic Press.

Alabouvette, C., P. Lemanceau, and C. Steinberg. 1993. Recent advances in the biological control of Fusarium wilts. *Pest. Sci.* 37:365–373.

Aritua, V., N. Parkinson, R. Thwaites, J.V. Heeney, D.R. Jones, W. Tushemereirwe, et al. 2008. Characterization of the *Xanthomonas* sp. causing wilt of enset and banana and its proposed reclassification as a strain of *X. vasicola*. *Plant Pathol.* 57:170–177.

Arzanlou, M., E.C.A. Abeln., G.H.J. Kema, C. Waalwijk, J. Carlier, I. de Vries, et al. 2007. Molecular diagnostics for the Sigatoka disease complex of banana. *Phytopathology* 97:1112–1118.

Aurore, G., B. Parfait, and L. Fahrasmane. 2008. Bananas, raw materials for making processed food products. *Trend Food Sci. Technol.* 20:78–91.

Bancroft, J. 1876. Report of the board appointed to enquire into the cause of disease affecting livestock and plants. Queensland, 1876. *Votes and Proceedings* 1877:1011–1038

Bentley, S., K.G. Pegg, and J.L. Dale. 1995. Genetic variation among a world-wide collection of isolates of *Fusarium oxysporum f. sp. cubense* analysed by RAPD-PCR fingerprinting. *Mycolog. Res.* 99:1378–1384.

Berg, L.A. and M. Bustamante. 1974. Heat treatment and meristem culture for the production of virus-free banana. *Phytopathology* 64:320–322.

Beveraggi, A., X. Mourichon, and G. Salle. 1995. Etude compare des premieres etapes de l'infection chez des bananiers sensible et resistants infectes par le *Cercospora fijiensis* (*Mycosphaerella fijiensis*) agent responsible de la maladie des raies noires. *Can. J. Bot.* 73:1328–1337.

Biruma, M., M. Pillay, L. Tripathi, G. Blomme, S. Abele, M. Mwangi, et al. 2007. Banana *Xanthomonas* wilt: A review of the disease, management strategies and future research directions. *Afr. J. Biotechnol.* 6:953–962.

Black, R. and A. Delbeke. 1991. Moko disease (*Pseudomonas solanacearum*) of *Musa* in Belize. *Trop. Sci.* 31:347–353.

Blomme, G. and E. Karamura. 2006. How farmers are stopping the spread of BXW. *MusAfrica* 17:12–14.

Blomme, G., L.F. Turyagyenda, H. Mukasa, F. Ssekiwoko, S. Mpiira, and S. Eden-Green. 2009. The effect of the prompt removal of inflorescence-infected plants and early debudding of inflorescences on the control of Xanthomonas wilt of banana. *Acta Hort.* 828:51–56.

Bouhida, M, B.E.L. Lockhart, and N.E. Olszewski. 1993. An analysis of the complete sequence of a sugarcane bacilliform virus genome infectious to banana and rice. *J. Gen. Virol.* 74:15–22.

Buddenhagen, I. 1960. Strains of *Pseudomonas solanacearum* in indigenous hosts in banana plantations of Costa Rica, and their relationship to bacterial wilt of banana. *Phytopathology* 50:660–664.

Buddenhagen, I.W. 1961. Bacterial wilt of bananas: History and known distribution. *Trop. Agric. Trinidad* 38:107–121.

Buddenhagen, I. 2009a. Understanding strain diversity in *Fusarium oxysporum* f. sp. *cubense* and history of introduction of 'Tropical Race 4' to better manage banana production. *Acta Hort.* 828:193–204.

Buddenhagen, I. 2009b. Blood bacterial wilt of banana: History, field biology and solution. *Acta Hort.* 828:57–68.

Buddenhagen, I.W. and T.A. Elsasser. 1962. An insect-spread bacterial wilt epiphytotic of Bluggoe banana. *Nature* 194:164–165.

Buddenhagen, I.W. and A. Kelman. 1964. Biological and physiological aspects of bacterial wilt caused by *Pseudomonas solanacearum*. *Annu. Rev. Phytopathol.* 2:203–230.

Buddenhagen, I.W. and L. Sequeira. 1958. Disinfectants and tool disinfection for prevention of spread of bacterial wilt of bananas. *Plant Dis. Repr.* 42:1399–1404.

Buregyeya, H., G. Tusiime, J. Kubiriba, and W.K. Tushemereirwe. 2008. Evaluation of distant transmission of banana bacterial wilt in Uganda. Paper presented at the conference on Banana and Plantain in Africa: Harnessing International Partnerships to Increase Research Impact, October 5–9, 2008, Mombasa, Kenya.

Burns, T.M., R.M. Harding, and J.L. Dale. 1995. The genome organization of banana bunchy top virus: Analysis of six ssDNA components. *J. Gen. Virol.* 76:1471–1482.

Burt, J.P.A., L.J. Rosenberg, J. Rutter, F. Ramírez, and O.H. Gonzales. 1999. Forecasting the airborne spread of *Mycosphaerella fijiensis*, cause of black Sigatoka disease on banana: Estimations of numbers of perithecia and ascospores. *Ann. Appl. Biol.* 135:369–377.

Carlier, J., E. Foure, F. Gauhl, D.R. Jones, P. Lepoivre, X. Mourichon, et al. 2000. Fungal diseases of the foliage. In: *Diseases of banana, abaca and enset*, D.R. Jones, ed., 37–79. Wallingford, UK: CAB International.

Cherian, A.K., R. Menon, A. Suma, S. Nair, and M.V. Sudheesh. 2002. Effect of banana varieties in Kerala. Paper presented at the Global Conference on Banana and Plantain, October 28–31, 2002. Kerala, India

Cook, D. and L. Sequeira. 1994. Strain differentiation of *Pseudomonas solanacearum* by molecular genetic methods. In: *Bacterial wilt: The disease and its causative agent, Pseudomonas solanacearum*, A.C. Hayward and G.L. Hartman, eds., 77–93. Wallingford, UK: CABI Publishing.

Cook, D., E. Barlow, and L. Sequeira. 1991. DNA probes as tools for the study of host-pathogen evolution: The example of *Pseudomonas solanacearum*. In: *Advances in molecular genetics of plant-microbe interactions*, Vol. 1, H. Henneke and D.P.S. Verma, eds., 103–108. The Netherlands: Kluwer.

Corden, M.E. and R.A. Young. 1959. The effects of fungicides on *Fusarium oxysporum* f. sp. *cubense* and other soil microorganisms. U.F.-Oregon State College Annual Reports.

Corden, M.E. and R.A. Young. 1960. The effects of fungicides on *Fusarium oxysporum* f. sp. *cubense* and the nature of wilt disease resistance. U.F.-Oregon State College Annual Reports.

Craenen, K. 1998. *Black Sigatoka disease of banana and plantain: A reference manual*. Ibadan, Nigeria: International Institute of Tropical Agriculture (IITA).

Crop Crisis Control Project (C3P). 2007. Final report submitted to USAID on October 15, 2007. http://c3project.iita.org/Doc/Final%20report%20c3P%20small.pdf (accessed 15 April 2010).

Crouch, J.H., D. Vuylsteke, and R. Ortiz. 1998. Perspectives on the application of biotechnology to assist in the genetic enhancement of plantain and banana (*Musa spp.*). *Elect. J. Biotechnol.* (1):1–2.

Crous, P.W. and X. Mourichon. 2002. *Mycosphaerella eumusae* and its anamorph *Pseudocercopsora eumusae* spp. nov.: Casual agent of eumusae leaf spot disease of banana. *Sydowia* 54:35–43.

Dabek, A.J. and J.M. Waller. 1990. Black leaf streak and other leaf streak—New banana diseases in East Africa. *Trop. Pest Manage.* 36(2):157–158.

Dahal, G., J. d'A. Hughes, G. Thottappilly, and B.E.L. Lockhart. 1998. Effect of temperature on symptom expression and reliability of banana streak badnavirus detection in naturally infected plantain and banana (*Musa spp.*) *Plant Dis.* 82:16–21.

Dale, J.L., D.A. Phillips, and J.N. Parry. 1986. Double-stranded RNA in banana plants with bunchy top disease. *J. Gen. Virol.* 67:371–375.

Daniells, J.W. 2009. Global banana disease management: Getting serious with sustainability and food security. *Acta Hort.* 828:411–416.

Davis, A.J., M. Say, A.J. Snow, and B.R. Grant. 1994. Sensitivity of *Fusarium oxysporum* f. sp. *cubense* to phosphonate. *Plant Pathol.* 43:200–205.

Davis, R.I., M. Fegan, B. Tjahjono, and S. Rahamma. 2001. An outbreak of blood disease of banana in Irian Jaya, Indonesia. *Australasian Plant Pathol.* 29:152.

De Beer, Z. 1997. Fusarium tolerance and horticultural characteristics of some FHIA hybrids. *Banana Growers Association of South Africa yearbook 2*. Sandton, South Africa: Studio Novell.

De Matos, A.P., Z.J.M. Cordeiro, S. De Oliveira e Silva, A.V. Trinidade, and D.M.V. Ferreira 1999. Research on Fusarium wilt of banana in Brazil: Achievements and current status. In: *Banana Fusarium wilt management: Towards sustainable cultivation*, A.B. Molina, N.H.N. Masdek, and K.W. Liew, eds., 95–102. Proceedings of the international workshop on banana Fusarium wilt disease, Genting Highlands Resort. Malaysia.

De Matos, A.P., Z.J.M. Cordeiro, S. de Oliveira e Silva, and D.M.V. Ferreira. 2009. Reaction of diploid (AA) and tetraploid (AAAB) banana hybrids to Fusarium wilt under field conditions. Paper presented at the ISHS/ProMusa Banana Symposium on Global Perspectives on Asian Challenge, September 14–18, 2009, Phoenix City, Guangzhou, China

Diekmann, M. and C.A.J. Putter. 1996. FAO/IPGRI Technical Guidelines for the Safe Movement of Germplasm. Rome: Food and Agriculture Organization of the United Nations/International Plant Genetic Resources Institute.

Dodds, J.A., T.J. Morris, and R.L. Jordan. 1984. Plant viral double-stranded RNA. *Ann. Rev. Phytopathol.* 24:151–168.

Dowson, W.J. 1943. On the generic names *Pseudomonas*, *Xanthomonas* and *Bacterium* for certain bacterial plant pathogens. *Trans. Br. Mycolog. Soc.* 26:4–14.

Drew, R.A., J.A. Moisander, and M.K. Smith. 1989. The transmission of banana bunchy-top virus in micropropagated bananas. *Plant Cell Tissue Organ Cult.* 16:187–193.

Eden-Green, S.J. 2004. How can the advance of banana Xanthomonas wilt be halted? *InfoMusa* 13(2):38–41.

Eden-Green, S.J. and H. Sastraatmadja. 1990. Blood disease of banana present in Java. *FAO Plant Prot. Bull.* 38:49–50.

Fegan, M. 2005. Bacterial wilt diseases of banana: Evolution and ecology. In: *Bacterial wilt disease and the Ralstonia solanacearum species complex*, C. Allen, P. Prior, and A.C. Hayward, eds., 379–386. St. Paul, MN: APS Press.

Fegan, M. and P. Prior. 2005. How complex is the *Ralstonia solanacearum* species complex? In: *Bacterial wilt disease and the Ralstonia solanacearum species complex,* C. Allen, P. Prior, and A.C. Hayward, eds., 449–461. St. Paul, MN: APS Press.

Fegan, M. and P. Prior. 2006. Diverse members of the *Ralstonia solanacearum* species complex cause bacterial wilts of bananas. *Australasian Plant Pathol.* 35:93–101.

Fiaboe, K.K.M., F. Beed, M. Mwangi, M. Katembo, and V. Ndungo. 2008. Survey of insects visiting banana male buds in eastern Democratic Republic of Congo and their contamination with the bacterium causing wilt. Paper presented at the conference on Banana and Plantain in Africa: Harnessing International Partnerships to Increase Research Impact, October 5–9, 2008, Mombasa, Kenya.

Fiaboe, K.K.M., J. Kubiriba, F. Beed, M. Mwangi, and W. Tushemereirwe. 2008. Comparative importance and epidemiology of BXW transmission through female and male inflorescences. Paper presented at the conference on Banana and Plantain in Africa: Harnessing International Partnerships to Increase Research Impact, October 5–9, 2008, Mombasa, Kenya.

Fouré, E. 1982. Les cercosporioses du bananier et leurs traitements. Comportement des varietés. 1. Incubation et évolution de la maladie. *Fruits* 37(12):749–766

Fouré, E. 1985. Les cercosporioses du bananier et leurs traitement. Etude de la sensibilité variétales des bananiers et plantain à *Mycosphaerella fijiensis* Morelet au Gabon. *Fruits* 40(60):393–399.

Fouré, E. 1987. Varietal reactions of bananas and plantains to black leaf streak disease. In: *Banana and plantain breeding strategies*, G.J. Persley and E.A. De Langhe, eds., 110–113. Proceedings of an international workshop held at Cairns, Australia, 13–17 October 1986, ACIAR Proceedings no. 21.

Fravel, D., C. Olivain, and C. Alabouvette. 2003. *Fusarium oxysporum* and its biocontrol. *New Phytologist* 157:493–502.

French, E.R. 1986. Interaction between strains of *Pseudomonas solanacearum*, its hosts and the environment. In: *Bacterial wilt diseases in Asia and the South Pacific. Proceedings of an international workshop held at PCARRD, Los Baños, Philippines, 8–10 October 1985*, G.J. Persley, ed., 99–104. Canberra, Australia: ACIAR Proceedings no. 13.

French, E.R. and L. Sequeira. 1970. Strains of *Pseudomonas solanacearum* from Central and South America: A comparative study. *Phytopathology* 60:506–512.

Frossard, P. 1980. Apparition d'une nouvelle et grave maladie foliaire des bananiers et plantains au gabon: la maladie des raies noires: *Mycosphaerella fijiensis* Morelet au Gabon. *Fruits* 35(9):519–527.

Fullerton, R.A. 1994. Sigatoka leaf diseases. In: *Compendium of tropical fruit diseases*, R.C. Ploetz, G.A. Zentmyer, W.T. Nishijinia, K.G. Rohrbach, and H.D. Oh, eds., 12–14. St. Paul, MN: American Phytopathological Society.

Gauhl, F. 1994. *Epidemiology and ecology of black Sigatoka (Mycosphaerella fijiensis Morelet) on plantain and banana (Musa spp.) in Costa Rica, Central America*. Montpellier, France: INIBAP.

Gauhl, F. and C. Pasberg-Gauhl. 1994a. Epidemiology of black Sigatoka disease in Nigeria. *Phytopathology* 84:100.

Gauhl, F. and C. Pasberg-Gauhl. 1994b. Virus symptoms in *Musa* germplasm: Report on observations from 1991–1994. Research Report, 40. Ibadan, Nigeria: International Institute for Tropical Agriculture (IITA).

Gauhl, F. and C. Pasberg-Gauhl. 1995. Temporal dynamics of banana streak badnavirus (BSV) symptoms in *Musa* clones in Southern Nigeria. *Phytomedizin* 25:27.

Gäumann, E. 1921. Onderzoekingen over de bloedziekte der bananen op Celebes I (Investigations on the blood-disease of bananas in Celebes I). *Mededeelingen van het Instituut voor Plantenziekten* No. 50.

Gäumann, E. 1923. Onderzoekingen over de bloedziekte der bananen op Celebes II (Investigations on the blood-disease of bananas in Celebes II). *Mededeelingen van het Instituut voor Plantenziekten* No. 59.

George, E.F. and P.D. Sherrington. 1984. *Plant propagation by tissue culture*. Bassingstoke: Exegetics Ltd.

Gerlach, K.S., S. Bentley, N.Y. Moore, E.A.B. Aitken, and K.G. Pegg. 1999. Investigation of non-pathogenic strains of *Fusarium oxysporum* for suppression of Fusarium wilt of banana in Australia. Paper presented at the Second International Fusarium Workshop, Dijon, France: INRA-CMSE.

Goods, R.D. and M. Tsirch. 1963. Greenhouse studies on the *Cercospora* leaf of bananas. *Trans. Br. Mycol. Soc*. 46:321–330.

Graham, K.M. 1969. A simple to distinguish black leaf streak from Sigatoka disease on bananas. *Bull. Dep. Agric*. 49:4.

Haase, A., J. Richter, and F. Rabenstein. 1989. Monoclonal antibodies for detection and serotyping of cucumber mosaic virus. *J. Phytopathol.* 127:129–136.

Harding, R.M., T.M. Burns, and D.L. Dale. 1991. Virus-like particles associated with banana bunchy top disease contain small single-stranded DNA. *J. Gen. Virol.* 72:225–230.

Harding, R.M., T.M. Burns, G. Hafner, R.G. Dietzgen, and J.L. Dale. 1993. Nucleotide sequence of one component of the banana bunchy top virus genome contains the putative replicase gene. *J. Gen. Virol.* 74:323–328.

Hayward, A.C. 2006. Fruit rots of banana caused by *Ralstonia solanacearum* race 2: Questions of nomenclature, transmission and control. *InfoMusa* 15(1–2):7–10.

Herbert, J.A. and D. Marx. 1990. Short-term control of Panama disease in South Africa. *Phytophylactica* 22:339–340.

Hu, J.S., H.P. Li, K. Barry, M. Wang, and R. Jordan. 1995. Comparison of dot blot, ELISA, and RT-PCR assays for detection of two cucumber mosaic virus isolates infecting banana in Hawaii. *Plant Dis.* 79:902–906.

Huang, J.W., S.K. Sun, and T.F. Hsieh. 1989. Characteristics of suppressive soil and its application to watermelon Fusarium wilt disease management. *Plant Prot. Bull.* (Taiwan, ROC) 31:104–118.

Huber, D.M. and R.D. Watson. 1974. Nitrogen form and plant disease. *Annu. Rev. Phytopathol.* 12:139–165.

Hunt, P. 1992. Moko research in Grenada: Broad aims of projects and preliminary findings. In: *Proceedings of a one-day seminar on control of Moko disease in Grenada*, N. Paul and M. Blanchard, eds., 13–16. Windward Islands Banana Growers' Association (WINBAN), July 1992.

Hwang, S.C. and W.H. Ko. 2004. Cavendish banana cultivars resistant to Fusarium wilt acquired through somaclonal variation in Taiwan. *Plant Dis.* 88:580–588.

INIBAP. 1997. *Networking banana and plantain.* Annual report 1996. Montpellier, France: International Network for the Improvement of Banana and Plantain.

Iskra, M.L., M. Garnier, and J.M. Bove. 1989. Purification of banana bunchy top virus (BBTV). *Fruits* 44:63–66.

Jacome, L.H. and W. Schuh. 1992. Effects of leaf wetness duration and temperature on development of black Sigatoka disease on banana infected by *Mycosphaerella fijiensis* var deformis. *Phytopathology* 82:515–520.

Jeger, M.J., S. Eden-Green, J.M. Thresh, A. Johanson, J.M. Waller, and A.E. Brown. 1995. Banana diseases. In: *Bananas and plantains*, S. Gowen, ed., 317–381. London: Chapman and Hall.

Jiménez, J.M., J.J. Galindo, and C. Ramirez. 1987. Estudios sobre combate biológico de *Mycosphaerella fijiensis* var difformis mediante bacterias epifitas. In: *Proc. ACORBAT meeting*, J.J. Galindo and R. Jaramillo, eds., 105–109. Turrialba, Costa Rica: Centro Agronomico Tropical de Investigacón y Enseñanza (CATIE).

Jones, D.R. 2000a. *Diseases of banana, abaca and enset.* Wallingford, UK: CABI Publishing.

Jones, D.R. 2000b. The distribution and importance of the *Mycosphaerella* leaf spot diseases of banana. In: *Proceedings of the second international workshop on Mycosphaerella leaf spot diseases of banana*, 25–42. Montpellier, France: INIBAP.

Jones, D.R, and B.E.L. Lockhart. 1993. Banana streak disease. In: *Musa disease fact sheet no.1. Banana streak disease.* Montpellier, France: INIBAP.

Kalyebara, M.R., P.E. Ragama, E. Kikulwe, F. Bagamba, C. Nankinga, and W. Tushemereirwe. 2006. Economic importance of the banana bacterial wilt in Uganda. *Afr. Crop Sci. J.* 14:93–104.

Karamura, E. 2006. Assessing the impact of the banana bacterial wilt, *Xanthomonas campestris pv. musacearum* (BXW), on household livelihoods in East Africa. DFID Crop Protection Programme Project R8437. http://www.research4development.info/PDF/Outputs/CropProtection/R8437_FTR.pdf (accessed 23 July 2010).

Kartha, K.K. 1986. Production and indexing of disease-free plants. In: *Plant tissue culture and its agricultural application*, L.A. Withers and P.G. Alderson, eds., 219–238. London: Butterworths.

Kelman, A. 1954. The relationship of pathogenicity in *Pseudomonas solanacearum* to colony appearance on a tetrazolium medium. *Phytopathology* 44:693–695.

Kelman, A. and L. Sequeira. 1965. Root-to-root spread of *Pseudomonas solanacearum. Phytopathology* 55:304–309.

Kubiriba, J. 2009. Banana bacterial wilt control in Uganda. Uganda presentation. Regional Lessons Learnt Workshop on the Control of BXW, Uganda, February 24–27, 2009 (Rwanda, Uganda, Tanzania, Kenya, DRC). http://km.fao.org/biosecwiki/images/d/de/BXW_Uganda.ppt (accessed 23 July 2010).

Kumar, N., T. Damodaran, and V. Krishnamoorthy. 2009. Breeding banana for combined resistance to Fusarium wilt and nematodes in Tamil Nadu, India. *Acta Hort.* 828:323–327.

Kung'u, J.N., A.A. Seif, and J.M. Waller. 1992. Black leaf streak and other foliage diseases in Kenya. *Trop. Pest Manage.* 38(4):359–361.

Lakshmanan, P., P. Selvaraj, and S. Mohan. 1987. Efficiency of different methods for the control of Panama disease. *Trop. Pest Manage.* 33:373–376.

Larkin, R. and D. Fravel. 1998. Efficacy of various fungal and bacterial biocontrol organisms for the control of Fusarium wilt of tomato. *Plant Dis.* 82:1022–1028

Lassoudière, A. 1974. La mosaïque dite "à tirets" du bananier 'Poyo' en Côte d'Ivoire. *Fruits* 29:349–357.

Lassoudière, A. 1979. Mise en évidence des répercussions économiques de la mosaïque en tirets du bananier en Côte d'Ivoire. Possibilités de lutte par éradication. *Fruits* 34:3–34.

Lee, S.Y., Y.U. Su, C.S. Chou, C.C. Liu, C.C. Chen, and C.P. Chao. 2009. Selection of new somaclone cultivar 'Tai-chiao No 5' with resistance to Fusarium wilt of banana in Taiwan. Paper presented at the ISHS/ProMusa banana symposium on Global Perspectives on Asian Challenge, September 14–18, 2009, Phoenix City, Guangzhou, China.

Lehmann-Danziger, H. 1987. The distribution of Moko disease in Central and South America and its control on plantains and bananas. In: *Seminar proceedings, improving citrus and banana production in the Caribbean through phytosanitation.* St Lucia, W.I., 2–5 September 1986, 153–155. Wageningen: CTA/CARDI.

Lemanceau, P. and C. Alabouvette. 1991. Biological control of Fusarium diseases by fluorescent *Pseudomonas* and non-pathogenic *Fusarium. Crop Prot.* 10:279–286

Lepoivre, P., P. Acuna, and A.S. Riveros. 1993. Screening procedures for improving resistance to banana leaf streak disease. In: *Breeding bananas for resistance to diseases and pests*, J. Ganry, ed., 213–220. Montpellier, France: CIRAD/INIBAP.

Lerch, B. 1987. On the elimination of plant virus multiplication by Ribavirin. *Antiviral Res.* 7:257–270.

Lian, J., Z. Wang, L. Cao, H. Tan, P. Inderbitzin, Z. Jiang, and S. Zhou. 2009. Artificial inoculation of banana tissue culture plantlets with indigenous endophytes originally derived from native banana plants. *Biologic. Control* 51:427–434.

Lockhart, B.E.L. 1986. Purification and serology of a bacilliform virus associated with banana streak disease. *Phytopathology* 76:995–999.

Lockhart, B.E.L. 1990. Evidence for a double-stranded circular DNA genome in a second group of plant viruses. *Phytopathology* 80:127–131.

Lockhart, B.E.L. 2000. Virus diseases of *Musa* in Africa: Epidemiology, detection and control. *Acta Hort.* 540:355–359.

Lockhart, B.E.L. and L.J.C. Autrey. 1988. Occurrence in sugarcane of bacilliform virus related serologically to banana streak virus. *Plant Dis.* 72:230–233.

Lockhart, B.E.L. and L.J.C. Autrey. 1991 Mealy bug transmission of sugar cane bacilliform and sugar cane clostero-like viruses. Paper presented at the third Sugarcane Pathology Workshop of the International Society of Sugarcane Technologists, July 22–26, 1991, Mauritius.

Lockhart, B.E.L. and D.R. Jones. 2000. Banana mosaic and banana streak disease caused by viruses. In: *Diseases of banana, abaca and enset*, D.R. Jones, ed., 241–293. Wallingford, UK: CABI Publishing.

Lockhart, B.E.L. and N.E. Olszewski. 1993. Serological and genomic heterogeneity of banana streak badnavirus: Implications for virus detection in *Musa* germplasm. In: *Breeding bananas and plantain for resistance to diseases and pests*, J. Ganry, ed., 105–113. Montpellier, France: CIRAD in collaboration with INIBAP.

Luis, J.G., F. Echeverría, W. Quinones, I. Brito, M. Lopez, F. Torres, et al. 1993. Irenolone and emenolone—Two new types of phytoalexin from *Musa paradisiaca. J. Org. Chem.* 58:4306–4308.

Magnaye, L.V. and R.R.C. Espino. 1990. Note: Banana bract mosaic, a new disease of banana I. Symptomatology. *Philipp. Agric.* 73:55–59.

Marin, D.H., R.A. Romero, M. Guzmán, and T.B. Sutton. 2003. Black Sigatoka: An increasing threat to banana cultivation. *Plant Dis.* 87:208–222.

Medberry, S.L., B.E.L. Lochhart, and N.E. Olszewski. 1990. Properties of Commelina yellow mottle virus' complete DNA sequence, genomic discontinuities and transcript suggest it is a pararetrovirus. *Nucl. Acids Res.* 18:5505–5513.

Meredith, D.S. and I.D. Firman. 1970. Banana leaf spot in Fiji. *Trop. Agric.* 47:127–130.

Meredith, D.S. and J.S. Lawrence. 1969. Black leaf streak disease of bananas (*Mycosphaerella fijiensis)*: Symptoms of disease in Hawaii, and notes on the conidial state of causal fungus. *Trans. Br. Mycol. Soc.* 52(3):457–476.

Mobambo, K.N. and M. Naku. 1993. Situation de la cercosporiose des bananiers et plantains (*Musa spp.*) sous différents systèmes de culture à Yangambi, Haut-Zaïre. *Tropiculturae* 11:7–10.

Mobambo, K.N., F. Gauhl, D. Vuylsteke, R. Ortiz., C. Pasberg-Gauhl, and R. Swennen. 1993. Yield loss in plantain from black Sigatoka leaf spot and field performance of resistant hybrids. *Field Crops Res.* 35:35–42.

Molina, A.B. 2006. Managing bacterial wilt/fruit rot disease of banana in southeast Asia. In: *Developing a regional strategy to address the outbreak of banana Xanthomonas wilt in East and Central Africa. Proceedings of a workshop held in Uganda, 14–18 February 2005*, E.B. Karamura, M. Osiru, G. Blomme, C. Lusty, and C. Picq, eds., 26–31. Montpellier: INIBAP.

Moore, N.Y., S. Bentley, K.G. Pegg, and D.R. Jones. 1995. Fusarium wilt of banana. *Musa disease fact sheet.* Montpellier, France: INBAP.

Moore, N.Y., K.G. Pegg, S. Bentley, and L.J. Smith. 1999. Fusarium wilt of banana: Global problems and perspectives. In: *Banana Fusarium wilt management: Towards sustainable cultivation*, A.B. Molina, N.H.N. Masdek, and K.W. Liew, eds., 11–30. Proceedings of the International Workshop on Banana Fusarium Wilt Disease, Kuala Lumpur, Malaysia: INIBAP.

Mourichon, X. 1986. Mise en évidence de la *Mycosphaerella fijiensis* Morelet, agent de la maladie des raies noires (black leaf streak) des bananiers et plantains au Congo. *Fruits* 41:371–374.

Mourichon, X. and R.A. Fullerton. 1990. Geographical distribution of the two species of *Mycosphaerella musicola Leach* (*Cercospora musae)* and *M. fijiensis* Morelet (*C. fijiensis)* respectively, agents of Sigatoka and black leaf streak diseases in bananas and plantains. *Fruits* 45:213–218.

Mourichon, X., J. Carlier, and E. Foure. 1997. Sigatoka leaf spot diseases. *Musa disease fact sheet no. 8.* Montpellier, France: INIBAP.

Mwangi, M. and V. Nakato. 2009. Key factors responsible for the Xanthomonas wilt epidemic on banana in East and Central Africa. *Acta Hort.* 828:395–404.

Mwangi, M., M. Pillay, R. Bandyopadhyay, W. Tushemereirwe, and P. Ragama. 2006. Progress in understanding mechanisms of host plant tolerance to banana bacterial wilt. Paper presented at the fourth International Bacterial Wilt Symposium, July 17–20, 2006, York, UK.

Mwangi, M., R. Bandyopadhyay, P.E. Ragama, and W. Tushemereirwe. 2007. Assessment of banana planting practices and cultivar tolerance in relation to management of soil-borne *Xanthomonas campestris* pv *musacearum*. *Crop Prot.* 26:1203–1208.

Mwebaze, J.M., G. Tushme, W. Tushemereirwe, and J. Kubiriba. 2006. The survival of *Xanthomonas campestris pv musacearum* in soil and plant debris. *Afr. Crop Sci. J.* 14:121–127.

Namu, F.N. 2008. The possible role of stingless bees in the spread of banana Xanthomonas wilt in Uganda and the nesting biology of *Plebeina hildebrandti* and *Hypotrigona gribodoi* (Hymenoptera-Apidae-Melamponini). PhD diss., University of Bonn.

Nankinga, C. and O. Okasaai. 2006. Community approaches used in managing BXW in Uganda. In: *Developing a regional strategy to address the outbreak of banana Xanthomonas wilt in East and Central Africa.* Proceedings of the Banana Xanthomonas Wilt Regional Preparedness and Strategy Development workshop held in Kampala, Uganda, 14–18 February 2005, 56–60. Montpellier: INIBAP.

Ndowora, T., G. Dahal, D. LaFleur, G. Harper, R. Hull, N.E. Olszewski, et al. 1999. Evidence that badnavirus infection in *Musa* can originate from integrated pararetroviral sequences. *Virology* 255:214–220.

Ndungo, V., K. Bakelana, S. Eden-Green, and G. Blomme. 2004. An outbreak of banana Xanthomonas wilt (*Xanthomonas campestris* pv. *musacearum*) in the Democratic Republic of Congo. *InfoMusa* 132:43–44.

Ndungo, V., K.K.M. Fiaboe, and M. Mwangi. 2008. Banana Xanthomonas wilt in the DR Congo: Impact, spread and management. *J. Appl. Biosci.* 1:1–7.

Nel, B., C. Steinberg, N. Labuschagne, and A. Viljoen. 2006. The potential of nonpathogenic *Fusarium oxysporum* and other biological control organisms for suppressing Fusarium wilt of banana. *Plant Pathol.* 55:217–223.

Nel, B., C. Steinberg, N. Labuschagne, and A. Viljoen. 2007. Evaluation of fungicides and sterilants for potential application in the management of Fusarium wilt of banana. *Crop Prot.* 26:697–705.

Nowak, J. 1998. Benefits of *in vitro* "biotization" of plant tissue cultures with microbial inoculants. *In Vitro Cell Dev. Biol. Plant* 34:122–130.

Nwauzoma, A.B., A. Tenkouano, J.H. Crouch, M. Pillay, D. Vuylsteke, and L.A. Daniel-Kalio. 2002. Yield and disease resistance of plantain (*Musa spp.*, AAB group) somaclones in Nigeria. *Euphytica* 123:323–333.

Nwauzoma, A.B., A. Tenkouano, and L.A. Daniel-Kalio. 2004. Disease incidence and symptom expression of banana streak virus on plantain (*Musa spp.*, AAB group) somaclones in Nigeria. *Nig. J. Plant Prot.* 21:63–72.

Nwauzoma, A.B., S. Uma, M.M. Mustaffa, and P. Durai. 2008. Response of banana hybrids to Sigatoka leaf spot disease under tropical conditions in southern India. *Acta Agronomica Nigeriana* 8(1):33–42.

Odoi, N.N. 2006. Growing bananas in Uganda: Reaping the fruit of participatory development communication, http://www.idrc.ca/openebooks/224–4/#page_129; and Communication tools in the hands of Ugandan farmers, http://www.idrc.ca/openebooks/224–4/#page_175. In: *People, land, and water: Participatory development communication for natural resource management*, G. Bessette, ed. London: Earthscan.

Ortiz, R. 1996. The potential of AMMI analysis for field assessment of *Musa* genotypes to virus infection. *HortScience* 31:829–832.

Ortiz, R. and D. Vuylsteke. 1998. 'Pita-14': A black Sigatoka-resistant tetraploid plantain hybrid with virus tolerance. *HortScience* 32(2):360–361.

Ortiz, R., R.S.B. Ferris, and D. Vuylsteke. 1995. Banana and plantain breeding. In: *Bananas and plantains*, S. Gowen, ed., 110–146. London: Chapman and Hall.

PBIP. 1997. *Plantain and banana improvement program, Crop Improvement Division.* 1995 Annual Report. Ibadan, Nigeria: International Institute for Tropical Agriculture.

Peterson, R., K. Grice, and R. Goebel. 2005. Eradication of black leaf streak disease from banana-growing areas in Australia. *InfoMusa* 14(2):2.

Phelps, R.H. 1987. The status of Moko disease in the Caribbean. In: *Seminar proceedings, improving citrus and banana production in the Caribbean through phytosanitation*, St. Lucia, W.I., 2–5 September, 1986, 100–107. Wageningen: CTA/CARDI.

Ploetz, R.C. 1990. *Fusarium wilt of banana*. St. Paul, MN: American Phytopathological Society.

Ploetz, R.C. 1994. Panama disease: Return of the first banana menace. *Int. J. Pest Manage.* 40:326–336

Ploetz, R.C. 2000. Panama disease: A classic and destructive disease of banana. Online. Plant Health Progress.

Ploetz, R.C. 2001. Black Sigatoka of banana. *Plant Health Instructor.* doi: 10.1094/PHI-I-2001–0126–01. www.epsnet.org/publications

Ploetz, R.C. 2005. Panama disease, an old enemy rears its ugly head: Pts 1, 2. In: *Plant health progress*, APSNET: doi:10.1094/PHP-2005–1221–01-RV.

Ploetz, R.C. and K.G. Pegg. 1997. Fusarium wilt of banana and Wallace's line: Was the disease originally restricted to his Indo-Malayan region? *Australasian Plant Pathol.* 26:239–249.

Ploetz, R.C. and K.G. Pegg. 2000. Fusarium wilt. In: *Diseases of banana, abacá and enset*, D.R. Jones, ed., 143–159. Wallingford, UK: CABI Publishing.

Ploetz, R.C., J. Herbert, K. Sebasigari, J.H. Hernandez, K.G. Pegg, J.A. Ventura, et al. 1990. Importance of *Fusarium* wilt in different banana-growing regions. In: *Fusarium wilt of banana*, R.C. Ploetz, ed., 9–26. St Paul, MN: American Phytopathological Society.

Ploetz, R.C., A.G. Channern, C.T. Chizala, D.L.N. Banda, P.W. Makina, and W.S. Braunworth Jr. 1992. A current appraisal of banana and plantain diseases in Malawi. *Trop. Pest Manage.* 38:36–42.

Ploetz, R.C., J.E. Thomas, and W.R. Slabaugh. 2003. Diseases of banana and plantain. In: *Diseases of tropical fruit crops*, R.C. Ploetz, ed., 73–134. Wallingford: CABI Publishing.

Prior, P. and M. Fegan. 2005. Diversity and molecular detection of *Ralstonia solanacearum* race 2 strains by multiplex PCR. In: *Bacterial wilt disease and the Ralstonia solanacearum species complex*, C. Allen, P. Prior and A.C. Hayward, eds., 405–414. St. Paul, MN: APS Press.

Promed. 2009. Undiagnosed disease, banana—Malaysia. http://www.promedmail.org (accessed 20 May 2010).

Quak, F. 1977. Meristem culture and virus-free plants. In: *Applied and fundamental aspects of plant cell, tissue and organ culture*, J. Reinert and Y.P.S. Bajaj, eds., 598–616. Berlin: Springer-Verlag.

Raemaekers, R. 1975. Black leaf streak-like disease in Zambia. *PANS* 21:396–400.

Rawal, R.D. 2000. Fungal diseases of banana: Current scenario in India. In: *Banana improvement, production and utilization*, H.P Singh and K.L. Chadha, eds., 355–363. Proceedings of the Conference on Challenges for Banana Production and Utilization in 21st Century, Association for the Improvement in Production and Utilization of Banana (AIPUB). Trichy, India: National Research Centre for Banana (NRCB).

Rhodes, P.L. 1964. A new disease in Fiji. *Commonw. Phytopathologic. News* 10:38–40.

Riechel, H., S. Bealalacazar, G. Munnera, E. Arévalo, and J. Narváez. 1996. First report of banana streak virus infecting plantains (*Musa spp.*) in Colombia. *Plant Dis.* 80:463.

Rijks, A.B. 1916. Rapport over een onderzoek naar de pisangsterfte op de saleiereilanden. *Mededeelingen van het Laboratorium voor Plantenziekten* No. 21.

Rillo, A.R. 1979. Bacterial wilt of banana in the Philippines. *FAO Plant Prot. Bull.* 27:105–108.

Rishbeth, J. 1955. Fusarium wilt of bananas in Jamaica. Some observations on the epidemiology of the disease. *Ann. Bot.* 19:293–329.

Rishbeth, J. and A.G. Naylor. 1957. Fusarium wilt of bananas in Jamaica. III. Attempted control. *Ann. Bot.* 21:599–609.

Rodoni, B.C., Y.S. Ahlawat, A. Varma, J.L. Dale, and R.M. Harding. 1997. Identification and characterisation of banana bract mosaic virus in India. *Plant Dis.* 81:669–672.

Rodoni, B.C., J.L. Dale, and R.M. Harding. 1999. Characterization and expression of the coat protein-coding region of banana bract mosaic potyvirus, development of diagnostic assays and detection of the virus in banana plants from five countries in southeast Asia. *Arch. Virol.* 144:1725–1737.

Rowe, P.R. and F.E. Rosales. 2000. Conventional banana breeding in Honduras. In: *Disease of banana, abacá and enset*, D.R. Jones, ed., 435–449. Wallingford, UK: CAB International.

Sági, L., S. Remy, and R. Swennen. 1997. *Genetic transformation for the improvement of bananas: A critical assessment*. Focus paper No. 2. In: INIBAP Annual Report 1997, 33–36, Montpellier, France: INIBAP.

Selvarajan, R., V. Balasubramanian, I. Ravi, T. Rajesh, S. Devi, R. Rajmohan, et al. 2007. *Molecular characterization and diagnostic techniques for plant viral and fungal pathogens*. Training manual. Tamil Nadu, India: National Research Centre for Banana.

Senula, A., E.R.J. Keller, and D.E. Leseman. 2000. Elimination of viruses through meristem culture and thermotherapy for the establishment of an in vitro collection of garlic (*Allium sativum*). *Acta Hort.* 530:121–128.

Sequeira, L. 1958. Bacterial wilt of bananas: Dissemination of the pathogen and control of the disease. *Phytopathology* 48:64–69.

Sequeira, L. 1962. Control of bacterial wilt of bananas by crop rotation and fallowing. *Trop. Agric. (Trinidad)* 39:211–217.

Sequeira, L. 1998. Bacterial wilt: The missing element in international banana improvement programs. In: *Bacterial wilt disease: Molecular and ecological aspects*, P. Prior, C. Allen, and J. Elphinstone, eds., 6–14. Berlin: Springer.

Sequeira, L. and C.W. Averre. 1961. Distribution and pathogenicity of strains of *Pseudomonas solanacearum* from virgin soils in Costa Rica. *Plant Dis. Repr.* 45:435–440.

Shehabu, M., A. Temesgen, L.F. Turyagyenda, T. Alemu, S. Mekonen, and G. Blomme. 2010. *The efficiency of air-drying pared corms of banana suckers in reducing the risk of soil-mediated Xanthomonas wilt infections in Ethiopia*. Tree and Forestry Science and Biotechnology—Special issue on banana, plantain and ensete. In press.

Shimelash, D., T. Alemu, T. Addis, F.L. Turyagyenda, and G. Blomme. 2008. Banana *Xanthomonas* wilt in Ethiopia: Occurrence and insect vector transmission. *Afr. Crop Sci. J.* (Special issue: Research advances in banana and enset in Eastern Africa) 16(1):75–87.

Sidi, R.K., D.H. Gelfand, S. Stoffel, S. Scharf, R.H. Higuchi, G.T. Horn, et al. 1988. Primer-detected enzymatic amplication of DNA with a thermostable DNA polymerase. *Science* 239:487–491.

Simmonds, N.W. 1962. *The evolution of bananas*. London: Longmans.

Simmonds, N.W. 1987. Classification and breeding of bananas. In: *Banana and plantain breeding strategies*, G. Persely and E.A. De Langhe, eds., 78–83. INIBAP and ACIAR Proceedings No. 21. Cairns, Australia.

Smith, J.J., D.R. Jones, E.B. Karamura, G. Blomme, and F.L. Turyagyenda. 2008. *An analysis of the risk from Xanthomonas campestris pv. musacearum to banana cultivation in Eastern, Central and Southern Africa*. Montpellier: Bioversity International.

Smith, M., S. Hamill, P. Langdon, J. Giles, V. Doogan, and K. Pegg. 2006. Towards the development of a Cavendish banana with resistance to race 4 of Fusarium wilt: Gamma irradiation of micropropagated Dwarf Parfitt (*Musa* spp., AAA group, Cavendish subgroup). *Aust. J. Exp. Agric.* 46:107–113.

Snyder, W.C. and H.N. Hanson. 1940. The species concept in *Fusarium*. *Am. J. Bot.* 27:64–67.

Soguilon, C.E., L.V. Magnaye, and M.P. Natural. 1994. Bugtok disease of cooking banana in the Philippines. *InfoMusa* 3(2):21–22.

Ssekiwoko, F., W.K. Tushemereirwe, M. Batte, P. Ragama, and A. Kumakech. 2006. Reaction of banana germplasm to inoculation with *Xanthomonas campestris* pv. *musacearum*. *Afr. J. Crop Sci.* 14:151–155.

Stover, R.H. 1962. *Fusarial wilt (Panama disease) of bananas and other musa species*. Kew, England: Commonwealth Mycological Institute.

Stover, R.H. 1972. *Banana, plantain and abaca diseases*. England: Commonwealth Mycological Institute.

Stover, R.H. 1980. Sigatoka leaf spot of bananas and plantains. *Plant Dis. Rep.* 64:750–756.

Stover, R.H. and S.E. Malo. 1972. The occurrence of fusarial wilt in normally resistant 'Dwarf Cavendish' banana. *Plant Dis. Repr.* 56:1000–1003.

Stover, R.H. and D.L. Richardson. 1968. 'Pelipita,' an ABB 'Bluggoe'-type plantain resistant to bacterial and fusarial wilts. *Plant Dis. Repr.* 52:901–903.

Su, H.J., S.C. Hwang, and W.H. Ko. 1986. Fusarial wilt of Cavendish bananas in Taiwan. *Plant Dis.* 70:814–818.

Subandiyah, S., S. Indarti, T. Harjaka, S.N.H. Utami, C. Sumardiyono, and Mulyadi. 2005. Bacterial wilt disease complex of banana in Indonesia. In: *Bacterial wilt disease and the Ralstonia solanacearum species complex*, C. Allen, P. Prior, and A.C. Hayward, eds., 415–422. St. Paul, MN: APS Press.

Sun, E.J. 1977. Survival of race 4 of *Fusarium oxysporum* f. sp. *cubense* in the soil. Master's thesis, National Taiwan University, Taipei.

Sun, S.K. and J.W. Huang. 1985. Formulated soil amendments for controlling Fusarium wilt and other soil borne disease. *Plant Dis.* 69:917–920.

Supriadi. 2005. Present status of blood disease in Indonesia. In: *Bacterial wilt disease and the Ralstonia solanacearum species complex*, C. Allen, P. Prior, and A.C. Hayward, eds., 395–404. St. Paul, MN: APS Press.

Thangavelu, R., P. Sundararaju, S. Sathiamoorthy, T. Raghuchander, R. Velazhahan, S. Nakkeeran, and A. Palaniswami. 2001. Status of Fusarium wilt of banana in India. In: *Banana Fusarium wilt management: Towards sustainable cultivation*, A.B. Molina, N.H. Nikmasdek, and K.W. Liew, eds., 58–63. Los Banos, Laguna, Philippines: INIBAP-ASPNET.

Thangavelu, R., A. Palaniswami, and R. Velazhahan. 2003. Mass production of *Trichoderma harzianum* for managing Fusarium wilt of banana. *Agric. Ecosytems Environ.* 103:259–263.

Thomas, J.E. and R.G. Dietzgen. 1991. Purification, characterization and serological detection of virus-like particles associated with banana bunchy top disease in Australia. *J. Gen. Virol* 72:217–224.

Thomas, J.E. and L.V. Magnaye. 1996. Banana bract mosaic disease. *Musa disease fact sheet, no. 7.* Montpellier, France: INIBAP.

Thomas, J.E., A.D.W. Geering, C.F. Gambley, A.F. Kessling, and M. White. 1997. Purification, properties and diagnosis of banana bract mosaic potyvirus and its distinction from abaca mosaic potyvirus. *Phytopathology* 87:698–705.

Thottappilly, G., G. Dahal, G. Harper, G. Hull, and B.E.L. Lockhart. 1997. Banana streak badnavirus: Development of diagnostics by ELISA and PCR. *Phytopathology* 87:S97.

Thwaites, R., S.J. Eden-Green, and R. Black. 2000. Diseases caused by bacteria. In: *Diseases of banana, abaca and enset*, D.R. Jones, eds., 213–239. Wallingford: CABI Publishing.

Tinzaara, W., C.S. Gold, F. Ssekiwoko, R. Bandyopadhyay, A. Abera, and S.J. Eden-Green. 2006. Role of insects in the transmission of banana bacterial wilt. *Afr. Crop Sci. J.* 14:105–110.

Tripathi, L., J. Odipio, J.N. Tripathi, and G. Tusiime. 2008. A rapid technique for screening banana cultivars for resistance to *Xanthomonas* wilt. *Eur. J. Plant Pathol.* 121:9–19.

Tripathi, L., J.N. Tripathi, and W.K. Tushemereirwe. 2008. Control of banana Xanthomonas wilt disease using genetic engineering. Paper presented at a conference on Banana and Plantain in Africa: Harnessing International Partnerships to Increase Research Impact, October 5–9, 2008, Mombasa, Kenya.

Tripathi, L., M. Mwangi, V. Aritua, W. Tushemereirwe, S. Abele, and R. Bandyopadhyay. 2009. Xanthomonas wilt. A threat to banana production in East and Central Africa. *Plant Dis.* 93:440–451.

Trujillo, I. and E. Garcia. 1996. Strategies for obtaining somaclonal variants resistant to yellow Sigatoka (*Mycosphaerella musicola*). *InfoMusa* 5:12–13.

Tumushabe, A., F. Ssekiwoko, and W. Tushemereirwe. 2006. Potential of infected banana parts to transmit *Xanthomonas campestris* pv. *musacearum*. *Afr. Crop Sci. J.* 14:137–142.

Turyagyenda, L.F., G. Blomme, F. Ssekiwoko, and S. Eden-Green. 2007. Determination of the appropriate fallow period to control *Xanthomonas* wilt following infection of banana. Paper presented at the ISHS/ProMusa symposium: Recent advances in banana crop protection for sustainable production and improved livelihoods, September 10–14, 2007, White River, South Africa.

Tushemereirwe, W.K. and J.M. Waller. 1993. Black leaf streak (*Mycosphaerella fijiensis*) disease in Uganda. *Plant Pathol.* 42:471–472.

Tushemereirwe, W., A. Kangire, J. Smith, F. Ssekiwoko, M. Nakyanzi, D. Kataama, et al. 2003. Outbreak of bacterial wilt on banana in Uganda. *InfoMusa* 12(2):6–8.

Waite, B.H. 1963. Wilt of *Heliconia* spp. caused by *Fusarium oxysporum* f. sp. *cubense* race 3. *Trop. Agric. (Trinidad)* 40:299–305.

Waite, B.H. and V.C. Dunlap. 1953. Preliminary host range studies with *Fusarium oxysporum* f. sp. *cubense*. *US Dept. Agric. Dis. Repr.* 37:79–80.

Walkey, D.G.A. and M.J.W. Webb. 1968. Virus in plant apical meristems. *J. Gen. Virol.* 3:311–313.

Wardlaw, C.W. 1961. *Banana diseases, including plantains and abaca*. London: Longmans, Green and Co. Ltd.

Welde Michael, G., K. Bobosha, G. Blomme, T. Addis, S. Mekonnen, and T. Mengesha. 2006. Screening banana cultivars for resistance to bacterial Xanthomonas wilt. *InfoMusa* 15(1–2):10–12.

Weller, D.M., J.M. Raaijmakers, B.B. McSpadden Gardener, and L.S. Thomashow. 2002. Microbial populations responsible for specific soil suppressiveness to plant pathogens. *Ann. Rev. Phytopath.* 40:309–48.

Wilson, G.F. and I. Buddenhagen. 1986. *The black Sigatoka threat to plantain and banana in West Africa*. IITA Research Briefs 7:3.

Wu, R.Y. and H.J. Su. 1990. Purification and characterization of banana bunchy top virus. *J. Phytopathol.* 128:153–160.

Wu, R.Y., L.R. You, and T.S. Soong. 1994. Nucleotide sequences of two circular single-stranded DNAs associated with banana bunchy top virus. *Phytopathology* 84:952–958.

Xie, W.S. and J.S. Hu. 1995. Molecular cloning, sequence analysis, and detection of banana bunchy top virus in Hawaii. *Phytopathology* 85:339–347.

Yirgou, D. and J.F. Bradbury. 1968. Bacterial wilt of enset (*Ensete ventricosum*) incited by *Xanthomonas musacearum* sp. *Phytopathology* 58:111–112.

Yirgou, D. and J.F. Bradbury. 1974. A note on wilt of banana caused by the enset wilt organism *Xanthomonas campestris* pv. *musacearum*. *East Afr. Agric. For. J.* 40:111–114.

Zimmerman, A. 1902. Uber einige tropischer Kulturpflanzen beobachtete Pilze (Abstract). *Zentralbl. Bakterriol. Parasitenk. Infektionskrankh. Hyg.* 8:219.

7 Integrated Pest Management of Banana

Thomas Dubois and Daniel L. Coyne

CONTENTS

7.1 INTRODUCTION

Bananas (*Musa* spp.) are plagued by a variety of nonmicrobial pests. Most attention has focused on the banana weevil *Cosmopolites sordidus* Germar and a complex of plant-parasitic nematodes, of which the burrowing nematode *Radopholus similis* Cobb Thorne has received the most attention. However, banana is grown widely across tropical and subtropical regions, attracting a wide range of associated pests. These can vary greatly according to geography and clone, while changes in cropping practices and the introduction of new or unfamiliar cultivars can introduce new pest species. In addition, banana serves varying purposes, ranging from the genetically diverse production systems of subsistence foods to commercially managed plantations of genetically uniform dessert bananas for export markets. For example, flower and fruit pests that cause cosmetic damage are of limited importance to subsistence cooking bananas but can result in refusal of export shipments when detected in even low numbers. Often, the management of pests is discussed in the context of integrated pest management (IPM). IPM, however, is often misused, referring instead to a plethora of ill-linked management options that can at times still be at the research stage. Within this chapter, therefore, the full spectrum of banana production systems will be taken into account when discussing the vast diversity of banana pests, while providing an important assessment on how IPM principles can be applied to manage them.

7.2 PLANT-PARASITIC NEMATODES

7.2.1 An Overview of Nematode Species and Their Distribution

Plant-parasitic nematodes are the most detrimental soil-borne pests of banana (Gowen et al., 2005). On a global basis, the key pest species are *Helicotylenchus multicinctus* (Cobb) Golden, root knot nematodes *Meloidogyne* spp. (Figure 7.1), the root-lesion nematode *Pratylenchus coffeae* Sher and Allen, *Pratylenchus goodeyi* Sher and Allen, and *R. similis* (Coyne, 2009). Other species not generally viewed as key pests may, however, be of local significance such as the reniform nematode *Rotylenchulus reniformis* Linford and Oliveira or *Hoplolaimus pararobustus* Schuurmans Stekhoven and Teunissen Sher. Virtually without exception, species occur in mixed communities.

Radopholus similis has been considered as the most damaging nematode affecting bananas worldwide, especially in lowland tropical areas (Sarah, 2000). However, this perception has essentially stemmed from the nuisance *R. similis* poses to commercial dessert banana plantations, where it has wreaked havoc and resulted in the substantial application of carbamate- and organophosphate-based pesticides (Cianco and Mukerji, 2009). Consequently, *R. similis* has traditionally been the main focus in breeding programs. In subsistence farming systems, though, the situation is less clearly defined. The nematode is thermophobic and in the tropics does not occur at high, cool altitudes, above 1400 m in the East African highlands (Price, 2006) where a substantial proportion of Africa's banana production is concentrated, nor does it occur at high latitudes, such as Taiwan and the Canary Islands (Jones, 2009). *R. similis* was previously the key nematode pest species in West Africa (Speijer and Fogain, 1999), but recent surveys show *P. coffeae* is often the most damaging species (Coyne, 2009). Since *P. coffeae* is also prevalent across the Pacific and Southeast Asia, it is

FIGURE 7.1 Root knot nematodes in banana (Courtesy of D. Coyne.)

of concern for banana and requires a greater attention in respect to pest management and resistance breeding.

Pratylenchus goodeyi, on the other hand, is viewed as thermophilic, and in the East African highlands, for example, replaces *R. similis* as the dominant species above 1400 m altitude (Speijer and Fogain, 1999). Its status as a pest of banana, however, is unclear. It can occur in extremely high densities, such as on banana in Tanzania (Speijer and Bosch, 1996) and enset in Ethiopia (Peregrine and Bridge, 1992), where it undoubtedly causes some damage. In Rwanda and Uganda, however, no correlation could be established between *P. goodeyi* and cooking banana losses (Gaidoshova et al., 2009; D. Coyne, unpublished). It is also interesting that *P. goodeyi* represents a major pest in commercial banana plantations in the Canary Islands (De Guiran and Vilardebo, 1962) and in Australia (Pattison et al., 2002) where prevailing temperatures tend to be higher than is optimal for this species. Recently, *P. goodeyi* was identified from bananas in Kenya. Further examination of *P. goodeyi* from toppled bananas on the Kenyan coast and the Canary Islands using molecular techniques demonstrated distinct molecular differences of these nematodes compared with *P. goodeyi* from the highlands of Uganda (Coyne and Waeyenberge, 2008). Results indicated that the "tropical" (Kenyan) *P. goodeyi* were more closely linked to *P. crenatus* Loof, *P. penetrans* Cobb, and *P. neglectus* Rensch than *P. goodeyi*, even though they physically resembled *P. goodeyi*. Within-species variability is a well-known phenomenon, which can explain differences in virulence and host range of some species (Starr et al., 2002). There are good reasons to separate certain strains into separate species, such as for the *P. coffeae* complex, which hitherto was a single species based on morphology (Duncan et al., 1999). For *R. similis*, studies have demonstrated that a series of different strains exist, with the Sri Lanka strain responsible for the severe damage to Ugandan bananas, amongst the most aggressive (Price, 2006), and able to overcome the resistance present in cv 'Yangambi Km5' (Plowright, 2000; Dochez, 2004). Such variability and diagnostic difficulties have significant implications to the development of management programs, especially for the use of resistance through breeding programs. Knowledge of the key pests and access to sources of resistance against these species and their variable strains are essential to make progress in managing these pests.

Helicotylenchus multicinctus is regularly associated with losses to banana, but almost exclusively in combination with other nematode species, especially *R. similis* and *Meloidogyne* spp. (Gowen et al., 2005). Its status has been subject to speculation, as determining the contribution of the individual nematode species to damage and losses is often difficult. Accumulating evidence, however, demonstrates that *H. multicinctus* is indeed responsible for large proportions of damage to banana production, even when other species are present (Ssango et al., 2004).

Meloidogyne spp. are amongst the most abundant nematode pests across all crops, with a global distribution. Their importance on bananas has been underestimated as they also regularly occur in combination with other damaging species (Gowen et al., 2005). However, in some instances they dominate the nematode populations and contribute significantly to production losses.

7.2.2 Nematode Damage

Nematodes generally cause damage through the destruction of root and rhizome tissue. Damaged tissue becomes necrotic and dies, reducing nutrient and water uptake, reducing bunch weights, and retarding harvest. Severe damage underscores plant anchorage, which can result in plant toppling (Sarah, 2000; Jones, 2009), while reduced plant turgidity can result in snapping of plant stems during periods of low water availability (D. Coyne, unpublished). Fruit on fallen plants generally have no value, resulting in extreme yield losses where infection levels and plant losses are high (Gowen et al., 2005). Common symptoms of severe nematode infection include stunting, poor plant growth, narrow and weak stems, foliar chlorosis, root rotting and galling, and plant toppling. Determining infestation levels can be difficult, especially to the untrained, as nematodes exist below ground and remain out of sight, until severe damage symptoms are observed. Nematodes almost always occur as species combinations that may be complex. Establishing the specific species contributions to damage is difficult, resulting in complications for developing management options that may be species specific.

7.2.3 Nematode Management

The discrepancy between management options for smallholders and commercial growers is vast. Nevertheless, nematodes remain a difficult group to manage effectively. However, because new infestations are primarily perpetuated through infected planting material, the use of clean, healthy, nematode-free planting material cannot be overemphasized for either system. Hot water treatment of suckers after removal of infected roots is a simple and effective technique for sanitizing material. For smallholder systems, this technique has been further adapted using a short 30 s exposure period in boiling water, which is less time and energy consuming, and more appealing to resource-poor farmers (Viaenne et al., 2006). For commercial systems, such as in Australia and Hawaii, hot water treatment is used to provide nematode-free material (Colbran, 1967). However, sterile plants produced using tissue culture and certified pest- and disease-free are ideal. Such material is now routinely used in commercial banana production but has yet to gain wider use by smallholder farmers (Dubois, Coyne, et al., 2006). The lack of virus indexing, suboptimal weaning procedures, accidental cultivar mixing during production, inappropriate farmer handling, and subsidization by governmental and nongovernmental organizations remain some of the major hurdles to overcome before tissue-culture technology can be widely rolled out among smallholder farmers.

In commercial plantations, postplanting nematicide applications continue to provide the most universal method of nematode management, primarily against *R. similis*, administered through granular applications or drip irrigation (Sarah, 2000; Jones, 2009). Soil sanitation can be achieved through a cleansing system based on glyphosate injection into banana plants before uprooting (Risède et al., 2009). However, many nematicides are being progressively removed from the market (Zum Felde et al., 2009). In the French West Indies, management of *R. similis* is based primarily upon the repeated application of carbamate or organophosphate nematicides; however, with increasing restrictions on their use, the search for alternative and environmentally responsible options has intensified. An environmentally sound scheme supported by three key pillars is being devised: use of tissue culture, fallow, and intercropping with nonhosts (Risède et al., 2009). In Hawaii's IPM scheme, incorporation of crop residue and fallowing fields for 6–8 months is common. Emphasis is also being increasingly placed on efforts to identify suitable biologically based solutions, such as mycorrhizae, endophytes, and biopesticides (Meyer and Roberts, 2002; Viaene et al., 2006; Sikora et al., 2008).

The use of healthy planting material is not only critical, but for smallholder farmers it also often is their only realistic option for nematode management. However, the use of locally grown, nematode-resistant cultivars, in combination with healthy planting material, is highly desirable (Coyne, 2009). Establishing nematode resistance is also a key target in banana breeding programs (Tenkouano and Swennen, 2004; Pillay and Tripathi, 2007; Lorenzen et al., 2009). Commercial dessert bananas are characterized by few landraces with an extremely narrow genetic base (Ortiz et al., 1995), while sources of resistance to nematode species are limited (De Waele and Elsen, 2002). To date no widely grown clone of export banana is known to be resistant to the important nematode species (Gowen et al., 2005). There are confirmed sources of resistance against *R. similis* but not necessarily against *Pratylenchus* spp. (De Waele and Elsen, 2002). Resistance to *R. similis* from cv 'Pisang Jari Buaya' has been incorporated into the widely used diploid parent cv SH-3142 of the Fundación Hondureña de Investigación Agrícola (FHIA) (Pinochet and Rowe, 1979). Recent successes have also been achieved in the breeding programs at the International Institute of Tropical Agriculture (IITA) (Pillay and Tripathi, 2007; Lorenzen et al., 2009) and the Centre de Coopération Internationale en Recherche Agronomique pour le Développement (CIRAD). At IITA, the diploid banana hybrids TMB2×5105-1 and TMB2×9128-3 have good combining ability and are resistant to *R. similis* (Tenkouano et al., 2003). At CIRAD, partial resistance to both *R. similis* and *P. coffeae* is reported within synthetic hybrids of *M. acuminata* (Quénéhervé et al., 2009). Meanwhile, the genetic modification of existing cultivars is also becoming a realistic option for nematode management (Roderick et al., 2009; Tripathi, 2009). Research efforts for biologically based solutions are equally being sought for smallholder farmers in Africa and India to compensate for the unsuitability and removal from use of nematicides.

7.3 INSECT AND MITE PESTS

Bananas can be attacked by a wide range of insect and mite pests. Rather than a taxonomic overview, it is best to group these into functional groups, as members from widely different taxa often pose similar problems and require similar management options.

7.3.1 Plant-Boring Pests

7.3.1.1 The Banana Weevil

The biology, distribution, and damage caused by *Cosmopolites sordidus* is comprehensively reviewed by Gold et al. (2001). Banana weevils feed only on bananas. Adults are most commonly found between leaf sheaths, in the soil at the base of the mat, or associated with crop residues. The banana weevil is nocturnally active and particularly susceptible to desiccation. As adults tend to have limited movement between mats and rarely fly, dissemination is primarily through infested planting material. The banana weevil is a typical k-selected insect with long life span and low fecundity. Adults normally survive for longer than 1 year, and oviposition has been estimated at 1 egg/week. Oviposition occurs in the leaf sheaths and rhizome surface, especially in flowered plants and in crop residues. Crop damage is inflicted by the larvae. The emerging larvae preferentially feed in the rhizome but will also attack the true stem and occasionally the pseudostem. Larval developmental rates are temperature dependent. Under tropical conditions, egg to adult development takes 5–7 weeks. Egg development does not occur below 12°C, restricting its distribution to lower altitudes. Adult banana weevils are attracted by volatiles emanating from host plants, explaining why cut rhizomes of fresh suckers for planting material are especially susceptible to attack (Gold et al., 2001).

The banana weevil has a cosmopolitan distribution, occurring in smallholder banana systems to commercial plantations. It is present in all banana and plantain production regions in the tropics and subtropics, and is generally considered the most important insect pest of banana (Jones, 2009). Beer, roasting, and cooking bananas are most susceptible, and therefore banana weevil problems appear to be most severe in smallholder systems and less in commercial cv Cavendish plantations (Gold and Messiaen, 2000).

Banana weevil attack has been reported to interfere with root initiation, kill existing roots, limit nutrient uptake, reduce plant vigor, delay flowering, and increase susceptibility to other pests and diseases. Yield reductions stem from both plant loss (plant death, rhizome snapping) and reduced bunch weights. Losses of more than 40% have been recorded (Gold and Messiaen, 2000; Gold et al., 2001). Young banana plants are most at risk because tunneling by the banana weevil can be fatal at this stage (Constantinides and McHugh, 2003).

As with nematodes, banana weevils are dispersed through contaminated planting material, emphasizing the importance of clean planting material as an essential prophylactic management measure. Rigorous field sanitation measures also take advantage of the adult's dependency on residues, lack of movement, and need for moisture. Despite numerous surveys, no known effective parasitoids of the banana weevil have been identified. In commercial systems, insecticides are applied. For resource-poor farmers, cultural management is the only means currently available to reduce banana weevil populations (Gold and Messiaen, 2000).

7.3.1.2 Stem Borers

Stem borers, such as the giant banana stem borer *Castniomera humboldti* Boisduval and the banana stem weevil *Odoiporus longicollis* Olivier, tunnel through the banana pseudostem (Jones, 2009). *Castniomera humboldti* occurs in Central and South America where it is a relatively minor pest, whereas *O. longicollis* can be a serious pest in Asia. The latter is among the main insect pests of quarantine importance for Australia (Pinese, 1999) and considered the most important insect pest in India. Eggs are laid inside air chambers through incisions made on the leaf sheath. In the advanced stage of infestation, severely affected plants break. Banana stem weevils often inflict total crop failures in susceptible cultivars (Jayanthi and Verghese, 1999). The pest survives in pseudostem stumps, which often remain as trash in the field after harvest. In India, the distribution of *O. longicollis* is aggravated when farmers cut the pseudostems at up to 1 m high from the ground level and allow them to decompose slowly until the establishment of the succeeding ratoon crop, which they believe transfers nutrition to subsequent ratoons (Padmanaban and Kandasamy, 2003).

7.3.2 Fruit and Flower Pests

Fruit and flower pests are especially important on exported banana. For example, in Hawaii, presence of the banana moth *Opogona sacchari* Bojer on fruit for export will result in their rejection (Constantinides, 2003). A small number of larvae of the banana scab moth *Nacoleia octasema* Meyrick may lead to the destruction of an entire bunch otherwise destined for export. The mere presence of *Bactrocera* spp. fruit flies, an insignificant pest of bananas, on shipments from Australia to New Zealand requires destruction of the fruit (Pinese, 1999; Ministry of Agriculture and Forestry, 2006). Most often, fruit and flower pests are managed using insecticide-treated bags that enclose the flower. Chlorpyrifos-treated bags are especially effective and used abundantly in commercial plantations, but environmentally safer alternatives, such as bifethrin, are increasingly sought (Chiquita Brands International, 2001).

7.3.2.1 Banana Moths

Opogona sacchari is endemic to Africa, where it is an insignificant pest. The pest has a wide host range and has been considered a serious pest of banana in the Canary Islands since the 1920s. In the 1970s, it was introduced into Brazil, where it has since become an important banana pest. The insect has also started to appear in a number of European countries on various tropical and subtropical glasshouse crops, and is now considered a serious quarantine pest (Smith et al., 1996; OEPP/EPPO, 2006). The banana moth oviposits on senescing flowers, decaying leaves, pseudostems, and fruits on which the larvae feed, although they will also feed on healthy adjacent tissue. Preventive management measures, such as the removal of plant debris and flowers, in addition to the application of insecticides to bunches prior to bagging, greatly reduces damage (Peña et al., 2002). Recently, a pheromone was discovered that attracts this pest, which may additionally aid the development of more efficient monitoring schemes (Wageningen University, 2009).

The banana fruit-piercing moth *Eudocima fullonia* Clercq attacks many fruits and vegetable crops, and can pose a serious banana risk. Unlike most moth and butterfly pests, the caterpillar stage is not the damaging stage. Instead, the adult moth punctures and feeds on ripening fruit, not only administering direct damage but also indirectly facilitating fungal and bacterial infections. High moth populations can result in premature ripening and fruit drop (CAPS online). Interestingly, in some endemic areas, such as Papua New Guinea, the pest is effectively managed below threshold levels by egg parasitoids (Sands and Liebregts, 2005).

The banana scab moth *Nacoleia octasema* Meyrick is one of the most serious pests in Malaysia, the southwest Pacific, and Queensland, Australia. Females lay eggs on flower bracts as the inflorescence emerges. Larvae feed on the surface of young fingers. They enter the flower and feed on the developing fruits within, gradually progressing down the maturing bunch. This causes brown scars, scabs, and severe cracking on the developing fruits. Cultural and biological control methods are not particularly effective due to the cryptic nature of the feeding larvae, and their management is based largely on injection of insecticides (Paine, 1964; Morton, 1987; Stover and Simmonds, 1987; Botha et al., 2000; Nelson et al., 2006).

7.3.2.2 Thrips

Numerous species of thrips of the family *Thripidae* feed on banana (CABI, 2005). Most thrips prefer sunny and dry areas, have a broad host range, and feed on flowers, fruits, or other young tissues, with both larvae and adults causing damage (Parker et al., 1995). Thrips cause superficial skin blemishes on immature banana fruits. Although severe infestations can cause peel splitting, the damage they cause is primarily cosmetic, and therefore only commercial banana systems require prophylactic management measures to meet stringent export requirements (Peña et al., 2002).

Banana is affected by several members of *Chaetanaphothrips*: the orchid thrips *Chaetanaphothrips orchidii* Moulton, the banana rust thrips *Chaetanaphothrips signipennis* Bagnall, and *Chaetanaphothrips leeuweni* Karny. These species are cosmopolitan pests, with most damage resulting from larval feeding. *Chaetanaphothrips signipennis* is a problem in Australia, while *C. orchidii* induces similar damage in Central and South America (Peña et al., 2002). Feeding on leaf sheaths results in damage on the outer surface of leaf petioles and is characterized by dark, V-shaped marks, while damage to the fruit initially presents a water-soaked appearance that later turns bronze- or rust-colored. The pest can split the fruit peel, exposing the flesh. It also feeds on the area where adjacent fingers touch, resulting in a reddish discoloration (Williams et al., 1990; CABI, 2005; Jones, 2009). The life cycle can be completed in 28 days. The insect is managed by spraying banana fruits with insecticide at bunch emergence and covering them with polyethylene bags prior to harvest (Morton, 1987; Hara et al., 2002; CABI, 2005).

The Hawaiian flower thrips, *Thrips hawaiiensis* Morgan and *T. florum* Schmutz, are similar, often confused with each other, and as a cosmopolitan species complex, feed on a wide variety of tropical flowers (Hollingsworth, 2003). The insect enters the developing fruit while the bracts remain present and oviposits on the young fruit. Feeding results in corky scabbing of the peel, flecked, spotted, or deformed flowers, and sometimes cracked fruits, especially during hot and dry weather. Infestations are lessened by removal of the terminal male bud, which tends to harbor the pest (Morton, 1987; CABI, 2005; Jones, 2009; Peña et al., 2002). Unlike most flower thrips, this species complex prefers wet and shady areas (Sakimura and Krauss, 1944).

The banded greenhouse thrips *Hercinothrips femoralis* Reuter is a cosmopolitan thrips species with a wide host range and has been recorded on bananas in various parts of the world. The closely related rind thrips *H. bicinctus* Bagnall, which is equally cosmopolitan, is considered a more important banana pest (Roditakis et al., 2006). Feeding by this insect causes unsightly silver and bronze fruit scars, reducing their marketability (Hawaiian Banana Industry, 2010a). The silvering usually occurs with small infestations. With larger infestations, especially in combination with the two-spotted spider mite *Tetranychus urticae* Koch, the fruits turn smoky-red in color, occasionally leading to skin cracks, further reducing the market value of the fruit (Lewis, 1997). *Hercinothrips* spp. are closely related to the rind thrips *Elixothrips brevisetis* Bagnall. *Elixothrips brevitis* is also a polyphagous foliage feeder and a common pest in commercial banana stands, and feeds on leaves, flowers, or stems. In Martinique, *E. brevisetis* has replaced *H. femoralis* as the predominant thrips pest (Rey, 2002). Flowers, buds, and the undersides of leaves become spotted with small black fecal specks. Injured tissue develops a silvery appearance and eventually turns dark brown, affecting banana marketing (Rey, 2002). *Elixothrips brevitis* also feeds on leaf tips, resulting in wilting and curling. When affected, buds may fail to open (Constantinides and McHugh, 2003; Hawaiian Banana Industry, 2010a).

Banana is also damaged by *Frankliniella* spp. The banana flower thrips *Frankliniella parvula* Hood pupates in the soil and only emerges during daylight hours to oviposit in the epidermis of young banana fruits. The host range of this thrips species seems restricted to banana plants (Harrison, 1963; Peña et al., 2002). The blossom thrips *Frankliniella insularis* Franklin mainly occurs in Central America (Mound and Marullo, 1996).

7.3.2.3 Peel-Scarring Beetles

Several species of *Colaspis* spp. are reported as banana pests, especially in Central and South America (Ostmark, 1975; Jones, 2009). *Colaspis hypochlora* Lefèvre in particular appears a troublesome pest in Venezuela, Guyana, and Mexico, where it invades young fruit on developing bunches, although this species is often confounded with other members of the genus. Severe outbreaks of this pest have been documented in Panama and Colombia (Ostmark, 1975; Morton, 1987). In the Philippines, several peel-scarring beetles belonging to *Philicoptus* spp. have also been reported as pests (Stephens, 1984).

7.3.2.4 Fruit Flies

Fruit flies only attack ripe banana fruits. Although minor pests, fruit flies, particularly of the genus *Batrocera* spp., can be highly significant quarantine pests (Nelson et al., 2006), disrupting international banana shipments. In October 2009, Mexico, for instance, halted all imports of fresh banana from regions where the banana fruit fly *Bactrocera musae* Tryon or the oriental fruit fly *B. dorsalis* Hendel occur (USDA, 2009). *Bactrocera musae* is considered among the most serious banana pests of Papua New Guinea (Kambuou, 2003). In Sri Lanka, severe outbreaks occurred of several *Bactrocera* spp. in the late 1990s (Ekanayake et al., 2002).

7.3.2.5 The Sugarcane Bud Moth Caterpillar

The sugarcane bud moth caterpillar *Decadarchis flavistriata* Walsingham is a localized pest in the Pacific, especially Hawaii. Caterpillars feed on decaying flowers, which can cause fruit scarring. Removing flowers prior to bagging reduces damage from this pest (Nelson et al., 2006).

7.3.3 Sucking Insects and Associated Arthropoda

7.3.3.1 The Banana Aphid

Colonies of the banana aphid *Pentalonia nigronervosa* Coquerel can occur anywhere on the plant but are most often found at the base (Robson et al., 2007). Young suckers are typically most heavily infested. The banana aphid is a phloem feeder, which causes plants to become deformed; the leaves become curled and shriveled, and in some cases galls form on the leaves (Metcalf, 1962). Direct damage from the banana aphid is normally negligible. More important is the role of *P. nigronervosa* as a virus vector. The alate form of the banana aphid is the sole vector of banana bunchy top virus (BBTV) disease, among the most serious of banana viruses in Asia, Africa, and the Pacific (Hu et al., 1996; Dale and Harding, 1998). With the exception of vector transmission, use of infected planting material is the only other mode of transmission of BBTV (Robson et al., 2006). Consequently BBTV management is highly dependent upon prophylactic vector management. Once BBTV contamination occurs, eradication is both difficult and costly, with vector management merely tending to reduce the rate of spread to healthy plants. In commercial plantations, pesticide applications are applied regularly, with newly released products, such as imidacloprid, being investigated (Robson et al., 2007). Utilization of disease-free planting material, windbreaks, horticultural oils, and detergents provide alternatives to pesticide treatments. Destruction of diseased plants immediately upon detection will also slow the spread. In Hawaii, eradication efforts continue to be conducted on an island-to-island basis. In Australia, a zero-tolerance policy and strict quarantine measures are established against the pest (Magee, 1967; Hawaii Banana Industry Association, 2010a, 2010b; Robinson, 1996; Robson et al., 2006).

7.3.3.2 Whiteflies

The spiraling whitefly *Aleurodicus dispersus* Russell is native to Central America but now has a cosmopolitan distribution. It is a polyphagous pest, including on banana. *Aleurodicus dispersus* is a sap-sucking insect that damages and discolors plant leaves during its nymphal stages but does not damage banana fruits directly (Waterhouse and Norris, 1989; Nelson et al., 2006). Whiteflies excrete honeydew, which serves as a substrate for mold fungi. Sooty mold blackens the leaves and decreases photosynthetic activity. During severe infestations in Costa Rica and Hawaii, high levels of sooty mold cause premature leaf drop and reduced yields (Botha et al., 2000; Nelson et al., 2006). The spiraling whitefly is not considered a principal threat to banana, as populations are usually maintained below economic thresholds by natural enemies, especially in the regions where it is endemic (Ramani et al., 2002), although in some countries it is considered a quarantine pest (Pinese, 1999). Other whiteflies have been reported as major pests locally, such as *Lecanoideus floccissimus* Martin in commercial banana greenhouses in the Canary Islands (Hernández-Suárez et al., 2006).

7.3.3.3 Scales and Mealybugs

The coconut scale *Aspidiotus destructor* Signoret is a common polyphagous pest of banana (Williams and Watson, 1988). The coconut scale usually occurs on the underside of leaves but can also affect petioles, peduncles, and fruits. The pest usually occurs in densely massed colonies on the lower leaf surface, except in extremely heavy infestations where it may be present on both sides. Mature scales are found on the older leaves. Their piercing and sucking mouthparts extract plant juices and inject toxic saliva, leading to discoloration and yellowing of plant tissue (Beardsley, 1970; Waterhouse and Norris, 1987; Wright and Diez, 2005). When attached to fruits, they pose a significant phytosanitary issue because of their quarantine status in many regions. The presence even of a single live coconut scale can effect the complete rejection of a California-bound banana shipment, causing substantial loss for Hawaii's growers and exports (Ming-yi, 2003; Nelson et al., 2006). Other species of scales have equally been reported as banana pests, such as the hemispherical scale *Saissetia coffeae* (Walker) and the green scale *Coccus viridis* De Lotto in the South Pacific (Ben-Dov, 1994; Nafus, 2000a, 2000b). Numerous species of mealybugs, which are closely related to scales, can be detrimental to bananas, such as the gray pineapple mealybug *Dysmicoccus neobrevipes* Beardsley in the Philippines, while severe outbreaks of *D. alazon* Williams, *Aspidiotus elaeidis* Marchal, and *A. hederae* (Vallot) are reported from organic banana plantations in Tenerife (Ben-Dov, 1994; Alvarez et al., 2001; Nafus 2000a, 2000b). The mango mealybug *Rastrococcus invadens* Williams, occasionally affecting bananas, has been identified as the primary exotic insect quarantine pest to banana production in Australia (Pinese, 1999). Other pests include members of

Planococcus spp., *Pseudococcus* spp., and *Ferrisia* spp. Although not a significant pest of banana in most locations, mealybugs have also been associated with transfer of banana streak virus (BSV) (Nelson et al., 2006).

7.3.3.4 Mites

Mites are generally considered a minor but frequent pest of bananas. However, several mites of the genus *Tetranychus* can cause significant damage to banana, such as *T. urticae* and especially the banana spider mite *T. lambi* Pritchard and Baker (Morton, 1987; Pinese and Piper, 1994). In West Bengal, India, *Oligonychus oryzae* Hirst was found to be the more damaging mite species (Karmakar and Dey, 2006). Mite activity and damage are mainly confined to localized, dry conditions, such as the underside of old leaves. In severe infestations, whole leaves turn brown-gray and wilt, resulting in sunburned bunches and a reduction in plant growth. However, in warm weather and during severe outbreaks, mites may migrate to the bunches and damage fruits. Dry and warm conditions under plastic bunch covers are particularly favorable for the buildup of banana spider mites. Fruit damage is characterized by a silver-gray discoloration of the fruit tip, and fruits may dry out and crack when serious infestations occur (Morton, 1987; Pinese and Piper, 1994). Mites are also implicated in fruit speckling, a disease with unknown etiology that, particularly during the rainy season, has caused up to 70% rejection of export banana in Central America (Pasberg-Gauhl, 2002).

7.3.4 Foliage Feeders

A large group of foliage-feeding insects, originating from several taxa, can cause damage to banana. The economic damage they cause is usually limited, with populations remaining below economic injury levels through natural predation and parasitism. However, serious crop losses can occur.

7.3.4.1 The Banana Skipper

The banana skipper *Erionota thrax* L. is considered to be the main insect pest in Papua New Guinea (Kambuou, 2003). In Australia, it is a quarantine pest (Pinese, 1999), where it is sometimes referred to as the banana leaf roller due to its habit of rolling leaves to make shelters. Caterpillars secrete a protective, white, waxy covering inside the rolled leaves. The feeding and rolling destroys the leaves and significantly reduces the plant's leaf area. Leaf defoliation can occur quickly with only three caterpillars per leaf (Queensland Horticulture Institute, 2000). In Asia, from where the banana skipper originates, the parasitic wasp *Cotesia erionotae* Wilkinson effectively manages banana skipper infestations, which were previously a serious problem. In Malaysia, populations are kept in check by at least five primary endoparasitoids (Queensland Horticulture Institute, 2000; Okolle et al., 2006; Jones, 2009), while in Papua New Guinea, parasitoids have been introduced to manage outbreaks of leaf rollers in areas of the country (Kambuou 2003).

7.3.4.2 The Chinese Rose Beetle

The Chinese rose beetle *Adoretus sinicus* Burmeister is a minor but common pest on all major banana-producing islands in Hawaii and in the Pacific. The larvae reside in the soil and litter, with damage caused only by adult feeding. The adult is nocturnal and feeds primarily on leaf and intervenal tissue. It most commonly attacks young plants (Nelson et al., 2006).

7.3.4.3 Other Foliage Feeders

Caterpillars from the genera *Antichoris*, *Caligo*, *Opsiphanes*, and *Sibene* have been reported to partially defoliate banana plants in Central and South America (Jones, 2009). The larval stages of *Opsiphanes tamarindi* Felder can consume large areas of leaf, making it a potentially serious pest (Uquillas, 2002). Especially the tamarind owlet *Opsiphanes tamarindis* Felder, the owl butterfly

Caligo mennon Felder, and *Antichloris viridis* Druce are considered economically important pests in countries such as Venezuela (Ramirez et al., 1999). Bagworm (*Oiketicus kirbyi* Guilding) can be a problem in Central America, such as in Costa Rica. Females live for only a maximum of 14 days but can produce over 6,500 eggs during their adult life span. However, natural parasites usually limit outbreaks (Stover and Simmonds, 1987).

7.4 PEST INTERACTIONS

Of the wide range of pests observed on banana, the level of damage inflicted depends on numerous factors. Banana pests are often highly interactive, occurring within a complex ecosystem that ultimately influences the damage they cause. As such, there is need to avoid pest management solutions that tend to focus on a single pest without considering its relation and interactions to other factors. It is necessary that pest management options be holistic in their approach.

7.4.1 Ants

Ants have at times been heralded as natural enemies for biological control in conservation programs. For example, encouraging colonies of ants has been suggested as a means to manage *C. signipennis* (CABI, 2005). Myrmicine ants such as *Tetramorium guinense* F. and the big-headed ant *Pheidole megacephala* F. have reportedly contributed to the successful management of banana weevils in plantain in Cuba and are even encouraged to nest in pseudostem sections that can then be used for their dissemination (Gold and Messiaen, 2000).

However, whereas ants are antagonistic to most other insect taxa, they can be highly protective of some honeydew-producing pest species, such as scales, whiteflies, and aphids. Ants will seek out honeydew sources to protect the supply, effectively farming the source, which may include their aggressive defense of the honeydew-producing insects. For example, honeydew produced by *A. dispersus* attracts ants, which, in turn, offer protection to the whiteflies, aggravating its damage and indirectly contributing to quarantine problems for export fruits (Waterhouse and Norris, 1989; Nelson et al., 2006).

Of particular concern is the intimate relationship of ants with the banana aphid, which produces honeydew. Aphid populations prosper in the presence of ant colonies, and thus ants indirectly aggravate BBTV incidence, increasing the probability of BBTV spread by the aphids. In Hawaii, *P. megacephala* and, more recently, the long-legged ant *Anoplolepis longipes* Jerdon are primarily associated with the banana aphid. By moving round aphids feeding on banana plants, they contribute to the spread of BBTV. Even directly, *A. longipes* feeds on the surface of the banana fruit, causing scarring of the fruit surface and reducing marketability (Brooks, 2003).

7.4.2 Natural Enemies

Several banana pests that require significant population densities before damage occurs are maintained below damage thresholds by natural enemies. Particularly good examples of this are demonstrated with the spiraling whitefly and the banana skipper (Ramani et al., 2002; Okolle et al., 2006). However, broad-spectrum insecticide applications, when relied upon for management of many pests simultaneously, may cause secondary outbreaks of otherwise minor pests, especially following the use of aerial or cover sprays (Pinese and Piper, 1994). Historical Lepidopteran outbreaks in banana have been associated with pesticide-induced disturbance of their natural enemies, such as the localized outbreaks of the banana skipper in Malaysia (Okolle et al., 2006). In a related study in Costa Rica, Hymenopteran parasitoid abundance and species richness were inversely related to application rates of nematicide and insecticide (Matlock and De La Cruz, 2002).

7.4.3 Weeds

Weeds not only compete with the banana crop for water and nutrients but also provide important pest havens, both by providing shelter and, more importantly, by serving as alternative hosts, especially for polyphagous thrips, banana moths, whiteflies, and mites. Consequently, weed management is an important component in many banana production areas for the indirect management of pests. In Hawaiian banana orchards, weed management strategies involve the prevention of weed seed formation and using pre-emergence herbicides, with emphasis on weed management prior to canopy closure (Hawaii Banana Industry Association, 2010b).

7.5 INTEGRATED PEST MANAGEMENT

7.5.1 History and Context of IPM

IPM is a term widely used but often misused in reference to banana. In relation to the literature, researchers and practitioners tend to equate IPM with a list of control options for a particular pest (often still at the research phase and biased towards biological control), which is not IPM. The term IPM was first coined in 1972 following a speech by President Nixon to the U.S. Congress, and originally defined in 1975 by the Food and Agricultural Organization (FAO, 1975). Since then, IPM has become a frequently used and misused term, often without the needful consideration of the subtleties and implications of its true meaning or impact on modern agriculture (Kogan, 1998). IPM was originally envisaged as a concept to counter the excessive applications of pesticides, particularly insecticides. Although pesticides can, and have, greatly increased crop productivity, their use has led to unintended adverse effects on human health and the environment. Furthermore, pesticide resistance among the target pests can result in secondary pest outbreaks through their ill effects on natural enemies (Stephenson, 2001). Originally, IPM was entomocentric, and only much later were weed science and plant pathology included in IPM principles (Kogan, 1998). Numerous definitions of IPM abound, with the concept of decision-making central to most (Bajwa and Kogan, 2002). Based on an analysis of the various definitions spanning the preceding 35 years, Kogan (1998) proposed a consensual definition of current thought: "IPM is a decision support system for the selection and use of pest control tactics, singly or harmoniously coordinated into a management strategy, based on cost/benefit analyses that take into account the interests of and impacts on producers, society and the environment."

7.5.2 Principles of IPM

The principles of IPM can be best explained by examining the terms of the acronym: integrated, pest, and management.

7.5.2.1 Integrated

"Integrated" refers to the harmonious use of multiple management methods to control single pests, as well as the impacts of these methods on multiple pests (Kogan, 1998). Management methods are traditionally categorized as chemical (for example, pesticides), cultural (such as intercropping), biological (for example, parasitoids), host plant resistance-based (such as breeding genetically modified organisms), and genetic (sterile insect release, for example). A mere list of categorized control options, however, does not necessarily enable the farmer to practice IPM. It is important to distinguish between preventive/prophylactic and curative management options. As IPM is founded on a decision-making process—namely, before economic damage levels are incurred—IPM implicitly relies on prophylactic management options (Bajwa and Kogan, 2002). The successful and harmonious integration of management options is a difficult, if not a near-impossible, task. Integration can be viewed as either vertical (that is, within a pest taxon, sometimes referred to as first level) or horizontal (that is, among pest taxa, sometimes referred to as second level). For example, an insecticide that affects both the target pest and its natural enemies represents a lack of vertical integration; similarly, application of a fungicide that is detrimental to the natural enemies of pests provides a lack of horizontal integration. Historically, the lack of such integration has been a major impediment to the implementation of IPM in agriculture (Ehler, 2006).

7.5.2.2 Pest

"Pest" refers to any organism causing crop damage, including invertebrate and vertebrate animals, pathogens, and weeds (Kogan, 1998). Pest is an anthropocentric term, and highly relative and dynamic. Any insect living in or on banana plants can, at some stage and in some locations, become a pest or cease to be a pest. As such, sampling and monitoring schemes are of paramount importance and are necessary components before IPM can be conducted. Even with a pest incurring identical levels of injury in different locations, the economic damage acceptance level can differ between production systems. This implies that sampling and monitoring schemes need to be adapted to be location, crop, crop system, and season specific (Stephenson, 2001).

7.5.2.3 Management

"Management," the most important term, refers to a series of decision rules based on ecological and economic considerations, and equipped with sound and specific information related to the pest and its management options. The key principle for this decision-making process is often the economic injury level (EIL) concept (Stern et al., 1959). EIL is based on economics: the study of the relationships between pest densities, host responses to injury, and resultant economic losses. EIL is a theoretical value that, if actually attained by a pest population, will result in economic damage. The EIL formula $[C/(V \times I \times D \times K)]$ is determined using five primary variables: cost of the management tactic per production unit (C), market value per production unit (V), injury units per pest (I), damage per injury unit (D), and the proportional reduction in pest attack (K). From the EIL, the economic threshold (ET) is calculated. The ET differs from the EIL in that it is a practical or operational rule, rather than a theoretical one. The ET is defined as the population density at which control action should be determined (initiated) to prevent an increasing pest population (injury) from reaching the economic injury level. The ET is effectively an action threshold and is more complex to calculate than the EIL. Besides information on the EIL, several other parameters need to be known to calculate the ET, such as pest and host phenology, population growth and injury rates, and time delays associated with the IPM tactics utilized. These parameters are also location, crop, crop system, and season specific, and require extensive research before their implementation.

7.5.3 Practical IPM as a Continuum

Decision making, based on pest populations, is the most critical element in any IPM program. Without the critical components of ET and EIL, there will be no decision making and hence no IPM. However, because ETs and EILs are difficult to calculate, the practical implementation of IPM has become less strict and is often applied as a continuum. For example, the U.S. Department of Agriculture (USDA) uses a four-tier approach. As a first line of defense, prophylactic cultural methods (rotation, resistant cultivars, and pest-free planting material) are encouraged. Once monitoring, identification, and action thresholds indicate that pest management is required and preventive methods are no longer effective, the least risky curative pest-management options are initially employed, including highly targeted pesticides, such as pheromones to disrupt pest mating, or mechanical control, such as trapping or weeding. If further monitoring, identification, and action thresholds indicate that less risky controls are not sufficiently effective, additional pest-management methods would be needed, such as targeted application of pesticides. Broadcast spraying of nonspecific pesticides is a last resort (Stephenson, 2001).

Practically, efficient and operational IPM programs exist for specific crops in specific locations. These programs take on a variety of formats: protocols, checklists, standards, and definitions. Many of these assign point values to each practice, facilitating use as a performance assessment tool (Green and Petzoldt, 2009).

7.6 IPM IN BANANA

7.6.1 IPM in Commercial Plantations

Pest management in commercial banana plantations is primarily chemical based, using nematicides with insecticidal activity and applying specific insecticides to the plant base or on bunches. Management of *R. similis* to date has essentially been achieved through the application of carbamates (aldicarb, carbofuran, and oxamyl) and organophosphates (fenamiphos, ethoprop, and terbufos) (Berg, 1991). Cyclodiene insecticides, once widely used but eventually abandoned following the development of pest resistance and the emergence of environmental concerns, are now replaced by less persistent organophosphates. However, pests such as the banana weevil have demonstrated the ability to develop resistance to most pesticide classes (Gold and Messiaen, 2000). In Hawaii, to avoid pesticide resistance, the industry is actively developing a pesticide resistance program. The organophosphate diazinon, the primary pesticide used for thrips management, is being replaced with low-risk pesticides such as imidacloprid and spinosad (Hawaii Banana Industry Association, 2010b). Compared to smallholder systems, much focus is directed towards managing fruit and flower pests in commercial banana plantations, because there tends to be a zero-tolerance policy on damage or even presence for export markets (Jones, 2009), leading to much lower EILs and ETs.

In commercial plantations, IPM is especially sought to substitute for the excessive use of nematicides particularly of late, following the imminent removal from use of many nematicides (http://www.pesticideinfo.org/) and the increasingly strict regulations on the maximum permitted residue levels of imported fruit, vegetable, and cereal products (European Commission, 2007). Furthermore, most systemic nematicides are short lived (2–5 weeks) (Zum Felde et al., 2009), with rapid microbe-enhanced biodegradation greatly reducing their effect, following their repeated and consistent use (Moens et al., 2004), leading to an ever increasing but untenable number of applications.

Examples of true banana IPM schemes are rare but can be found in Hawaii. Their banana IPM protocol uses a combination of guidelines and point values to determine the level of IPM being utilized on a particular farm, which is constantly subject to change with new IPM developments. Pest management practices are grouped according to five categories (cultural, physical, mechanical, biological, and chemical) and each category is assigned a point value. Those practices that require more active management decisions or present reduced environmental risks receive higher point values. In practicality, points are low for pesticide-dependent practices while high for biologically dependent ones. A grower is certified as an IPM practitioner if he or she enrolls in the program and provides documentation that at least 70% of the total possible points is achieved. Although not applied for all pests, binomial pest sampling plans have been developed to allow for informed decision making regarding pesticide applications (Robson et al., 2006).

Currently, the public demand for quality products grown with environmentally responsible methods is strongly encouraging organic banana farming. Nonpesticide pest-management methods are paramount for the promotion and adoption of such cropping systems (Roditakis et al., 2006). There is the erroneous tendency to associate the production of organic banana with smallholder farmers, possibly because of the perception that they can least afford the pesticides used on conventional crops and are more likely to use naturally occurring inputs. Current price premiums for organic produce, only enjoyed in niche Western markets and required to offset increased production costs, are also decreasing, perhaps limiting organic banana production in the long run (Reid, 2000; Abele et al., 2007).

7.6.2 IPM in Smallholder Systems

7.6.2.1 Problems

Compared to commercial systems, smallholder banana systems are intrinsically different, making it virtually impossible to implement IPM, while adoption of IPM has remained low in the developed world, contrary to original expectations (Vereijken, 1989; World Bank, 2005; Ehler, 2006). Pest problem recognition, and more specifically species identification, is required as a basis for IPM. However, in resource-poor situations, the obstacles to pest problem recognition and species identification are often much greater than for commercial situations. For example, in Kenya, only 15% of the farmers had knowledge on the damage caused by banana weevils and none for nematodes or their damage symptoms, mostly mistaking their damage for that of banana weevils. Importantly, and of greater concern, was that both farmers and extension officers made this mistake (Seshu Reddy et al., 1999). In addition, pest profiles and their management cannot be simply transferred from developed countries, as banana clones are highly variable and differ from the commercial cv Cavendish types. For example, in East Africa, based on a comprehensive farm baseline study, several banana pests, such as banana weevils and plant-parasitic nematodes, may not be as important as traditionally assumed (CIALCA, 2009).

Government support and resources for IPM implementation at the national level in less-developed countries are often scarce or absent. Through the Cooperative Global Program, sponsored by the Food and Agricultural Organization (FAO) and the United Nations Environmental Program (UNEP), IPM strategies were targeted at cropping systems and regions, which included implementation, research, training, and education, and which have led to some success stories (such as the reduction of pesticide use against the brown plant hopper *Nilaparvata lugens* Stål on rice (*Oryza sativa* L.) (FAO, 1995). Since the 1990s, the Consultative Group for International Agricultural Research (CGIAR) has increased its attention towards crop protection in general and IPM in particular, away from host-plant-resistance breeding (Kogan, 1998).

Smallholder banana systems are typically low input/low output, or resource poor, meaning that pest-management practices are often limited at best. In Cameroon, for example, insecticide applications among smallholder banana farmers are negligible, because the returns are too low to allow meaningful investment into pest-management measures. Only management options that offer less capital investment, therefore, offer long-term sustainability (Tomekpe and Sadom, 2008).

7.6.2.2 Focus on Cultural Management

Mainly because of limited resources and availability of other options, pest management in smallholder systems is heavily dependent on cultural options. However, cultural control options against insect pests in smallholder systems differ markedly from those of commercial systems. For example, one of the characteristics of smallholder banana systems is the perennial nature of banana, creating difficulties for short-term rotation options, while annual or single-cycle cropping characterize some commercial systems. In addition, a combination of banana cultivars with varying levels of resistance is often present in the same field in smallholder systems (Seshu Reddy et al., 1999).

Most of the cultural management options available to smallholder farmers are simple and rely on good crop husbandry and habitat management—practices that encourage vigorous crop growth, which leads to less damage. These practices include deep planting, weeding, mulching, and the application of organic manure. For example, systematic trapping with pseudostem or rhizome pieces, as well as removal of crop residues, can also be effective in reducing populations of adult banana weevils (Masanza et al., 2006). Using crop residues as mulch is effective at decreasing nematode populations, especially when applied to low fertility systems (McIntyre et al., 2000). One of the most critical management options for borers and nematodes is the use of clean, healthy planting material. Tissue culture plantlets are widely used in commercial banana plantations for pest and disease prevention but less available or understood in smallholder systems. Where tissue culture is not available, removal of roots and emersion of cleaned (pared) suckers in hot water treatments can be highly effective against banana weevils and nematodes (Speijer et al., 1999; Gold and Messiaen, 2000), while adapting the system to a simpler system of using 30 s periods of immersion in boiling water can be equally effective against nematodes (Tenkouano et al., 2006; Viaenne et al., 2006).

Besides cultural management options, much research currently focuses on biologically based management options. How feasible or economical these will be to smallholder farmers remains to be seen. However, based on interviews, Mugisha-Kamatenesi (2008) observed that subsistence farmers around the Lake Victoria basin in East Africa commonly use botanical pesticides. Botanical compounds especially are seen as substitutes for costly pesticides. Applications of neem (*Azadirachta indica* A. Juss) in the field, such as neem oil for treating planting material or pseudostem traps, protect bananas from weevil and nematode attack, inhibiting weevil larvae development by up to 14 days (ICIPE, 1997).

7.6.2.3 Technology Transfer

As a consequence of the intrinsic differences between smallholder and commercial systems, the focus for pest management for smallholder farmers is on transfer of the basic technologies, mostly cultural based, to the farming communities and extension personnel. For example, in Zanzibar, pest management includes formation of farmers' groups, training of trainers, establishment of plots used for participatory action, and farmer field schools, which are used for demonstrating basic technologies, such as good crop husbandry (Rajab and Fundi, 1999). In Kenya, mobile training workshops were initiated on a trial basis, which proved very effective in information dissemination (Seshu Reddy et al., 1998). Global Plant Clinics is a recent initiative to link smallholder farmers with pest information. The initiative aims to improve access to effective plant health services by adopting similar approaches used in human health, through regular advisory services made available in local communities (Boa, 2007).

7.7 ADVANCES IN SEED-BASED MICROBIAL MANAGEMENT

Endophytes are organisms that, at some time during their life cycle, live within plant tissues yet cause no disease symptoms to their host (Petrini, 1986). Endophytes are natural and integral components of all plants. The relationship can be mutualistic: Endophytes protect the host plant against pests and diseases, and increase plant growth and vigor. Endophytes occupy a niche with relatively low competition from other microorganisms, provided they gain access initially. As such, endophytes have received increasing attention as biological control organisms in vegetatively multiplied crops, such as banana (Sikora et al., 2008). Tissue-culture banana plants are becoming increasingly used in banana production, even in smallholder systems, because of the advantages offered by these plants. Tissue-culture plants are free from pests and pathogens, simple and quick to multiply in larger numbers, and exhibit faster and more uniform growth in the field than sucker-planted fields (Vuylsteke, 1989; Mateille et al., 1994; Robinson, 1996; Dubois, Coyne, et al., 2006; Pocasangre, 2006).

However, because tissue-culture plants are propagated under sterile conditions, they are void of all beneficial organisms, including endophytes (Pocasangre, 2006). By selecting the best performing endophytic strains and reintroducing them early into tissue-culture plantlets, the natural equilibrium is somewhat restored, extending the benefits of clean planting material. Research into enhancing banana with endophytes started in the early 1970s (Sikora and Schlosser, 1973; Sikora and Schönbeck, 1975).

The use of endophyte-enhanced tissue-culture plants is a unique form of microbial pest control, mainly because it is seed based and thus circumvents many of the obstacles normally associated with augmentative biological control. Thus tissue-culture plantlets can be supplied to farmers already fortified with endophytes, eliminating the need for farmers to apply the biopesticides. Costs and know-how associated with formulation, distribution, application, and storage of endophytes can be transferred to commercial tissue-culture laboratories (Dubois and Coyne, 2006). As endophytes exist *in planta*, they offer great potential to manage cryptic pests and diseases. Furthermore, endophytes escape the rhizosphere community, where they would otherwise compete with the native flora and where they would be exposed to environmental factors that may adversely affect their efficacy. Avoiding this competition and exposure allows for low initial inoculation levels, improving consistency of endophyte performance and substantially reducing costs.

Endophyte-enhanced tissue-culture technology is sought after both in smallholder as well as commercial systems. In smallholder systems, the early stages of plant growth can be challenging because banana tissue-culture plantlets need higher levels of care and attention than conventional planting material. Where soils are depleted and pests and diseases abundant, tissue culture is only superior to conventional planting material if accompanied by significant field maintenance. Especially in the smallholder banana production systems, where high-input field maintenance routines are largely absent, endophyte enhancement creates more robust tissue-culture plants. In commercial systems, endophytes are also being investigated as replacements for nematicides (Zum Felde et al., 2009).

How endophytes protect bananas is only just beginning to be understood. The primary mode of protection of the endophyte *Fusarium oxysporum* Schlecht.: Fries against *R. similis* appears to involve a number of mechanisms, including induced resistance (Athman, 2006). Induced resistance is the activation of defense mechanisms in plants after contact with biotic initiators, such as endophytes. The endophyte triggers pathways that induce physiological changes in the plant, enabling a susceptible cultivar to express similar properties as a resistant cultivar (Dubois, Gold, et al., 2006). This mode of action is economically interesting because it may transfer host resistance across a broad range of pest groups. Also, endophytic inoculum can be further reduced and may not necessarily need to persist for long periods, as long as the resistance remains triggered. Furthermore, *F. oxysporum* seems to prime the banana plant against pests and diseases rather than inducing a constitutive response. The priming of plants in this way thus avoids waste and helps optimize the use of resources, a prerequisite for the implementation of the plant-enhancement technology (Heil et al., 2000).

Several groups of endophytes are currently under investigation worldwide. A first group is comprised of mostly hyphomyceteous fungi that are nonobligate endophytes, which have a saprophytic stage in the rhizosphere. *Fusarium oxysporum* is the most predominant endophytic taxon in banana (Hallman and Sikora, 1996; Dubois, Gold, et al., 2006) and offers great commercial potential, mainly due to the relative ease of inoculum production (Dubois, Coyne, et al., 2006). Endophytic protection of tissue-culture banana plants has been demonstrated in the field under both commercial and smallholder settings. In Panama, inoculation with *Trichoderma atroviride* P. Karst. protected tissue-culture banana plants from *R. similis* better than two applications of ethoprop and temephos nematicides, reducing *R. similis* population levels by 30–50% (Pocasangre et al., 2006; Pocasangre et al., 2007). In Kenya, inoculation with *F. oxysporum* reduced nematode population densities by > 45% and damage by > 20% over one growth cycle (JKUAT, 2008; Waithira, 2009). Mass multiplication mechanisms of promising strains are currently being researched, in coordination with private industry in East Africa and Central America.

A second group is the arbuscular mycorrhizal fungi (AMF), obligate symbionts of almost all higher plants, including most cultivated plant species (Abbott and Robson, 1984; Sikora et al., 2003). AMF not only antagonize banana pests but also improve plant growth and survival through water and nutrient uptake. Tissue-culture plants enhanced with *Glomus fasciculatum* Thaxter and *G. mosseae* Thaxter have demonstrated suppression of nematodes, such as *R. similis* (Umesh et al., 1988) and *P. goodeyi* (Jaizme-Vega and Pinochet, 1997). However, to obtain the inoculum needed for application, AMF need to be produced on living plants (Sikora et al., 2003), creating difficulties and expense in their production and application. Until recently, the use of AMF was not viewed to be commercially viable, although products are now beginning to appear on the market. As intercropping is such a common aspect of subsistence farming, some intercrops that favor AMF inoculum buildup, such as sorghum (*Sorghum* spp.), could be promoted (Elsen et al., 2003; Elsen et al., 2009).

Several entomopathogenic fungal products based on *Metarrhizium* spp. and *Beauveria* spp. are commercially available for use as insect biopesticides. Despite near 100% efficacy *in vitro* against pests such as the banana weevil (Kaaya et al., 1993), their efficacy in the field tends to be slow, erratic, and ultimately an expensive option. The development of efficient and cost-effective field delivery systems currently hampers their use in smallholder and commercial banana systems. Recently, it was demonstrated that such fungi can be applied as artificial endophytes in banana plants, reducing banana weevil populations and damage (Akello et al., 2007; Akello, Dubois, et al., 2008a, 2008b; Akello, Coyne, et al., 2008).

7.8 CONCLUSION

This chapter provides an overview of nonmicrobial pests of bananas. Pest profiles are highly variable and depend on region, clone, and crop system. Of particular importance is that recommendations for their management, and research leading to these, vary greatly between smallholder systems and commercially managed plantations. Whereas IPM is necessary and can lead to reduced pesticide reliance, especially nematicides, in commercial systems, the situation contrasts markedly with smallholder systems. For smallholder systems, ill-linked management options, some that are often unsuitable, should be avoided for use in IPM, and a focus on key basic issues, such as pest identification, cultural management options, and deployment of basic training for farmers and extension workers should prevail.

REFERENCES

Abbot, L.K. and A.D. Robson. 1984. The effect of VA mycorrhizae on plant growth. In: *VA mycorrhizae*, C.L. Powell and D.J. Bagyaraj, eds., 113–130. Boca Raton: CRC Press.

Abele S., T. Dubois, E. Twine, K. Sonder, and O. Coulibaly. 2007. Organic agriculture in Africa: A critical review from a multidisciplinary perspective. *J. Agric. Rural Dev. Trop. Subtrop.* 89:143–166.

Akello J., T. Dubois, C.S. Gold, J. Nakavuma, and P. Paparu. 2007. *Beauveria bassiana* Balsamo (Vuillemin) as a potential endophyte in tissue culture banana (*Musa* spp.). *J. Invertebrate Pathol.* 96:34–42.

Akello J., T. Dubois, D.L. Coyne, and S. Kyamanywa. 2008a. Effect of endophytic *Beauveria bassiana* on populations of the banana weevil, *Cosmopolites sordidus*, and their damage in tissue-cultured banana plants. *Entomologia Experimentalis et Applicata* 129:157–165.

Akello J., T. Dubois, D.L. Coyne, and S. Kyamanywa. 2008b. Endophytic *Beauveria bassiana* in banana (*Musa* spp.) reduces banana weevil (*Cosmopolites sordidus*) fitness and damage. *Crop Prot.* 27:1437–1441.

Akello, J., D.L. Coyne, A. Wasukira, and T. Dubois. 2008. Endophytic *Beauveria bassiana* controls *Radopholus similis* in tissue culture banana. In: *Fifth International Congress of Nematology.* 13–18 July 2008, Brisbane, Australia. M. Hodda and A. Ciancio, eds., 254. Canberra: CSIRO.

Alvarez, C.E., A. Ortega, M. Fernández, and A.A. Borges. 2001. Growth, yield and leaf nutrient content of organically grown banana plants in the Canary Islands. *Fruits* 56:17–26.

Athman, S.Y. 2006. Host-endophyte-pest interactions of endophytic *Fusarium oxysporum* antagonistic to *Radopholus similis* in banana (*Musa* spp.). PhD diss., University of Pretoria, South Africa.

Bajwa, W.I. and M. Kogan. 2002. Compendium of IPM definitions—What is IPM and how is it defined in the worldwide literature? IPPC publication no. 998. Corvallis: Oregon State University.

Beardsley, J.W. 1970. *Aspidiotus destructor* Signoret, an armored scale pest new to the Hawaiian Islands. *Proc. Hawaiian Entomolog. Soc.* 20:505–508.

Ben-Dov, Y. 1994. *A systematic catalogue of the mealybugs of the world (Insecta, Homoptera, Coccoidea, Pseudococcidae and Putoidae) with data on geographical distribution, host plants, biology and economic importance.* Andover: Intercept.

Berg, G.L. 1991. *Farm chemicals handbook '91.* Willoughby: Meister Publishing Company.

Boa, E. 2007. Global Plant Clinic 2005–2008. *2006 annual report.* Surrey: CAB International.

Botha, J., D. Hardie, and G. Power. 2000. *Fact sheet no. 45/2000: Banana scab moth* Nacoleia octasema. *exotic threat to western Australia.* Bentley Delivery Centre: Agriculture Western Australia.

Brooks, F. 2003. Crop profile for bananas in the American Pacific. National information system for the regional IPM centers. http://www.ipmcenters.org/cropprofiles/docs/APbananas.pdf (accessed 1 November 2009).

CAB International. 2005. *Crop protection compendium*. Wallingford: CAB International.

CAPS. 2010. Fruit piercing moth *Eudocima fullonia*. http://www.dem.ri.gov/programs/bnatres/agricult/pdf/fruitpiercingmoth.pdf (accessed 1 February 2010).

Chiquita Brands International. 2001. *2001 corporate responsibility report*. Cincinatti: Chiquita Brands International.

CIALCA. 2009. *CIALCA progress reports 05. Final report phase I-CIALCA*. January 2006–December 2008. Kampala: CIALCA.

Cianco, A. and K.G. Mukerji. 2009. *Integrated management of fruit crops and forest nematodes*. Heidelberg: Springer.

Colbran, R.C. 1967. *Hot water tank for treatment of banana planting material*. Advisory leaflet Division of Plant Industry. no. 924. Brisbane: Queensland Department of Primary Industries.

Constantinides, N.L. and J.J. McHugh. 2003. *Pest management strategic plan for banana production in Hawaii*. Workshop summary, March 3, 2003, Pearl City Urban Garden Center. Honolulu: University of Hawaii at Manoa.

Coyne, D.L. 2009. Pre-empting plant-parasitic nematode losses on banana in Africa: Which species do we target? *Acta Hort.* 828:227–235.

Coyne, D.L. and L. Waeyenberge. 2008. Plant parasitic nematodes associated with *Musa* in Africa. In: *Fifth International Congress of Nematology*, 13–18 July 2008, Brisbane, Australia. M. Hodda and A. Ciancio, eds., 64. Canberra: CSIRO.

Coyne, D.L., O. Rotimi, P. Speijer, B. De Schutter, T. Dubois, A. Auwerkerken, et al. 2005. Effects of nematode infection and mulching on plantain (*Musa* spp., AAB-group) yield of ratoon cycles and plantation longevity in southeastern Nigeria. *Nematology* 7:531–543.

Dale, J.L. and R.M. Harding. 1998. Banana bunchy top disease: Current and future strategies for control. In: *Plant virus disease control*, A. Hadidi, R.K. Khetarpal, and H. Koganezawa, eds., 659–669. St. Paul: APS Press.

De Guiran, G. and A. Vilardebo. 1962. Le bananier aux Iles Canaries. IV. Les nématodes parasites. *Fruits* 17:263–277.

De Waele, D. and A. Elsen. 2002. Migratory endoparasities: *Pratylenchus* and *Radopholus* species. In: *Plant resistance to parasitic nematodes*, J.J. Starr, R. Cook, and J. Bridge, eds., 175–206. Wallingford: CAB International.

Dochez, C. 2004. *Breeding for resistance to* Radopholus similis *in East African Highland bananas (Musa spp.)*. PhD diss., Katholieke University Leuven, Belgium.

Dubois, T. and D.L. Coyne. 2006. Endophytes: Natural biodiversity bolstered to combat banana pests. *Geneflow* 2006:52.

Dubois, T., D.L. Coyne, E. Kahangi, L. Turoop, and E.N. Nsubuga. 2006. Endophyte-enhanced banana tissue culture: Technology transfer through public-private partnerships in Kenya and Uganda. *Afr. Technol. Dev. Forum J.* 3:18–23.

Dubois, T., C.S. Gold, P. Paparu, S. Athman, and S. Kapindu. 2006. Tissue culture and the *in vitro* environment. Enhancing plants with endophytes: Potential for ornamentals? In: *Floriculture, ornamental and plant biotechnology: Advances and topical issues*, J. Teixeira Da Silva, ed., 397–409. London: Global Science Books.

Duncan, L.W., R.N. Inserra, W.K. Thomas, D. Dunn, I. Mustika, M.L. Frisee, et al. 1999. Molecular and morphological analysis of *Pratylenchus coffeae* and closely related species. *Nematropica* 29:61–80.

Ehler, L.E. 2006. Perspective integrated pest management (IPM): Definition, historical development and implementation, and the other IPM. *Pest Manage. Sci.* 62:787–789.

Ekanayake, H.M.R.K., W.W.M.S.N. Wekadapola, and K.A.N.P. Bandara. 2002. Studies on fruit fly infestation in banana cultivars in Sri Lanka. *Ann. Sri Lanka Dep. Agric.* 4:269–274.

Elsen, A., H. Baimey, R. Swennen, and D. De Waele. 2003. Relative mycorrhizal dependency and mycorrhiza-nematode interaction in banana cultivars (*Musa* spp.) differing in nematode susceptibility. *Plant Soil* 256:303–313.

Elsen, A., L. Van Der Veken, and D. De Waele. 2009. AMF-induced bioprotection against migratory plant-parasitic nematodes in banana. *Acta Hort.* 828:91–100.

European Commission. 2007. *Part 1, Monitoring of pesticide residues in products of plant origin in the European Union, Norway, Iceland and Liechtenstein 2005*. Commission staff working document. Brussels: European Commission.

Food and Agriculture Organization. 1975. *FAO panel of experts on integrated pest control*. October 15–25, 1974. Meeting rep. 1975/M/2. Rome: FAO.

Food and Agriculture Organization. 1995. *Intercountry programme for the development and application of integrated pest control in rice in South and South-East Asia*. Rome: FAO.

Gaidashova, S.V., P. Van Asten, D. De Waele, and B. Delvaux. 2009. Relationship between soil properties, crop management, plant growth and vigour, nematode occurrence and root damage in East African Highland banana-cropping systems: A case study in Rwanda. *Nematology* 11:883–894.

Gold, C.S. and S. Messiaen. 2000. *The banana weevil* Cosmopolites sordidus. *Musa* pest fact sheet no. 4. Montpellier, France: INIBAP.

Gold, C.S., J.E. Peña, and E.B. Karamura. 2001. Biology and integrated pest management for the banana weevil *Cosmopolites sordidus* (Germar) (Coleoptera: Curculionidae). *Integrated Pest Manage. Rev.* 6:79–155.

Gowen, S.C., P. Quénéhervé, and R. Fogain. 2005. Nematode parasites of bananas and plantains. In: *Plant parasitic nematodes in subtropical and tropical agriculture.* 2nd ed., M. Luc, R.A. Sikora, and J. Bridge, eds., 611–643. Wallingford: CAB International.

Green, T.A. and C. Petzoldt. 2009. *Guide to IPM elements and guidelines*. State College, PA: Northeastern IPM Center.

Hallman, J., and R.A. Sikora. 1996. Toxicity of fungal endophyte secondary metabolites to plant parasitic nematodes and soil-borne plant pathogenic fungi. *Eur. J. Plant Pathol.* 102:155–162.

Hara, A.H., R.F.L. Mau, R. Heu, C. Jacobsen, and R. Niino-Duponte. 2002. *Banana rust thrips damage to banana and ornamentals in Hawaii*. Honolulu: University of Hawaii at Manoa.

Harrison, J.O. 1963. Notes on the biology of the banana flower thrips, *Frankliniella parvula*, in the Dominican Republic (Thysanoptera: Thripidae). *Ann. Entomolog. Soc. Am.* 56:664–666.

Hawaiian Banana Industry Association. 2010a. Banana IPM protocol supporting documentation. http://www.extento.hawaii.edu/IPM/Certification/banana/banprotocol1.pdf (accessed 1 February 2010).

Hawaiian Banana Industry Association. 2010b. EPA pesticide environmental stewardship program. Industry background. http://www.extento.hawaii.edu/IPM/Certification/banana/banbackground.pdf (accessed 1 February 2010).

Heil, M., A. Hilpert, W. Kaiser, and E. Linsenmair. 2000. Reduced growth and seed set following chemical induction of pathogen defence: Does systemic acquired resistance (SAR) incur allocation costs? *J. Ecol.* 88:645–654.

Hernández-Suárez, E., C. Ramos-Cordero, and A. Carnero. 2006. Laboratory screening of pesticides against *Lecanoideus floccissimus*. *IOBC/WPRS Bull.* 29:19.

Hollingsworth, R.G. 2003. Life history observations on *Thrips florum* (Thysanoptera: Thripidae) infesting gardenia in Hawaii, and a comparison of the humidity requirements for *T. florum* and *Frankliniella occidentalis*. *Proc. Hawaiian Entomolog. Soc.* 36:79–88.

Hu, J.S., M. Wang, D. Sether, W. Xie, and K.W. Leonhardt. 1996. Use of polymerase chain reaction (PCR) to study transmission of banana bunchy top virus by the banana aphid (*Pentalonia nigronervosa*). *Ann. Appl. Biol.* 128:55–64.

ICIPE. 1997. *Annual scientific report 1995–1997*. Nairobi: ICIPE.

Jaizme-Vega, M.C. and J. Pinochet. 1997. Growth response of banana to three mycorrhizal fungi in *Pratylenchus goodeyi* infested soil. *Nematropica* 27:69–76.

Jomo Kenyatta University of Agriculture and Technology (JKUAT). 2008. *Managing microorganisms to enhance plant health for sustainable banana production in Eastern Africa: Biological control of banana nematodes with fungal endophytes*. Summary report. Nairobi: JKUAT.

Jones, D.R. 2009. Diseases and pest constraints to banana production. *Acta Hort.* 828:21–36.

Kaaya, G.P., K.V. Seshu Reddy, E.D. Kokwaro, and D.M. Munyinyi. 1993. Pathogenicity of *Beauveria bassiana, Metarrhizium anisopliae* and *Serratia marcescens* to the banana weevil, *Cosmopolites sordidus*. *Biocontrol Sci. Technol.* 3:177–187.

Kambuou, R.N. 2003. Current situation in banana R&D in Papua New Guinea. In: *Advancing banana and plantain R&D in Asia and the Pacific*. Vol. 12, *Proceedings of the 2nd BARNET Steering Committee meeting. Jakarta*. Los Bãnos: INIBAP.

Karmakar, K., and S. Dey. 2006. Studies on seasonal incidence of phytophagous mite species on selected germplasms of banana in West Bengal. *Indian J. Crop Sci.* 1:138–139.

Kogan, M. 1998. Integrated pest management: Historical perspectives and contemporary development. *Annu. Rev. Entomol.* 43:243–70.

Lewis, T. 1997. *Thrips as crop pests*. Wallingford: CAB International.

Lorenzen, J., A. Tenkouano, R. Bandyopadhyay, B.I. Vroh, D.L. Coyne, and L. Tripathi. 2009. The role of crop improvement in pest and disease management. *Acta Hort.* 828:305–314.

Magee, C.J. 1967. *The control of banana bunchy top*. New Caledonia: Government of Western Samoa South Pacific Commission, New Caledonia.

Masanza, M., C.S. Gold, A. Van Huis, and P.E. Ragama. 2006. Effects of crop sanitation on banana weevil, *Cosmopolites sordidus* (Germar) (Coleoptera: Curculionidae), populations and crop damage in Uganda. *Afr. Entomol.* 14:267–275.

Mateille, T., P. Quénéhervé, and R. Hugon. 1994. The development of plant-parasitic nematode infestations on micro-propagated banana plants following field control measures in Côte d'Ivoire. *Ann. Appl. Biol.* 125:147–159.

Matlock, R.B. and R. De La Cruz. 2002. An inventory of parasitic Hymenoptera in banana plantations under two pesticide regimes. *Agric. Ecosystems Environ.* 93:147–164.

McIntyre, B.D., P.R. Speijer, S.J. Riha, and F. Kizito. 2000. Effects of mulching on biomass, nutrients, and soil water in banana inoculated with nematodes. *Agron. J.* 92:1081–1085.

Metcalf, R.L. 1962. *Destructive and useful insects*. New York: McGraw-Hill.

Meyer, S.L.F. and D.P. Roberts. 2002. Combinations of biocontrol agents for management of plant-parasitic nematodes and soilborne plant-pathogenic fungi. *J. Nematol.* 34:1–8.

Ming-Yi, C. 2003. Occurrence and control of coconut scale (*Aspidiotus destructor* Signoret) in bananas. MsC diss., University of Hawaii at Manoa.

Ministry of Agriculture and Forestry. 2006. *Import health standard commodity sub-class: Fresh fruit/vegetables. Bananas (Musa spp) from Australia.* Issued pursuant to section 22 of the Biosecurity Act 1993. Date issued: 2 June 2006. Wellington: Ministry of Agriculture and Forestry.

Moens, T., M. Arya, R. Swennen, and D. De Waele. 2004. Enhanced biodegradation of nematicides after repetitive applications and its effects on root and yield parameters in commercial banana plantations. *Biol. Fertility Soils* 39:407–414.

Morton, J.F. 1987. *Fruits of warm climates*. Miami: Florida Flair Books.

Mound, L.A. and R. Marullo. 1996. The thrips of Central and South America: An introduction (Insecta: Thysanoptera). *Memoirs Entomol.* 6:1–488.

Mugisha-Kamatenesi, M., A.L. Deng, J.O. Ogendo, E.O. Omolo, M.J. Mihale, M. Otim, et al. 2008. Indigenous knowledge of field insect pests and their management around Lake Victoria basin in Uganda. *Afr. J. Environ. Sci. Technol.* 2:342–348.

Nafus, D. 2000a. *Green scale (*Coccus viridis *De Lotto).* Honolulu: Agricultural Development in the American Pacific.

Nafus, D. 2000b. *Hemispherical scale (*Saissetia coffeae *(Walker)).* Honolulu: Agricultural Development in the American Pacific.

Nelson, S.C., R.C. Ploetz, and A.K. Kepler. 2006. *Species profiles for Pacific Island agroforestry.* Musa *species (banana and plantain).* Holualoa: Traditional Tree Initiative.

OEPP/EPPO. 2006. Diagnostics. *Opogona sacchari*. PM 7/71(1). *OEPP/EPPO Bull.* 36:171–173.

Okolle, J.N., M. Mansor, and A.H. Ahmad. 2006. Seasonal abundance of the banana skipper, *Erionota thrax* (Lepidoptera: Hesperiidae) and its parasitoids in a commercial plantation and a subsistence farm in Penang, Malaysia. *Int. J. Trop. Insect Sci.* 26:197–206.

Ortiz, R., R.S.B. Ferris, and D.R. Vuylsteke. 1995. Banana and plantain breeding. In: *Bananas and plantains*, S. Gowen, ed., 110–146. London: Chapman and Hall.

Ostmark, H.E. 1975. Banana pests in the genus *Colapsis*, including description of a new species, Coleoptera: Chrysomelidae. *Fla. Entomol.* 58:1–8.

Padmanaban, B. and S. Sathiamoorthy. 2001. *The banana stem weevil* Odoiporus longicollis. Musa *pest fact sheet no. 5*. Montpellier: INIBAP.

Padmanaban, B. and M. Kandasamy. 2003. Survival of banana weevil borers in banana plant residues. *Indian J. Entomol.* 65:424–425.

Paine, R.W. 1964. *The banana scab moth,* Nacoleia octasema *(Meyrick): Its distribution, ecology and control.* Technical paper. Noumea: South Pacific Commission.

Parker, B.L., M. Skinner, and T. Lewis. 1995. *Thrips biology and management.* NATO science series. New York: Springer-Verlag.

Pasberg-Gauhl, C. 2002. How can we reduce losses of banana bunches due to "fruit speckling" in commercial banana plantations? In: *XV reunión international ACORBAT 2002*, 477–480. Medellin: Asociación de Bananeros de Colombia.

Pattison, A.B., J.M. Stanton, J.A. Cobon, and V.J. Doogan. 2002. Population dynamics and economic threshold of the nematodes *Radopholus similis* and *Pratylenchus goodeyi* on banana in Australia. *Int. J. Pest Manage.* 48:107–111.

Peña, J., J. Sharp, and M. Wysoki. 2002. *Tropical fruit pests and pollinators: Biology, economic importance, natural enemies and control.* Wallingford: CAB International.

Peregrine, W.T.H. and J. Bridge. 1992. The lesion nematode *Pratylenchus goodeyi*, an important pest of ensete in Ethiopia. *Trop. Pest Manage.* 38:325–326.

Petrini, O. 1986. Taxonomy of endophytic fungi of aerial plant parts. In: *Microbiology of the phyllosphere*, N.J.J. Fokkema and J. Van Den Heuvel, eds., 175–187. Cambridge: Cambridge University Press.

Pillay, M. and L. Tripathi. 2007. Banana breeding. In: *Breeding major food staples*, M.S. Kang and P.M. Priyadarshan, eds., 393–428. New York: Blackwell Science.

Pinese, B. 1999. Insects of quarantine importance to Australian banana production. In: *Third Australian Banana Industry Congress*. Brisbane: Australian Banana Growers' Council.

Pinese, B. and R. Piper. 1994. *Bananas: Insect and mite management*. Brisbane: Department of Primary Industries.

Pinochet, J. and P.R. Rowe. 1979. Progress in breeding for resistance to *Radopholus similis* on bananas. *Nematropica* 9:76–77.

Plowright, R. 2000. *Analysis of the pathogenic variability and genetic diversity of the plant-parasitic nematodes Radopholus similis, Pratylenchus coffeae and Pratylenchus goodeyi on bananas and plantains in East and West Africa.* Final technical report. Egham: CAB International.

Pocasangre, L.E. 2006. Banana improvement based on tissue culture propagation, bioenhancement and agronomic management for sustainable production. In: *Tropentag 2006. International research on food security, natural resource management and rural development. Prosperity and poverty in a globalised world—Challenges for agricultural research. Book of abstracts*, F. Asch and M. Becker, eds., 75. Bonn: Rheinische Friedrich-Wilhelms-Universität Bonn.

Pocasangre, L.E., R.D. Menjivar, A. Zum Felde, A.S. Riveros, F.E. Rosales, and R.A. Sikora. 2006. Hongos endifíticos como agentes biológicos de control de fítonemátodos en banana. In *XV reunión international ACORBAT 2002*, 249–254. Joinville: ACORBAT.

Pocasangre, L.E., A. Zum Felde, C. Canizares, J. Munoz, P. Suarez, C. Jimenes, et al. 2007. Field evaluation of the antagonistic activity of endophytic fungi towards the burrowing nematode, *Radopholus similis*, in plantain in Latin America. Paper presented at the ISHS/ProMusa symposium: Recent advances in banana crop protection for sustainable production and improved livelihoods, September 10–14, 2008, White River, South Africa.

Price, N.S. 2006. The banana burrowing nematode, *Radopholus similis* (Cobb) Thorne in the Lake Victoria region of East Africa: Its introduction, spread and impact. *Nematology* 8:801–817.

Queensland Horticulture Institute. 2000. *Banana quarantine threats: The banana skipper*. Brisbane: Queensland Horticulture Institute.

Quénéhervé, P., F. Salmon, P. Topart, and J.P. Horry. 2009. Nematode resistance in bananas: Screening results on some new *Mycosphaerella* resistant banana hybrids. *Euphytica* 165:123–136.

Rajab, K. and H. Fundi. 1999. Banana and plantain in IPM Zanzibar. In: *Mobilizing IPM for sustainable banana production in Africa.* Proceedings of a workshop on banana IPM held in Helspruit, South Africa 23–28 November 1998, E.A. Frison, C.S. Gold, E.B. Karamura, and R.A. Sikora, eds., 209–213. Montpellier: INIBAP.

Ramani, S., J. Poorani, and B.S. Bhumannavar. 2002. Spiraling whitefly, *Aleurodicus dispersus*, in India. *Biocontrol News Inf.* 23:55–62.

Ramírez, R., O.E. Dominguez-Gil, O. Liscano, M. Vilchez, and R. Urdaneta. 1999. Importancia de *Antichloris viridis* Druce como lepidóptero defoliador del plátano (*Musa* AAB cv. 'Hartón') en la zona sur del Lago de Maracaibo, Venezuela. *Revista de la Facultad de Agronomía. Universidad del Zulia* 16:88–94.

Reid, E.D. 2000. Status of banana production in the Windward Islands and the prospects for organic banana production. In: *Organic banana 2000: Towards an organic banana initiative in the Caribbean.* Report of the international workshop on the Production and Marketing of Organic Bananas by Smallholder Farmers 31 October–4 November 1999, Santo Domingo, Dominican Republic. M. Holderness, S. Sharrock, E. Frison, and M. Kairo, eds., 41–54. Montpellier: INIBAP.

Rey, F. 2002. Rouille argentée des bananes martiniquaises: Distribution et méthode de lutte. *Fruits* 57:3–10.

Risède J.M., C. Chabrier, M. Dorel, B. Rhino, K. Lakhia, C. Jenny, and P. Quénéhervé. 2009. Recent and upcoming strategies to counter plant-parasitic nematodes in banana cropping systems of the French West Indies. *Acta Hort.* 828:117–127.

Robinson, J.C. 1996. *Bananas and plantains*. Wallingford: CAB International.

Robson, J.D., M.G. Wright, and R.P.P. Almeida. 2006. Within-plant distribution and binomial sampling plan of *Pentalonia nigronervosa* (Hemiptera, Aphididae) on banana. *J. Econ. Entomol.* 99:2185–2190.

Robson, J.D., M.G. Wright, and R.P.P. Almeida. 2007. Effect of imidacloprid foliar treatment and banana leaf age on *Pentalonia nigronervosa* (Hemiptera, Aphididae) survival. *New Zealand J. Crop Hort. Sci.* 35:415–422.

Roderick, H., P.E. Urwin, H.J. Atkinson, and L. Tripathi. 2009. Nematode resistant plantain for subsistence farmers. In: *Book of abstracts. Tropical crop biotechnology conference, 22–25 July 2009*, 74. Stellenbosch: University of Stellenbosch.

Roditakis, E., L.A. Mound, and N.E. Roditakis. 2006. Note: First record in Crete of *Hercinothrips femoralis* in greenhouse banana plantations. *Phytoparasitica* 34:488–490.

Sakimura, K. and N.L.H. Krauss. 1944. Thrips from Maui and Molokai. *Proc. Hawaiian Entomolog. Soc.* 12:113–122.

Sands, D. and W. Liebregts. 2005. Biological control of fruit piercing moth (*Eudocima fullonia* [clerck]) (Lepidoptera: Noctuidae) in the Pacific: Exploration, specificity, and evaluation of parasitoids. In: *Second International Symposium on Biological Control of Arthropods.* 12–16 September 2005, Davos, Switzerland, M.S. Hoddle, ed., 267–276. Washington, DC: USDA Forest Service.

Sarah, J.L. 2000. Burrowing nematode. In: *Diseases of banana, abacá and enset*, D.R. Jones, ed., 295–303. Wallingford: CAB International.

Seshu Reddy, K.V., J.W. Ssennyonga, and L. Ngode. 1998. Banana integrated pest management information transfer through training. In *Second Triennial Crop Protection Conference, 16–17 September 1998.* Nairobi: KARI.

Seshu Reddy, K.V., L. Ngode, J.W. Ssenyonga, M. Wabule, M. Onyango, T.O. Adede, et al. 1999. Management of pests and diseases of banana in Kenya: A status report. In *Mobilizing IPM for sustainable banana production in Africa.* Proceedings of a workshop on banana IPM held in Nelspruit, South Africa 23–28 November 1998, E.A. Frison, C.S. Gold, E.B. Karamura, and R.A. Sikora, eds., 215–223. Montpellier: INIBAP.

Sikora, R.A. and E. Schlösser. 1973. Nematodes and fungi associated with root systems of banana in a state of decline in Lebanon. *Plant Dis. Rep.* 57:615–618.

Sikora, R.A. and F. Schönbeck. 1975. Effect of vesicular-arbuscular mycorrhizae (*Endogone mosseae*) on the population dynamics of the root-knot nematodes *Meloidogyne incognita* and *Meloidogyne hapla.* In *Proceedings of the VIII international congress on plant protection*, 158–166. Moscow.

Sikora, R.A., B.A. Niere, and A.J. Kimenju. 2003. Endophytic microbial biodiversity and plant nematode management in African agriculture.. In *Biological control in IPM systems in Africa*, P. Neuenschwander, C, Borgemeister, and U. Langewald, eds., 179–192. Wallingford: CAB International.

Sikora, R.A., L. Pocasangre, A. Zum Felde, B. Niere, T.T. Vu, and A.A. Dababat. 2008. Mutualistic endophytic fungi and *in-planta* suppressiveness to plant parasitic nematodes. *Biologic. Control* 46:15–23.

Smith, I.M., D.G. McNamara, P.R. Scott, and M. Holderness. 1996. *Quarantine pests for Europe.* 2nd ed. Wallingford: CAB International.

Speijer, P.R. and C.H. Bosch. 1996. Susceptibility of *Musa* cultivars to nematodes in Kagera region, Tanzania. *Fruits* 51:217–222.

Speijer, P.R. and R. Fogain. 1999. *Musa* and *Ensete* nematode pest status in selected African countries. In: *Mobilizing IPM for sustainable banana production in Africa.* Proceedings of the International Workshop on Banana IPM, November 1998. Nelspruit, South Africa, E. Frison, C. Gold, E. Karamura, and R.A. Sikora, eds., 99–108. Montpellier: INIBAP.

Speijer, P.R., C. Kajumba, and W.K. Tushemereirwe, T. 1999. Dissemination and adaptation of a banana clean planting technology in Uganda. *InfoMusa* 8:11–13.

Ssango, F., P.R. Speijer, D.L. Coyne, and De Waele. 2004. Path analysis: A novel approach to determine the contribution of nematode damage to East African Highland banana (*Musa* spp., AAA) yield loss under two crop management practices in Uganda. *Field Crops Res.* 90:177–187.

Starr, J.J., R. Cook, and J. Bridge. 2002. *Plant resistance to parasitic nematodes.* Wallingford: CAB International.

Stephens, C.S. 1984. Bionomics of three *Philicoptus* banana pests (Coleoptera: Curculionidae) and notes on other weevils in Mindanao, Philippines. *Philipp. Agric.* 67:243–253.

Stephenson, J.B. 2001. *Report to the chairman, subcommittee on research, nutrition, and general legislation, Committee on Agriculture, Nutrition, and Forestry, U.S. Senate. Agricultural pesticides management improvements needed to further promote integrated pest management.* Washington, DC: US Government Printing Office.

Stern, V.M., R.F. Smith, R. Van Den Bosch, and K.S. Hagen. 1959. The integrated control concept. *Hilgardia* 29:81–101.

Stover, R.H. and N.W. Simmonds. 1987. *Bananas.* Harlow: Longmans.

Tenkouano, A. and R.L. Swennen. 2004. Progress in breeding and delivering improved plantain and banana to African farmers. *Chronica Hort.* 44:9–15.

Tenkouano, A., D. Vuylsteke, J. Okoro, D. Makumbi, R.L. Swennen, and R. Ortiz. 2003. Diploid banana hybrids TMB2x5105–1 and TMB2x9128–3 with good combining ability, resistance to black Sigatoka and nematodes. *Hort. Sci.* 38:468–472.

Tenkouano, A., S. Hauser, D.L. Coyne, and O. Coulibaly. 2006. Clean planting materials and management practices for sustained production of banana and plantain in Africa. *Chronica Hort.* 46:14–18.

Tomekpe, K. and L. Sadom. 2008. A new generation of dwarf plantain hybrids with resistance to black Sigatoka and high yield. In: *Banana 2008. Banana and plantain in Africa: Harnessing international partnerships to increase research impact*, 154. Ibadan: IITA.

Tripathi, L. 2009. Biotechnology and nematodes. *R4D Review* 2. Ibadan: IITA.

Umesh, K.C., K. Krishnappa, and D.J. Bagyaraj. 1988. Interaction of burrowing nematode, *Radopholus similis* (Cobb 1893) Thorne 1949, and VA mycorrhiza, *Glomus fasciculatum* (Thaxt.) Gerd. and Trappe in banana (*Musa acuminata* Colla.). *Indian J. Nematol.* 18:6–11.

Uquillas, C.M. 2002. Cicle de vida del Lepidoptero (*Opsiphanes tamarindi*) criado en laboratorio y el consumo de follaje en sus diversos instares. In: *XV reunión international ACORBAT 2002*, 301–305. Medellin: Asociación de Bananeros de Colombia.

U.S. Department of Agriculture. 2009. *Banana pest regulation modifies conditions for internal transit*. Mexico City: USDA-FAS.

Vereijken, P. 1989. From integrated control to integrated farming, an experimental approach. *Agric. Ecosystems Environ.* 26:37–43.

Viaene, N., D.L. Coyne, and B.R. Kerry. 2006. Biological and cultural control. In: *Plant nematology*, M. Moens and R. Perry, eds., 346–369. Wallingford: CAB International.

Vuylsteke, D. 1989. Shoot-tip culture for the propagation, conservation and exchange of *Musa* germplasm. *Practical manual for handling crop germplasm in vitro*. Rome: IBPGR.

Wageningen University. 2009. Early detection *Opogona* now possible. http://www.pri.wur.nl/UK/newsagenda/news/opogona140909.htm (accessed 1 November 2009).

Waithira, W.B. 2009. Screening selected isolates of endophytic Fusarium oxysporum for biological control of banana nematodes in tissue culture banana. Master's thesis, Nairobi: JKUAT.

Waterhouse, D.F. and K.R. Norris. 1987. *Biological control: Pacific prospects*. Melbourne: Inkata Press.

Waterhouse, D.F. and K.R. Norris. 1989. *Biological control: Pacific prospects. Supplement 1*. Canberra: Australian Center for International Agriculture Research.

Williams, D.J. and G.W. Watson. 1988. *Aspidiotus destructor* Signoret. In: *The scale insects of the tropical South Pacific region.* Part 1, *The armored scales (Diaspidae)*, D.J. Williams and G.M. Watson, eds., 53–56. Wallingford: CAB International.

Williams, M.H., M. Vesk, and M.G. Mullins. 1990. Development of the banana fruit and occurrence of the maturity bronzing disorder. *Ann. Bot.* 65:9–19.

World Bank. 2005. *Sustainable pest management: Achievements and challenges*. Report 32 714-GBL. Washington, DC: World Bank.

Wright, M.G. and J.M. Diez. 2005. Coconut scale *Aspidiotus destructor* (Hemiptera: Diaspididae) seasonal occurrence, dispersion and sampling on banana in Hawaii. *Int. J. Trop. Insect Sci.* 25:80–85.

Zum Felde, A., A. Mendoza, J.A. Cabrera, A. Kurtz, A. Schouten, L. Pocasangre, and R.A. Sikora. 2009. The burrowing nematode of banana: Strategies for controlling the uncontrollable. *Acta Hort.* 828:101–108.

8 Reproductive Biology

Jeanie Anne Fortescue and David William Turner

CONTENTS

8.1 FLORAL BIOLOGY

Bananas and plantains (Musaceae) are a unique group of plants within the Zingiberales. They contribute to the nourishment and culture of millions of people, especially in tropical regions, and are exported to temperate regions. This chapter discusses the reproductive biology of the *Musa* spp. because it is central to working with the breeding system and the relative success of banana breeding schemes.

In recent years knowledge of the sexual reproduction of plants has progressed considerably, but few studies have been made on members of the Musaceae, despite their economic importance and scientific interest. Many of the key references on floral biology of *Musa* spp. were published between 1930 and 1980, especially those involving ovules and the female gametophyte.

8.1.1 Development of the Inflorescence

The vegetative stage of each shoot ends when the apical meristem produces an inflorescence that undergoes much of its early development inside the pseudostem. Extension of the internodes below it thrusts the inflorescence to the top of the plant and clear of the surrounding leaf sheaths. Until its emergence, the changes from a vegetative to floral apex and even the development of the inflorescence is unmarked by any external sign. The only indication is leaf number and size. Approximately 26 to 50 laminate leaves are produced, increasing in size until the last two, which are much reduced in size, just before the inflorescence emerges (Simmonds, 1962; Stover and Simmonds, 1987). The vegetative apical meristem of *Musa* is a flattened dome. The main meristem lies deep under the center of the dome. As in other monocotyledons, vegetative growth is delegated to lateral organs; leaves are differentiated in regular succession from the outer flanks of the corpus. The transition to flowering is accompanied by a broadening of the apex by both cell division and cell expansion. After floral initiation, the meristem becomes convex and rises above the surrounding leaf bases. Flower bracts appear instead of leaves.

Initiation of the inflorescence is accompanied by a great increase in mitotic activity deep in the corpus and thickening of the tunica. The products of all this activity and growth are a stem with elongated internodes, non-encircling bracts in place of encircling sheaths, and a regular system of axillary lateral branches—the flowers (Simmonds, 1966; Stover and Simmonds, 1987). In quantitative terms, before floral initiation, the meristem produces a leaf and a lateral bud (phytomer) every 10 days, but after floral initiation it produces a bract and up to 20 flower initials every 1 to 2 days.

At the earliest stages of development, the floral parts are undifferentiated mounds of tissues. The leaf and following bracts are homologous organs and have the same morphology during early floral initiation. The major difference between the two is that the bracts have flower primordia in their axils. The early floral meristem produces secondary meristems that then initiate the individual flowers along the central axis. Swellings soon appear at the base of the flower bracts that enlarge and differentiate into female flowers. Subsequent bracts subtend male flowers (White, 1928). In tree crops the determinate floral meristems are axillary and the terminal apex remains vegetative; this situation is typical in tropical crops such as avocado, a dicotyledon (Sedgley and Griffin, 1989). In *Musa* it is not known whether the apex truly changes from vegetative to reproductive or whether the lateral activity around it becomes reproductive (White, 1928).

The inflorescence axis is the distal part of the aerial stem, which is terminal on the corm. It is a raceme or spike of cymose clusters of flowers at nodes covered by colored bracts. The female flowers are within the basal (proximal) bracts and the male flowers in the apical (distal) bracts; the intermediary clusters or neuters are of transitional structure. *Musa* flowers are predominantly unisexual and the plant monoecious. However, bananas grow in a clump with several generations connected to the underground corm and geitonogamous pollination may occur between inflorescences on the one clump or mat. Within an inflorescence, transition from nodes with female flowers to nodes with male flowers is marked in the juvenile inflorescence by a sudden decline in ovary length from one node to another. This is the first sign of gross sex differentiation. The underlying biochemical processes must take place much earlier in the sequence of floral differentiation. The major difference between the male and female ovary is the latter are larger, have a massive style that exceeds the perianth in length, and the stamens are reduced to staminodes. In the male flowers the ovary is small and in many cultivars and species they develop an abscission zone at their base and are shed a few days after anthesis. The female flowers develop no such abscission zone; however, the style and staminodes may abscise, leaving a calloused scar at the top of the ovary (Simmonds, 1966; Stover and Simmonds, 1987).

The inflorescence bears 1 to 30 nodes (or hands) of pistillate female flowers depending on genotype, environment, and edaphic conditions, followed by 0 to 4 hands of neuter flowers or pseudohermaphrodite hands. The remainder of the inflorescence contains staminate flowers, of which there are from 150 to 300 hands. Spikes of cv 'Gros Michel' (AAA) can contain over 100 male hands representing 2,500 male flowers compared with fewer than 200 female flowers (White, 1928). The apex may continue to produce male flowers long after the female fruits have rotted. In some clones, especially among plantains, the apex is short lived so that growth ceases soon after the bunch emerges from the top of the pseudostem; Horn plantains are characterized by the absence of male flowers at maturity (Simmonds, 1966; Stover and Simmonds, 1987; Swennen et al., 1995; De Langhe et al., 2005).

The basal nodes of the banana inflorescence bear the female flowers, and the upper nodes the male flowers. When the ovaries of the female flowers are developing into fruit, the axis continues to grow. As the stem grows, the bracts open to expose the male flowers and then both bract and male flowers usually abscise after a day or two. The distribution of the female and male flowers and the continued growth of the distal portion of the axis confer on the mature inflorescence a characteristic appearance. At the basal end is a mass of fruit, then a length of axis to which the male flowers and their subtending bracts may adhere, or abscise, and a "bell" of new male bracts and flowers at the distal end. The whole inflorescence may be 1 to 3 m long.

8.1.2 Development of the Female Flower

Musa has an inferior trilocular ovary with axile placentation. The ovules, of which there may be 300 to 1,500 per ovary, spring as relatively late outgrowths from the placenta and are two or four rowed in each locule. This character is expressed differently by the *acuminata* (A) and *balbisiana* (B) genomes and is diagnostic for determining the genomic origins of the edible bananas (Simmonds and Shepherd, 1955; Simmonds, 1966; Stover and Simmonds, 1987). The development of the female gametophyte of *Musa* is typical of angiosperms.

8.1.2.1 Pre-Anthesis

The ovary has septal nectaries with copious nectar and a tough outer skin that splits open longitudinally. The stigma is trilobate with a constantly wet and papillate surface that is sticky when receptive. It has two to four rows of ovules in each locule. The ovules are numerous, embedded in a strip of mucilage, and axile to the placenta. Development from floral initiation to ovule primordium to megaspore mother cell occurs while the inflorescence is inside the pseudostem (White, 1928; Juliano and Alcala, 1933; Dodds, 1945; Bouharmont, 1963; Dahlgren et al., 1985; Krishnamoorthy et al., 2004). As the inflorescence moves upward, the flowers arise in each node as a double row of closely grouped protuberances numbering 12 to 20 in 'Gros Michel' (White, 1928). When the length of the bract that envelops the inflorescence is twice its width, the fascicular primordium, which produces the flowers, appears as a mamillate hump at the axis. Development of the floral organs of the three kinds of flowers is acropetal. They arise in the following sequence: outer perianth lobes, inner perianth lobes, stamen, and pistil (Juliano and Alcala, 1933).

When it is midway up the pseudostem, the inflorescence of cv 'Gros Michel' (AAA) is about 120 mm long and the staminate and pistillate flowers are distinguishable. The ovary of the female

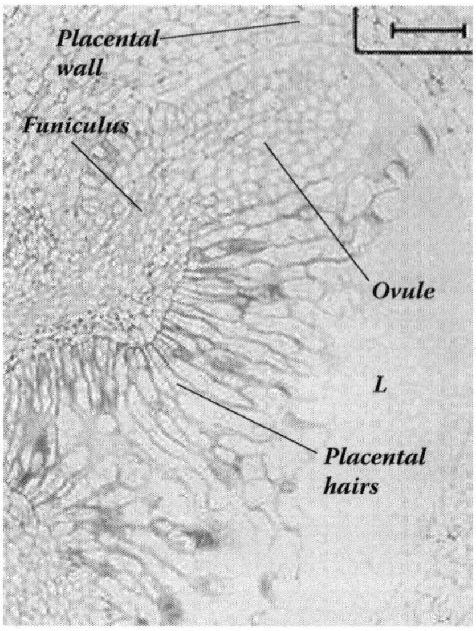

FIGURE 8.1 *Musa acuminata* ssp. (AA), an undescribed seeded banana similar to *microcarpa* but with longer pedicels. Also in Figures 8.3, 8.4, and 8.8. Bar = 50 μm. The ovule is small and orthotropous in orientation. It is only a fingerlike projection from the placenta. The outer and inner integuments, nucellus, and nucellar cap have not differentiated. L, locule. (Reprinted from *Sci. Hort.* 104, J.A. Fortescue, and D.W. Turner, Growth and development of ovules of banana, plantain and ensat (Musaceae), 463–478, Copyright 2005, with permission from Elsevier.)

flowers is about 10 mm long and 2.5 mm in diameter. The megasporangia are differentiating and appear as a rounded protuberance growing at right angles from the placental wall. Initially the ovule primordia are slender, fingerlike projections not more than five cells in width and two to four times longer (Figure 8.1) (White, 1928; Bouharmont, 1963). The ovule is at first atropous and by differential growth becomes anatropous. The micropyle points towards the placental wall.

The inner integument has already formed when the archesporium arises from any subepidermal cell near the summit of the nucellus. The differentiation of the archesporia takes place before the differentiation of the outer integument and just at the time the megasporange is half anatropous. It is easily distinguished from the surrounding cells by its relatively large size and great affinity for stains (Dodds, 1945; Bouharmont, 1963). The ovules attain almost maximum size when the megaspore mother cell begins to divide. When the inflorescence emerges from the top of the pseudostem, the embryo sac has differentiated and the nuclei are in their respective positions.

The stage when the ovule is half anatropous is critical in its development because changes are rapid and profound. The outer and inner integuments have differentiated but not the nucellar cap; however, the integuments are not large enough to form a micropyle. A large cell is visible in the nucellus and is most probably the megaspore mother cell (Figure 8.2). Within a few days the ovule is circular and the inner and outer integuments have enlarged to encircle the nucellus and form the micropyle. A nucellar cap is evident and an elongated and vacuolated megaspore has formed (Figure 8.3). Environmental conditions at the stage between archesporia and embryo sac are critical, because a cold snap can produce malformations only inside the ovule and not the fruit (Fortescue and Turner, 2005c).

8.1.2.2 Anthesis (Opening of the Female Flowers)

In *Musa* spp. the ovule is anatropous, bitegmic, and crassinuclear. The external integument is thick, distinct from the internal integument, and the funicle is partly united to the external integument.

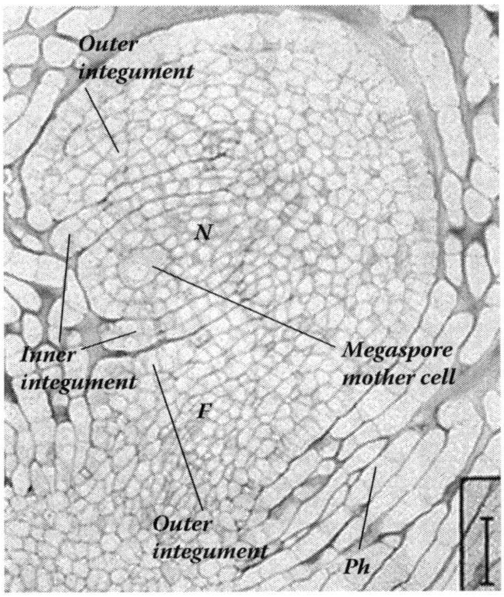

FIGURE 8.2 *Musa* spp. Cavendish subgroup, (AAA), sterile edible cultivar, also in Figures 8.5, 8.6, and 8.7. Bar = 50 μm. The ovule is now anatropous, the outer and inner integuments have formed but not the nucellar cap. The megaspore mother cell has formed with a central vacuole. F, funiculus; N, nucellus; Ph, placental hairs. (Reprinted from *Sci. Hort.* 104, J.A. Fortescue and D.W. Turner, The anatomy of ovule development of banana, plantain and ensat (Musaceae), 479–492, Copyright 2005, with permission from Elsevier.)

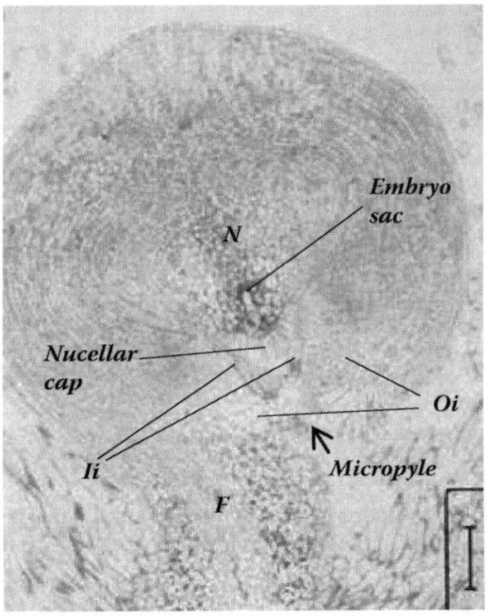

FIGURE 8.3 *Musa acuminata* ssp. (AA), bar = 100 μm. All the components of the ovule have now differentiated. F, funiculus; Ii, inner integument; N, nucellus; Oi, outer integument. (Reprinted from *Sci. Hort.* 104, J.A. Fortescue and D.W. Turner, The anatomy of ovule development of banana, plantain and ensat (Musaceae), 479–492, Copyright 2005, with permission from Elsevier.)

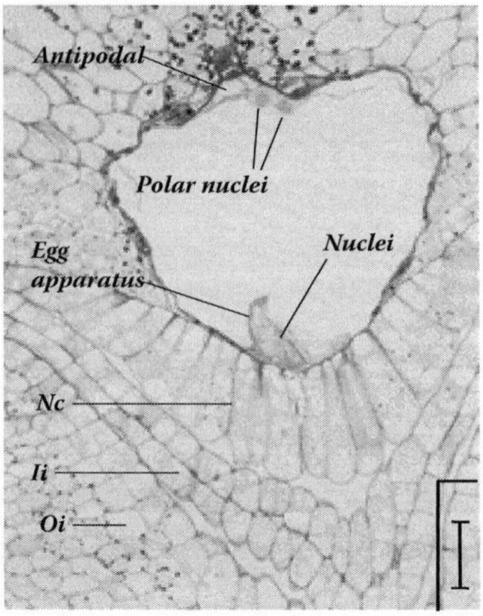

FIGURE 8.4 *Musa acuminata* ssp. (AA), bar = 25 µm. The diploid embryo sac at anthesis. It is very large, round, and lies flush against the nucellar cap. All the constituent embryo sac components are present. Two polar nuclei are visible over the antipodal pit where one antipodal is in view. A nucleus is visible in the egg apparatus. Ii, inner integument; Nc, nucellar cap, Oi, outer integument. (Reprinted from *Sci. Hort.* 104, J.A. Fortescue and D.W. Turner, The anatomy of ovule development of banana, plantain and ensat (Musaceae), 479–492, Copyright 2005, with permission from Elsevier.)

The chalazal mass is a transitional zone between the integuments, the funicle, and the base of the nucellus, which is massive. The nucellus consists of two parts; on the outside there are parenchymatous cells with very small nuclei, and towards the center there are large cells with voluminous and very colored nuclei. The micropyle is formed by both integuments and points in all directions towards the placental wall. At anthesis, the ovule of seeded *Musa* ssp. is 0.6 mm wide and 0.7 mm long. The embryo sac is large and located next to the nucellar cap; it is eight nucleate and has a characteristic bell shape (Figure 8.4). Its broad base is embedded in the nucellus and it has two large unfused polar nuclei, usually immediately over the antipodal pit in which lie the three antipodals. Its tapered end is flush with the columnar cells of the micropylar cap and contains the egg apparatus and two large and active synergids (White, 1928; Dodds, 1945; Maheshwari, 1950; Bouharmont, 1963; Fortescue and Turner, 2005a, 2005b). Figure 8.4 shows a median cross-section of an embryo sac of *Musa* sp. at anthesis; in view are two antipodals, two polar nuclei, and the egg apparatus with a nucleus.

The primary vascular bundles are found in the placental wall and outer integuments, in which they appear to be in a functioning state at anthesis and probably during the maturation of the seed. The chalazal mass consists of two doughnut-shaped discs positioned one above the other. Before anthesis, the chalaza is not very apparent but as anthesis approaches it grows larger and becomes very obvious after fertilization. In the unfertilized ovule, the chalaza remains thin and lightly stained.

The developmental progression of the embryo sac of the edible triploids is very similar to that of the seeded diploids; examples can be found in all stages of development from tetrad to embryo sac. At anthesis the majority of the embryo sacs are not found flush with the nucellar cap and their contents appear in disarray, particularly in the AAA genotypes but also in the ABB genotypes. They are often oval shaped and vacuolated. However, many embryo sacs still showed internal order with one or more nuclei present (Figure 8.5) (Fortescue and Turner, 2005a, 2005b).

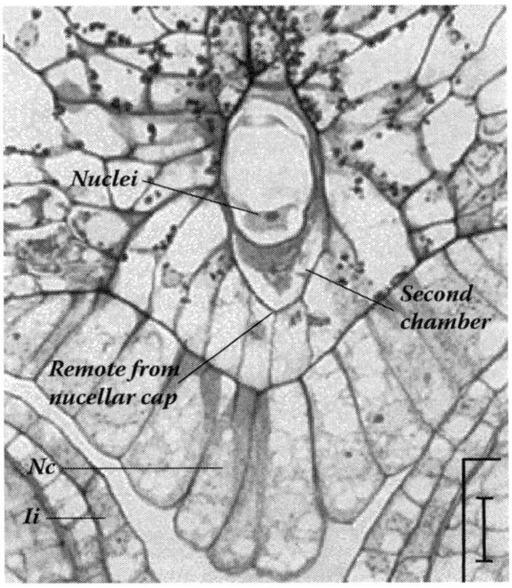

FIGURE 8.5 *Musa* spp. Cavendish subgroup, (AAA), bar = 25 μm. At anthesis the triploid embryo sac is present, usually with its contents in disarray, but with some order and one or more nuclei present. Note that this sac is not flush with the nucellar cap. Ii, inner integument; Nc, nucellar cap.

8.1.2.3 Post-Anthesis

A few days after anthesis, the nucellus, lying subadjacent to the integument and extending from the walls of the embryo sac outward and downward to the chalaza, begins to break down to form an inverted funnel-shaped cavity in the center of the main mass of the nucellus. This autolytic process is independent of the development of the embryo sac. Without the division of the endospermic nucleus to fill the cavity, the unfertilized ovule shrinks, collapsing on itself. After fertilization the internal part of the nucellus entirely disintegrates; after 25 days none of the inner cells remain and a large internal cavity has formed. The thickness of the larger parenchymatous cells does not decrease. By 55 days after fertilization, these parenchymatous cells are almost empty, becoming the perisperm, which later disintegrates when the endosperm becomes cellular (Bouharmont, 1963; Fortescue and Turner, 2005a, 2005b).

In fertilized ovules the development of the endosperm and zygote begins, and the endospermic nucleus commences to divide almost immediately. Three days after fertilization, the endosperm is 20 to 40 nucleate. Divisions proceed rapidly, resulting in a convoluted vesicle of free nuclei devoid of walls, situated immediately over the antipodal pit. The divisions are not synchronous, forming isolated vesicles and nuclei that wander the periphery of the embryo sac. The nucellar autolysis begins at the outer basal regions of the embryo sac and endosperm nuclei pass into the resultant cavity. After 22 days, the endosperm becomes cellular. The cells are large, separate from one another, uninucleate, and do not divide further. Later they increase in volume a little and crush the perisperm against the integuments. After 32 to 40 days, the endosperm thickly carpets the embryo sac.

After fertilization the zygote possesses one or two nucleoli and the synergids disappear. It does not begin to divide until after endosperm formation is well underway. It develops slowly, the multicellular bi- or quadra-cellular proembryo can be observed after 13 days. After 25 days, the proembryo is a small multicellular, undifferentiated, and elongated mass measuring 65 by 35 μm. After 40 days it measures 110 by 40 μm and has partially emerged from the hollow of the micropylar cap. After 55 days the proembryo measures 120 to 250 by 80 to 170 μm. After 62 days, what is now the embryo is fixed to the nucellar cap by the suspensor and nearly fills the space inside what is now the micropylar collar. The embryo continues to grow until it fills the micropylar cavity as a short cylindrical body and finally pushes into the main nucellar cavity where it encounters the dense nutritive

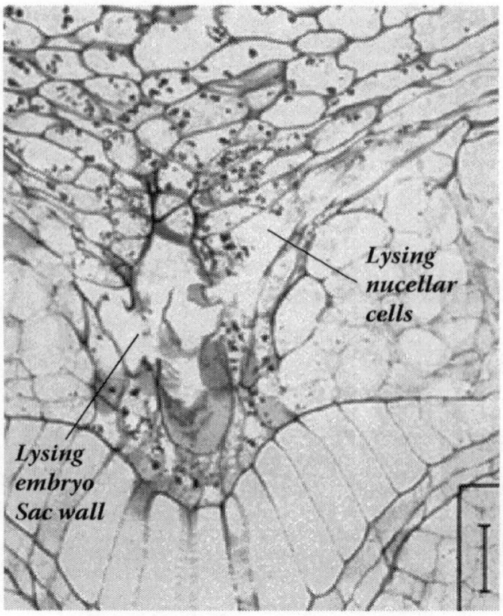

FIGURE 8.6 *Musa* spp. Cavendish subgroup, (AAA), bar = 25 μm. The post-anthesis embryo sac and surrounding nucellus tissues begin to lyse. The nucellus cells at the posterior end and to the lateral sides of the embryo sac have begun to lyse, the cells have become compressed, and their walls are disrupted. The embryo sac also has begun to lyse, the walls have ruptured, splitting the sac and spilling the contents into the surrounding nucellus tissue. (Reprinted from *Sci. Hort.* 104, J.A. Fortescue and D.W. Turner, The anatomy of ovule development of banana, plantain and ensat (Musaceae), 479–492, Copyright 2005, with permission from Elsevier.)

endosperm. The protruding portion flattens against this to form the fungiform digestive pad characteristic of the embryos of *Musa*. After 68 days the meristems of the shoot and root appear. The mature embryo is very small and does not contain a well-defined suspensor. The fertilized ovule is enormous, and the outer integuments enlarge to close the micropyle and form the operculum. The nucellar cap cells alter their arrangement to form a large cradle in which the proembyron rests (Fortescue and Turner, 2005c).

In the edible triploids the embryo sac begins to show deterioration soon after anthesis; the cells surrounding and in contact with the sac from the posterior end lyse along with the embryo sac walls and their contents fill the space the sac had occupied (Figure 8.6). The lysing cells continue to spread laterally to create an internal chamber (Figure 8.7) very similar to the conditions seen in the ovules of seeded diploid bananas immediately after fertilization (Figure 8.8).

8.1.2.4 Ovule Growth in Cross Section

At anthesis the ovules of *Musa* ssp. measure 0.49±0.03 mm^2 in cross-sectional area. At 47 days after fertilization, the size increases eightfold to 3.8±0.5 mm^2. Unfertilized ovules remained a similar size of 0.48±0.06 mm^2 after 47 days. At anthesis the triploid ovules are approximately twice as large, in cross-sectional area, as the diploid ovules, averaging 0.82±0.08 mm^2. Within *Musa*, ovule size increases with ploidy, the ovules of triploids are larger than diploids and those of tetraploids are larger again. The *acuminata*/*balbisiana* content does not appear to significantly affect ovule size or shape. In *Ensete* sp., the ovules at anthesis are up to 11 times larger than those of *Musa*, measuring up to 5.0 mm^2 (Fortescue and Turner, 2005b). This difference is reflected later in the size of mature seeds.

8.1.2.5 Embryology of *Musa*

The megasporogenesis and gametogenesis of *Musa* is typical of angiosperms with a few minor exceptions. The megasporocyte undergoes the usual meiotic divisions to form a tetrad. The tetrads are usually linear, but isobilateral T and inverted T shapes can occur in the same species. The

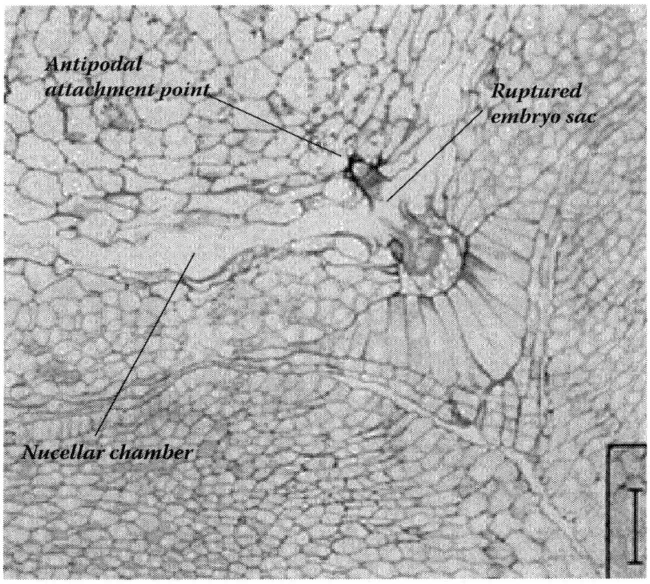

FIGURE 8.7 *Musa* spp. Cavendish subgroup, (AAA), bar = 50 μm. The lysis has spread laterally to create an internal chamber, very similar to the early postfertilized circumstances found in the fertilized diploid. (Reprinted from *Sci. Hort.* 104, J.A. Fortescue and D.W. Turner, The anatomy of ovule development of banana, plantain and ensat (Musaceae), 479–492, Copyright 2005, with permission from Elsevier.)

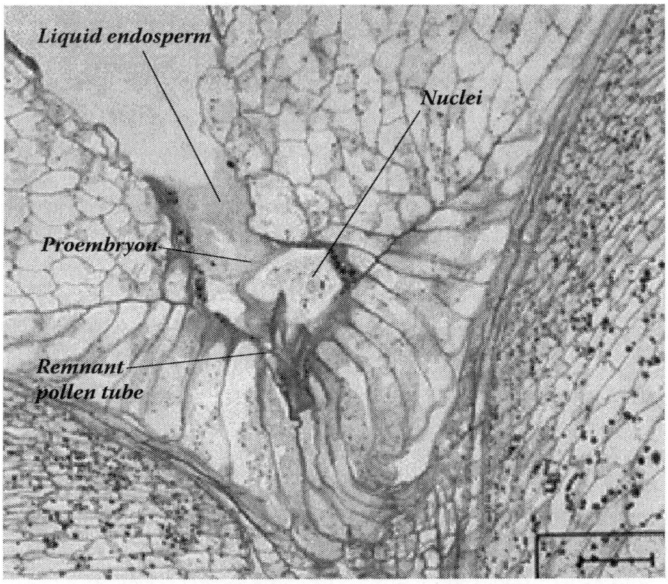

FIGURE 8.8 *Musa acuminata* ssp. (AA), bar = 25 μm. The fertilized ovule 20 days post-anthesis. The fertilized ovule is large, the nucellus has completely lysed, creating a large central cavity, the micropyle is closed, and the nucellar cap has formed a large cradle in which sits the proembryo. The endosperm is beginning to form but is not yet cellular. A nucleus is present inside the proembryo and the remnants of the pollen tube can be seen—it has pushed apart the nucellar cap cells and pierced the embryo sac with an apparent forked tip. (Reprinted from *Sci. Hort.* 104, J.A. Fortescue and D.W. Turner, The anatomy of ovule development of banana, plantain and ensat (Musaceae), 479–492, Copyright 2005, with permission from Elsevier.)

embryo sac is large, with a characteristic bell shape and located next to the columnar cells of the micropylar cap. It is eight nucleate, containing the egg apparatus with two large and active synergids, two large unfused polar nuclei, and three antipodals. The antipodals are ephemeral but sometimes persistent (White, 1928; Dodds, 1945; Maheshwari, 1950; Bouharmont, 1963; Fortescue and Turner, 2005a, 2005b). The polar nuclei fuse at the time of fertilization. The development of the endosperm is the nuclear type. In *Musa*, not only is nuclear division nonsynchronous (which in itself is not unusual) but some nuclei divide more actively, forming isolated vesicles or nodules that develop into separate endosperm masses. Usually one large vesicle is formed over the antipodal pit and lesser free nuclei are dispersed about the sac. Later the endosperm becomes cellular beginning at the micropylar region. The embryology of *Musa* has not been studied in sufficient detail. Bouharmont (1963) describes the embryology as of the Asterad type. According to Johri et al. (1992), Asterad is the common type of embryology in the order Zingiberales. It is reasonable to assume, from lack of evidence to the contrary, that the embryology of *Musa* is also the Asterad type. The mature embryo is capitate, more or less basal, relatively small, and restricted to the lower part of the seed, distally expanded and in copious endosperm. The seed is medium to large.

8.1.3 Development of Male Organs and Gametes

In angiosperms the stamen differentiates into an anther and filament. The anther usually develops into two lobes with two pollen sacs with four groups of archesporial tissue. The archesporial tissue differentiates into a mass of pollen mother cells surrounded by various wall layers. The pollen mother cells become surrounded by callose, undergo meiosis, and produce four haploid cells. These haploid cells subsequently develop into pollen grains surrounded by the pollen grain wall. At maturity the size, shape, surface pattern, and stickiness of the pollen grains are highly characteristic of species (Sedgley and Griffin, 1989). Bananas do not differ from the normal angiosperm example, except in gross morphological appearance. Five of the six stamens are functional and the anthers are elongate to linear, basifixed, and tetrasporangiate. The pollen grains are two locular and nonaperturate. The anthers of Musaceae have a glandular-secretory tapetum and the pollen grains have a thin or virtually nonexistent exine but a thick intine (Dahlgren et al., 1985; Fortescue and Turner, 2004). The tapetum synthesizes substances that contain fibrogranular proteins and carotenoid lipid droplets that help form the exine (Johri et al., 1992). As in most angiosperms, the anthers dehisce by two longitudinal slits; each microsporangium opens by a common slit with the other microsporangium.

8.1.3.1 Microsporogenesis

Microsporogenesis and male gametogenesis in Musaceae are not well studied and there is insufficient data from each of the families of Zingiberales to draw close comparisons. However, from the literature it does appear to be a uniform order (Dahlgren et al., 1985). The following account relates to the anther development in *Musella lasiocarpa* (Musaceae) by Xue et al. (2005). The anthers are tetrasproangiate. The formation of the anther wall is of the basic type. The mature anther wall consists of an epidermis, an endothecium, many middle layers, and a two-layered glandular tapetum with uninucleate cells. The old anther wall consists of an epidermis with annular and helical thickening and reduced endothecium. Successive cytokinesis follows meiosis of the microspore mother cells, forming a T-shaped or isobilateral tetrad of microspores. Pollen grains are two celled. The generative cell nucleus is clavate in shape (Xue et al., 2005). The archesporial cells are recognized by their dense cytoplasm and conspicuous nuclei. A row of sporogenous cells produced by the archesporia gives rise to a mass of microspore mother cells by several mitotic divisions. While changes take place in the anther walls, the primary sporogenous cells undergo mitosis forming secondary sporogenous cells from which the microsporocytes are derived. Microspores are separated from the tetrad as uninucleate free microspores. Each microspore has a dense cytoplasm with a prominent and centrally placed nucleus that moves to a peripheral position as the vacuole develops. The first mitotic division of the microspore results in the formation of two unequal cells, the large vegetative and smaller generative cell. The pollen grains are two celled and nonaperturate at the time of anther dehiscence. The callose surrounding the microsporocytes is thin, and the pollen grains have a thin or virtually nonexistent exine and a thick intine (Xue et al., 2005).

The tapetal cells are uninucleate throughout their development. At the time of microsporocyte meiosis, the walls of the tapetal cells become indistinct and the tapetal cells begin to degenerate. The endothecium reduces and does not develop fibrous thickenings as in most angiosperms. At the mature pollen grain stage, the tapetal cells have degenerated completely. During maturation the epidermal cells enlarge and thicken. Thus the mature anther wall is composed of the fibrous thickened epidermis and the reduced endothecium (Xue et al., 2005).

8.1.3.2 Pollination

In terms of pollination, no fewer than about 4,000 pollen grains are needed to cover the stigmatic surface of a female flower of a *Musa* diploid (Dodds, 1945). This is approximately 20 to 40 times the number of ovules in an ovary. About 12 hr after pollination, the pollen tubes transverse the entire length of the style. The style is approximately 30 mm long, so growth rate is 0.33 mm/hr. There appears to be no inhibition of the pollen tube growth and more tubes enter the ovary than there are embryo sacs awaiting fertilization. Pollen tubes enter the ovule only through the micropyle. The styles abscise about 30 hr after the maturation of their receptive surfaces. Ovules must be fertilized within 24 hrs of flower opening, after which they begin to disintegrate.

8.1.4 Floral Factors and Breeding Systems

8.1.4.1 Outbreeding or Inbreeding?

Bananas are monoecious: The female and male reproductive organs are separated into different floral structures on the same inflorescence. This is the dominant condition among gymnosperms, also frequent in angiosperm monocotyledons and some dicotyledonous timber and crops species (Sedgley and Griffin, 1989). If the monoecious condition is complete, then self-pollination of an individual flower is impossible. In the monoecious condition the flowers are functionally unisexual but still structurally hermaphrodite. Both carpel and stamen primordia are initiated, and female and male flowers are identical in early stages. Further development results in female flowers with functional carpel and vestigial sterile staminodes and male flowers with functional stamens but small and underdeveloped carpels. Most monoecious plants produce considerably more male than female flowers; the proportions range from 0.1% female flowers in *Aesculus* sp. to 60% in lychee, oil palm, and mangoes (Sedgley and Griffin, 1989). In *Musa* cv 'Gros Michel' (AAA), 7% of the flowers within an inflorescence are female (White, 1928).

Bananas are also dichogamous: There is temporal as well as spatial separation of the sexes within an inflorescence, and the female organs mature before the male organs. Dichogamy is particularly common in monoecious plants, especially the gymnosperms. In wild bananas the first bunch thrown by a new plant would be out-pollinated. But bananas are clump forming; therefore, later bunches thrown by suckers arising from the parent rhizome could be pollinated by male flowers from the preceding bunches. Therefore selfing cannot be uncommon. Since bananas tend to be gregarious plants of transient habitats, sib pollination must be frequent. According to Simmonds (1962) out-pollinations do occur as evidenced by the occasional occurrence of interspecific hybrids in batches of open pollinated seeds.

Female flowers produce considerable quantities of nectar, and male flowers tend to produce less or none at all. The female flowers of *M. salaccensis*, which has a vertical inflorescence, produce during the day 17–33 μL/h of nectar that contains 18–19% sugar (Itino et al., 1991). Male flowers produce less (17–19 μL/h) with similar sugar concentrations (19–20%). Male flowers of *M. acuminata* ssp. *halabanensis* that occur on the pendant section of the inflorescence produce 8 μL/h of nectar during the night. This nocturnal nectar contains 23–25% sugar (Itino et al., 1991). Similar quantities of nectar are produced by the flowers of *M. paradisiaca* ssp. *sapientum* Kuntze (possibly AAA, cv 'Pisang Masak Hijau' or 'Red-Green Red,' or AAB, 'Silk Fig') but the male flowers produce about four times more (35 μL/h) nectar than the female flowers (8–9 μL/h) (Fahn and Benouaiche, 1979). In contrast, the male flowers of *Musa* spp. AAA Cavendish subgroup 'Dwarf Cavendish' do not produce any nectar because the epithelial nectar cells disintegrate before the nectary matures (Fahn and Benouaiche, 1979). In *M. itinerans*, female flowers produce 6 μL/h and male flowers a little less at 4 μL/h, but the sugar concentration is similar in each flower type (21–22%) (Liu, Li, et al., 2002). The female flowers of *Musella lasiocarpa* produce nectar at a rate of 7 μl/h and the male flowers 2 μl/hr, with a sucrose content of 15% (Liu, Kress, et al., 2002). Among species and cultivars there is considerable variation in the rate of production of nectar from female flowers. The reasons for these differences among *Musa* spp. have yet to be determined.

Within the Musaceae, some species have a proportion of functionally hermaphrodite flowers and presumably would be frequently, if not regularly, self-pollinated. *Musa acuminata* ssp. *banksii*, *M. ingens* and *M. schizocarpa* are examples (Simmonds, 1962). In addition, Nur (1976) excluded pollinators from inflorescences of *M. velutina*, but flowers set seed because of self-pollination. The species with hermaphrodite flowers occur at the outer margins of the geographical distribution and represent a mechanism that favors selfing by isolated plants at the limit of their range (Simmonds, 1962). Bananas are intermediate between the two extremes of inbreeding and outbreeding. If they were highly outbred, there should be significant loss of vigor on selfing, but this is not so. Bananas are moderately outbred and can tolerate an occasional generation of close inbreeding without significant harm. This strategy is consistent with their status as "jungle weeds" that provide a useful understory when forest vegetation is damaged by extreme weather events.

8.1.4.2 Reproductive Abnormalities

Abnormalities of the physiological features of the reproductive process can influence the breeding systems of plants. They cause a breakdown in the sequence of flower and fruit development and cause marginal or substantial effects on the breeding system. Significant abnormalities can affect the genetic composition of the subsequent generation. Parthenocarpy is defined as the formation of fruit without fertilization; the phenomenon reduces the unreliability of fruit set and results in fruit without seeds that consumers may prefer. There are two types: stimulative parthenocarpy, which requires pollination before the fruit develops, and vegetative parthenocarpy, which requires no external stimulus. *Musa* has vegetative parthenocarpy (Dodds and Simmonds, 1948; Ortiz and Vuylsteke, 1995a). Parthenocarpy is sometimes inaccurately used to describe fruit without mature seeds.

Ploidy affects the breeding system. It often results in reduced fertility from faulty chromosome pairing at meiosis. However, polyploids, especially triploids, are very vigorous and produce few seeds. Triploids are uniformly highly sterile and have an expectation of only 0.05% gametic fertility. Triploids can still be slightly fertile through chromosome restitution. Tetraploids are moderately pollen fertile and make effective male parents. In the tetraploid breeding system, the tetraploids must be very highly female sterile to minimize the risk of seediness in fruit for consumption. Male sterility is not confined to triploid bananas, as Dumpe and Ortiz (1996) showed that half of their diploid accessions were male sterile. Structural heterozygosity in the diploids is responsible for moderate to nearly complete male sterility. Chromosome pairing is generally low and variable with genotype and environment. The diploids 'Sucrier' (AA), 'Bande' (AA), and 'Palembang' (AA) are nearly completely male sterile, while 'Pisang Lilin' (AA, allied to *M. acuminata* ssp. *malaccensis*) is only 50% sterile.

Female sterility is less often reported as being selected against in cultivation. The implications of female sterility are more important than that of male sterility since a totally female sterile cultivar will not grow fruit unless it is parthenocarpic. However, flowers can still contribute to the breeding system and overall fertility by augmenting the floral display and pollen production. Bananas are probably the most conspicuously sterile of all cultivated fruits. Female meiosis of bananas shows a similar behavior range that characterizes male meiosis, and there is no reason for the sexes to differ in the mechanisms of meiosis. Irregularities notwithstanding, parthenocarpic bananas can produce a few morphologically normal embryo sacs and fertilized embryo sacs with young endosperm when suitably pollinated.

8.1.4.3 Female Sterility

Errors in development of seed can occur in the sporophyte, the gametophyte, and female × male interactions. Sporophytic mutations cause defects in a discrete aspect of ovule development such as initiation of integuments or development of the nucellus. They can also cause a defect in embryo sac development by disrupting meiosis. Gametic mutations are determined by the haploid genotype of the embryo sac and usually affect the number of mitotic divisions, that is, multiple functional egg cells and embryos within a sac. Gametic mutations are not well understood or documented. The gametophyte can contain up to five micropylar cells as either eggs or synergids, and seeds have been obtained with twin, triplet, and quadruplet embryos formed by the fertilization of multiple egg cells (Haig, 1990; Reiser and Fischer, 1993). Unlike pollen, the sporophyte surrounds the gametophyte and maintains physical contact throughout its development. In *Musa* none of the sporophyte problems such as the *Arabidopsis sin1*, *ovm2*, *ovm3*, and *bel1* mutations have been reported. The literature does not mention any defects in ovule development. The published works, particularly of Dodds (1945) and Dodds and Simmonds (1948), focused on problems of female × male interactions. Dodds (1945), Dodds and Simmonds (1948), Simmonds (1962), and Shepherd (1999) have reported detailed studies of the cytogenetics of wild and cultivated bananas.

Female infertility can be divided into two basic causes: those that inhibit the development of the embryo sac and those that prevent its fertilization. In the first case, the two well-defined causes of sterility in *Musa* spp. are multiple archesporia, in frequencies of 0–45%, and tetrahedral shaped tetrads (Dodds, 1945; Dodds and Simmonds, 1948; Simmonds, 1962). In the second case, the sporophytic cause of infertility is failure of the pollen tube to reach the ovule. This is apparently common despite a great excess of pollen supplied to the stigma and the number of tubes in the style far exceeding that necessary to fertilize all the ovules (Dodds, 1945; Shepherd, 1954, 1960b, 1999). In female cytology, besides structural hybridity and deficiencies in chromosome homology, there are other causes of sterility, such as abnormal spore development from spindle irregularities and deficient wall formation, sometimes leading to diploid or tetraploid spores from genetic restitution. Failures in embryo sac development resulted in morphological errors and errors of polarity and failure of fertilization and undefined derangement of postfertilization events (Dodds, 1945; Dodds and Simmonds, 1948; Simmonds, 1962; Shepherd, 1999). Embryo sac failure accounts for most sterility in edible banana clones. By far the greatest cause is gametogenic, ill-defined irregularities of embryo sac formation caused by mismatching of chromosomes. Simmonds (1962) considered that all bananas, whether diploid or triploid, suffer one or more of these defects, which may affect as many as 50% of ovules. Early irregularities notwithstanding, edible diploids can produce a few percent of morphologically normal embryo sacs whether genetically balanced or not (Fortescue and Turner, 2005b).

8.1.4.4 Female–Male Interaction

The first step in the female–male interaction is pollination, the transfer of pollen from male reproductive structures to receptive stigmas. In the wild-seeded diploid bananas, pollination is essential for fruit development. The result is mature fruits that contain a mass of hard black or brown seeds surrounded by a scanty sweetish pulp. The parthenocarpic edible bananas develop as mass of pulp without pollination and without seeds, and pollination has no detectable effect on the development of the edible banana fruit (Simmonds, 1966).

8.1.4.5 Pollen–Pistil Interaction

The second step in the female–male interaction, after the arrival of pollen, is the extracellular secretions of the stigma and pollen grains. The angiosperm stigma is covered by extracellular secretions that may contain carbohydrates, proteins, enzymes, phenolics, and amino acids. The primary recognition of species occurs at the stigma and its secretions. The pollen tube's journey from stigma to embryo sac is long. It grows through the stigma secretions and enters the stigmatic tissue between the papilla cells. Its route through the stigma, style, and ovary is entirely confined to the extracellular secretions produced by the cells of the transmitting tract (Sedgley and Griffin, 1989). In angiosperms the embryo sac is generally mature and receptive at the time of anthesis and the pollen tube reaches the ovary and fertilizes the egg cell within days of pollination—a relatively short proportion of the overall time span of the reproductive process. The male gametes are transferred from the male to female organs via the pollen tubes, a highly complex process consisting of a number of sequential steps. At all stages of the process, there is potential for acceptance or rejection of incompatible pollen.

8.1.4.6 Self-Incompatibility

Many plants have prezygotic self-incompatibility (SI), in which the growth of self-pollen tubes is inhibited and fertilization prevented. Most prezygotic SI is a gametophytic mechanism with pollen tube growth inhibited in the style. The tubes generally cease growth in the upper portion of the style and have a characteristic appearance of swollen tips, terminal deposition of callous, and often discharge their contents into the intercellular matrix. Thus self-pollen tubes show clear signs of inhibited growth thought to be controlled by the stylar tissue. SI is an out-crossing mechanism that reduces inbreeding and promotes heterozygosity in natural populations. A disadvantage is that it reduces the amount of possible crosses available to the plant breeder.

Interspecific and intergeneric hybridizations are important in many plant breeding programs where a combination of characteristics from different species is required. The ease of these hybridizations varies greatly between species and genera. Research has concentrated on developing methods to overcome barriers. Methods of overcoming SI involve exploration of the physiology of the pistil and the utilization of the recognition mechanism between pollen and pistil. Some techniques employed for overcoming SI at the breeder's disposal include bud pollination, temperature, and hormone manipulation. The concentrations of glycoproteins in the style that cause incompatibility increase immediately before anthesis. Immature flowers and old flowers produced near the end of the flowering season have weaker SI control. This may be less relevant in *Musa* spp. where flowering can occur at any time of the year. Temperature can affect SI. Thus, in almond and cherry the optimum for selfing is lower than that for crossing, 15°C compared with 25°C. In contrast, high temperatures in apple, 32–60°C, results in seed set following self-pollination presumably due to denaturing of the glycoprotein. Similar effects are achieved by γ radiation. Applications of auxins, gibberellins, boric acid, or succinic acid to the base of the pistil before pollination affect SI. The female–male interaction is still a difficult area of research; the major events occur within the female structures, making it difficult to locate the organs of interest. In the literature there has been much work conducted on pollen germination and tube growth, though little specifically on bananas. Most effort has been placed on the pollen–pistil interactions.

In *Musa* there are two events that could be interpreted as SI. In the seeded diploid, over half of normal ovules are unfertilized, despite a great excess of pollen applied to the stigma and the tube number at the base of the style far in excess of that necessary to fertilize all the ovules. There are frequent failures of fertilization despite normal pollen tube growth. In 'Gros Michel' (AAA), erratic pollen tube growth contributed to sterility by delayed or slow but otherwise normal growth of the tubes or abnormal growth manifested by arrest and swelling of pollen tube tip (Dodds, 1945; Shepherd, 1954, 1960a, 1960b). Secondly, in relation to apical bias in fertilization within the fruit, Shepherd showed that the more fertile the banana fruit and the earlier the pollination, the less the bias.

8.2 POLLEN PRODUCTION, VIABILITY, AND GERMINATION

For use as a male parent in conventional *Musa* breeding, the genotype must produce ample pollen that is viable, can germinate, and can fertilize the female gametophyte. A deficiency in any of these requirements precludes selection, even if the plant possesses the characteristics required in the progeny, such as resistance to disease. The amount of pollen can be assessed by counting and expressing the results on a per anther basis (Krishnamoorthy and Kumar, 2005a, 2005b) or by ranking the amount of pollen using an hedonic scale (0 = none, 6 = extremely high; Ssebuliba et al., 2008). Pollen viability is often assessed using vital stains; the most common methods are the use of nuclear and vital dyes such as acetocarmine glycerol jelly (Marks, 1954) and Alexander's procedure (Alexander, 1969). However, different stains do not always give the same answer either within or across species (Rodriguez-Riano and Dafni, 2000) and so results are indicative, depending on the stains used. In *Musa* there is not necessarily a significant correlation between viability of pollen and its capacity to produce seed ($r = 0.25$, $P = 0.36$; Ortiz et al., 1998). Germination, expressed as percent, is usually evaluated *in vitro* and may equal or more usually be less than the percent viability indicated by vital stains (Rodriguez-Riano and Dafni, 2000). This may mean that the stains are not indicative of the capacity of the pollen to germinate, or it may be because the germination is expressed on the basis of the whole population rather than the proportion of viable pollen.

8.2.1 Pollen Production

Pollen production is influenced by genetic and environmental factors. In *Musa*, male meiosis is essentially regular, and sterility in certain cultivars is caused by chromosome misbehavior at meiosis. Structural complexities met at meiosis may be overcome by genetic restitution in mitosis. In essentially sterile triploids with very low levels of structural hybridity, genetic restitution could account for pollen present at anthesis.

Musa pollen grains are almost visible to the naked eye and are rather sticky. The amount can vary from 200 to more than 40,000 pollen grains per anther (Sathiamoorthy, 1994; Krishnamoorthy and Kumar, 2005a). Sathiamoorthy (1994) presents data on pollen counts for a range of diploid, triploid, and tetraploid banana genotypes grown in India. All genotypes produced pollen, at least 2,500 per anther, even cv 'Matti,' a tall edible diploid (AA) that is male sterile. Different cultivars contain more viable pollen than others, either within or between genomes or groups. Between ploidy groups, the diploids contained three times more pollen than the tetraploids and 11 times more than the triploids. Krishnamoorthy and Kumar (2005b) measured the pollen production of 18 tetraploid (AABB) banana hybrids in India. The number of pollen grains per anther ranged from 400 to 12,000. Most triploid accessions do not produce enough pollen for use in breeding programs. Low pollen production in triploids is mostly attributed to meiotic abnormalities, which may be under genetic control (A or B) and influenced by environment (Simmonds, 1966; Ortiz, 1995; Dumpe and Ortiz, 1996; Ortiz et al., 1998).

To account for the variation in seed set observed during conventional cross-breeding of *Musa*, many researchers have assessed the viability of pollen used in pollination. A reliable method of detecting pollen viability and germination allows not only for increased efficiency in breeding but raises the possibility of *in vitro* pollination of ovules, especially since banana breeding is hampered by pre- and postzygotic barriers of incompatibility. Here, from the published literature, we summarize data on the pollen viability of species, clones, and cultivars of *Musa* grown in Australia (Fortescue and Turner, 2004), India (Sathiamoorthy and Madhava Rao, 1980; Sathiamoorthy, 1994), tropical America and the Caribbean (Landa et al., 1999; Perea Dallos, 1998; Roman et al., 2004; Soares et al., 2008), and Africa (Dumpe and Ortiz, 1996; Ortiz et al., 1998; Adeleke et al., 2004; Nyine and Pillay, 2007; Ssebuliba et al., 2008). The data on pollen viability were sorted into ploidy and genomic groups (Table 8.1).

Diploids produce pollen with high viability (80–90%), compared with triploids (40–49%), and these differences are significant at $P = 0.05$ level of probability (Table 8.1). The pollen viability of tetraploids was variable but the numbers available were too small to make a comparison with other ploidy levels. Within the diploids and triploids, there was no significant effect of the proportion of A or B genome on pollen viability. A feature of pollen viability is the large range within each

TABLE 8.1
Pollen Viability in *Musa* spp. in Relation to Ploidy and Genomic Constitution

	Genome		Viability, %		
Ploidy		***N***	**Mean**	**SE**	**Range**
Diploid	AA	81	81a	2	28–100
	BB	12	90a	5	44–100
Diploid total		93	83A	2	
Triploids	AAA	39	44b	5	3–92
	AAB	15	37b	9	0–91
	ABB	23	49b	6	5–90
Triploid total		77	44B	3	
Tetraploids	AAAA	1	31		
	AAAB	4	56	16	28–94
	AABB	1	14		
	ABBB	2	33	4	30–37
Tetraploid total		8	42B	10	

Note: Mean values of genomes followed by the same lowercase letter are not significantly different by t-test at $P = 0.05$. Ploidy means followed by the same uppercase letter are not significantly different ($P = 0.05$). SE = standard error. The tetraploids had too few data and were excluded from the t-test to compare genomes within ploidy levels.

Source: Data from Sathiamoorthy, S. and V.N. Madhava Rao, 1980, Pollen production in relation to genome and ploidy in banana clones, In: *National seminar on banana production technology*, C.R. Muthukrishnan, and J.B.M.Md. Abdul Khader, eds., 46–49, Chennai, India, Tamil Nadu Agricultural University; Sathiamoorthy, S., 1994, *Musa* improvement in India, In: *The improvement and testing of Musa: A global partnership*, D.R. Jones, ed., 188–200, Montpellier, France, INIBAP; Dumpe, B.B. and R. Ortiz, 1996, Apparent male fertility in *Musa* germplasm, *HortScience*, 31, 1019–1022; Ortiz, R., F. Ulburghs, and J.U. Okoro, 1998, Seasonal variation of apparent male fertility and 2*n* pollen production in plantain and banana, *HortScience*, 33, 146–148; Landa, R., A. Rayas, T. Ramirez, J. Ventura, J. Albert, and O. Roca, 1999, Study of the pollen fertility in the INIVIT genetic improvement programme, *InfoMusa*, 8(1), 27; Perea Dallos, M., 1998, Pollen and anther culture in *Musa* spp., *Acta Hort.*, 490, 493–497; Adeleke, M.T.V., M. Pillay, and B.E. Okoli, 2004, Relationship between meiotic irregularities and fertility in diploid and triploid *Musa* L., *Cytologia*, 69, 387–393; Fortescue, J.A. and D.W. Turner, 2004, Pollen fertility in *Musa*: Viability in cultivars grown in Southern Australia, *Aust. J. Agric. Res.*, 55, 1085–1091; Roman, M.I., M. Alonso, X. Xiques, C. Gonzalez, and I. Sanchez, 2004, Estudio del numero cromosomico y la fertilidad del pollen especies y clones diploids de platano fruta (*Musa* ssp.), *Cultivos Tropicales*, 25, 71–73; Nyine, M. and M. Pillay, 2007, Banana nectar as a medium for testing pollen viability and germination in *Musa*, *Afr. J. Biotechnol.*, 6, 1175–1180; Soares, T.L., S.O. Silva, M.A.P.C. Costa, J.A. Santos-Serejo, A.S. Souza, L.S.M. Lino, et al., 2008, *In vitro* germination and viability of pollen grains of banana diploids, *Crop Breed. Appl. Biotechnol.*, 8, 111–118; Ssebuliba, R.N., A. Tenkouano, and M. Pillay, 2008, Male fertility and occurrence of 2*n* gametes in East African Highland bananas (*Musa* ssp.), *Euphytica*, 164, 53–62.

genomic group (Table 8.1), suggesting that some genotypes contain high pollen viability irrespective of their ploidy level. However, viability below 30% is more common among triploid genotypes than diploids.

This broad picture contains exceptions when applied at the local level. For example, in a range of genotypes grown across southern Australia, diploids had 88% viable pollen compared with 29% for tetraploids. Tetraploid cultivars contained three times more viable pollen than the triploids AAA (9%), ABB (10%), and four times more than AAB cultivars (6%) (Fortescue and Turner, 2004). Similar results were found previously with cultivars grown in India where the diploids had 50 to 66% viable pollen, triploids had 21–29%, and the tetraploids had 28% (Sathiamoorthy and Madhava Rao, 1980; Sathiamoorthy, 1994). Adeleke et al. (2004) found that the viability of pollen in homogenomic triploids (AAA) was 30–48%, while in heterogenomic triploids (AAB, ABB) it was much less and averaged 8–30%. *Musa balbisiana* (diploid, section *Eumusa*) and *M. ornata* (section *Rhodochlamys*) have more viable pollen than *M. acuminata* (diploid, section *Eumusa*). Within triploids the cv 'Gros Michel' has more viable pollen than its dwarf mutant cv 'Highgate' (AAA) but similar to the Cavendish subgroup (AAA). Shepherd (1960a, 1960b) found that among *M. acuminata*, ssp. *burmannica* had much more viable pollen than ssp. *malaccensis*.

Flower age, expressed as node position in relation to anthesis, can affect the viability of pollen. The day between pre-anthesis and anthesis can see a decrease of 7% per node in viable pollen (Fortescue and Turner, 2004). Shepherd (1960a) compared pollination in the afternoon of the day before flower opening with the morning of the day after flower opening. Late pollination always resulted in lower yields of seed. Ssebuliba et al. (2008) demonstrated a significant relationship between the node from which the pollen was obtained and variation in pollen viability. All nodes were at anthesis and all collection was done early in the morning. Pollen viability was highest in the mid nodes 30 to 39 of the male rachis, especially in the triploids and certain diploids. Therefore, choice of pollen source should consider the position of the node from which pollen grains are harvested as well as time of day and day of anthesis. Genotype, node number, and time of day do not account for all the variation seen by different authors, especially when the same cultivar gives different results in different locations. It is known that pollen viability is influenced by temperature and humidity (Ortiz et al., 1998). Variation in pollen viability caused by organ ontogeny, season, or genotype will influence the practical aspects of pollen storage and use in a breeding scheme.

8.2.2 Polyspory

Polyspory, in this context, is the production of pollen grains of different ploidy levels within the same anther. This is a feature of plants that have a hybrid origin and is linked to the production of univalents during pairing of chromosomes in meiosis (Penland, 1923; Griffiths et al., 2000; Adeleke et al., 2004). The different levels of ploidy in pollen are usually determined by measuring the size of the pollen grains: Larger grains have a higher ploidy level than smaller grains.

Sathiamoorthy (1994) found that in 8 diploid cultivars, 7 showed polyspory with 39±4% haploid pollen and 11±3% diploid. Among 17 triploid cultivars, 16 showed polyspory with 16±2% pollen being haploid, 15±3% being diploid, and 5±2% being tetraploid. Among 4 tetraploid cultivars, 3 showed polyspory with 7±2% pollen being haploid and 23±4% pollen being diploid. The two seeded species, *M. acuminata* and *M. balbisiana*, produced only haploid pollen (66±4%). In contrast, Shepherd (1999) compared pollen samples from 11 plants from the one cross, of which four produced entirely haploid pollen with 95% viability. Only one uncommonly large grain was seen and no irregular ones except in two plants where the pollen was extremely variable in size representing haploid, diploid, and tetraploid pollen.

Gametes with the sporophytic chromosome number are known as $2n$ gametes. The production of $2n$ gametes is uncontrolled and fairly rare in diploid clones (Dodds, 1943). However, Sathiamoorthy (1994) found 11% of $2n$ gametes in seven diploid banana genotypes grown in India. Ortiz (1997) examined $2n$ pollen production in a range of *Musa* spp. and genotypes in Africa and found that those genotypes that produced $2n$ pollen were also parthenocarpic. The formation of $2n$ pollen has been described by Simmonds (1962), Ortiz (1995), and Shepherd (1999). Diploid spore formation results from either the failure of the first meiosis division followed by division restitution mechanisms or the failure of the second division following a successful first division. The usual mode of $2n$ spore production in bananas is due to second division restitution. Ortiz (1997) found that not all genotypes can produce $2n$ pollen and that at least one locus may be involved in the inheritance of $2n$ pollen production in *Musa*.

Pollen diameter is a function of genome size since both the nucleus and cytoplasm increase as chromosome number increases. Diploid or $2n$ pollen is approximately 25% larger than haploid (n) pollen, and the diameter of $3n$ and $4n$ pollen is considered to be 44 and 59% greater than that of haploid pollen, respectively (Ssebuliba et al., 2008). In *Musa*, haploid pollen diameter ranges between

TABLE 8.2
Proportion of Pollen with Different Levels of Ploidy from AA or AAA Cultivars

	Diploids, AA			East African Highland Bananas, AAA		
Pollen Ploidy	**Mean**	**SE**	**Range**	**Mean**	**SE**	**Range**
n	96.0	2.0	86–100	82.0	2.0	68–95
2*n*	2.9	1.5	0–9	14.0	1.8	4–24
3*n*	0.8	0.7	0–4	2.5	0.6	0–7

Note: SE = standard error.
Source: Data from Ssebuliba, R.N., A. Tenkouano, and M. Pillay, 2008, Male fertility and occurrence of 2*n* gametes in East African Highland bananas (*Musa* ssp.), *Euphytica*, 164, 53–62.

78 and 128 μm (Shepherd, 1999; Ortiz et al., 1998; Ssebuliba et al., 2008). In the diploid species *M. balbisiana*, the haploid pollen has a diameter of 94±2 μm compared with 105±1 μm for pollen of *M. acuminata*. There are also slight differences in mean pollen diameter within ploidy level between landraces and hybrids: ABB cooking bananas have a slightly smaller pollen than AAB landraces. Darlington (1937) indicated than pollen with a diameter of 129 μm or less was haploid while the average diameter of 2*n* pollen was 148 microns. In *Musa*, pollen grains with a diameter of 160 μm are generally considered to be 2*n* pollen grains.

Triploid genotypes produce less haploid pollen than diploid genotypes (Table 8.2). The range of pollen that is haploid is large and so there is variation among genotypes within ploidy levels. In East African Highland bananas, the proportions of haploid (*n*), diploid (2*n*), and triploid (3*n*) pollen grains differ greatly, the proportion of haploid pollen being significantly higher than that of diploid and triploid pollen (Ssebuliba et al., 2008). Haploid pollen grains make up 95% of the population in an anther; diploid pollen constitutes 3–5%, and triploid pollen from 0–1% of the population. This is consistent with the findings of Ortiz (1997) that the frequency of genotypes that can produce 2*n* pollen was 22% in diploid *M. acuminata*, 56% in AAA, 17% in AAB, and 14% in ABB triploids. The higher frequency of 2*n* pollen production in AAA triploids compared with diploids is consistent with the occurrence of sexual polyploidization in bananas (Ortiz, 1997).

8.2.3 Pollen Germination

There is much information in the broader plant literature on pollen germination and pollen tube growth; however, there is still little information available for banana. Pollen grains need to be alive when they reach the stigma (viable), be able to germinate and grow towards the ovule (functional competence), and be able to complete double fertilization to produce a healthy zygote. The vital stains used to indicate viability cannot determine which pollen is functionally competent. This is best observed by pollinating fertile female flowers. Krishnamoorthy and Kumar (2005b) measured pollen germination, *in vitro*, of 18 tetraploid (AABB) banana hybrids in India. Germination ranged from 4 to 17%. There was no relationship between pollen production and germination and so the amount of germinable pollen produced by a genotype could not be predicted from the amount produced. In diploids grown in Uganda, pollen germination was 84% (Nyine and Pillay, 2007).

The method of *in vitro* germination uses various combinations of sucrose and micronutrients; however, Nyine and Pillay (2007) suggested that better results could be obtained using diluted banana nectar harvested from male flowers from newly opened bracts as the main nutritive source. Pollen germination increased from 47% (range 17–50%) with sucrose to 84% (range of 60–97%) with nectar. Pollen germination was better for *M. acuminata* (83%) than *M. balbisiana* (44%). Soares et al. (2008) achieved germination rates of 16–20% with a range of 2–90% on regular pollen media. Banana pollen germinates readily, many within 1 hour from the start of incubation. The main factors that affect germination are the genotype, culture medium and incubation regime, sampling time, physiological development, and flower age. To a lesser extent, the nutritional state of the parent, temperature, humidity, and photoperiod may be contributing factors (Soares et al., 2008).

8.3 SEED PRODUCTION

8.3.1 Seed Morphology and Anatomy

The family Musaceae contains three genera—*Musa*, *Ensete*, and *Musella*—although the generic status of *Musella* is not universally accepted (De Langhe et al., 2009). All seeds of the Musaceae contain two chambers, similar to the spiral gingers—*Costus* sp. (*Zingiberaceae*; Humphrey, 1896). The larger, proximal chamber contains the embryo and endosperm and the smaller, a chalazal mass. The outer integument is thick, has a silicified exotesta (Graven et al., 1996) and internally, the middle layers are sclerotic cells. The innermost layer of the outer integument is an endotesta of thickened cells with a U shape. The tegmen or inner integument has two layers and a thick cuticle on its inner surface. The cuticle originates partly from the inner layer of the tegument and partly from the nucellus (Graven et al., 1996). The outer integument encloses the whole seed while the inner integument encloses the large chamber (McGahan, 1961). Within the large chamber, the embryo is wedged into the micropylar collar and its cotyledonary haustorium extends towards the center of the seed and is surrounded by endosperm. The nucellar pad lies between the embryo and the micropylar plug. Within the micropylar plug, which forms an operculum, the micropyle can be detected. The external surface of the micropylar plug becomes the hilum. The micropylar plug, which fills the only opening in the seed, is easily removed, at least in the seed of *Musa balbisiana* studied by McGahan (1961). The chalazal mass protrudes towards the center of the seed, creating the doughnut shape of the endosperm. The aril is trichomatous and in the ovular stage surrounds the seed coat. Later it is reduced and may not always be detected on the mature seed (Humphrey, 1896).

Broadly, *Ensete* has large smooth seeds with a rimmed hylar depression. *Musella* and *Musa* have small seeds but the coats have a different texture. Seeds of *Musella* are smooth compared with the rough coat of *Musa* (Graven et al., 1996). De Langhe (2009) presents a key to genera, species, sections and some subspecies based on seed shape, size, and the nature of the surface of the seed coat.

The seeds of the Musaceae are either cylindrical (for example, *Musa violascens*, section *Callimusa*) or subglobular, angular, and flattened (section *Eumusa*) (Chin, 1996). Seed weight, adjusted to 10% water content, varies from 27 mg for *Musa ornata* (section *Rhodochlamys*) to 71 mg for *Musa violascens*. Seeds of *Musa acuminata*, as recorded by Chin (1996), weigh about 45 mg. The seeds of *Ensete glauca* are about 10 to 22 times heavier than the seeds of *Musa* spp. Wattanachaiyingcharoen (1990) measured seed weight within a population of more than 2,000 seeds of an undescribed *Musa acuminata* ssp. (similar to *microcarpa*; Turner and Hunt, 1984). Mean seed weight was 55 to 56 mg over two seasons but the range was from 10 to 72 mg. Seed weight was not normally distributed, with both skewness (mode > mean) and kurtosis (peaked > normal curve) being significant ($p = 0.05$). Seed size can vary significantly within a population of seeds of *Musa* sp.

The main features of mature dried seed of *Musa balbisiana* (McGahan, 1961) are similar to those described by Humphrey (1896) for *M. ornata*. In a number of studies, summarized by McGahan (1961), although different species of Musaceae were examined, the characteristics of the seed of the family are clear. Graven et al. (1996) examined seeds of several *Musa* species (*M. acuminata*, *M. balbisiana*, *M. mannii*, and *M. paradisiaca* [possibly French Plantain, AAB triploid clone, Simmonds 1966], *M. textilis*, *M. velutina*), *Ensete* (*E. ventricosum*, *E. glaucum*), and *Musella lasiocarpa*. They concluded that the seed coats were very similar in structure across the three genera and so they presented only their data for *Musa*.

8.3.2 The Seed Coat

Musaceae grow in a range of environments in the tropics and subtropics that experience extremes of water supply and a range of temperatures, although few below 0°C. In the wild species, the seed carries the plant propagule, the embryo, through time and space. The complexity of the seed coat of *Musa* was identified more than a century ago (Humphrey, 1896) and has been studied by numerous people since, including a detailed study of *M. balbisiana* by McGahan (1961). More recently, Graven et al. (1996) examined the structure and macromolecular properties of the testa of *Musa*, *Ensete*, and *Musella* with the objective of determining the features that contributed to its ability to protect the embryo as well as allow germination.

Under selection pressure, the seed coat has characteristics that increase the probability of survival of the new propagule and that allow germination. Graven et al. (1996) consider that survival and germination place competing demands on the seed coat. In *Musa*, the fruit of wild species are usually eaten by animals, mainly monkeys, birds, and bats (Simmonds, 1959, 1962). The seed itself may not have appeal for animals, but the sweet flesh of the surrounding pulp tissue would be attractive, thus consumption of seeds may be accidental. Graven et al. (1996) thought that the presence of silica on the outer seed coat may assist the seed to pass through the alimentary tract of animals. Simmonds (1959) fed ripe, seeded fruit of *M. acuminata* and *M. balbisiana* to adult pigs and domestic fowl. The gastric acid of a pig has a pH of 1–2 (Manners, 1976). Passage through the gut destroyed 90–96% of seeds fed to pigs and 99% of seed fed to fowl. Simmonds (1962) concluded that survival of seed in the alimentary tract was possible, but not good. Passage of seed through the gut of birds has different effects, depending on the species of bird and the species of seed (Traveset et al., 2001), and so the results of Simmonds (1959) are indicative only.

Acid is used experimentally to scarify seed. Scarification with acid destroyed most seeds of *M. acuminata* and *M. balbisiana* (Simmonds, 1952), more than halved the germination of *M. balbisiana* (Stotzky et al., 1962), and halved the already low (18%) germination of *Ensete* seed (Tesfaye, 1992). A pH of 3.7 in the medium reduced germination of seed of *M. balbisiana* by 88%, compared with germination at pH 5.2 to 6.3 (Perry and Boodley, 1980). Seeds of Musaceae are not particularly resistant to acids, despite the presence of silica in the exotesta that might assist them to survive such conditions (Graven et al., 1996). The animal gut applies high selection pressure to seed and the fact that many don't survive the experience suggests that over time, a significant proportion of seed are dispersed by other means.

For germination, water needs to enter the seed and gas exchange needs to occur. Graven et al. (1996) suggested that the silica in the exotesta could restrict water loss if the silica gel was hydrated. They also suggest that the thick cuticle on the innermost side of the seed coat may resist the diffusion of water, in addition to the outer layers. For interpreting experiments on scarification, for example those of Stotzky et al. (1962), it is worthwhile distinguishing those techniques that affect mainly the exotesta (such as acids and alkali; Stotzky et al., 1962) and those that penetrate the inner cuticle (chipping; Stotzky et al., 1962), allowing more rapid access of water to the endosperm and embryo. Chipping increased germination of *Musa balbisiana* from 0 to 81%, whereas the other scarification techniques increased germination from 0 to 34% at most (Stotzky et al., 1962). These data imply that the inner cuticle is a significant barrier for the flow of water into the seed.

Removing the micropylar cap did not increase germination, and so it is unlikely that water penetrates at this part of the seed, at least in *M. balbisiana*. Removal of the chalazal mass, which lies between the inner and outer integuments, increased germination from 0 to 16%. Wattanachaiyingcharoen (1990) measured imbibition of seeds of *M. acuminata* ssp. by weighing seed sequentially over several days. He compared intact seed with that which had either been scarified using sandpaper or pierced. The scarification ruptured some of the exotesta and piercing created a hole in the outer and inner integuments, allowing water to enter the endosperm. Scarifying the seed coat did not significantly ($p = 0.05$) increase the rate of water absorption or the final amount absorbed (23.4 mg/seed; Figure 8.9), but piercing the

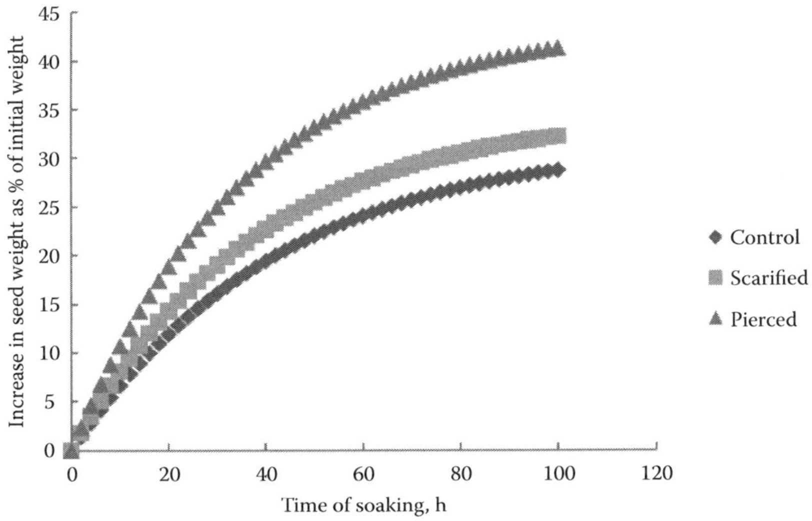

FIGURE 8.9 Imbibition of seed of *Musa acuminata* ssp. that is either intact (control) or has been scarified or pierced. Scarification ruptured the exotesta, piercing penetrated the inner cuticle allowing access of water to the endosperm. The rate of increase in water content for the pierced seeds and the maximum weight reached are significantly greater than the values for the intact or scarified seed, which do not differ ($p = 0.05$). (Data from Wattanachaiyingcharoen, D., 1990, Viability, germination and dormancy of banana seed (*Musa acuminata* ssp.), master's thesis, The University of Western Australia.)

seed did increase the rate and the amount of water absorbed (29.6 mg/seed; Wattanachaiyingcharoen, 1990). Over the time span of this experiment (8 days) germination had not commenced.

8.3.3 Pollinators

Pollinators place selection pressure on the pollen source. Therefore, if there is variation between pollinators within *Musa*, there will be differences in plant form and function (Inito et al., 1991). Apart from birds and bats, pollinators include tree shrews (*Tupaia* sp.) and bees (*Trigona* sp.), but ants, butterflies, flies, and spider-hunters (*Arachnothera* sp.) were unimportant in the experiments conducted by Nur (1976). *Musella* is pollinated by bumblebees (*Bombus* sp.), honeybees (*Apis* sp.), and wasps (*Vespa* sp.). Indeed, the insect pollination of *Musella* has contributed to its reproductive isolation (Liu, Kress, et al., 2002). Features of *Musa* that favor pollination are: female flowers have a longer flowering time than male flowers; the banana produces flowers all year, meaning that vertebrate pollinators will find it attractive; and there are more male flowers than female flowers in a population of *Musa* plants.

Itino et al. (1991) studied the bat and bird pollinators of *Musa acuminata* ssp. *halabanensis* (chiropterophily) and *Musa salaccensis* (ornithophily) in the field. The timing of anthesis and the peaks in nectar production by flowers are consistent with pollinators being present during the day (birds) or at night (bats). Itino et al. (1991) concluded that seed set in these two species was pollinator limited and that this indicated competition among nectar-producing plants for the available vertebrate pollinators. *Musa itinerans* in southern China is pollinated equally by birds and bats (Liu, Li, et al., 2002). *Musa itinerans* has a peak of nectar production near midday and midnight but most flower opening occurs near dawn and sunset. The total resource available to pollinators will be a function of the rate of nectar production and the number of flowers open. Assuming that flowers stay open for 1 day, the availability of the resource throughout the day coincides with the number of visits from birds (day) or bats (night). The data of Liu, Li, et al. (2002) show that the greatest amount of resource is available from 8 a.m. to 12 noon and from 10 p.m. to 2 a.m. This matches the timing of most visits by birds or bats. Male flowers show a similar pattern of resource availability throughout the day as do female flowers, but they provide less resource at the peak times. This is reflected in a lower frequency of visits by pollinators, compared with female flowers.

Stephens and Tyson (1975) recorded the visits of nectar bats to bunches of *Musa* spp. (AAA, Cavendish subgroup) cv 'Valery' in Panama. Almost all bunches within the large commercial plantation were located by bats, as indicated by scratches on the young fruit. Bats did not visit bunches in the first few nights of their opening, and so most of the scratches appear on hands towards the bunch apex. The cv 'Valery' is sterile but if bats visited the bunches of fertile seeded *Musa* species in the same way, then this may disadvantage the first few hands on a bunch in terms of pollination and seed production.

Despite structural self-incompatibility, Nur (1976) observed that fruit of *Musa velutina* (section *Rhodochlamys*) set seed even though all pollinators were excluded from the inflorescence. He concluded that *M. velutina* was self-pollinating as it had hermaphrodite basal flowers, or some other mechanism was responsible for seed set.

8.3.4 Factors Affecting Seed Set

For breeding, seed production is important because it provides the population from which desirable progeny can be selected. On the other hand, the advantage of banana fruit for human consumption is that they are sterile and parthenocarpic. An ongoing difficulty in breeding of *Musa* has been the low seed set, and several studies have had the objective of identifying the factors contributing to seed production. Many of these studies have been empirical (Table 8.3). Looking deeper into the problem, Shepherd (1954) investigated the significance of pollen tube growth in contributing to the variation in seed fertility. More recently Ortiz and Vuylsteke (1995b) gathered indirect evidence that seasonal conditions affect the ploidy of the gametes of the female parents in French Plantains. Ortiz (1997) obtained direct evidence that seasonal conditions affected the ploidy of pollen derived from 'Pisang Lilin.' The value of these studies for breeding is that since seed set is very low, even a small increase in seed production can significantly increase the population of progeny available for selection.

8.3.4.1 Maximum Seed Set

In Jamaica, the 'Gros Michel' (AAA) × 'Pisang Lilin' (AA) crosses generally produced only 1 to 3 seeds per bunch but the maximum recorded was 60 seeds/bunch (Shepherd, 1954). In Nigeria, the 'Bobby Tannap' AAB × 'Calcutta 4' (AA derived from *M. acuminata* ssp. *burmannicoides*) cross produces about 22 seeds per bunch, but has produced 219 seeds/bunch (Swennen et al., 1991). The maximum values can be taken as indicating what is possible, and lower numbers of seeds encourage the search for reasons for reduced seed fertility.

8.3.4.2 Bunch and Fruit Factors

Most cultivars of edible triploid bananas are largely male-sterile, but almost all will set seed if pollinated with suitable pollen (Shepherd, 1954). If ovaries of cv 'Gros Michel' (AAA) are hand-pollinated with material from the edible diploid 'Pisang Lilin' (AA), then bunches containing thin fruit at maturity are more likely to have more seed than bunches containing well-developed fruits. Bunches with mature fruit of 10 cm circumference had 1.7 seeds per bunch and those with a circumference of 15 cm or more had only 0.6 seeds per bunch. Shepherd (1954) thought this difference was associated with the impact of growth regulators that suppressed seed fertility but promoted growth in the larger fruit. In these experiments, differences in fruit diameter were associated with differences in seasonal conditions and so the impact of environmental factors on seed fertility may be confounded with the effects of fruit size.

In hand-pollinated bunches of cv 'Gros Michel' (AAA) pollinated with 'Pisang Lilin' (AA), the proximal hands of female flowers are more fertile than the distal hands. In bunches that had

TABLE 8.3
Plant and Environmental Factors That Affect Seed Set in *Musa* spp.

Factor	Magnitude of Effect	Ref.
Bunch size	Small bunches, 6–8 hands, produce fewer seeds (1.3 seeds/100 fruit) than large bunches, 12–13 hands (5.2 seeds/100 fruit), cv 'Gros Michel'	2
Position of hands within a bunch	66–79% of seed within a bunch are located in the three proximal hands in bunches with 7–8 hands, cv 'Gros Michel'	1
	42–45% of seed in a bunch located in the three proximal hands, cvv 'Bobby Tannap,' 'Obino l'Ewai'	8
Fruit size	Thin fruit at maturity produce more seeds (1.7 seeds/bunch) than full fruit (0.6 seeds/bunch), cv 'Gros Michel'	1
Position within the fruit	In fruit 150 mm long, seeds tend to be located 30 mm from flower end. In fruit 230 mm long, seeds tend to be 50 mm from flower end, cv 'Gros Michel.'	1
Style length	Increasing length of style negatively correlated with number of seed set/fruit, 78 genotypes of East African Highland bananas	4
Style receptivity	Pollination at stage 3 of stigma development was correlated with the greatest number of seeds per hand, five genotypes of East African Highland bananas	5
Season of pollination across locations	Consistent seasonal variation in seed fertility at two locations in Jamaica (range 0–3 seeds/bunch), one location having high seed fertility from November to February (4–6 seeds/bunch) compared with 1 seed/bunch at other locations, cv 'Gros Michel'	1
Season of pollination at one location	July better than September/October or January/February in Jamaica, cv 'Gros Michel'	3
	Bunches pollinated in February contain 65–70 seeds but at other times of the year the number falls to 2–30 seeds, cv 'Bobby Tannap' (French Plantain) with pollen from 'Calcutta 4'	6
Time of day of pollination	Morning pollination (6–10 a.m.) increased seed set by 12–100% compared with afternoon (1 p.m.), cv 'Gros Michel'	3
	Pollination at 7 a.m. better (1.95 seeds/bunch) than later in the day (0.6–0.8 seeds/bunch), cv 'Gros Michel'	1
Altitude	Seeds occur in fruit of *M. fehi* grown at 900 to 1100 m in Tahiti, but not in the lowlands	7
Fields at one location	1.83 seeds/bunch in one field compared with 6.82 seeds/bunch in another field at the one location, cv 'Gros Michel'	3
Soil fertility	More seeds/bunch in section of field with higher P and K than other sections, cv 'Gros Michel'	1
Pruning of suckers	No pruning produced 73% more seed/bunch than pruning to one sucker per mat, cv 'Gros Michel'	1

Sources: Data from (1) Shepherd, K., 1954, Seed fertility of the Gros Michel banana in Jamaica, *J. Hort. Sci.*, 29, 1–11. (2) Shepherd, K. 1960a. Seed fertility of edible bananas. *J. Hort. Sci.* 35:6–20. (3) Shepherd, K. 1960b. Seed fertility of 'Gros Michel' bananas. *Trop. Agric.*, Trinidad 37:211–221. (4) Ssebuliba, R.N., P. Rubaihayo, A. Tenkouano, D. Makumbi, D. Talengera, and M. Magambo, 2005, Genetic diversity among East African Highland bananas for female fertility, *Afr. Crop Sci. J.*, 13, 13–26. (5) Ssebuliba, R.N., M. Magambo, D. Talengera, D. Makumbi, A. Tenkouano, P. Rubaihayo, et al., 2006, Biological factors affecting seed production in East African Highland bananas, *J. Crop Improvement*, 16:67–79. (6) Swennen, R., D. Vuylsteke, and K. De Smet, 1991, Season dependent seed set in plantain, *Banana Newslett.*, 14, 35–36. (7) Baker, J.G., 1893, A synopsis of the genera and species of *Museae*, *Ann. Bot.*, 7, 189–222. (8) Swennen, R. and D. Vuylsteke, 1990, Aspects of plantain breeding at IITA, in: *Sigatoka leaf spot diseases of bananas*, R.A. Fullerton and R.H. Stover, eds., 252–266, Montpellier, INIBAP.

eight hands and were from several locations, 67–79% of the seed was in the proximal three hands (Shepherd, 1954). This pattern of seed distribution within the bunch differs from that in the French Plantain cvv 'Bobby Tannap' (AAB) and 'Obino l'Ewai' (AAB), pollinated with Calcutta 4 (AA), where only 42 to 45% of the seeds in a bunch occur in the three proximal hands (Swennen and Vuylsteke, 1990). Within fruit of cv 'Gros Michel' (AAA) that are pollinated by hand, seeds may set as close to the stylar end of the fruit as 13 mm or as distant as 250 mm. However, the mean distance is within 30 to 58 mm from the stylar end and the distance tends to increase with fruit size (1.8 mm deeper into the fruit for each 10 mm increase in fruit length). These observations caused Shepherd (1954) to examine pollen tube growth. Penetration of the style was assumed to take 1 hour and subsequent growth 3 mm h^{-1}, based on observations on *M. acuminata*. The tubes of pollen from cv 'Pisang Lilin' (AA edible diploid allied to *M. acuminata* ssp. *malaccensis*; Simmonds, 1966) applied by hand to ovaries of cv 'Gros Michel' generally grew at a slower rate than might be expected from the rate observed in *Musa acuminata*. No growth occurred in 26% of ovaries. Pollen tube growth varied greatly between hands and between fruit (ovaries) within hands. In some hands pollen grew in all fruit, while in others pollen tubes grew in only one or two fruit. The data available on pollen tube growth seem consistent with the observed distribution of seeds within fruit and the low seed fertility of cv 'Gros Michel' (AAA) (Shepherd, 1954). There is a need to discover why so many pollen tubes fail to grow, especially in the light of recent knowledge about this process (Higashiyama and Hamamura, 2008).

8.3.4.3 Environmental Factors

Environmental conditions influence seed fertility in *Musa*. Baker (1893) observed that *M. fehi* (*Australimusa*) in Tahiti is seedless when grown at low altitudes but contains seeds when grown at 900–1100 m. A number of environmental components change with altitude, such as temperature, humidity, and cloudiness but this observation suggests a strong environmental component in the fertility system of *M. fehi*. Alternatively, pollinators may be absent at low altitudes. Firmer evidence supporting the influence of environment on the success of pollination comes from hand crosses between 'Gros Michel' (AAA) with pollen from 'Pisang Lilin' (AA, parthenocarpic) from different locations in Jamaica (Shepherd, 1954). At an exposed coastal site, 80% of pollinated bunches contained fruit with seeds, whereas at a sheltered site in a nearby valley, only 20% of pollinated bunches contained seeds. Maintaining humidity around bunches immediately after pollination increased the proportion of fertile bunches from 30 to 48%, but had no significant effect on the number of seeds per bunch (Shepherd, 1954). Despite large variation in seediness of bunches between localities and seasons, Shepherd was unable to detect significant correlations with temperature, rainfall, or humidity. Therefore, standard meteorological data could not be used to predict bunch fertility, under these conditions.

A stronger link between environment and seed fertility has been reported by Swennen and Vuylsteke (1990) and Swennen et al. (1991, 1992). Swennen et al. (1991) found a very distinct effect, repeated over 2 years, of month of pollination on the seed fertility of French Plantain cv 'Bobby Tannap' (AAB) pollinated with 'Calcutta 4' (AA, *M. acuminata* ssp. *burmannicoides*). Bunches pollinated in February contained 65–70 seeds per bunch. Bunches pollinated from March to June produced fewer than 10 seeds/bunch and from July to January seed number varied from 15 to 35 seeds/bunch. These data were further collected for 5 years (Figure 8.10) and the data analyzed to determine any correlations between meteorological conditions and seed set. For the AAB plantains, variations in humidity accounted for 32–51% of the variation in seed set. Over the same time period, variations in rainfall, solar radiation, temperature, and humidity accounted for 63% of the variation in seed set in the AAA and ABB bananas (Ortiz and Vuylsteke, 1995b). Similarly, Ssebuliba et al. (2009) found that there were two major seed-set periods in East African Highland banana within a year with a major peak occurring in March–April and a minor peak occurring in September. Optimum seed set appeared to be related to climatic variables.

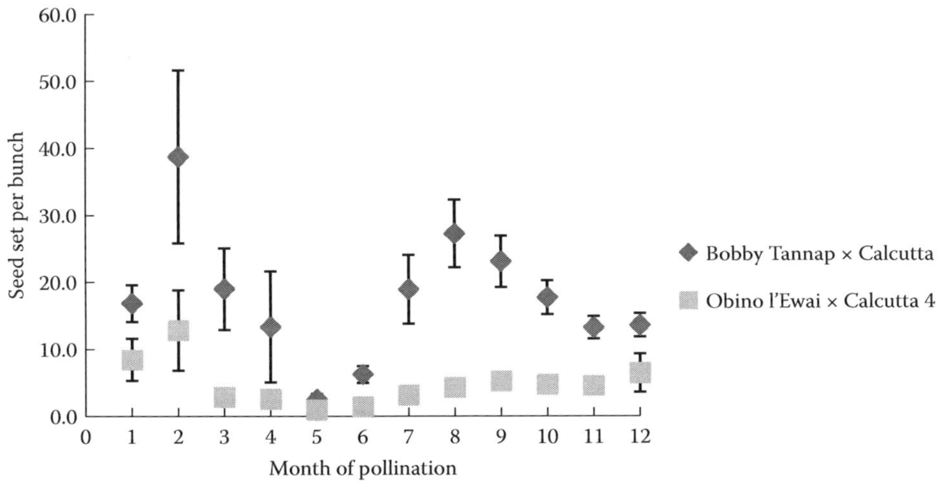

FIGURE 8.10 Seed set per bunch in two French Plantain (AAB) cultivars grown at Onne, Nigeria, and pollinated every day, but mean shown for each month, with pollen from 'Calcutta 4' (derived from *Musa acuminata* ssp. *burmannicoides*), a wild-seeded diploid. The rainy season is from March to December. Vertical bars are standard errors across years, $n = 5$. (Data from Ortiz, R. and D. Vuylsteke, 1995, Factors influencing seed set in triploid *Musa* spp. L. and production of euploid hybrids, *Ann. Bot.*, 75, 151–155.)

While these correlations are useful, they are specific for location and cultivar, or group of cultivars. The process of pollination and seed set is complex and season may influence any one or more components. Deeper studies are required to identify which components are sensitive to environmental perturbations. An example of this is the study of $2n$ pollen production by Ortiz (1997) who found that the proportion of $2n$ pollen produced by 'Pisang Lilin' throughout the year in Nigeria was positively correlated with the amount of solar radiation ($r^2 = 0.72$). Nonetheless, the empirical relationships tell us that planting dates of female parents can be managed so that seed set can be maximized, thus increasing the efficiency of breeding *Musa* spp. (Swennen and Vuylsteke, 1990). Such information, if available, could also be used to select sites for breeding programs in bananas.

Time of day when pollination occurs strongly influenced bunch fertility in 'Gros Michel' in Jamaica (Shepherd, 1954). When bunches were pollinated at 0700 h (compared with 1000, 1300, or 1600 h), 60% produced seed, compared with 30–40% at other times. As well as increasing bunch fertility, pollinating at 0700 h increased the fertility per bunch by two- to threefold. Whether these effects are caused by the environmental conditions at that time of day or the receptivity of the flowers remains to be determined.

8.3.5 Seed Growth

Seeds stimulate the growth of the pulp in wild *Musa* species, partially parthenocarpic and, if seeds are present, in parthenocarpic genotypes, according to Simmonds (1960). He investigated this by applying plant growth regulators, or their antagonists, to growing fruit. In the wild-seeded species, *M. acuminata* and *M. balbisiana*, each seed produced 0.23 cm^3 of pulp, about four times the size of the seed itself. In parthenocarpic fruit, Simmonds (1960) estimated the stimulatory effect of the seed on fruit volume to be 23 times the size of the seed. He concluded that the stimulatory effect of the seed increased with increasing parthenocarpy. However, these calculations ignore the two autonomous synthetic systems (Simmonds, 1960) that contribute to growth of the pulp in partially or fully parthenocarpic fruit. A more likely scenario is that seeds across the genetic spectrum in *Musa* have a more limited stimulatory effect on growth of the pulp but in parthenocarpic fruit this is complemented by the independent growth of pulp linked to the genes in the maternal tissues that contribute to parthenocarpy (Simmonds, 1953b). In seedless mandarin (*Citrus reticulata*), endogenous gibberellins in the ovaries are linked to the parthenocarpic development of fruit (Talon et al., 1992). Parthenocarpy in tomato (*Lycopersicon esculentum*) is controlled by the deregulated production of gibberellins in the seedless fruit (Olimpieri et al., 2007). These two examples indicate that the seed is not the only source of growth regulator for fruit development, and it is likely that similar systems operate in *Musa* fruits.

The size of mature fruit of wild bananas (*M. acuminata* and *M. balbisiana*) is linearly proportional to the number of seeds in the fruit (Simmonds, 1953a, 1960). The regression coefficients are 0.23–0.24 cm^3 per seed. A similar, but more limited, data set is available in the studies of Inito et al. (1991) on pollination of *M. salaccensis*. Their data show a nonlinear relationship between seed number per fruit and fruit size such that the contribution of each seed to fruit growth decreases with increasing seed number. When a fruit contains fewer than 15 seeds, the contribution of the seed to fruit weight is about 0.37 g/seed; but when there are between 15 and 62 fruit per seed, the contribution falls to 0.10 g/seed. Inito et al. (1991) measured whole-fruit weight, not the mass of the pulp, and so the differences between the two coefficients may be linked to the high proportion of peel weight likely to be present in small fruit where the pulp has not developed.

If insufficient ovules contain embryo sacs, then seed fertility will be affected, especially in triploid cultivars. Shepherd (1954) measured the number of embryo sacs in ovaries obtained from plantings of 'Gros Michel' (AAA) in Holland and Potosi, two locations in east Jamaica. At Holland, 18% of ovules contained embryo sacs but only 10% contained embryo sacs at Potosi. While this is consistent with the observations on seed fertility observed at these locations 3 years earlier, at Holland there were many embryo sacs with abnormal constitution. Shepherd (1954) concluded that it was impossible to estimate the importance of embryo sac formation in controlling seed fertility, based on the evidence available.

8.3.6 Seed Storage

In the forest the ripe fruit of seeded wild bananas are usually eaten by animals or birds and so to recover seed from the field, mature green fruit need to be harvested as long as the seeds are black and hard. Seed can be extracted from ripe fruit, washed, and cleaned. Initially the seed water content may vary from 30 to 45%. For storage, the seed need to be dried to 8–10% water content, placed in a sealed container, and then they can be stored for 1 to 2 years at 5°C or at –18°C (Chin, 1996). Banana seeds are thought to be intermediate in their storage behavior, although seeds of *Ensete* may be orthodox (Seed Information Database, 2010). Seed water content strongly influences the viability of banana seed in storage; seed with high water content loses its viability much more rapidly than dried seed (Figure 8.11).

8.4 SEED GERMINATION

8.4.1 Dormancy

The issues about dormancy in seeds of the Musaceae center on seed structure, imbibition, viability, proportion of germinable seeds, experimental evidence for dormancy, changes in the definition of dormancy, and differences between and within *Musa*, *Ensete*, and *Musella*.

Opinions about the nature of dormancy in the seed of Musaceae vary. Simmonds (1962) states that banana seed are not inherently dormant as they will germinate as soon as they are extracted from ripe fruit (Figure 8.12), and this is supported by the application of embryo rescue technology (Cox et al., 1960; Vuylsteke et al., 1990). However, in the field they may lie viable for many years and often germinate at the same time and in large numbers when the soil is disturbed or when vegetation is removed (Simmonds, 1962). Simmonds (1959) interpreted the delay in germination, under

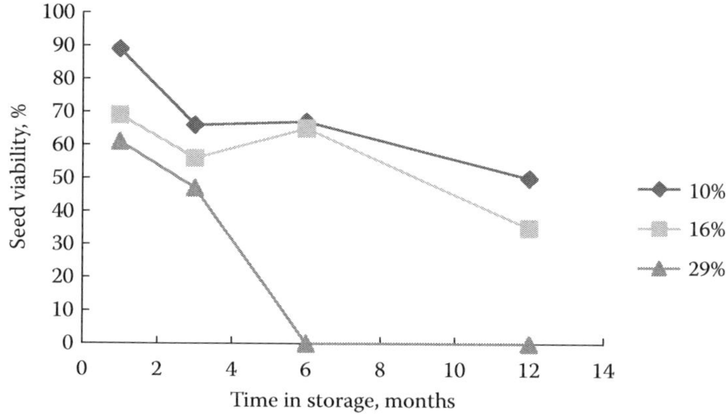

FIGURE 8.11 Effect of seed water content (10, 16, or 29%) on the viability of seed of *Musa violascens* during 12 months' storage at 22°C. (Data from Chin, H.F., 1996, Germination and storage of banana seed, in *New frontiers in resistance breeding for nematode, Fusarium and Sigatoka*, E.A. Frison, J. Horry, and D. De Waele, eds., 218–227, Montpellier, France, INIBAP.)

apparently favorable conditions, as a storage phenomenon, rather than dormancy. He thought that the carbon dioxide concentration in the soil was sufficient to prevent germination and retain seed viability, and he provided some experimental evidence to support this contention. Simmonds (1959) also tested storage of seeds of *M. balbisiana* at different depths in the soil, compared with storage of seed on the soil surface. Seed stored at depth maintained viability longer than seed stored on the soil surface. Chin (1996) linked dormancy with seed water content, especially the desiccation of the chalazal mass. If the chalazal mass was large, then seed usually germinated, but if it was desiccated, then the seeds became dormant. For example, freshly harvested seeds of *M. gracilis* (46% water content) germinated (70%) after exposure to moist heat for 7 days. When dried to 12% water content, these seed did not germinate after heat treatment, even after 3 months. Chin (1996) linked germination of large numbers of banana seeds in the field after soil disturbance with the exposure of these seed to diurnal changes in soil temperature. This is the interpretation reached by Stotzky and Cox (1962) in their studies on the effect of alternating temperature on the germination of seed

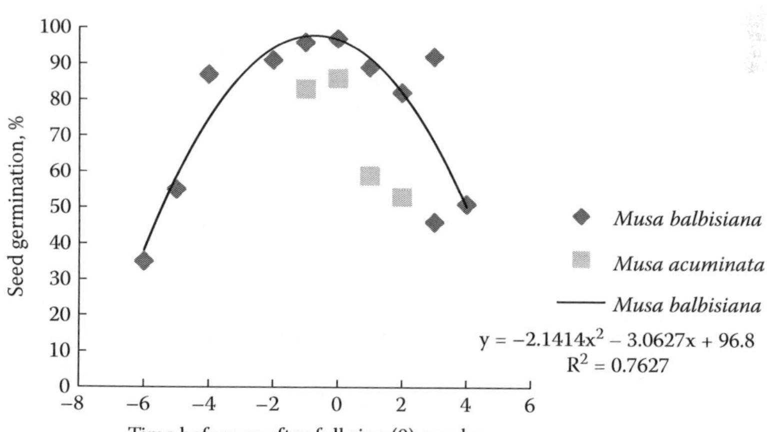

FIGURE 8.12 Relationship between the maturity of the bunch, expressed as weeks before or after the fruit are fully ripe (week 0) and the germination of the seeds of *Musa balbisiana* and *Musa acuminata*. Seeds were extracted and sown fresh and undried. The fitted equation is only for the *M. balbisiana* data. (Data from Simmonds, N.W., 1952, The germination of banana seeds, *Trop. Agric., Trinidad,* 29, 35–49.)

of *M. balbisiana*. Removal of vegetation will also expose the soil to large variations in temperature, without necessarily disturbing the soil.

Stotzky et al. (1962) state that difficulties in germination of seed of *M. balbisiana* lie in the seed coat and that the embryo exhibits no dormancy because it is easy to culture aseptically (Cox et al., 1960). On the other hand, Graven et al. (1996) believe that seeds of Musaceae show embryo dormancy. De Langhe (2009) stated that seeds possess dormancy but its nature is unknown. In Uganda, seeds of *M. balbisiana* were observed to germinate around the parental plants from which they fell, while this was not observed in *M. acuminata* 'Calcutta 4,' suggesting that there was some inherent dormancy required for 'Calcutta 4' (M. Pillay, personal communication). Whether seeds of the Musaceae possess dormancy or not, they are often notoriously difficult to germinate. In the work on banana breeding in Africa, usually only 1% of the hybrid seeds germinated (Swennen et al., 1992). This can be increased to 12% if the embryos are extracted and cultured aseptically (Vuylsteke et al., 1990).

There are a number of reasons why authors may have different opinions about dormancy in seeds of the Musaceae. The concept of dormancy in relation to seeds has undergone revision during the years that people have been working on banana seeds. Vleeshouwers et al. (1995) suggest that dormancy should not be viewed simply as the absence of germination. Dormancy reflects a changing internal status of the seed and it can vary on a continuous scale. The level of dormancy determines the conditions needed for germination. If the environment external to the seed meets those requirements, then the seed germinates. They conclude that dormancy is a dynamic state, influenced by the response of the seed to changes in the environment.

Baskin and Baskin (2004b) proposed a classification system for dormancy. Seeds may possess dormancy that is physical, physiological, morphological, or combinations of these. In addition, physiological dormancy and some of the combinations can have different levels and different types of dormancy. Baskin and Baskin (2004a) present a dichotomous key, allowing the type of dormancy to be identified providing the following are known: seed coat permeability, response to scarification, embryo differentiation, and response to temperatures that simulate the seasonal variation in the habitat. Using the classification of Baskin and Baskin (2004b) and bearing in mind the concepts of dormancy proposed by Vleeshouwers et al. (1995), we can examine the experimental data available for seeds of the Musaceae.

Baskin and Baskin (1998) included Musaceae in a list of families that contained at least some species that had physical dormancy. This was based on the observation of Stotzky et al. (1962) that scarification of seeds of *M. balbisiana* promoted germination and that the seed coat was impermeable to water (Bhat et al., 1994). However, Baskin et al. (2000) removed Musaceae from the group of families in which physical dormancy occurred because the seed coat was permeable, according to their interpretation of the comments of Stotzky and Cox (1962), and because a water-impermeable layer of palisade cells could not be detected in the seed coat (Humphrey, 1896; Graven et al., 1996).

Physical dormancy is characterized by an impermeable seed coat (Baskin et al., 2000). When the seed coat becomes permeable to water through the action of factors such as high or fluctuating temperature, freezing/thawing, drying, fire, or passage through the alimentary tract of animals, then the embryo germinates under a wide range of temperatures and in the light as well as the dark (Baskin et al., 2000). The feature that creates the high resistance to water flow across the seed coat is a layer(s) of palisade cells. In addition, the comments of Stotzky and Cox (1962) on imbibition, which in intact seed of *M. balbisiana* took 3 days but in chipped seed took 2 days, and both intact and chipped seed absorbed the same amount of water, appear not to support physical dormancy in *Musa*.

A seed may functionally express physical dormancy, especially if a layer other than the palisade layer prevented water entering the endosperm and embryo. Graven et al. (1996) suggest such a role for the thick cuticle on the inner side of the tegument in seed of Musaceae. The data of Wattanachaiyingcharoen (1990) (Figure 8.9) suggest that the inner cuticle prevents water movement into the chamber containing the endosperm and embryo in *M. acuminata* ssp. Piercing the cuticle allows an extra 7 mg of water to flow into the seed, and this difference was maintained over the 8 days of the experiment. Thus the thick inner cuticle provides the barrier to water flow to the embryo, providing a functional expression of physical dormancy. However, Stotzky and Cox (1962) state that chipped seed absorbed as much water as the intact seed. Chipping was done with a scalpel and differs from the technique of piercing with a needle used by Wattanachaiyingcharoen (1990). Chipping cut through to the inner cuticle and so would have removed a section of the seed coat. This loss of part of the seed coat would reduce the amount of water that the seed could imbibe into the seed coat, but in chipped seed it could be compensated with the water entering the chamber containing the endosperm and embryo. Thus the comment by Stotzky and Cox (1962) that chipped and intact seeds absorbed the same amount of water is not necessarily inconsistent with the measurements of Wattanachaiyingcharoen (1990) showing that piercing the cuticle increases the amount of water absorbed.

The seed of *Musa* may well have a functional physical dormancy, where the barrier to water flow is located at the inner surface of the seed coat, rather than the exterior as in *Canna* sp. (Graven et al., 1997). Once the seed coat is imbibed, alternating temperatures increase the permeability of the inner cuticle, allowing germination to proceed in *M. balbisiana* (Stotzky and Cox, 1962) and *M. acuminata* ssp. (Wattanachaiyingcharoen and Turner, 1989a). Stotzky and Cox (1962) established that alternating temperatures promote the germination of seed of *M. balbisiana* (Figure 8.13). In their experiments, no germination occurred at constant temperature. Germination of *M. balbisiana* increased as the difference between the maximum and minimum temperatures increased, until the difference reached 15°C and beyond that germination was unaffected (Figure 8.13). Stotzky and Cox (1962) excised the embryos of seeds of *M. balbisiana* and exposed them to constant and alternating temperatures. In contrast to intact seeds, the excised embryos germinated at constant temperature and did not require alternating temperature for germination or growth. Stotzky and Cox (1962) concluded that the factor(s) that promoted germination that were affected by alternating temperatures did not lie in the embryo. Thus the embryo of *M. balbisiana* meets another requirement of Baskin

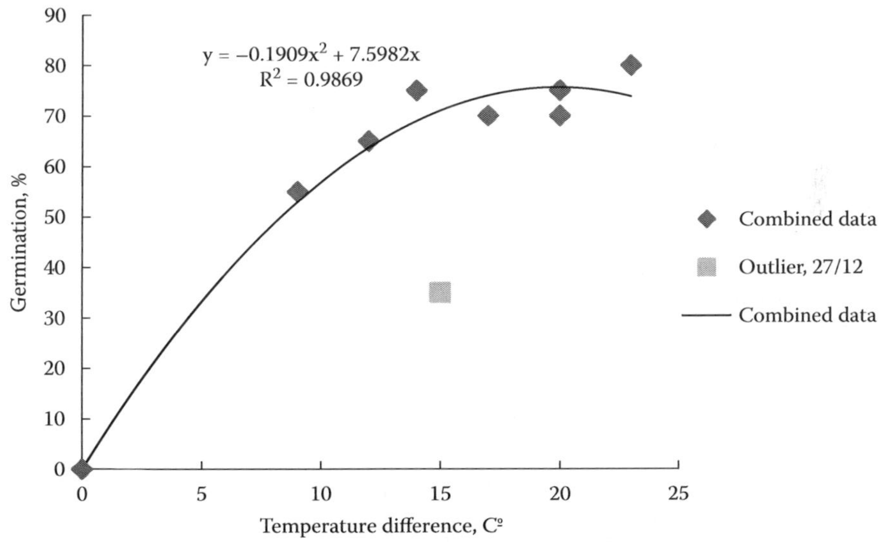

FIGURE 8.13 Effect of the difference between maximum and minimum temperature on the germination of seeds of *M. balbisiana*. Maximum temperatures were 27°C, 32°C, or 35°C, minimum temperatures ranged from 12°C to 35°C. The response of the seeds to the 27/12°C treatment is about half that expected and is presented as an outlier. (Data from Stotzky, G. and E.A. Cox, 1962, Seed germination studies in *Musa*, Pt. 2, Alternating temperature requirement for the germination of *Musa balbisiana*. *Am. J. Bot.* 49:763–770.) The fitted equation was forced through zero because no germination occurred at constant temperature.

et al. (2000) for physical dormancy: that the embryo can grow over a range of temperatures once physical dormancy is broken.

Baskin et al. (2000) point out that for species with physical dormancy, the mechanism for breaking dormancy needs to be fine-tuned to the environment so that individuals may germinate, establish, and eventually reproduce. For *M. acuminata* ssp., the seed requires at least 4 days to reach maximum water absorption (Figure 8.9) and for *M. balbisiana*, more seed germinates as the number of alternating temperature cycles increases and while the maximum temperatures remain at about 32°C (Stotzky and Cox, 1962). In addition, Stotzky and Cox (1962) compared alternating temperatures where the high temperature was maintained for 5 or 19 h and the minimum was 19 or 5 h, respectively. The longer time at the maximum temperature reduced germination to less than 10% and the shorter time at maximum temperature promoted seed germination (30–60%). In tropical conditions, it is more important for a seed to detect the wet or dry season, since temperatures are reasonably moderate; nonetheless, the response of banana seed to temperature is quite precise (Stotzky and Cox, 1962). Simmonds (1962) described the wild bananas as seral plants functioning as "jungle weeds" that spring up in response to disturbance in vegetation or soil. Disturbance may be clearing of forest by humans, land slips, or sections of forests damaged by storms. Exposure of soil to sunlight and high temperature for a short time would more likely be a feature of the edge of disturbed vegetation, rather than its center. Being able to detect this difference would allow banana seeds to germinate adjacent to the forest but in a disturbed area.

8.4.2 Viability

Fresh wild banana seed can germinate at a high level (Figure 8.11) but as they dry seeds become dormant (Chin, 1996). In studies on germination, knowledge of the viability of seed is important for interpreting the response of seeds to environmental cues. If seeds do not germinate, are they viable, or is there some other reason for the inability to germinate? Earlier studies on banana seed usually separated them into "good" or "bad" seed depending on their appearance, hardness, and endosperm content (Shepherd, 1954, 1960a, 1960b). However, in the analysis of data "good" and "bad" seeds were combined because it was difficult to predict the behavior of the seed based on this classification. Triphenyl tetrazolium salts have been widely used to determine viability of embryos in seed and they are straightforward to use on seed of *Musa* spp. (Wattanachaiyingcharoen and Turner, 1989b). The tetrazolium salts form a red color when they react with dehydrogenases in living tissue. If seed is killed by autoclaving, no red color appears. Embryos of *Musa* sp. can be taken as viable if the red color exists on 90% of the embryo. The tetrazolium test for viability is destructive and other ways of determining seed viability are worth exploring. In addition to problems of viability, a seed may not germinate because it lacks an embryo or endosperm, or there is no functional connection between the embryo and the endosperm.

Wattanachaiyingcharoen (1990) found that seed weight was a good predictor of viability, as determined by the tetrazolium test. In *M. acuminata* ssp., seed below 20 mg fresh weight contained no viable embryos and the increase in the proportion of seed in the population with viable embryos ($P\%$) increased linearly with seed weight (W) so that the largest seeds of 75 mg had 100% viable embryos. The relationship was $P\% = 1.56W - 27.21$, $r^2 = 0.92$. Floating was unable to separate viable from nonviable seed. Most seed of less than 30 mg weight floated. Among the seeds that sank, viability varied from 25 to 100% (Wattanachaiyingcharoen 1990). Tesfaye (1992) used floating to separate seeds of *Ensete* sp.

8.4.3 Germination

The sequence of germination in *Musa* sp., as recorded by Simmonds (1959), is: exudation of a brownish fluid from the micropyle, the micropylar lid or operculum is extruded, and the primary root emerges from the micropylar canal. One week after the primary root emerges, lateral roots dominate; after 2 weeks the endosperm has disappeared, after 4 weeks seedlings show nutrient deficiency symptoms if germinated in sand. If the seed coat is ruptured, then the expanding embryo will emerge from the seed through the injury, whether it is at the chalazal end of the seed (Stotzky and Cox, 1962) or through a hole pierced in the coat (Wattanachaiyingcharoen, 1990). This implies that in intact seed, the expanding embryo needs to push the micropylar cap from the seed in order to emerge and will use an alternative route if it is available. This would provide a point for selection pressure because in the wild, embryos that were not able to dislodge the micropylar plug would not germinate. Selection based on *in vitro* culture of excised embryos, such as is used in banana breeding technology, would avoid this selection and may therefore include embryos that are less fit than those germinated from seed.

Fresh seed germinates (Simmonds, 1952; Figure 8.12). On drying, the seed enters functional physical dormancy and this can be broken in *M. balbisiana* by alternating temperature, after the seed coat has imbibed (Stotzky and Cox, 1962). The embryos can readily grow after excision from the seed using embryo rescue technology (see, for example, Cox et al., 1960; Pancholi et al., 1995) and are not dormant, although Graven et al. (1996) thought that they were. In the study of Pancholi et al. (1995) on *M. velutina* seeds, sowing the seed in a propagating medium allowed 82% to germinate, but the germination was spread over 10 months. Excising the embryos allowed the seed to germinate (74%) in 14 days. Ellis et al. (1985) provide a list of the various treatments that either do or do not promote germination in Musaceae. Alternating temperatures promoted germination in intact seed of *M. balbisiana* (Stotzky and Cox, 1962) but did not promote germination in *Ensete* sp. (Tesfaye, 1992). However, soaking in water at 40°C for 24–48 h did promote germination in *Ensete* (Tesfaye, 1992). On the other hand, soaking seeds of *M. acuminata* or *M. balbisiana* in water usually had deleterious effects (Simmonds, 1952). Seeds from different species within the Musaceae do not appear to have the same requirements for germination and may not have the same mechanism of dormancy, despite the similarities in the structure of the seed and seed coat (Graven et al., 1996). In *Ensete*, the embryo is small (8 mm) in relation to the size of the seed (20–30 mm) and so it needs to grow considerably within the seed before it can emerge (Tesfaye, 1992).

8.5 CONCLUSION

The inflorescence of most *Musa* species and cultivars is monoecious. The female flowers are first formed and precede the male flowers in anthesis. Ovules are formed and are similar in the wild-seeded species and the edible, sterile, and parthenocarpic landraces, at least up until anthesis. Beyond this, numerous events combine to minimize seed development in cultivars. While many of these are known, the importance of each in any situation is uncertain. There is a need to know more about why so few seeds are produced, when there are records of high seed production, even in cultivated genotypes. Several factors, such as receptivity of the female organs, pollen tube growth, ovary position within the bunch, season of pollination, time of day, and location contribute to seed production in banana. There is also an interaction between environment and fertility of female and male components. These factors can be managed to manipulate seed production so that the immediate objective of crosses (such as production of triploids or tetraploids) can be met.

The clumped nature of the plant means that crossing in the wild-seeded species can be achieved between inflorescences within the one clump, and outbreeding can occur if other clumps of bananas are nearby. Heterozygosity is high among the edible diploids, which have been dispersed by people, but somewhat less among the seeded wild species that tend to be geographically isolated, one species from another.

Pollen production varies considerably between genotypes and locations, as does its viability. There is not necessarily a correlation between the viability of pollen and the number of seeds produced by receptive flowers. The production of $2n$ pollen is used to increase the production of tetraploids in breeding schemes and it is believed that this was the process that produced the triploid landraces used widely today.

Hybridization lies at the heart of breeding. There has been considerable effort to discover the likely route by which the diploid and triploid landraces were produced. This opens the possibility of determining which crosses might best be used to not only incorporate disease or pest resistance in progeny but to also provide new cultivars that retain the desirable qualities that people selected in the first place.

Selection of bananas for edibility has been a high priority for people for thousands of years and the primary aim was to select plants with fruit that were seedless. Breeding bananas for resistance to disease, tolerance of edaphic and environmental constraints, and desirable postharvest qualities is a more recent activity spread over several decades. Breeding requires seed. Once seed are produced, few of them germinate and so seed production and germination are bottlenecks in banana breeding. Seeds of the wild bananas germinate readily when they are fresh and as they dry they enter a functional physical dormancy that can be broken by alternating temperatures when the seed is hydrated. Many banana seeds do not germinate because they do not contain embryos. Many embryos are not viable, further reducing germination. Viable embryos can be excised and cultured aseptically to produce plants for selection.

In recent decades, knowledge of the reproductive biology of plants has progressed, but work on banana has been slow. There is a need to reengage research on reproductive biology in Musaceae, bringing to bear the new knowledge gained on "model plants" so that new, disease-resistant cultivars can contribute to sustaining the world's population.

ACKNOWLEDGMENTS

We are grateful to Dr. David Merritt, Professor Rony Swennen, and Jeff Daniells who made constructive comments on the text.

REFERENCES

Adeleke, M.T.V., M. Pillay, and B.E. Okoli. 2004. Relationship between meiotic irregularities and fertility in diploid and triploid *Musa* L. *Cytologia* 69:387–393.

Alexander, M.P. 1969. Differential staining of aborted and nonaborted pollen. *Stain Technol.* 44:117–122.

Baker, J.G. 1893. A synopsis of the genera and species of Museae. *Ann. Bot.* 7:189–222.

Baskin, C.C. and J.M. Baskin. 1998. *Seeds: Ecology, biogeography, and evolution of dormancy and germination*. San Diego: Academic Press.

Baskin, C.C. and J.M. Baskin. 2004a. Germinating seeds of wildflowers, an ecological perspective. *Hort. Technol.* 14:467–473.

Baskin, J.M. and C.C. Baskin. 2004b. A classification system for seed dormancy. *Seed Sci. Res.* 14:1–16.

Baskin, J.M., C.C. Baskin, and X. Li. 2000. Taxonomy, anatomy and evolution of physical dormancy in seeds. *Plant Species Biol.* 15:139–152.

Bhat, S.R., K.V. Bhat, and K.P.S. Chandel. 1994. Studies on the germination and cryopreservation of *Musa balbisiana* seed. *Seed Sci. Technol.* 22:637–640.

Bouharmont, J. 1963. Evolution de l'ovule feconde chez *Musa acuminata* Colla subsp. *burmannica* Simmonds. *La Cellule* 63:261–283.

Chin, H.F. 1996. Germination and storage of banana seed. In: *New frontiers in resistance breeding for nematode, Fusarium and Sigatoka*, E.A. Frison, J. Horry, and D. De Waele, eds., 218–227. Montpellier, France: INIBAP.

Cox, E.A., G. Stotzky, and R.D. Goos. 1960. *In vitro* culture of *Musa balbisiana* Colla embryos. *Nature* 185:403–404.

Dahlgren, R.M.T., H.T. Clifford, and P.F. Yeo. 1985. *The families of the monocotyledons. Structure evolution and taxonomy*. Berlin: Springer-Verlag.

Darlington, C.D. 1937. *Recent advances in cytology*, 2nd ed. London: Churchill.

De Langhe, E. 2009. Relevance of banana seeds in archaeology. *Ethnobotany Res. Applns.* 7:271–281.

De Langhe, E., M. Pillay, A. Tenkouano and R. Swennen. 2005. Integrating morphological and molecular taxonomy in *Musa*: The African plantains (*Musa* spp. AAB group). *Plant Syst. Evol.* 255:225–236.

De Langhe, E., L. Vrydaghs, P. de Maret, X. Perrier, and T. Denham. 2009. Why bananas matter: An introduction to the history of banana domestication. *Ethnobot. Res. Applns.* 7:165–177.

Dodds, K.S. 1943. Genetical and cytological studies of *Musa* 5. Certain edible diploids. *J. Genetics* 45:113–138.

Dodds, K.S. 1945. Genetical and cytological studies of *Musa* 6. The development of female cells of certain edible diploids. *J. Genetics* 46:161–179.

Dodds, K.S. and N.W. Simmonds. 1948. Sterility and parthenocarpy in hybrids of *Musa*. *Heredity* 2:101–117.

Dumpe, B.B. and R. Ortiz. 1996. Apparent male fertility in *Musa* germplasm. *HortScience* 31:1019–1022.

Ellis, R.H., T.D. Hong, and E.H. Roberts. 1985. *Handbook of seed technology for genebanks*, Vol. 2, *Compendium of specific germination information and test recommendations*. Chapter 49 Musaceae. Rome: IBPGR. http://www2.bioversityinternational.org/publications/Web_version/52/ch34.htm (accessed 17 February 2010).

Fahn, A. and P. Benouaiche. 1979. Ultrastructure, development and secretion in the nectar of banana flowers. *Ann. Bot.* 44:85–93.

Fortescue, J.A. and D.W. Turner. 2004. Pollen fertility in *Musa*. Viability in cultivars grown in Southern Australia. *Aust. J. Agric. Res.* 55:1085–1091.

Fortescue, J.A. and D.W. Turner. 2005a. Growth and development of ovules of banana, plantain and ensat (Musaceae). *Sci. Hort.* 104:463–478.

Fortescue, J.A. and D.W. Turner. 2005b. The anatomy of ovule development of banana, plantain and ensat (Musaceae). *Sci. Hort.* 104:479–492.

Fortescue, J.A. and D.W. Turner. 2005c. The association between low temperatures and anatomical changes in preanthetic ovules of *Musa* (Musaceae). *Sci. Hort.*104:433–444.

Graven, P., C.G. De Koster, J.J. Boon, and F. Bouman. 1996. Structure and macromolecular composition of the seed coat of the Musaceae. *Ann. Bot.* 77:105–122.

Graven, P., C.G. De Koster, J.J. Boon, and F. Bouman. 1997. Functional aspects of mature seed coat of the *Cannaceae*. *Plant Syst. Evol.* 205:223–240.

Griffiths, A.J.F., J.H. Miller, D.T. Suzuki, R.C. Lewontin, and W.M. Gelbart. 2000. *An introduction to genetic analysis*, 7th ed. New York: W.H. Freeman.

Haig, D. 1990. New perspectives on the angiosperm female gametophyte. *Bot. Review* 56:236–274.

Higashiyama, T. and Y. Hamamura. 2008. Gametophytic pollen tube guidance. *Sexual Plant Reprod.* 21:17–26.

Humphrey, J.E. 1896. The development of the seed in Scitaminae. *Ann. Bot.* 10:1–47.

Itino, T., M. Kato, and M. Hotta. 1991. Pollination ecology of the two wild bananas, *Musa acuminata* ssp. *halabanensis* and *M. salaccensis*: Chiropterophily and ornithophily. *Biotropica* 23:151–158.

Johri, B.M., K.B. Ambegaokar, and P.S. Srivastava. 1992. *Comparative embryology of angiosperms*, Vols. 1, 2. Berlin: Springer-Verlag.

Juliano, J.B. and P.E. Alcala. 1933. Floral morphology of *Musa errans* (Blanco) Teodoro var. Botoan Teodoro. *Philipp. Agric.* 22:91–121.

Krishnamoorthy, V. and N. Kumar. 2005a. Preliminary evaluation of diploid banana hybrids for yield potential, male fertility and reaction to *Radopholus similus*. *PGR Newslett.* 141:39–43.

Krishnamoorthy, V. and N. Kumar. 2005b. Breeding potential of new synthetic banana hybrids. *PGR Newslett.* 144:63–66.

Krishnamoorthy, V., N. Kumar, and K. Sooriyanathasundaram. 2004. Influence of male and female parent on parthenocarpy. *InfoMusa*. 13(1):7–9.

Landa, R., A. Rayas, T. Ramirez, J. Ventura, J. Albert, and O. Roca. 1999. Study of the pollen fertility in the INIVIT genetic improvement programme. *InfoMusa* 8(1):27.

Liu, A-Z., W.J. Kress, H. Wang, and D-Z. Li. 2002. Insect pollination of *Musella* (Musaceae), a monotypic genus endemic to Yunnan, China. *Plant Syst. Evol.* 235:135–146.

Liu, A-Z., D-Z. Li, H. Wang, and W.J. Kress. 2002. Ornithophilous and chiropterophilous pollination in *Musa itinerans* (Musaceae), a pioneer species in tropical rain forests on Yunnan, Southwestern China. *Biotropica* 34:254–260.

Maheshwari, P. 1950. *An introduction to the embryology of the angiosperms*. New York: McGraw Hill.

Manners, M.J. 1976. The development of digestive function in the pig. *Proc. Nutr. Soc.* 35:49–55.

Marks, G.E. 1954. An acetocarmine glycerol jelly for use in pollen fertility counts. *Stain Technol.* 29:277.

McGahan, M.W. 1961. Studies on the seed of banana. Pt. 1, Anatomy of the seed and embryo of *Musa balbisiana*. *Am. J. Bot.* 48:230–238.

Nur, N. 1976. Studies on pollination in Musaceae. *Ann. Bot.* 40:167–177.

Nyine, M. and M. Pillay. 2007. Banana nectar as a medium for testing pollen viability and germination in *Musa*. *Afr. J. Biotechnol.* 6:1175–1180.

Olimpieri, I., F. Siligato, R. Caccia, L. Mariotti, N. Ceccarelli, G.P. Soressi, and A. Mazzucato. 2007. Tomato fruit set driven by pollination or by the parthenocarpic fruit allele are mediated by transcriptionally regulated gibberellin biosynthesis. *Planta* 226:877–888.

Ortiz, R. 1995. *Musa* genetics. In: *Bananas and plantains*, S. Gowen, ed., 84–109. London: Chapman and Hall.

Ortiz, R. 1997. Occurrence and inheritance of 2n pollen in *Musa*. *Ann. Bot.* 79:449–453.

Ortiz, R. and D. Vuylsteke. 1995a. Effect of the parthenocarpy gene P_1 and ploidy on fruit and bunch traits of plantain-banana hybrids. *Heredity* 75:460–465.

Ortiz, R. and D. Vuylsteke. 1995b. Factors influencing seed set in triploid *Musa* spp. L. and production of euploid hybrids. *Ann. Bot.* 75:151–155.

Ortiz, R., F. Ulburghs, and J.U. Okoro. 1998. Seasonal variation of apparent male fertility and 2*n* pollen production in plantain and banana. *HortScience* 33:146–148.

Pancholi, N., A. Wetten, and P.D.S. Calgari. 1995. Germination of *Musa velutina* seeds: Comparison of *in vivo* and *in vitro* systems. *In Vitro Cell Dev. Biol.—Plant* 31:127–130.

Penland, C.W.T. 1923. Cytological behavior in *Rosa*. *Bot. Gaz.* 76:403–410.

Perea Dallos, M. 1998. Pollen and anther culture in *Musa* spp. *Acta Hort.* 490:493–497.

Perry, L.P. and J.W. Boodley. 1980. Germination of foliage plant seeds in response to sowing media, depths of sowing, pH levels, and medium temperature. *HortScience* 15:194–196.

Reiser, L. and R.L. Fischer. 1993. The ovule and the embryo sac. *Plant Cell* 5:1291–1301.

Rodriguez-Riano, T. and A. Dafni. 2000. A new procedure to assess pollen viability. *Sexual Plant Reprod.* 12:241–244.

Roman, M.I., M. Alonso, X. Xiques, C. Gonzalez, and I. Sanchez. 2004. Estudio del numero cromosomico y la fertilidad del pollen especies y clones diploids de platano fruta (*Musa* ssp.). *Cultivos Tropicales* 25:71–73.

Sathiamoorthy, S. 1994. *Musa* improvement in India. In: *The improvement and testing of Musa: A global partnership*, D.R. Jones, ed., 188–200. Montpellier, France: INIBAP.

Sathiamoorthy, S. and V.N. Madhava Rao. 1980. Pollen production in relation to genome and ploidy in banana clones. In: *National seminar on banana production technology*, C.R. Muthukrishnan, and J.B.M.Md. Abdul Khader, eds., 46–49. Chennai, India: Tamil Nadu Agricultural University.

Sedgley, M. and A.R. Griffin. 1989. *Sexual reproduction of tree crops*. Kent: Academic Press.

Seed Information Database. 2010. http://data.kew.org/sid/SidServlet?Clade=&Order=&Family=MUSACEAE &APG=off&Genus=&Species=&StorBehav=0 (accessed 19 March 2010)

Shepherd, K. 1954. Seed fertility of the Gros Michel banana in Jamaica. *J. Hort. Sci.* 29:1–11.

Shepherd, K. 1960a. Seed fertility of edible bananas. *J. Hort. Sci.* 35:6–20.

Shepherd, K. 1960b. Seed fertility of 'Gros Michel' bananas. *Trop. Agric., Trinidad* 37:211–221.

Shepherd, K. 1999. *Cytogenetics of the genus* Musa. Montpellier, France: INIBAP.

Simmonds, N.W. 1952. The germination of banana seeds. *Trop. Agric., Trinidad* 29:35–49.

Simmonds, N.W. 1953a. The development of the banana fruit. *J. Exp. Bot.* 4:87–105.

Simmonds, N.W. 1953b. Segregations in some diploid bananas. *J. Genetics* 51:458–469.

Simmonds, N.W. 1959. Experiments on the germination of banana seeds. *Trop. Agric., Trinidad* 36:259–273.

Simmonds, N.W. 1960. Experiments on banana fruit development. *Ann. Bot.* 24:212–222.

Simmonds, N.W. 1962. *The evolution of the bananas*. London: Longman.

Simmonds, N.W. 1966. *Bananas*. 2nd ed. London: Longman.

Simmonds, N.W. and K. Shepherd. 1955. The taxonomy and origins of the cultivated bananas. *J. Linn. Soc. London (Bot.)* 55:302–312.

Soares, T.L., S.O. Silva, M.A.P.C. Costa, J.A. Santos-Serejo, A.S. Souza, L.S.M. Lino, et al. 2008. *In vitro* germination and viability of pollen grains of banana diploids. *Crop Breed. Appl. Biotechnol.* 8:111–118.

Ssebuliba, R.N., P. Rubaihayo, A. Tenkouano, D. Makumbi, D. Talengera, and M. Magambo. 2005. Genetic diversity among East African Highland bananas for female fertility. *Afr. Crop Sci. J.* 13:13–26.

Ssebuliba, R.N., M. Magambo, D. Talengera, D. Makumbi, A. Tenkouano, P. Rubaihayo, et al. 2006. Biological factors affecting seed production in East African Highland bananas. *J. Crop Improvement* 16:67–79.

Ssebuliba, R.N., A. Tenkouano, and M. Pillay. 2008. Male fertility and occurrence of 2*n* gametes in East African Highland bananas (*Musa* ssp.). *Euphytica* 164:53–62.

Ssebuliba, R.N., D. Makumbi, and M. Pillay. 2009. Patterns of seed set in East African Highland banana (*Musa* sp.) hybrids. *J. New Seeds* 10:160–170

Stephens, C.S. and E.L. Tyson. 1975. The nectar bat, *Glossophaga soricina*, as a pest of bananas in Panama. *Trop. Agric., Trinidad* 52:85–89.

Stotzky, G. and E.A. Cox. 1962. Seed germination studies in *Musa*. Pt. 2, Alternating temperature requirement for the germination of *Musa balbisiana*. *Am. J. Bot.* 49:763–770.

Stotzky, G., E.A. Cox, and R.D. Goos. 1962. Seed germination studies in *Musa*. Pt. 1, Scarification and aseptic germination of *Musa balbisiana*. *Am. J. Bot.* 49:515–520.

Stover, R.H. and N.W. Simmonds. 1987. *Bananas*. 3rd ed. London: Longman.

Swennen, R. and D. Vuylsteke. 1990. Aspects of plantain breeding at IITA. In: *Sigatoka leaf spot diseases of bananas*, R.A. Fullerton and R.H. Stover, eds., 252–266. Montpellier: INIBAP.

Swennen, R., D. Vuylsteke, and K. De Smet. 1991. Season dependent seed set in plantain. *Banana Newslett.* 14:35–36.

Swennen, R., D. Vuylsteke, and S.K. Hahn. 1992. The use of simple biotechnological tools to facilitate plantain breeding. In: *Biotechnology: Enhancing research on tropical crops in Africa*, G. Thottappilly, L.M. Monti, D.R. Mohan Raj, and A.W. Moore, eds., 69–74. Ibadan, Nigeria: CTA/IITA co-publication.

Swennen, R., D. Vuylsteke and R. Ortiz. 1995. Phenotypic diversity and patterns of variation in West and Central African Plantains (*Musa* spp., AAB group Musaceae). *Econ. Bot.* 49:320–327.

Talon, M., L. Zacarias, and E. Primo-Millo. 1992. Gibberellins and parthenocarpic ability in developing ovaries of seedless mandarins. *Plant Physiol.* 99:1575–1581.

Tesfaye, M. 1992. The effect of soaking, temperature and other pretreatments on the germination of "enset" seed. *Seed Sci. Technol.* 20:533–538.

Traveset, A., N. Riera and R.E. Mas. 2001. Passage through bird gut causes interspecific differences in seed germination characteristics. *Functional Ecol.* 15:669–675.

Turner, D.W. and N. Hunt. 1984. *Growth, yield and leaf nutrient composition of 30 banana varieties in subtropical New South Wales*. Technical Bulletin 31. Sydney, Australia: Department of Agriculture, NSW.

Vleeshouwers, L.M., H.J. Bouwmeester, and C.M. Karssen. 1995. Redefining seed dormancy: An attempt to integrate physiology and ecology. *J. Ecol.* 83:1031–1037.

Vuylsteke, D., R. Swennen, and E. De Langhe. 1990. Tissue culture technology for the improvement of African plantains. In: *Sigatoka leaf spot diseases of bananas*, R.A. Fullerton and R.H. Stover, eds., 316–337. Montpellier: INIBAP.

Wattanachaiyingcharoen, D. 1990. Viability, germination and dormancy of banana seed (*Musa acuminata* ssp.). Master's thesis, The University of Western Australia.

Wattanachaiyingcharoen, D. and D.W. Turner. 1989a. Improving germination and breaking dormancy of banana seed (*Musa acuminata* ssp.) *Banana Newslett.* 12:4–5.

Wattanachaiyingcharoen, D. and D.W. Turner. 1989b. The staining pattern of tetrazolium and its association with the number of viable and non-viable seeds of *Musa acuminata* ssp. *Banana Newslett.* 12:5–7.

White, P.R. 1928. Studies on the banana. An investigation on the floral morphology and cytology of certain types of the genus *Musa* L. *Zeitschrift. Fuer. Zellforschung und Mikroskopische Anatomie. Bd* 7:673–733.

Xue, C-Y., H. Wang, and D-Z. Li. 2005. Microsporogenesis and male gametogenesis in *Musella* (Musaceae), a monotypic genus from Yunnan, China. *Ann. Bot. Fennici* 42:461–467.

9 Breeding Techniques

Abdou Tenkouano, Michael Pillay, and Rodomiro Ortiz

CONTENTS

9.1 INTRODUCTION: SHORT HISTORY OF BREEDING PROGRAMS

Banana improvement formally started in response to the decimation of large commercial plantations of dessert bananas by Panama disease caused by *Fusarium oxysporum* f. sp. *cubense* (Ploetz, 1992). No chemical or cultural method was available to control this fungal disease, which rapidly spread from its area of first occurrence in Surinam to Honduras, the largest commercial banana producer, in the early 1920s. By the 1950s, the disease had virtually wiped out the prevailing 'Gros Michel' cultivar and threatened the mere existence of commercial dessert banana production. This prompted the search for new cultivars, initially through exploration of existing accessions, to replace the 'Gros Michel' cultivar. These were found in the Cavendish subgroup of AAA bananas, which became and remain to date the predominant type of commercial dessert banana. The outbreak of fungal Sigatoka diseases in the early 1970s and the financial and environmental cost of chemical treatment gave additional impetus for genetic improvement because no existing cultivar had natural resistance to the new diseases.

The earliest attempts to breed bananas were at the Imperial College of Tropical Agriculture (Trinidad), but this program has virtually ceased to operate. The oldest breeding program in operation is that of the Fundación Hondureña de Investigación Agrícola (FHIA, http://honduras.com/fhia/banana.htm, accessed December 27, 2009), which was established in 1958 by the United Fruit

Company in La Lima, Honduras. FHIA's program initially focused on dessert bananas, but it gradually expanded to plantain and cooking banana for which it is most known. Other major programs in existence were established from the mid 1970s to the mid 1990s. This includes the programs of the Centre de Coopération Internationale en Recherche Agronomique pour le Développement (CIRAD, based in Guadeloupe for dessert bananas), the Centre Africain de Recherches sur les Bananiers et Plantains (CARBAP, based in Cameroon for plantains), and the International Institute of Tropical Agriculture (IITA, with two main programs established in Nigeria for plantains and in Uganda for the East African Highland bananas).

Relatively smaller or domestically focused programs have been established in Austria (International Atomic Energy Agency [IAEA]), Brazil (Empresa Brasileira de Pesquisa Agropecuaria [EMBRAPA]), India (National Center for Research on Banana [NCRB], Tamil Nadu Agricultural University [TNAU], and Kerala Agricultural University [KAU]), Taiwan (Taiwan Banana Research Institute [TBRI]), Uganda (National Research Organization [NARO]), and Côte d'Ivoire (Centre National de Recherches Agronomiques [CNRA]). The program based in Taiwan is focused on the search for somaclonal variants using induced mutagenesis as the main avenue for creating genetic diversity. All other programs, except that of CNRA, use the array of tools available for creating diversity, with a preponderance of cross-pollination.

In 1985, the International Network for Improvement of Banana and Plantains (INIBAP, France) was established to coordinate the efforts of the various institutions involved in the development and dissemination of new cultivars and to facilitate access to genetic stocks and new cultivars through the International *Musa* Testing Program (IMTP). INIBAP established offices in Africa, Asia, and Latin America and coordinates several networks of banana researchers involved in various disciplines contributing to genetic improvement. The INIBAP International Transit Center operating at the Katholieke Universitat Leuven (KULeuven, Belgium) remains as the main *ex situ, in vitro* gene bank of bananas in the world. To date, INIBAP, nowadays included in one of the programs of Bioversity International (Italy), constitutes the largest source of information on banana research. The number of breeding programs remains small, despite the place of banana in world trade and as a global staple crop (Escalant and Panis, 2002).

9.2 BREEDING OBJECTIVES IN AFRICA

Banana and plantain (*Musa* spp. L.) constitute major export crops in some countries, predominantly in Latin America and the Caribbean and a few countries in Western and Central Africa. However, only about 10% of world production enters international trade. Banana and plantain are major staple foods for rural and urban consumers and an important source of income for rural populations in the production areas, particularly in the equatorial belt of Africa, where more than 70 million people derive in excess of 25% of their daily calorie intake from plantains in West and Central Africa (Robinson, 1996). Seventy-three percent of the world crop of plantain is grown and consumed in Western and Central Africa (WCA), although increasing quantities of plantain are being exported to the United States and Europe (Marin et al., 2003).

Despite their introduction to Africa only about 3,000 years ago (De Langhe, 1995), a remarkable morphological diversity now exists for both banana and plantain in this region, with a multitude of vernacular names for the prevailing cultivars, superimposing linguistic diversity to genetic and ecological diversity of the cultivars in any given area (Rossel, 1998). Similar interactions between linguistics and genecology occur throughout Asia, the geographical origin and primary diversification center of the Musaceae. Many indigenous cultivars are susceptible to a range of pests and diseases. The most serious threat to the production of plantains is caused by the Sigatoka leaf diseases: yellow Sigatoka (*Mycosphaerella musicola*) in the highlands and black Sigatoka (*M. fijiensis*) in the lowlands. Black Sigatoka is the most damaging and costly disease of bananas and plantains because its control accounts for 27% of total production costs in commercial plantations (Stover, 1980). Other long-standing threats to banana and plantain production are caused by Fusarium wilt (*Fusarium oxysporum* f. sp. *cubense* [E.F. Smith] Snyder and Hansen), banana weevil (*Cosmopolites sordidus* Germar), and a complex of plant parasitic nematodes (*Pratylenchus goodeyi* Sher and Allen, *Helicotylenchus multicinctus* [Cobb] Golden, and *Radopholus similis* [Cobb] Thorne), but bacterial wilt (*Xanthomonas campestris* pv. *musacearum*) has recently emerged as the most damaging disease of bananas in the highlands and Great Lakes region of East Africa while banana bract mosaic virus has become a cause of concern in Central Africa and to a lesser extent in West Africa.

Black Sigatoka was accidentally imported into Africa in the late 1970s (Wilson and Buddenhagen, 1986) and quickly reached epidemic proportions. When unimpeded, the disease induces leaf decay, thereby reducing the photosynthetic area, and considerable reduction in yield or complete crop failure may ensue (Mobambo et al., 1993). Because none of the indigenous cultivars had natural resistance, black Sigatoka has since reduced the yields of these cultivars to less than half what they were before the arrival and spread of this disease.

Black Sigatoka has been especially devastating to the food security for the people in the affected regions, essentially for two reasons: (1) there were inadequate supplies of food even before this new disease further reduced production, and (2) prior to the arrival and spread of this disease, plantains and bananas were one of the most dependable foods since these crops were not as vulnerable to production uncertainties (droughts, floods, diseases, insects, and storage losses) as the basic grains.

Black Sigatoka turned plantain into a delicacy in urban Nigeria and other countries in Africa, due to the high cost of the fruits subsequent to the reduced production caused essentially by the disease. Chemical protection is relatively effective but expensive and detrimental to the environment, leaving host-plant resistance as the most practical option for sustainable control of the disease for resource-poor farmers of the developing world where the bulk of the production takes place. In particular, the thousands of small-scale plantain farmers in Africa do not have access to, and cannot afford, the fungicides used by banana exporters and commercial producers for control of black Sigatoka; the only sustainable solution is to plant hybrids with genetic resistance to this disease.

It is against this background that *Musa* breeders aim to develop new disease- and pest-resistant cultivars that also retain the organoleptic properties of the traditional cultivars. Besides resistance to biotic threats, genetic improvement of *Musa* also aims at developing hybrids that are high yielding per unit area and time (Buddenhagen, 1996). Hence, improved hybrids should be photosynthetically efficient, early to mature in the first production cycle, and display minimum delay between consecutive harvests (Eckstein et al., 1995; Ortiz and Vuylsteke, 1994a). Other desirable characteristics include short stature and strong roots for optimal nutrient uptake and greater resistance to wind damage. These breeding objectives are summarized in Figure 9.1.

9.3 BREEDING SCHEMES

There are two basic steps in plant breeding: (1) access natural variation or artificially create genetic diversity, and (2) select individuals with the desirable gene combinations from existing or artificially created populations. These principles have been successfully applied to the improvement of sexual, seed-propagated crop species using conventional breeding techniques, which are commonly viewed as inappropriate for banana and plantain improvement. It seems that the distinction between "conventional" or "classical" and "nonconventional" breeding lies in differences in the methods of gene shuffling and in the origin of the genes being manipulated.

Nevertheless, the methodological principles applied for sexually propagated crops can be, and have been, applied to banana and plantain breeding, largely with the aid of nonfield methods of recovering viable progeny and identifying those progenies with putatively desirable gene combinations. Thus, banana and plantain breeding is neither strictly conventional nor strictly non-conventional, which will be illustrated in the subsequent sections of this chapter.

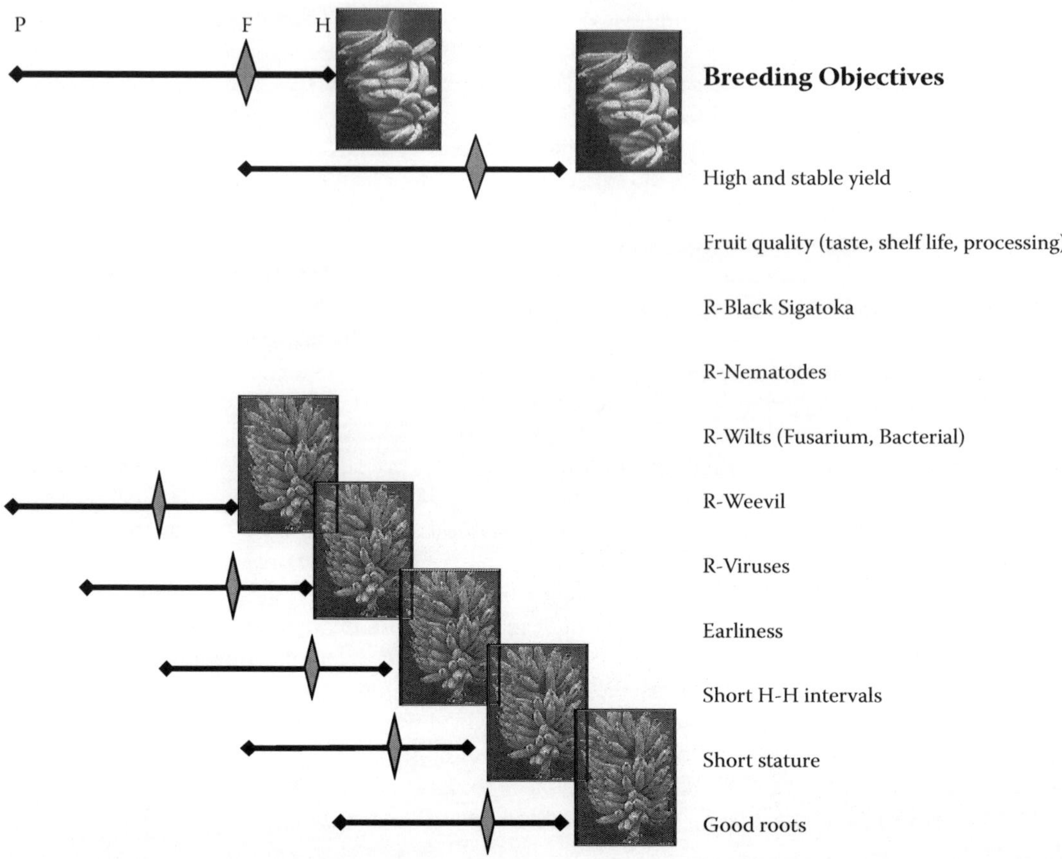

FIGURE 9.1 Illustration of breeding objectives for banana and plantain: increased frequency of harvests of high and stable yield with good fruit quality through resistance to an array of pests and diseases and greater biological efficiency. The letters P, F, and H indicate planting, flowering, and harvesting events, respectively. The letter R indicates resistance.

9.3.1 Pollination and Seed Production

Breeding starts with effective pollination to produce seeds. Pollination is the transfer of pollen from the anther to stigma either within a flower or between flowers of the same genotypes (autogamy) or between flowers of different genotypes (allogamy). This mechanism guarantees genetic recombination and preserves variability within populations. An efficient pollination scheme can be measured by the proportion of seed set (Cruden, 1977) and the diversity of genes that can be incorporated into individual plants within limited time and resources.

In banana and plantain, the male gametophyte is easily affected by the environment because the pollen grains must be transferred from one plant to another by pollinating agents. This is compensated for by the production of a large number of pollen grains. In contrast, the female gametophyte develops within the ovary and is less affected by the environment. Hence, smaller numbers of female gametophytes are observed across species. Pollination success also depends on the ability of the male gamete to access the ovule and the ability of the female to provide adequate resources for seed and fruit development. The duration of flowering, flower arrangement, and the pattern of anther dehiscence are also important aspects that determine pollination success.

A common practice in hand pollination in most crops is to start in the morning as soon as pollen is shed, because pollen viability declines with time. Thus, the male and female flowers of bananas and plantains are covered shortly before anthesis to prevent contamination by alien pollen. At anthesis, pollen grains from male flowers are collected and brushed against female flowers, usually between 7 a.m. and 10:30 a.m. (Swennen and Vuylsteke, 1993). Pollination can be carried out over a period of 7 to 15 days depending on the number of hands present in the bunch under pollination. It usually takes from 60 to 90 days for the pollinated bunch to reach maturity, following which the fruits are subjected to rapid ripening before seeds are mechanically extracted.

Pollination can be carried out year round, but Ortiz and Vuylsteke (1995) found that periods of high temperature, high solar radiation, and low relative humidity were most favorable for seed production in banana and plantain. Hand pollination is a tedious and relatively inefficient process, considering the time required, the poor seed set, and the quality of the seed produced. The potential of open pollination for improvement in *Musa* has been suggested (Ortiz et al., 1995). Among the open pollination methods, the polycross approach has been widely used in forage and maize to develop superior genetic recombinants (Allard, 1960). Although attractive, this scheme requires effective control over prospective parents, notably by choosing prospective parents on the basis of their combining ability and floral synchrony prior to growing them in isolated crossing blocks where natural agents would effect pollination.

At maturity, the seeds are black or dark brown stony bodies of variable size and shape with a rough seed coat and a complex structure that makes germination very difficult to achieve (Kiew, 1987; Chin, 1995; Graven et al., 1996).

9.3.2 Seed Germination

Regardless of the approach, most cultivated cultivars display relatively high infertility levels, which is the most important obstacle to overcome in order to create genetic variability. Cross-pollination usually produces sexual embryos that can develop into an array of genetically varied offspring, but the embryos often abort or develop into seeds that lack enough autonomy to develop into seedlings. The embryos must be rescued using tissue culture, which has become an integral component of banana and plantain breeding (Vuylsteke, 1989). With the small number of seeds recovered from crosses (typically a few per pollinated bunch), *Musa* breeders disproportionately allocate their resources for producing, rather than evaluating, progenies.

9.3.3 Somaclonal Variation

Tissue culture not only serves as a tool for producing progenies from crosses but also stands as a diversity-generating option on its own. Tissue culture has the potential of generating somaclonal variation, which is genetic in nature and can be stably passed onto progeny (Larkin and Scowcroft, 1981). Vuylsteke et al. (1988) have reported that variations in inflorescence morphology occur in populations of 'False Horn' plantains derived from micropropagation, with a reversion to a typical French plantain bunch type accounting for 40–100% of the total variability. Somaclonal variation prevents clonal uniformity and can be a serious nuisance, particularly for populations used to establish commercial banana plantations (Damasco et al., 1996; Grillo et al., 1998). However, this phenomenon may also provide a useful source of genetic variability to the plant breeder as was first demonstrated in sugarcane with the *in vitro* selection of a commercial cultivar resistant to Fiji disease (Heinz and Mee, 1969). Likewise, Hwang and Ko (1987) obtained somaclonal variants of bananas with putative field resistance to Fusarium wilt (*Fusarium oxysporum* Schlect. f. sp. *cubense* [E.F. Smith] Snyder and Hansen) while Trujillo and Garcia (1996) selected a variant of the Cavendish subgroup resistant to yellow Sigatoka (*M. musicola* Leach).

The resistant somaclones often had inferior horticultural characteristics, but these examples suggest that somaclonal variation arising in culture may produce useful variants for the genetic improvement of *Musa* spp. However, to date all studies concerning somaclonal variation of plantains suggested that this source of variation may not readily provide useful agronomic improvements in this crop (Vuylsteke et al., 1991; Vuylsteke, 1998; Nwauzoma et al., 2002). Understanding the molecular basis of such variations could provide a more deterministic approach for using somaclonal variation as a breeding tool. A recent study at IITA assessed differences in genomic regions varying within clones grown naturally or *in vitro*, using 52 simple sequence repeats (SSR) markers and 6 *Eco*RI/*Msp*I-AFLP primer combinations (Vroh et al., 2010). Sequencing the polymorphic regions at one SSR locus with significant similarity to an arcelin gene revealed a deletion in a subculture regenerant. Out of 390 amplified fragment length polymorphism (AFLP) bands, 24 (6.15%) accounted for within-clone variations, among which 0.5% and 5.65% occurred in conventionally and micro-propagated plants, respectively. Homology searches revealed that most polymorphic AFLP sequences were related to cytochrome P450, cell-wall biosynthesis, and senescence genes. Thus, somaclonal variation may result from mutagenic events occurring in hot spots that might not be associated with the expression of traits of horticultural value.

Tissue culture also offers opportunities for using physical or chemical mutagenesis as well as genetic transformation based on site-directed or unspecific gene insertion, disruption, or substitution to create genetic variability, with varying degrees of success (Mohan Jain and Swennen, 2004; Tripathi et al., 2008). There has been no assessment of the comparative efficiency of alternative schemes of generating, managing, and exploiting genetic diversity (induced mutation breeding, fortuitous somaclonal variation, genetic transformation, and biotechnology-assisted crossbreeding). However, each approach has produced lines that are now reaching farmers, either as experimental materials in the case of genetic transformation or as confirmed cultivars for the other approaches. The rest of this chapter will focus on crossbreeding.

9.4 BREEDING PHILOSOPHIES

Breeding for resistance to *M. fijiensis* is a fairly simple process in theory, but there are two competing philosophies that are not mutually exclusive. Both are based on a good understanding of the evolution of banana and plantain from interspecific hybridization and polyploidization involving derivatives of *M. acuminata* Colla and *M. balbisiana* Colla that are the diploid ($2n = 2x = 22$) ancestors of present-day cultivars to which they contributed the A and B genomes, respectively (Simmonds and Shepherd, 1955).

Two critical events subtend the evolutionary pathway leading to the emergence of cultivated cultivars from the seminiferous ancestors. The first event is the appearance of vegetative parthenocarpy and female sterility in *M. acuminata*, allowing for production of pulp without seeds as evidenced by the occurrence of parthenocarpic and seedless diploid *M. acuminata*. The second event includes crosses within *M. acuminata* or between *M. acuminata* and *M. balbisiana* coupled with female restitution and haploid fertilization, which produced two groups of natural hybrids. One group comprises autoploid and homogenomic hybrids that are essentially AAA dessert and East African Highland bananas. The other group contains the alloploid and heterogenomic AAB plantains and the ABB cooking bananas. Since these cultivars produce no seeds and cannot reproduce sexually, only vegetative propagation is possible, implying that survival in nature and geographical dispersal do not occur without human intervention. Consequently, secondary diversification in areas devoid of wild *Musa* plants must be due to somatic mutations of the introduced materials. It is generally accepted that edibility of mature fruits in interspecific hybrids comes from the A genome while starchiness and acid taste come from the B genome, causing AAB plantain to be starchier but less sweet and less palatable when raw than the AAA dessert bananas (Simmonds, 1962).

9.4.1 Evolutionary Breeding

One school of thought, which we may call evolutionary breeding, attempts to mimic the evolutionary development of the *Musa* species complex (Rowe and Rosales, 1996; Ortiz, 1997a). In this process, female fertile triploid ($2n = 3x$) landraces of banana (AAA genomes) or plantain (AAB genomes) are crossed to diploid ($2n = 2x$) accessions of *M. acuminata* (AA genomes) or *M. balbisiana* (BB genomes) that are resistant to black Sigatoka, having coevolved with *M. fijiensis* in their area of origin in Southeast Asia. The $3x \times 2x$ crosses produce primary hybrids with considerable phenotypic variation subtended by ploidy and genome polymorphisms, among which $2x$ and tetraploid ($2n = 4x$) progenies that are disease resistant and agronomically desirable are selected and intermated via $4x \times 2x$ crosses to produce high-yielding secondary $3x$ hybrids (Vuylsteke et al., 1993a). When $4x$ and $2x$ offspring take into consideration the genetic diversity of their $3x$ progenitors, this process allows for the establishment of genetic bridges between banana or plantain landraces that cannot be crossed directly due to reproductive barriers (Tenkouano, 2001). Ideally, therefore, plantain genes from different sources may be brought together in a $3x$ background where the resistance genes lacking in the traditional cultivars will have been restored, resulting in an improved plantain hybrid that retains the preferred processing properties of the landraces (Tenkouano, 2001). Completing this scheme is recurrent $2x$ breeding to produce $2x$ stocks that retain the disease resistances of their $2x$ progenitors with improved agronomic characteristics. It is widely accepted that improved $2x$ would be better parents for the improvement of $3x$ landraces (Tezenas du Montcel et al., 1996). It has been reported that traits of economic importance in banana and plantain—for example, pest resistance, increased bunch weight, and reduced time interval between flowering and harvest—are more predictably inherited from a $2x$ male background than from parents with a higher ploidy status (Tenkouano et al., 1998a, 1998b). Furthermore, fewer chromosomal imbalances occur during meiosis in $2x$ compared to polyploids due to the disomic nature of inheritance in $2x$, thereby making genetic analysis more attractive in $2x$ than in polyploids. This fact gives in retrospect a justification for past and current investment in the development of $2x$ breeding stocks with a balanced combination of resistance and good agronomic features, as pursued by major programs worldwide (Ortiz and Vuylsteke, 1996). Breeders following this school of thought have used a limited pool of accessions in their work, due to reproductive barriers caused by the low fertility of existing triploid cultivars used as candidate parents (base generation) for improvement. At IITA, for example, the base generation consisted mainly of a few triploid landraces ('Bobby Tannap,' 'Mbi Egome,' and 'Obino l'Ewai') of the Medium French subgroup of AAB plantains (Swennen, 1990; Swennen et al., 1995). The sources of resistance were the diploid AA accessions *M. acuminata* ssp. *burmannicoides* 'Calcutta 4,' *M. a.* ssp. *malaccensis* 'Pisang Lilin,' and *M. a.* ssp. *microcarpa* 'Tjau Lagada.'

The accession 'Calcutta 4' is a nonedible wild accession from Myanmar (De Langhe and Devreux, 1960) that produces many true-seeded fleshless fruits because it lacks one of the three dominant complementary genes for parthenocarpy (Simmonds, 1953). 'Calcutta 4' is abundantly polleniferous and resistant to black Sigatoka and several nematode species, including *R. similis* and *P. coffeae* (Viaene et al., 2000). 'Calcutta 4' is morpho-taxonomically related to plantain (De Langhe, 1969), justifying its extensive use in crosses to improve plantain landraces (Swennen and Vuylsteke, 1993; Vuylsteke et al., 1993b; Vuylsteke and Ortiz, 1995). The accession 'Pisang Lilin' originated in Malaysia (Stover and Simmonds, 1987) and is described as a translocation heterozygote with edible parthenocarpic fruits and high resistance to black Sigatoka. It is moderately female fertile but highly male fertile and known to frequently produce $2n$ or $4n$ pollen, depending on the season (Ortiz, 1997b). Both 'Calcutta 4' and 'Pisang Lilin' are reference accessions of the International *Musa* Testing Program (Orjeda, 2000). 'Tjau Lagada' was originally collected in the Java island of Indonesia (Stover and Simmonds, 1987) with an incomplete resistance to black Sigatoka that may best be described as a slow-lesion development type. It is moderately female and male fertile, with a characteristic long bunch with many hands, and moderately parthenocarpic fruits. We denote the triploid cultivars as $3x^0$ and the diploid accessions as $2x^0$. The primary diploid descendants derived from crossing these accessions to triploid cultivated landraces or to other $2x^0$ diploid accessions constitute the first generation, which we refer to as $2x^1$ hybrids. Likewise, primary $3x$ or $4x$ descendants from the $3x^0 \times 2x^0$ crosses are denoted as $3x^1$ and $4x^1$, respectively. However, the most interesting descendants in the first generation were $2x^1$ and $4x^1$ hybrids. The second generation is obtained by crossing the primary generation hybrids in various combinations ($2x^1 \times 2x^1$, $4x^1 \times 2x^1$, or $2x^1 \times 4x^1$). The most frequent combinations have been the $2x^1 \times 2x^1$ and $4x^1 \times 2x^1$ crosses, producing

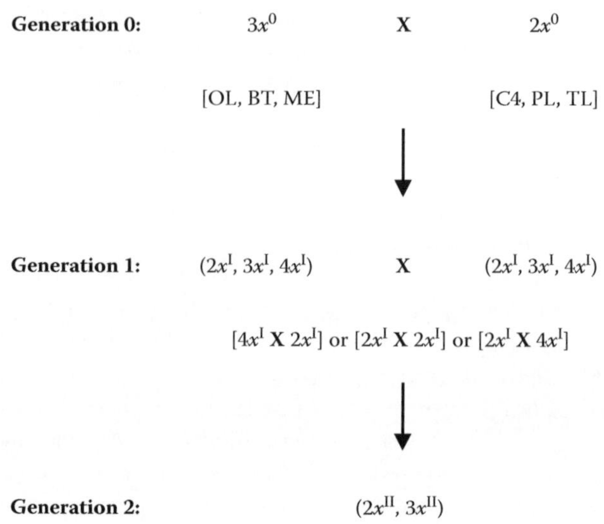

FIGURE 9.2 Schematic representation of the plantain breeding process whereby initial crosses are carried out between triploid ($2n = 3x$) landraces (predominantly 'Obino L'Ewai' [OL], 'Bobby Tannap' [BT], and 'Mbi Egome' [ME]) and diploid ($2n = 2x$) sources of resistance (predominantly 'Calcutta 4' [C4], 'Pisang Lilin' [PL], and 'Tjau Lagada' [TL]), followed by intermating best primary diploid, triploid, and tetraploid ($2n = 4x$) progenies using $2x \times 2x$, $4x \times 2x$, and $2x \times 4x$ crosses to produce secondary diploid ($2x^{II}$) or triploid ($3x^{II}$) descendants. Diploid improvement can also be concomitantly carried out as a complementary scheme.

predominantly $2x^{II}$ and $3x^{II}$ descendants, respectively. The $2x^I \times 4x^I$ crosses have been less frequent, but they produce predominantly $2x^{II}$ descendants (Oselebe et al., 2006).

9.4.2 Reconstructive Breeding

The second school of thought may be described as reconstructive breeding, whereby breeders attempt to reenact the landraces by first determining their most likely ancestors from the pool of $2x$ species and subsequently using only such putative ancestors or closely related derivatives in crosses aiming at synthesizing improved variants of the susceptible landraces (Carreel et al., 2002). Proponents of this school of thought advocate the use of colchicine to induce tetraploidization of $2x$ accessions prior to crossing with another $2x$ (Tezenas du Montcel et al., 1996). Colchicine prevents the formation of mitotic spindles, causing mitotic restitution in treated cells. However, colchicine treatment may not affect uniformly all cells in multicellular meristems, causing cytochimeras that may not be easy to dissociate (Roux et al., 2001). Thus, efficient methods for *in vitro* dissociation of chimeras and selection of the desired cells are required before such cells can be cultured to regenerate a plant that will now be used for crossbreeding. Furthermore, the use of colchicine could result in increased inbreeding, reduced vigor, and reduced genetic variability (Ortiz et al., 1992).

Despite these limitations, Bakry et al. (2007) have successfully derived stable, nonchimerical, colchicine-induced $4x$ variants from several diploid *M. acuminata* accessions. Crossing the induced $4x$ to $2x$ clones resulted in the production of $3x$, as anticipated (Bakry et al., 2007). This approach now forms one of the pathways used by the breeding program of the Centre de Coopération Internationale en Recherche Agronomique pour le Développement (CIRAD). CIRAD's program aims at creating commercial cultivars of AAA dessert bananas for the banana industry operating in the Caribbean islands of Guadeloupe and Martinique. Whether this approach could be used for other *Musa* groups is an attractive idea that needs to be explored.

However, it is conceivable that the identification of putative $2x$ ascendants of the present-day cultivars could also be followed by a breeding process that could be the same as for evolutionary breeding, albeit with initial crosses of the landraces with their putative ascendants, as described above. With the tremendous developments in the use of molecular or cytological tools, it is possible to clarify the phylogenetic relationships in the Musaceae, allowing for inferences about gene pools and the search for novel genes. Thus, AFLP analysis of total DNA revealed that the *M. acuminata* complex of about 50 morphological subspecies can be ascribed to only three genetic subspecies (*burmannica*, *malaccensis*, and *microcarpa*) while *M. balbisiana* was shown to contain two forms ('Singapuri,' 'I-63'), suggesting that there might be at least three A genomes and two B genomes (Ude et al., 2002a, 2002b). Furthermore, it seems that the A genomes in the plantains came from the subspecies *microcarpa* while the B genome was from the 'I-63' form of *M. balbisiana*. More recently, Ge et al. (2005) also traced variation in *M. balbisiana* to two main clades, based on SSRLP and cpDNA analysis.

A major retroactive justification for this school of thought was the discovery in the early 1990s of banana streak badnavirus (BSV) and its association with genetically variable but otherwise quiescent integrated forms in the B genomes that are activated by factors located in the A genomes (Lockhart and Olszewski, 1993; LaFleur et al., 1996; Dahal et al., 1998). Thus, interspecific hybridization may trigger genome interactions that are favorable to the transcriptional recombination of dispersed BSV sequences into virulent episomes. Therefore, breeding emphasis would be on avoiding bringing compatible genomes together in hybrid cultivars (Tenkouano et al., 2001).

9.5 OCCURRENCE AND POTENTIAL OF 2*N* GAMETES IN BANANA BREEDING

Triploid breeding can take advantage of the occurrence of $2n$ gametes via chromosome doubling during microsporogenesis, which has been observed in some diploid *Musa* accessions and is apparently under genetic control (Ortiz, 1997b). Fertilization of a normal egg ($n = x = 11$) from a $2x$ individual by a $2n$ pollen grain ($n = 2x = 22$) from another $2x$ plant produces a $3x$ progeny ($2n = 3x = 33$), which is a process termed unilateral sexual polyploidization (USP). A major advantage of the USP-based breeding is that genetic analyses will be essentially carried out at the $2x$ level, avoiding complex inheritance patterns involving ploidy polymorphisms within and across generations.

The analysis of breeding populations at IITA revealed that nearly 21% of these population produced $2n$ pollen with highly significant ($\chi^2 = 597.8$, $P = 0.001$) differences among genotypes. Furthermore, $3x$ progenies were obtained from $2x \times 2x$ crosses. Preliminary data suggests that such $3x$ hybrids may prove to be agronomically superior to their $2x$ counterparts derived from same or similar crosses. For example, the cross 'Wh-o-gu' ($2x$) × 'Calcutta 4' ($2x$) has produced a $3x$ genotype '1968-2' which is early maturing and has shown field resistance to black Sigatoka, weevils, and nematodes. Another example is 25287-S28 (1448-1 × 1448-1) that displayed a bigger bunch associated with increased fruit length and circumference, compared to the diploid hybrids (Table 9.1). This could have been a consequence of higher order interactions, leading to better exploitation of recombinative heterosis.

In practice, breeding $3x$ hybrids via USP requires (1) population improvement at the $2x$ level to accumulate favorable alleles in the $2x$ background, (2) screening of the $2x$ breeding populations for $2n$ pollen production, and (3) synchronizing pollination with periods that are favorable for $2n$ pollen production. Hence, $4x \times 2x$ crosses, coupled with recurrent $2x$ breeding to make better $2x$ sources of resistance or other desirable traits, remain the predominant $3x$ breeding scheme.

9.6 SELECTION

Selection is the process whereby individuals carrying the genes of interest are identified on the basis of the level of expression of the traits subtended by the genes. While this may seem straightforward for many crop species, efficient selection in banana and plantain breeding requires sieving through a diversity of ploidy and genome configurations.

TABLE 9.1
Performance of USP-Derived Triploid Hybrid (25287-S28) Compared to Diploid Hybrids in a Preliminary Yield Trial at IITA High Rainfall Station, Onne, Southeastern Nigeria

	Bunch Weight (kg)		Fruit (no. of hands)		Fruit (no. of fingers)		Length (cm)		Girth (cm)	
Hybrid	P	R	P	R	P	R	P	R	P	R
Triploid										
25287-S28	17.0	14.4	9	8	182	179	17.8	15.4	9.7	9.6
Diploids										
25291–1A	6.5	15.8	8	11	167	291	10.7	14.4	7.5	9.4
25291-S26	10.2	–	8	–	177	–	16.8	–	9.1	–
25291-S32	15.1	14.5	13	14	262	287	16.8	14.0	8.3	9.0
25291-S41	5.6	10.3	10	12	151	226	12.2	14.2	8.1	8.0
25291-S62	9.1	14.0	10	12	168	246	18.6	17.2	9.2	8.6
25447-14	6.9	15.6	9	11	193	257	12.1	14.4	7.6	9.3
LSD ($P = 0.05$)	1.7	3.9	1	2	27	43	6.7	2.8	1.2	1.3

Note: P = plant crop; R = ratoon crop; LSD = least significant difference.

The multiploidy and heterogenomic structure of breeding populations results in unpredictable variation in genome size and structure across and within generations (Tenkouano, 2001). This causes complex inheritance patterns and complicates phenotypic selection for most yield and growth traits (Ortiz and Vuylsteke, 1996). Breeders would gain in efficiency if they were able to assign segregating progenies to ploidy and genome classes putatively indicative of their potential use as plantain, dessert, or cooking banana, prior to field evaluation.

9.6.1 Ploidy and Genome Analysis

Methods for unambiguous detection of the ploidy and genome status of *Musa* breeding populations have been extensively described. The most accurate and user friendly of these methods are flow cytometry (Dolezel et al., 1994, 1997) and genome-specific DNA markers (Pillay et al., 2000, 2004; Nwakanma et al., 2003). The application of these methods is best carried out at the nursery stage in juvenile plants, allowing for sorting the progenies into ploidy and genome classes, prior to field establishment for performance evaluation, which typically takes place approximately 9 to 10 months following the date the cross was made.

9.6.2 Phenotypic Evaluation

Field evaluation follows a rather classical approach of phenotypic selection. Thus, seedlings from crosses usually undergo a series of trials, starting with early evaluation trials (EET) that are mostly nonreplicated trials in which different numbers of plants per genotype are evaluated during two production cycles (1 cycle = year). This is followed by two cycles of clonal evaluation in preliminary yield trials (PYT). At IITA, the PYT are replicated trials with each test genotype being represented by four plants in each of three replications. Implication of farmers in field evaluation and selection at this stage followed by on-farm validation of selected clones under farmers' field conditions, together with replicated multilocational reference evaluation trials (MET) at three to five locations, completes the hybrid development process. The MET follows the same experimental details as the PYT.

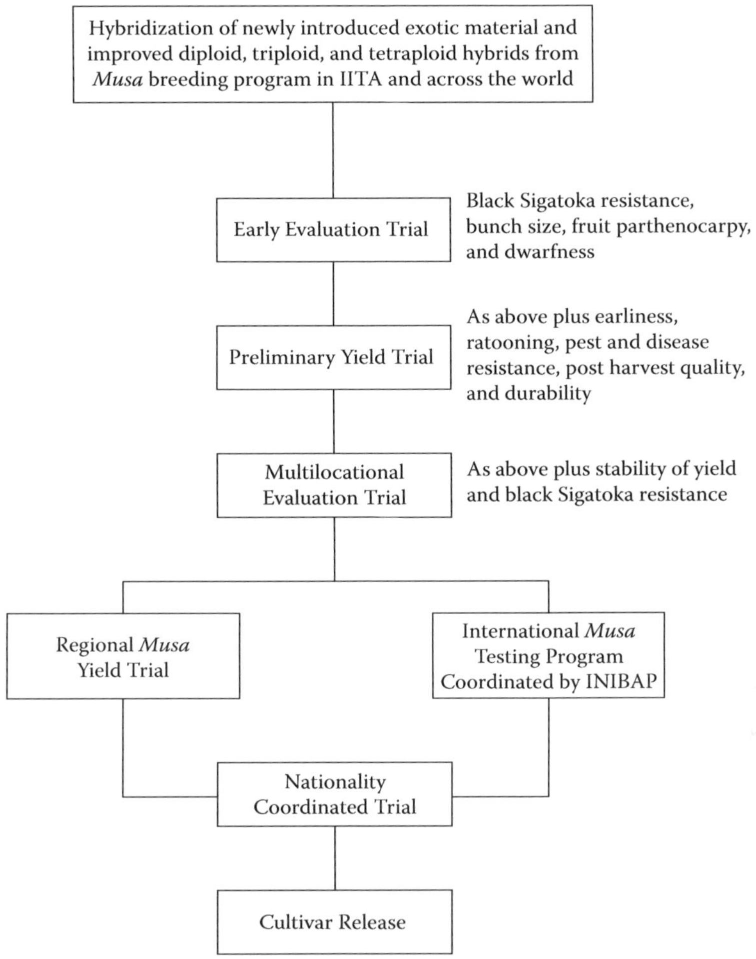

FIGURE 9.3 Typical flow diagram of the banana and plantain selection process: Each step may take approximately 2 years, so that about 15 years may be required from crossing to release of an improved cultivar. Furthermore, each plant occupies about 6 m^2 of field space and less than 1% of the plants in the EET typically make it to the next evaluation steps.

A variant of the MET is the evaluation of the best bet selections, usually through a farmer-participatory approach outside research stations, to comply with requirements for official cultivar release in nationally coordinated trials (NCT). The characteristics of the NCT are country specific. Optional steps are regional or international yield trials that are carried out outside the countries where the new cultivars have been bred. Figure 9.3 summarizes the salient features of the selection process followed by the breeding programs operating in Africa, notably those of IITA.

The selection criteria are earliness, short plant stature, resistance to black Sigatoka, resistance to nematodes, bunch weight significantly higher than the mean of the triploid breeding population, and good plant vigor. Both 'Calcutta 4' and 'Pisang Lilin' are routinely included in the EET as resistant controls. For 4*x* breeding, the EET and PYT also include the most popular landrace (for example, 'Agbagba' for plantain) as susceptible control and the best first-generation tetraploid hybrid (for example, 'PITA14' for plantain) as the reference control. Specifically, breeders aim to maintain or improve yield and resistance to black Sigatoka above the levels of the best tetraploid hybrids, reduce plant stature to enhance the ability of the plant to withstand wind damage and reduce the need for propping, and improve fruit quality (size, shape, and color) and the nutritional value of fresh and processed fruits, using the most popular landrace as the target reference.

Genotype response to black Sigatoka infection is assessed under natural field conditions using the youngest leaf spotted (YLS) method of Vakili (1968). Increasing YLS values indicate the presence of more healthy leaves on the plant and, hence, greater resistance to black Sigatoka. To adjust for genetic differences in number of standing leaves (NSL), the index of nonspotted leaves (INSL) is calculated as INSL = 100(YLS − 1)/NSL.

Nematode-resistance screening is carried out in parallel field trials using a method that is based on the inoculation of individual roots (De Schutter et al., 2001). Screening trials usually include two reference cultivars with known responses to *R. similis*, namely, 'Valery' (*Musa* AAA, Cavendish subgroup), which is very susceptible to *R. similis* and 'Yangambi Km 5' (*Musa* AAA group), which is resistant to *R. similis* (Sarah et al., 1992; Price, 1994; Fogain and Gowen, 1998). 'Yangambi Km5' is also used as a reference cultivar of the International *Musa* Testing Program (Orjeda, 2000). Based on the assumption that susceptible plants allow a higher reproduction ratio of the nematodes than resistant plants, test genotypes are classified as resistant (resp. susceptible) when the final nematode population density is not significantly different from that of the resistant (resp. susceptible) check but significantly different from that of the susceptible (resp. resistant) check. Otherwise, the test genotype is considered partially resistant when its final nematode population density is significantly different from those observed on both checks (Dochez et al., 2009).

Growth and yield variables are evaluated at flowering and at harvest, as described by Swennen and De Langhe (1985). For each plant, data are recorded on bunch weight (BWT, kg) and its components, days to flowering (DTF), days for fruit filling (that is, the number of days elapsed from flowering to harvest [DFF]), plant height (PHT, cm), height of tallest sucker (HTS, cm), and total number of leaves (TNL) at flowering. Three other variables are often calculated from collected data in 6 m^2 plots: yield (YLD = 1.667 × 365 × BWT/[DTF + DFF], t ha^{-1} yr^{-1}), sucker growth index (SGI = HTS/PHT), and leaf emission rate (LER = DTF/TNL, the number of days required for the production of one leaf). The number of days to flowering is calculated as the time interval between planting and flowering in the plant crop, and the interval between harvest of the previous crop and flowering of the next in ratoon crops.

Postharvest characteristics are assessed using the methods of Dadzie and Orchard (1996). Of recent, breeders have been interested in biofortification of banana and plantain by assessing micronutrient content of fresh and processed fruits from field-grown plants, including total carotenoids, iron, and zinc (Adeniji and Tenkouano, 2007; Honfo et al., 2007).

The completion of an evaluation cycle may take about 15 years from crossing to release of an improved cultivar. Furthermore, each plant occupies about 6 m^2 of field space and less than 1% of the plants in the EET typically make it to the next evaluation steps. Thus, traditional banana and plantain breeding has been quite inefficient in utilizing resources, and breeders have developed a keen interest in improving the breeding process.

9.6.3 Improving Selection Efficiency

Essentially due to the lack of large segregating populations, early breeding efforts focused on hybrid development rather than genetic analysis, except for a few qualitative traits. However, knowledge of genetic parameters is essential in (1) designing crosses targeting the improvement of specific traits, based on combining ability and heterosis, and (2) predicting the expected advance from selection for these traits, based on heritability and genetic correlations among traits. However, several advances have been made in understanding the inheritance of important traits in banana and plantain and it is now possible to design crosses on the basis of combining ability (Ortiz, 1997a; Tenkouano et al., 1998a). Likewise, information on the meiotic behavior of prospective parental stocks, their inbreeding status, and genetic relatedness have been estimated using a combination of pedigree and molecular analyses, increasing the prospects for designing crosses with maximum recombinative heterosis (Tenkouano et al., 1999; Tenkouano, 2005). For example, it was observed that nongenetic effects accounted for a large proportion of the phenotypic variation observed among hybrids from $4x \times 2x$ crosses for most traits, except bunch weight (Tenkouano et al., 1998b). General combining ability (GCA) effects were exclusively associated with the female parents for plant height, number of leaves, suckering behavior, and fruit circumference, suggesting that selection for these traits should be carried out in the prospective $4x$ female parents. Conversely, GCA effects were primarily of paternal origin for fruit filling time, bunch weight, and fruit length. Thus, breeding for these traits would be more efficient when carried out in prospective males at the $2x$ level. Specific combining ability (SCA) effects were observed for all traits, except fruit filling time, suggesting that additional genetic gain could be achieved through recombinative heterosis for these traits.

Most growth and yield characteristics of *Musa* display complex inheritance and genetic association patterns and are subject to genotype-by-environment interactions. Powerful statistical genetics tools have been used to further our knowledge of the phenotypic and genetic correlations between indirectly and directly selected traits, in order to help determine whether an indirectly selected trait will increase, decrease, or remain constant in advanced cycles of selection (Tenkouano et al., 2002). It was suggested that selecting for specific adaptation based on site-specific selection indices constructed from site-specific genetic correlations may significantly enhance *Musa* breeding. A neutral genotype-by-environment (GxE) interaction effect on the relationships between bunch weight and yield components was reported, suggesting that indirect selection for bunch weight could be achieved by selecting for yield components. Conversely, the predictive value of phenological traits for yield was considered negligible, the only exception being the relatively high genetic correlations between bunch weight and the number of leaves, which was not surprising given the role of leaves in photosynthesis and in reaction to black Sigatoka.

Selection for specific rather than broad adaptation has been credited with the potential of achieving greater genetic gains (Simmonds, 1981). This is of particular relevance for farmers who often are more interested in crop cultivars that ensure reliable performance from year to year in their local environment and bother less about what happens at other locations.

Clearly, selection for most of these traits, particularly those that are expressed late in the plant cycle or strongly influenced by the environment, would benefit from the use of molecular markers, but progress in developing trait-specific molecular markers has been very limited. Nevertheless, a considerable amount of genomic resources is becoming available (Vilarinhos et al., 2003; Kaemmer et al., 2002; Safar et al., 2002).

9.7 BREEDING ACHIEVEMENTS

9.7.1 Substitutes for Commercial or Subsistence Production

The initial focus of *Musa* breeding was to develop improved dessert banana cultivars of commercial value, but progress has been rather limited, partly due to genetic drag from the sources of resistance to disease used in the breeding process. However, recent breakthroughs have been obtained by the CIRAD program, based on their reconstructive breeding approach (Bakry, personal communication). Nonetheless, it is speculated that the banana industry would massively embrace the new cultivars only if this does not impose drastic modification of the production to delivery logistics that have been so well tuned to the existing Cavendish cultivars.

In contrast, past investments in breeding stand to benefit smallholder producers of banana and plantain, with several new cultivars becoming available, particularly in the last decade (Tenkouano and Swennen, 2004; Tenkouano et al., 2006). Undoubtedly, the most successful breeding program has been that of the Fundación Hondureña de Investigación Agrícola (FHIA) based in Honduras. The most publicized cultivars from this program are 'FHIA-01,' 'FHIA-21,' and 'FHIA-25,' which

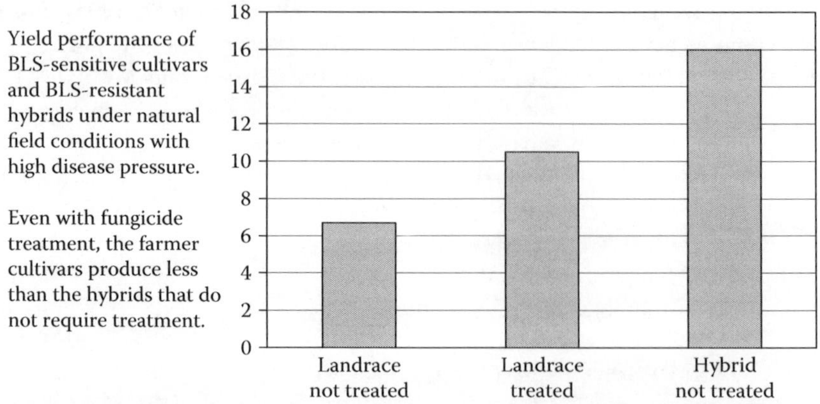

FIGURE 9.4 Illustration of progress in breeding for resistance to black Sigatoka.

have been tested worldwide and have been released for cultivation in several countries, notably in Ghana and Uganda.

The oldest breeding programs in Africa are those of CARBAP and IITA, which both started by assembling a considerable pool of local cultivars (more than 100 from various countries) and sources of resistance to black Sigatoka (more than 200 from Southeast Asia where bananas and the disease originated from) at their research stations in Cameroon and Nigeria, respectively. The first IITA cross to transfer resistance to the local cultivars was made in 1987 (Swennen, personal communication). A decade later, several improved plantain-derived hybrids with resistance to black Sigatoka had been produced and tested in many countries, revealing the superior performance of the hybrids under situations of high black Sigatoka (BLS) prevalence (Figure 9.4).

These efforts earned IITA the CGIAR's 1994 King Baudouin Award. The most promising of the first-generation plantain-derived hybrids has been 'PITA14' in Nigeria and neighboring countries. First-generation hybrids of the East African bananas did not perform better than the landraces. Secondary 3*x* derived from the first-generation plantain or East African banana hybrids were obtained, displaying lesser propensity to forming seeds and greater culinary resemblance to the prevailing landraces (Tenkouano et al., 2006), but adoption prospects only increased with the distribution of the new cultivars through a scheme that was nondisruptive to the farmers' practice of cultivar mixtures and through the promotion of novel postharvest processing options. Apart from the hybrids' high yields and disease resistance, farmers appreciate their capacity for rapid multiplication (a result of excellent proliferation of the rhizomes, or "suckers," used for propagation), their good taste and cooking qualities, and their resistance to other diseases and pests. Some farmers who have adopted the hybrids are generating significant income from the sale of suckers, in addition to boosting returns from the sale of fruit (Tenkouano et al., 2006). This is discussed in more detail in Chapter 16 of this book. The CARBAP program has also developed a series of first-generation hybrids, among which the most promising has been 'CRBP39,' but these 4*x* hybrids presented several defects that prevented their diffusion, and some of them are being used as intermediate products for further improvement (Noupadja et al., 2007).

9.7.2 Genetic Stocks for Breeding

The limited success of the first-generation hybrids was largely attributed to defects they inherited from their wild 2*x* parents. Thus, major efforts were devoted to developing improved 2*x* by stacking desirable alleles while decreasing the extent of wildness. Such improved 2*x* could then be used for transfer of resistance genes to the landraces without the genetic drag of undesirable traits.

Notorious progress in this area has been achieved (Ortiz and Vuylsteke, 1995; Krishnamoorthy and Kumar, 2005), but the products from these efforts have not been widely distributed outside the program that developed them. To date, only the improved 2*x* stocks from the IITA program have been placed in the public domain and have been accessible worldwide.

Genetically, many of the 2*x* stocks of the first generation displayed excellent levels of resistance to black Sigatoka but had unappealing agronomic or fruit characteristics (Figure 9.5 top). However, it is possible to correct this situation and bring together productivity and resistance traits (Figure 9.5 bottom) to produce 2*x* stocks with greater potential for use in the improvement of 3*x* cultivars.

9.8 BREEDING CONSTRAINTS

Despite these advances, many problems specific to the biology of plantains and bananas impede rapid breeding progress, including low reproductive fertility, triploidy, and slow propagation (Vuylsteke et al., 1997). Thus, breeders may not be able to access all available variation, largely due to the high sterility of many cultivated cultivars. The conundrum of banana breeding is that seed set is required to produce seedless cultivars. Improving female or male sterile accessions can only be achieved by circumventing reproductive barriers, for example, by deliberate mutagenesis or by direct gene transfer through genetic engineering. Ideally, genetic engineering would provide a more deterministic approach for substituting undesirable genetic materials with those that would enhance the performance of the recipient line or cultivar with respect to a particular constraint.

Additionally, the available germplasm may not harbor the desired genes with respect to an emerging threat, which is clearly the case for the bacterial wilt caused by *Xanthomonas campestris* pv. *musacearum* that is threatening livelihoods in the Great Lakes region of Central and Eastern Africa. There is little that crossbreeding can do to address such a threat in the near future, leaving genetic engineering as the best alternative to control the bacterial wilt epidemics through host resistance.

The crossbreeding approach is time consuming, given the long generation time of the crop; it is also technically complex, since it requires both ploidy and genome selection. With the advent of flow cytometric analysis of nuclear DNA content (Dolezel, 1997) and molecular markers for the A and B genomes (Pillay et al., 2000), ploidy and genome selection have become easier. However, predicting progeny performance remains a major challenge when ploidy and genome variation occurs within and across generations (Tenkouano et al., 1999a, 1999b). Until very recently, there were few assessments on parent–offspring relationships so that parental contribution to offspring and ensuing patterns of ploidy segregation were not well understood.

Microsporogenis studies (Oselebe et al., 2001) revealed that 2*x* individuals produced essentially haploid gametes ($n = 1x = 1C$), whereas 4*x* individuals produced mostly hypoploid gametes with half the chromosome content of the haploid class that would have been expected with normal meiosis, which would only result from double reduction ($2n = 4x = 4C \rightarrow n = 2x = 2C \rightarrow 0.5n = 1x = 1C$). Other findings by the IITA research team were that euploid gametes with two or four times the basic chromosome set were also produced by the 2*x* individuals, presumably through first divisional restitution (FDR) or second divisional restitution (SDR) or both (double restitution), the first two events producing 2*n* gametes and the last event 4*n* gametes. Progeny analysis showed that 4*x* × 2*x* crosses mostly produced 3*x* progenies, against mostly 2*x* progenies for 2*x* × 4*x* and 2*x* × 2*x* crosses, with very low frequencies of 4*x* and 5*x* progenies. It was concluded that parental contribution to progenies was unequal with respect to 3*x* progeny ($n = 2x = 2C$ by the 4*x* parent plus $n = 1x = 1C$ by the 2*x* parent), but equal with respect to the parents, both contributing half of their endowment in 4*x* × 2*x* crosses. Conversely, parental contribution was equal with respect to the progeny ($n = 1x = 1C$ plus $n = 1x = 1C$) but unequal with respect to the parents in 2*x* × 4*x* crosses, with the 2*x* parent donating half of its endowment and the 4*x* parent a quarter of its endowment. In 2*x* × 2*x*, 2*x* progeny would result from equal parental contribution with respect to both parental and progeny endowment, whereas, 3*x* progeny obtained by unilateral sexual polyploidization (2*n* pollen) would be attributed to unequal parental contributions with respect to both parental and progeny endowments. These studies further

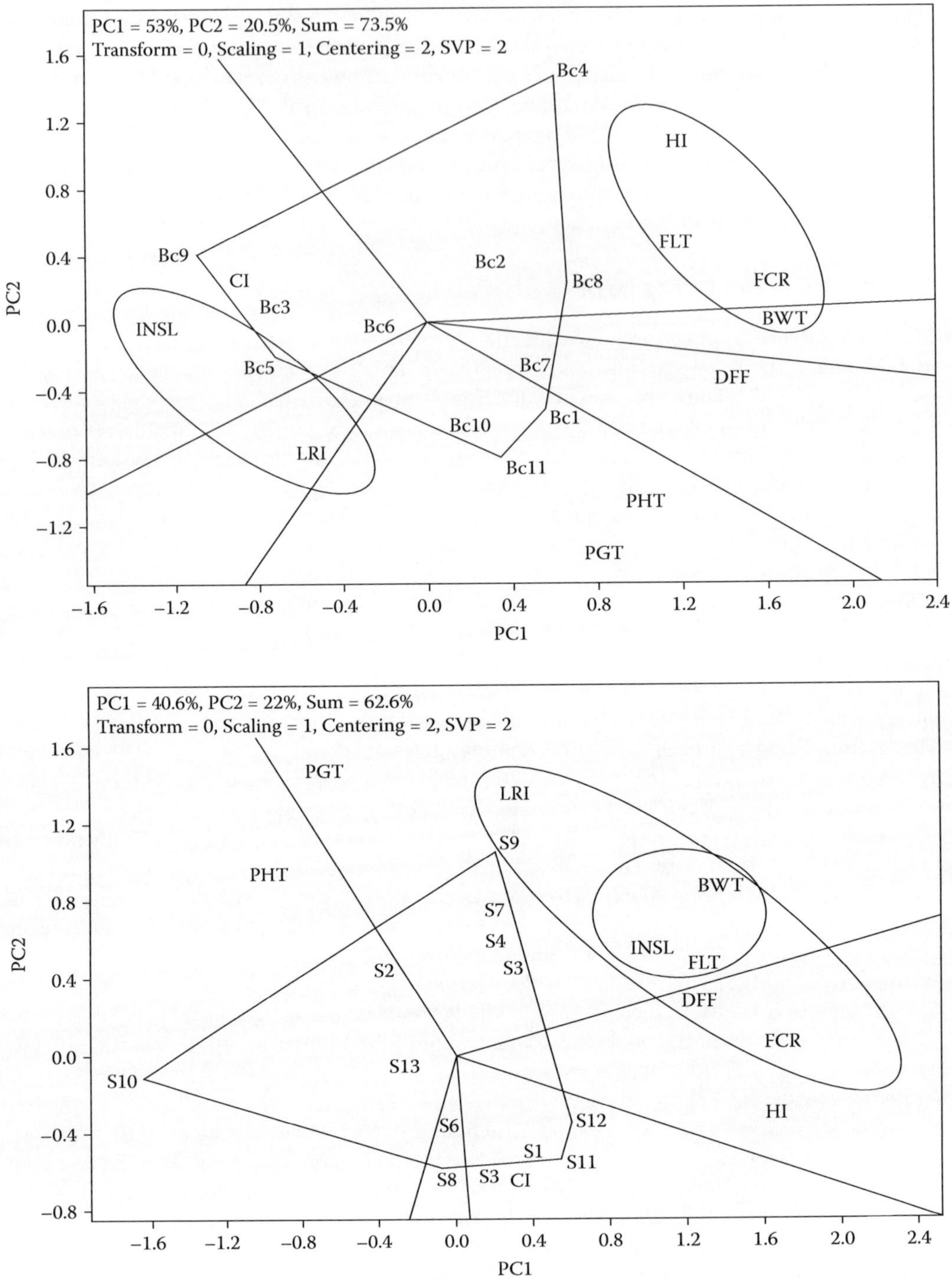

FIGURE 9.5 GGE biplot analysis of trait association in diploid banana and plantain hybrids. Primary hybrids derived from the cross 'Bobby Tannap' × 'Calcutta 4' often display resistance traits in opposition relation to productivity traits (top). Crossing selected primary hybrids with unrelated stocks from the same generation (e.g., 'Tjau Lagada' × 'Calcutta 4') produces secondary hybrids that combine resilience and productivity traits (bottom). (From Daniel N. Igili. Progress in Breeding Diploid Genetic Stocks of Banana with Resistance to Black Sigatoka, master's thesis, University of Nigeria, Nsukka.)

suggested that $3x$ progeny could also be derived from $4x \times 4x$ crosses, a scheme that has not been frequently used but is now being considered for the improvement of banana and plantains (Pillay, unpublished). Macrosporogenesis was not directly examined but has been assumed to be essentially normal in both $2x$ and $4x$ individuals, attested by the predominantly $2x$ nature of progenies from $2x \times 2x$ crosses and the reduced frequency of $2x$ progenies from $4x \times 2x$ crosses. Knowledge of the meiotic behavior of breeding stocks could thus allow for designing informed crosses with highest probability of generating $3x$ lines with putative plantain, dessert, or cooking banana labels.

9.9 FUTURE BREEDING GOALS

Despite the evolving nature of biotic constraints, black Sigatoka remains at the forefront of the constraints to commercial and subsistence production of banana and plantain. However, overcoming this problem through genetics is no longer a challenge. Breeders have been looking at refined ways of utilizing the genes embedded in the available improved stocks. Likewise, the focus of breeding is gradually shifting from addressing existing constraints to assessing the risk potential of emerging threats and preparing to respond to these.

For decades, banana and plantain research has centered mainly on the development of technologies that can help boost and stabilize production. As farmers gain better access to new germplasm and knowledge generated by that research, they will be better able not only to enhance crop productivity but also to strengthen their ties with markets for surplus production. New opportunities to accomplish both ends are rapidly emerging. A major challenge is to find ways of ensuring that small farmers are able to capture benefits from such developments through stronger market links.

Thus, breeding is now conceived as a component of a larger, integrated effort aimed at developing and promoting market-preferred cultivars. This genetic-improvement undertaking is done through the following options, which constitute emerging orientations of the existing programs:

- Preemptive breeding (germplasm diversification) to counteract emerging and shifting biological and environmental constraints, in close association with:
  - Development of upstream prebreeding methods (e.g., anther or ovule culture for production of haploid lines, haplo-diploidization with support from cytogenetics) to recover useful progenies while minimizing wildness
  - Search for novel genes (drought, bacterial, and viral diseases) through judicious assembly and mining of agro-biodiversity of crop species and biotic threats

- Deployment of pest and disease management and monitoring schemes (increased alertness to threats), coupled with:
  - Development of cost-effective alternatives that combine environmental friendliness with human health safety for prevalent production systems
  - Design of early alert systems based on predictive analysis of biological shifts in pest and pathogen populations

- Seed systems, including technical and policy-related issues, combined with germplasm (source populations and genetically stabilized cultivars) testing through networks
  - Development of standards and protocols for cost-effective dissemination of improved germplasm, including efficient cross-border exchange of plant propagules (seeds, vegetative materials)
  - Promotion of traceability mechanisms that protect intellectual property and ensure visibility, e.g., facilitating cultivar release processes that are more agile to ensure effective delivery, facilitating the emergence of a commercial seed system

- Crop utilization and product development, including health-related issues (mycotoxins, nutrition) and trade-related issues (commodity policy) with emphasis on:
  - Development of sustainable systems for intensification of production as more effort is directed to industrial use
  - Assessment of the distribution of the toxic organisms and promotion of remedial technologies and standards

The purpose of *Musa* breeding has been to develop new cultivars resistant to diseases. However, *Musa* breeders very well know that, even with widespread adoption of new cultivars by farmers, decreases in soil fertility would still remain an obstacle to the profitable cultivation of banana and plantains.

REFERENCES

Adeniji, T.A. and A. Tenkouano. 2007. Micronutrient content of chips from plantain and banana hybrids. *Fruits* 62:345–352.

Allard, R.W. 1960. *Principles of Plant Breeding*. New York: John Wiley & Sons.

Bakry, F., N.P. de la Reberdiere, S. Pichot, and C. Jenny. 2007. In liquid medium colchicine treatment induces non chimerical doubled-diploids in a wide range of mono- and interspecific diploid banana clones. *Fruits* 62:3–12

Buddenhagen, I.W. 1996. Banana research needs and opportunities. In: Banana improvement: Research challenges and opportunities, G.J. Persley and P. George, eds., 1–20. Washington, DC: The World Bank.

Carreel, F., D. Gonzalez de Leon, P.J. Lagoda, C. Lanaud, C. Jenny, J.-P.Horry, and H. Tézenas du Montcel. 2002. Ascertaining maternal and paternal lineage within *Musa* by chloroplast and mitochondrial DNA RFLP analyses. *Genome* 45:679–692.

Chin, H.F. 1995. Germination and storage of banana seeds. In: *New frontiers in resistance breeding for Nematode, Fusarium and Sigatoka*, Frison, E.A., J.P. Horry, and D. De Waele, eds., Proc. workshop held in Kuala Lumpur, Malaysia, 2–5 October 1995, 218–227. Los Baños, Philippines: INIBAP.

Cruden, R.W. 1977. Pollen–ovule ratios: A conservative indicator of breeding systems in flowering plants. *Evolution* 31:32–36.

Dadzie, B.K. and J.E. Orchard. 1996. *Postharvest criteria and mehtods for routine screening of banana and plantain hybrids*. Montpellier, France: International Network for the Improvement of Banana and Plantain,. 47 pp.

Dahal, G., J. d'A. Hughes, G. Thottappilly, and B.E.L. Lockhart. 1998. Effect of temperature on symptom expression and reliability of banana streak badnavirus detection in naturally-infected plantain and banana (Musa spp.). *Plant Disease* 82:16–21.

Damasco, O.P., G.C. Graham, R.J. Henry, S.W. Adkins, M.K. Smith, and I.D. Godwin. 1996. Random amplified polymorphic DNA (RAPD) detection of dwarf off-types in micropropagated Cavendish (*Musa* spp. AAA) bananas. *Plant Cell Reports* 16:188–123.

De Langhe E. and M. Devreux. 1960. Une sous-espèce nouvelle de *Musa acuminata* Colla. *Bull. Jard. Bot. Brux*. 30:375–388.

De Langhe, E. 1969. Bananas (*Musa* spp.). In: *Outlines of perennial crop breeding in the tropics*, F.P. Ferwarda and F. Wit, eds., 53–78. Miscellaneous Papers No. 4, Agricultural University, Wageningen, The Netherlands.

De Langhe, E. 1995. Is plantain the oldest fruit crop in the world? pp 44–47. In The King Baudouin Award to IITA. Proceedings of the Celebration at Leuven, 6 April 1995.

De Schutter, B., P.R. Speijer, C. Dochez, A. Tenkouano, and D. De Waele. 2001. Evaluating host plant reaction of *Musa* germplasm to Radopholus similis by inoculation of single primary roots. *Nematropica* 31:295–299.

Dochez, C., A. Tenkouano, R. Ortiz, J. Whyte, and D. De Waele. 2009. Host plant resistance to *Radopholus similis* in a diploid banana hybrid population. *Nematology* 11:329–335

Dolezel, J. 1997. Application of flow cytometry for the study of plant genomes. *J. Appl. Genetics* 38:285–302.

Dolezel, J., M. Dolezelova, and F.J. Novak. 1994. Flow cytometric estimation of nuclear DNA amount in diploid bananas (*Musa acuminata* and *M. balbisiana*). *Biol. Plant.* 36:351–357.

Dolezel, J., M.A. Lysak, I. Van den Houwe, M. Dolezelova, and N. Roux. 1997. Use of flow cytometry for rapid ploidy determination in *Musa* species. Info*Musa*. 6:6–9.

Ecktein, K., J.C. Robinson, and S.J. Davie. 1995. Physiological responses of banana (*Musa* AAA; Cavendish sub-group) in the subtropics. III. Gas exchange, growth analysis and source-sink interaction over a complete crop cycle. *J. Hort. Sci.* 70:169–180.

Escalant, J.V. and B. Panis. 2002. Biotechnologies toward the genetic improvement in *Musa*. pp 68–85. In: *Memorias XV Reunión Internacional.* Proc. ACORBAT 2002. Cartagena de Indias, Colombia, 27 October–2 November 2002. Augura, Medellin, Colombia.

Fogain, R. and S.R. Gowen. 1998. 'Yangambi Km5' (*Musa* AAA, Ibota subgroup): A possible source of resistance to *Radopholus similis* and *Pratylenchus goodeyi. Fund. Appl. Nematol.* 21:75–80.

Ge X.J., M.H. Liu, W.K. Wang, B.A. Schaal, and T.Y. Chiang. 2005. Population structure of wild bananas, *Musa balbisiana*, in China determined by SSR fingerprinting and cpDNA PCR-RFLP. *Molec. Ecol.* 14:933–944.

Graven, P., C.G. De Koster, J.J. Boon, and F. Bouman. 1996. Structure and macromolecular composition of the seed coat of the Musaceae. *Ann. Bot.* 77:105–122.

Grillo, G. S., M.J.G. Martin, and A.M. Domínguez. 1998. Morphological methods for the detection of banana off-types during the hardening phase. *Acta Horticulturae* 490:239–245.

Heinz D.J. and G.W.P. Mee. 1969. Plant differentiation from callus tissue of Saccharum species. *Crop Sci.* 9:346–348.

Honfo, F.G., P.A.P. Kayodé, O. Coulibaly, and A. Tenkouano. 2007. Relative contribution of banana and plantain products to the nutritional requirements for iron, zinc and vitamin A of infants and mothers in Cameroon. *Fruits* 62:1–11

Hwang S.C. and W.H. Ko. 1987. Somaclonal variation of bananas and screening for resistance to Fusarium wilt. In: *Banana and plantain breeding strategies*, G.J. Persley and E.A. De Langhe, eds., 151–156. Proc. Int. Workshop held at Cairns, Australia. Canberra, Australia: Australian Centre for International Agricultural Research.

Kaemmer D., L. Peraza-Echeverrýa, B. Canche´, A. Arroyo, E. Ortiz, H.-B. Zhang, and A. James. 2002. Characterization and utilization of the first banana BAC library of *Musa acuminata* spp. *burmannicoides* type 'Calcutta IV'. In: *Program and abstracts of the 3rd international symposium on molecular and cellular biology of bananas*, Leuven, Belgium, 9–11 September 2002, INIBAP, p 14.

Kiew, R. 1987. Notes on the natural history of the Johore Banana, *Musa gracilis* Holtum. *Malayan Nat. J.* 41:239–248.

Krishnamoorthy, V. and N. Kumar. 2005. Preliminary evaluation of diploid banana hybrids for yield potential, male fertility and reaction to *Radopholus similis. Plant Genet. Res. Newslr.* 141:39–43.

La Fleur, D.A., B.E.L. Lockhart, and N.E. Olszewski 1996. Portions of banana streak badnavirus genome are integrated in the genome of its host, *Musa. Phytopathology* 86 (Supplement 11):100.

Larkin, P.J. and W.R. Scowcroft, 1981. Somaclonal variation: A novel source of variability from cell culture for plant improvement. *Theor. Appl. Genet.* 58:197–214.

Lockhart B.E.L. and N.E. Olszewski. 1993. Serological and genomic heterogeneity of banana streak badnavirus: Implications for virus detection in *Musa* germplasm. In: *Breeding banana and plantain for resistance to diseases and pests,* J. Ganry, ed., 105–113. International Network for the Improvement of Banana and Plantain, Montpellier, France.

Marin, D.H., R.A. Romero, M. Guzmán, and T.B. Sutton. 2003. Black Sigatoka: An increasing threat to banana cultivation. *Plant Disease* 87:208–222.

Mobambo, K.N., F. Gauhl, D. Vuylsteke, R. Ortiz, C. Pasberg-Gauhl, and R. Swennen. 1993. Yield Loss in plantain from black Sigatoka leaf spot and field performance of resistant hybrids. *Field Crop Res.* 35:35–42.

Mohan Jain, S. and R. Swennen R. (eds.). 2004. *Banana improvement: Cellular, molecular biology, and induced mutations*. Enfield, NH, Plymouth, UK: Science Publishers, 394 pp.

Noupadja, P., K. Tomekpe, and E. Youmbi. 2007. Évaluation d'hybrides tétraploïdes de bananiers plantains (*Musa* spp.) résistants à la maladie des raies noires au Cameroun. *Fruits* 62:77–88.

Nwakanma, D.C., M. Pillay, B.E. Okoli, and A. Tenkouano. 2003. PCR-RFLP of the ribosomal DNA internal transcribed spacers (ITS) provides markers for the A and B genomes in *Musa* L. *Theor. Appl. Genet.* 108:154 –159.

Nwauzoma, A.B., A. Tenkouano, J.H. Crouch, M. Pillay, D.Vuylsteke, and L.A. Daniel-Kalio. 2002. Enhanced yield and disease resistance of plantain (*Musa* spp., AAB group) somaclones in Nigeria. *Euphytica* 123:323–331.

Orjeda, G. 2000. Evaluating bananas: *A global partnership. Results of IMTP Phase II.* International Network for the Improvement of Banana and Plantain, Montpellier, France, 466 pp.

Ortiz, R. 1997a. Secondary polyploids, heterosis and evolutionary crop breeding for further improvement of the plantain and banana (*Musa* spp. L.) genome. *Theor. Appl. Genet.* 94:1113–1120.

Ortiz, R. 1997b. Occurrence and inheritance of 2n pollen in *Musa*. *Ann. Bot.* 79:449–453.

Ortiz, R. and D. Vuylsteke. 1994a. Inheritance of black Sigatoka resistance in plantain–banana (*Musa* spp.) hybrids. *Theor. Appl. Genet.* 89:146–152.

Ortiz, R. and D. Vuylsteke. 1994b. Genetics of apical dominance in plantain (*Musa*, AAB group) and improvement of suckering behavior. *J. Amer. Soc. Hort. Sci.* 119:1050–1053.

Ortiz, R. and D. Vuylsteke. 1995. Factors influencing seed set in triploid *Musa* spp. L. and production of euploid hybrids. *Ann. Bot.* 75:151–155.

Ortiz, R. and D. Vuylsteke. 1996. Recent advances in *Musa* Genetics, breeding and biotechnology. *Plant Breed. Abstr.* 66:1355–1363.

Ortiz, R., R.S.B. Ferris, and D. Vuylsteke. 1995. Banana and plantain breeding. In: *Bananas and plantains*, S. Gowen, ed., 110–146. London: Chapman & Hall.

Oselebe, H.O., A. Tenkouano, I.U. Obi, and M.I. Uguru. 2001. Prospects for breeding agronomically superior triploid plantain and banana hybrids using 2n gametes. Paper presented at the 5th African Crop Science Society Conference, 21–26 October, 2001, Lagos, Nigeria.

Oselebe, H.O., A. Tenkouano, M. Pillay, I.U. Obi, and M.I. Uguru. 2006. Ploidy and genome segregation in *Musa* breeding populations assessed by flow cytometry and random amplified polymorphic DNA markers. *J. Amer. Soc. Hort. Sci.* 131:780–786.

Pillay, M., A. Tenkouano, G. Ude, and R. Ortiz. 2004. Molecular characterization of genomes in *Musa* and its applications. In: *Banana improvement: Cellular, molecular biology, and induced mutations*. M.S. Jain and R. Swennen, eds., 271–286. Enfield, NH, Plymouth, UK: Science Publishers.

Pillay, M., D.C. Nwakanma, and A. Tenkouano. 2000. Identification of RAPD markers linked to A and B genome sequences in *Musa*. *Genome* 43:763–767.

Ploetz, R.C. 1992. *Fusarium* wilt of banana (Panama disease). In: Plant diseases of international importance, Vol. III, A.N. Mukhopadhyay, H.S. Chaube, J. Kumar, and U.S. Singh, eds., 270–282. Englewood Cliffs, NJ: Prentice Hall.

Price, N.S. 1994. Field trial evaluation of nematode susceptibility within *Musa*. *Fund. Appl. Nematol.* 17:391–396.

Robinson, J.C. 1996. Crop production in horticulture 5: Bananas and plantains. Centre for Agriculture and Bioscience International, Wallingford, Oxon, UK: CAB International.

Rossel, G. 1998. Taxonomic-linguistic study of plantain in Africa. PhD thesis, Wageningen Agricultural University, Wageningen. The Netherlands, 277 pp.

Roux, N., J. Dolezel., R. Swennen, and F.J. Zapata-Arias. 2001. Effectiveness of three micropropagation techniques to dissociate cytochimeras in *Musa* spp. *Plant Cell Tiss. Organ Culture* 66:189–197.

Rowe, P. and F. Rosales. 1996. Bananas and plantains. In: *Fruit breeding, Volume I: Tree and tropical fruits*, J. Janick and J.N. Moore, eds., 167–211. New York: John Wiley & Sons.

Safar, J., P. Pifanelli, J. Glaszmann, and J. Dolezel. 2002. Construction of BAC library for the B genome of banana (*Musa balbisiana*). In: *Program and abstracts of the 3rd international symposium on molecular and cellular biology of bananas*, Leuven, Belgium, 9–11 September 2002, INIBAP, pp. 18–19.

Sarah, J.L., F. Blavignac, C. Sabatini, and M. Boisseau. 1992. Une méthode de laboratoire pour le criblage variétal des bananiers vis-à-vis de la résistance aux nématodes. *Fruits* 47:559–564.

Simmonds, N.W. 1953. Segregations in some diploid bananas. *J. Genet.* 51:458–469.

Simmonds, N.W. 1962. *The evolution of the bananas*. Tropical Science Series. London, UK: Longmans.

Simmonds, N.W. 1981. Genotype (G), environment (E) and GE components of crop yields. *Exper. Agric.* 17:355–362.

Simmonds, N.W. and K. Shepherd. 1955. The taxonomy and origins of the cultivated bananas. *J. Linn. Soc. Lond. Bot.* 55:302–312.

Stover, R.H. 1980. Sigatoka leaf spot of banana and plantain. *Plant Dis.* 64:750–756

Stover, R.H. and N.W. Simmonds. 1987. *Bananas*. 3rd ed. London, UK: Longman Scientific and Technical.

Swennen, R. 1990. *Plantain cultivation under West African conditions: A reference manual*. Ibadan, Nigeria: IITA.

Swennen, R. and E. De Langhe. 1985. Growth parameters of yield of plantain (*Musa* cv. AAB). *Ann. Bot.* 56:197–204.

Swennen, R., D. Vuylsteke, and R. Ortiz. 1995. Phenotypic diversity and patterns of variation in West and Central African Plantains (*Musa* spp., AAB group Musaceae). *Econ. Bot.* 49:320–327.

Swennen. R. and D. Vuylsteke. 1993. Breeding black Sigatoka-resistant plantains with a wild banana. *Trop. Agric.* (Trinidad) 70:74–77.

Tenkouano, A. 2001. Current issues and future directions for Musa genetic improvement research at the International Institute of Tropical Agriculture. In: Advancing banana and plantain R&D in Asia and the Pacific, Vol 10, A.B. Molina, V.N. Roa, and M.A.G. Maghuyop, eds., 11–23. Proceedings of the 10th INIBAP-ASPNET Regional Advisory Committee meeting, Bangkok, Thailand, 10–11 November 2000. International Network for the Improvement of Banana and Plantain-Asia and the Pacific Network, Los Baños, Laguna, Philippines.

Tenkouano, A. 2005. Breeding banana and plantain: Integrating molecular techniques and conventional approaches to explore genetic polymorphisms and predict progeny performance. In: *Genetic resources and biotechnology*, Vol. 3, D. Thangadurai, T. Pullaiah, and L. Tripathi, eds., 153–168. New Delhi, India: Regency Publications.

Tenkouano, A. and R.L. Swennen. 2004. Progress in breeding and delivering improved plantain and banana to African farmers. *Chronica Hort.* 44:9–15.

Tenkouano, A. J.H. Crouch, H.K. Crouch, and R. Ortiz. 1998a. Genetic diversity, hybrid performance and combining ability for yield in *Musa* germplasm. *Euphytica* 102:281–288.

Tenkouano, A., J. Hughes, and M. Pillay. 2001. Prospects for breeding agronomically superior *Musa* with resistance or tolerance to BSV. Paper presented at a Conference on Plant Virology in Sub-Saharan Africa, 4–8 June 2001, International Institute of Tropical Agriculture, Ibadan, Nigeria.

Tenkouano, A., J.H. Crouch, H.K. Crouch, and D. Vuylsteke. 1998b. Ploidy determination in *Musa* germplasm using pollen and chloroplast characteristics. *HortSci.* 33:889–890.

Tenkouano, A., J.H. Crouch, H.K. Crouch, D. Vuylsteke, and R. Ortiz. 1999a. Comparison of DNA marker and pedigree-based methods of genetic analysis in plantain and banana (*Musa* spp.) Clones: I. Estimation of genetic relationships. *Theor. Appl. Genet.* 98:62–68.

Tenkouano, A., J.H. Crouch, H.K. Crouch, D. Vuylsteke, and R. Ortiz. 1999b. Comparison of DNA marker and pedigree-based methods of genetic analysis in plantain and banana (*Musa* spp.) Clones: II. Predicting hybrid performance. *Theor. Appl. Genet.* 98:69–75.

Tenkouano, A., R. Ortiz, and D. Vuylsteke. 1998b. Combining ability for yield and plant phenology in plantain-derived populations. *Euphytica* 104:151–158.

Tenkouano, A., R. Ortiz, and K.P. Baiyeri. 2002. Phenotypic and genetic correlations in *Musa* populations in Nigeria. *African Crop Science J.* 10:121–132.

Tenkouano, A., S. Hauser, D. Coyne, and O. Coulibaly. 2006. Clean planting materials and management practices for sustained production of banana and plantain in Africa. *Chronica Hort.* 46:14–18.

Tezenas du Montcel, H., F. Carreel, and F. Bakry. 1996. Improve the diploids: The key for banana breeding. In: *New Frontiers in Breeding for Nematode, Fusarium and Sigatoka*. E.A. Frison, J.-P. Horry, and D. De Waele, eds., 119–127. Proceedings of workshop held in Kuala Lumpur, Malaysia. International Network for the Improvement of Banana and Plantain, Montpellier, France.

Tripathi, L., J.N. Tripathi, A. Tenkouano, and P. Bramel. 2008. Banana & Plantain. In: *A compendium of transgenic crop plants*, Vol. 5, C. Kole and T.C. Hall, eds., 77–108., Oxford, UK: Blackwell.

Trujillo, I. and E. Garcia, 1996. Strategies for obtaining somaclonal variants resistant to yellow Sigatoka (*Mycosphaerella musicola*). *Infomusa* 5:12–13.

Ude, G., M. Pillay, D. Nwakanma, and A. Tenkouano. 2002a. Analysis of genetic diversity and sectional relationships in *Musa* using AFLP markers. *Theor. Appl. Genet.* 104:1239–1245.

Ude, G., M. Pillay, D. Nwakanma, and A. Tenkouano. 2002b. Genetic diversity in *Musa acuminata* Colla and *Musa balbisiana* Colla and some of their natural hybrids using AFLP markers. *Theor. Appl. Genet.* 104:1246–1252.

Vakili, N.G. 1968. Responses of *Musa acuminata* species and edible cultivars to infection by *Mycosphaerella musicola. Tropical Agric. (Trinidad)* 45:13–22.

Viaene, N., L.F. Durán, J. Dueñas, J.M. Rivera, D. DeWaele, and P. Rowe. 2000. Reaction of hybrids and natural genotypes of *Musa* to *Pratylenchus coffeae* and *Radopholus similis* in shadehouse conditions. Abstract. P77. ACORBAT 2000, 31 Jul–4 Aug., 2000, San Juan, Puerto Rico.

Vilarinhos, A.D., P. Pifanelli, P. Lagoda, S. Thibivilliers, X. Sabau, F. Carrel., and A. D'Hont. 2003. Construction and characterization of a bacterial artificial chromosome library of banana (*Musa acuminata* Colla). *Theor. Appl. Genet.* 106:1102–1106.

Vroh-Bi, I., C. Anagbogu, S. Nnadi, and A. Tenkouano. 2010. Genomic characterization of natural and somaclonal variations in bananas (*Musa* spp.). *Plant Mol. Biol. Rep.* (in press).

Vuylsteke D., R. Swennen, G.F. Wilson, and E. De Langhe. 1988. Phenotypic variation among in vitro propagated plantain (*Musa* spp. Cv. AAB). *Scientia Hort.* 36:79–88.

Vuylsteke, D. 1989. Shoot-tip culture for the production, conservation and exchange of *Musa* germplasm. *Practical manuals for handling crop germplasm* in vitro 2. Rome, Italy: International Board of Plant Genetic Resources.

Vuylsteke, D. 1998. Field performance of banana micropropagules and somaclones. In: *Somaclonal variation and induced mutation in crop improvement*, S.M. Jain, D.S. Brar, and B.S. Ahloowalia, eds., 219 – 231, Dordrecht, The Netherlands: Kluwer Academic.

Vuylsteke, D. and R. Ortiz. 1995. Plantain-derived diploid hybrids (TMP2x) with black Sigatoka resistance. *HortSci.* 30:147–149.

Vuylsteke, D., R. Ortiz, and S. Ferris. 1993a. Genetic and agronomic improvement for sustainable production of plantain and banana in sub-Saharan Africa. *Afri. Crop Sci. J.* 1:1–8.

Vuylsteke, D., R. Ortiz, R.S.B. Ferris, and J.H.Crouch. 1997. Plantain Improvement. *Plant Breed. Rev.* 14:267–320.

Vuylsteke, D., R. Swennen, and E. De Langhe. 1991. Somaclonal variation in plantains (*Musa* spp., AAB group) derived from shoot-tip culture. *Fruits* 46:429–439.

Vuylsteke, D., R. Swennen, and R. Ortiz. 1993b. Registration of 14 improved tropical Musa plantain hybrids with black Sigatoka resistance. *HortScience* 28:957–959

Wilson, G.F. and I. Buddenhagen. 1986. The black Sigatoka threat to plantain and banana in West Africa. *IITA Research Highlights* 7:3.

10 Mutations and Cultivar Development of Banana

Shri Mohan Jain, Bradley Till, Prasnna Suprasanna, and Nicolas Roux

CONTENTS

10.1 INTRODUCTION

Spontaneous genetic variation has contributed to the origin of present-day *Musa*. Spontaneous mutations are considered to have played a role in the origin of almost all of the edible banana and plantain varieties (Buddenhagen, 1987). A spontaneous Cavendish mutant, resistant to Fusarium wilt (race 1), that originated in Vietnam replaced 'Gros Michel' in 1950s–60s (Ploetz, 1994) as the major export banana and saved the commercial banana industry from collapse. Development of new banana clones with improved agronomic features has been slow due to the complex and polyploid nature of the *Musa* genome as well as sterility barriers and other obstacles to conventional breeding approaches (Pillay and Tripathi, 2006, 2007). Breeding efforts have produced only a few cultivars through selection of improved dessert and cooking bananas and plantains (Rowe, 1984; Vuylsteke et al., 1995). Concerted efforts are being made to develop new banana cultivars by a number of breeding programs.

In vegetatively propagated crops like banana that have a narrow genetic base (Pillay et al., 2001), it is essential to generate additional genetic variability to facilitate selection of desirable traits. Many techniques involving cellular and molecular biological tools such as *in vitro* plant regeneration, mutagenesis, transgenics, and molecular markers have been deployed for crop improvement (Jain and Swennen, 2004). Induced mutagenesis and *in vitro* selection for desired traits offer several advantages such as mutagenizing the plant parts, uniform mutagen treatment, handling large number of samples in a short time span, rapid production of large populations to separate chimeras, and facilitating *in vitro* selection (Van Harten, 1998). Currently, it is believed that *in vitro* mutation breeding programs can deliver more promising results (Smith et al., 2005, 2006).

There is rapid accumulation of genome sequence information for many crop species, including banana. This has allowed the development of hypotheses for gene function based on homologies between different organisms. Reverse-genetic techniques target disruptions in specific loci, providing *in vivo* testing of gene function and the development of crops with novel traits (for example, Enns et al., 2005; Slade et al., 2005). Targeting induced local lesions in genomes (TILLING) is a general reverse-genetics strategy utilizing traditional mutagenesis and high-throughput mutation discovery that has been applied to many species (McCallum et al., 2000; Colbert et al., 2001). TILLING provides a method to combine the power of induced mutations with expanding sequence information for functional genomics and crop improvement projects. Challenges exist for the efficient adaptation of TILLING for banana, and several approaches are being evaluated. While the development of large mutant populations in banana can be considered a bottleneck, reverse-genetics may provide a practical means for genetic studies and plant improvement.

10.2 MUTATION INDUCTION

The combination of mutation breeding and *in vitro* culture (also called *in vitro* mutagenesis) makes the induction and selection of induced somatic mutations more effective. This method can also aid in further propagation, which will ensure the formation of periclinal chimeras or homohistont individuals. Furthermore, increased recovery of plantlets through decreased somatic competition can be obtained by modifying culture conditions. Plant growth regulators, and in particular a cytokinin-enriched medium, can increase the recovery rate of mutated cells. Hence, the combined use of mutation induction and *in vitro* technology is more efficient because it speeds up the production of mutants as a result of an increased propagation rate and a greater number of generations per unit time and space (Morpurgo et al., 1997).

10.2.1 Methods on Mutation Induction

Mutagens cause random changes in the nuclear or cytoplasmic DNA, resulting in gene, chromosomal, or genomic mutations (Jain and Maluszynski, 2004). Both physical and chemical mutagens have been employed for mutagenesis in banana for generating useful mutations (see Table 10.1 and Table 10.2). Physical mutagens such as gamma- and x-rays have been useful for mutation induction (Kulkarni et al., 1997; Roux et al., 2004). Physical mutagens, especially gamma rays, show reasonable reproducibility and have a high and uniform penetration of multicellular systems. However, selection of a mutagen is dependent on the type of plant material, such as organs (shoot tips, meristems, axillary buds), cell suspension cultures, or protoplasts. Although gamma rays have been commonly used for physical mutagenesis, ion-beam techniques have recently been described. Ion beams integrate the factors of mass, energy, and charge, inducing damage to the biological materials, thereby displacing, recombining, and compounding the biological molecules and atoms. Ion beams can frequently produce large DNA alterations such as inversions, translocations, and large deletions rather than point mutations. Reyes-Borja et al. (2007) were the first to report the use of ion-beam irradiation for mutation breeding in banana by selecting lines tolerant to black Sigatoka. The effect of irradiation doses on the regeneration of plantlets was investigated in Japan, and the variation of the black Sigatoka response under field conditions was evaluated in Ecuador (Reyes-Borja et al., 2007). Chemical mutagens can also be used, producing a high density of predominantly point mutations. The problems involved in using physical mutagens are the high cost of radiation source requirement and the simultaneous induction of chromosomal and gene mutations. Chemical mutagens are carcinogenic in nature and penetrate multicellular tissues nonuniformly.

TABLE 10.1
Studies on Mutation Induction Using Chemical and Physical Mutagens in Banana

Target Material	Cultivar/ Genomic Group	Mutagen Dose	Observed Responses/ Mutation(s)	Mutant/Clones Developed	Reference
Excised shoot tips	SH-3362 (AA); GN-60A (mutant of Grande Naine, AAA)	EMS 24.69 mM	Number of newly initiated adventitious buds decreased with increased EMS concentrations	—	1
Shoot tips of *in vitro* grown cultures	Highgate, AAA Group	-Sodium azide (NaN_3) 2.3 mM -diethyl sulphate (DES)-20 mM -EMS 200 mM.	Tolerance to *Fusarium oxysporum* f. sp. *cubense*	Tolerant clones for field screening	2
Shoot tips	Grande Naine, AAA	Gamma rays	Earliness	Early flowering putative mutant designated GN-60A	3
Shoot tips		Gamma rays 25 Gy	Bunch size and cylindrical shape	Klue Hom Thong KU1	4
Shoot tips	Dwarf Parfitt, an extra-dwarf Cavendish banana	Gamma rays 20 Gy	Improved agronomic characteristics (taller plant size, increased yield and no choking)	Improved lines with productivity and resistance	5
Shoot tips	Highgate, AAA	Gamma rays 8–20 Gy	Tolerance to *Fusarium oxysporum* f. sp. *cubense*	Tolerant clones for field screening	6
Protocorm like bodies	Nanicao, AAA	Gamma rays 2kR	Aluminum tolerance	Tolerant selections	7
In vitro cultured shoot tips	Diploid and tetraploid clones	Gamma rays 25 Gy	Diploid clones were more sensitive than tetraploids	Plants with morphological and physiological traits	8
Tissue cultured shoot tips	Dwarf Cavendish, AAA	Gamma rays 8–20 Gy	Morphological traits	22 clones for different morphological traits	9
In vitro cultured shoot tips	Basrai, AAA	Gamma rays	Morphological traits	Clones for different morphological traits	10
Excised shoot tips *in vivo* and *in vitro*	Basrai, AAA	Gamma rays	Differential response of *in vivo* and *in vitro* plant materials	—	11
In vitro propagated shoot tips	Williams and Cavendish Enano	Carbon ion beam 0, 0.5, 1, 2, 4, 8, 16, 32, 64 and 128 Gy	Fungal disease resistance	Resistant plants to black Sigatoka	12

(continued)

TABLE 10.1 (CONTINUED)
Studies on Mutation Induction Using Chemical and Physical Mutagens in Banana

Target Material	Cultivar/ Genomic Group	Mutagen Dose	Observed Responses/ Mutation(s)	Mutant/Clones Developed	Reference
In vitro shoot cultures	Basari, AAA, Chakkarakela, AAB, and Rasthali, AAB	Gamma rays 30Gy (recurrent dose)	Morphological traits	Morphological variations: thick shiny dark green leaves, ovate leaves and a dwarf with a rosette of leaves	13

Sources: Data from (1) Omar, M.S., F.J. Novak, and H. Brunner, 1989, *In vitro* action of ethyl-methanesulphonate on banana shoot tips, *Sci. Hort.*, 40, 283–295. (2) Bhagwat, B. and E.J. Duncan, 1998, Mutation breeding of banana cv. Highgate (*Musa acuminata*, AAA group) for tolerance to *Fusarium oxysporum* f. sp. *cubense* using chemical mutagens, *Sci. Hort.*, 73, 11–22. (3) Roux, N., R. Afza, H. Brunner, R. Morpurgo, and M. van Duren, 1994, Complementary approaches to cross-breeding and mutation breeding for *Musa* improvement, 213–218, Proceedings of the first global conference on International *Musa* Testing Program, FHIA, Honduras. (4) List of new mutant cultivars; *Musa* sp. (banana), 1990, *Mutation Breed. Newslett.*, 35, 32–41. (5) Smith, M.K., S.D. Hamill, P.W. Langdon, J.E. Giles, V.J. Doogan, and K.G. Pegg, 2006, Towards the development of a Cavendish banana resistant to race 4 of Fusarium wilt: Gamma irradiation of micropropagated Dwarf Parfitt (*Musa* spp., AAA group, Cavendish subgroup), *Aus. J. Exp. Agri.*, 46, 107–113. (6) Bhagwat, B. and E.J. Duncan, 1998, Mutation breeding of Highgate (*Musa acuminata*, AAA) for tolerance to *Fusarium oxysporum* f. sp. *cubense* using gamma irradiation, *Euphytica*, 101, 143–150. (7) Matsumoto, K. and H. Yamaguchi, 1990, Selection of aluminium-tolerant variants from irradiated protocorm-like bodies in banana, *Trop. Agric. (Trinidad)*, 67, 229–232. (8) Novak, F.J., R. Afza, M. Van Duren, and M.S. Omar, 1990, Mutation induction by gamma irradiation of *in vitro* cultured shoot-tips of banana and plantain (*Musa* cvs), *Trop. Agric. (Trinidad)*, 67, 21–28. (9) Miri, S.M., A. Mousavi, R. Mohammad, A. Naghavi, M. Mirzaii, A.R. Talaei, et al., 2009, Analysis of induced mutants of salinity resistant banana (*Musa acuminata* cv. Dwarf Cavendish) using morphological and molecular markers, *Iranian J. Biotech.*, 7, 86–92. (10) Kulkarni, V.M., T.R. Ganapathi, P. Suprasanna, V.A. Bapat, and P.S. Rao, 1997, Effect of gamma irradiation on *in vitro* multiple shoot cultures of banana (*Musa species*), *J. Nucl. Agric. Biol.*, 26, 232–240. (11) Karmarkar, V.M., V.M. Kulkarni, P. Suprasanna, V.A. Bapat, and P.S. Rao, 2001, Radio-sensitivity of *in vivo* and *in vitro* cultures of banana cv. Basrai (AAA), *Fruits*, 56, 67–74. (12) Reyes-Borja, W.O., I. Sotomayor, D. Garzón, M. Vera, B. Cedeño, A. Castillo, et al., 2007, Alteration of resistance to black Sigatoka (*Mycosphaerella fijiensis* Morelet) in banana by *in vitro* irradiation using carbon ion-beam, *Plant Biotech.*, 24, 349–353. (13) Mishra, P.J., T.R. Ganapathi, P. Suprasanna, and V.A. Bapat, 2007, Effect of single and recurrent gamma irradiation on *in vitro* shoot cultures of banana, *Int. J. Fruit Sci.*, 7, 47–57.

10.2.2 Choice of Plant Material

Both *in vivo* and *in vitro* cultures have been used to induce mutations in banana. Earlier studies used seeds and suckers of plantains and bananas, including the diploid *Musa balbisiana*, and demonstrated that gamma rays affected the rates of seed germination and seedling survival (Stotzky et al., 1964). The chemical mutagen ethyl methanesulphonate (EMS) was used in seeds of diploid *M. acuminata* (Menendez, 1973). Irradiation of suckers, prior to excision of shoot tips for *in vitro* culture, generally gives a low yield of mutagenized material for further screening. De Guzman et al. (1976) reported that viable shoots could be obtained from suckers irradiated with a dose of up to 100 Gy. *In situ* sucker irradiation before initiation of tissue culture was not effective and yielded only a low number of regenerants in the M_1V_1 generation. In a comparative study on the radiosensitivity of different *in vivo* and *in vitro* planting materials, Karmarkar et al. (2001) observed a decrease in percent survival of irradiated material with increasing irradiation dose. The hardened plants were more sensitive than suckers, while individual *in vitro* shoots were least sensitive to increases in doses of

TABLE 10.2
Examples of Desirable Variants/Putative Mutants Identified for Release or Further Confirmation Trials

Country	Parent/Selection	Traits	Technique	Place of Induction
Cuba	SH3436 (AAAB)/'SH3436	Reduced height	Gamma rays	Cuba
	Parecido al Rey (AAA)/ Parecido al Rey 6.44	Reduced height	Gamma rays	IAEA
Malaysia	Pisang Rastali (AAB)/Mutiara	Tolerance to *Foc* race 4	Somaclones	United Plantation Bhd., Malaysia
	Grande Naine GN-GoA (AAA)/Novaria	Tolerance to *Foc* race 4	Somaclones	
	Pisang Berangan	Early flowering and reduced height	Somaclones	
	Pisang Berangan	Tolerance to *Foc* race 4	Gamma rays	IAEA
	Pisang Mas	Tetraploid	Colchicine	
Philippines	Lakatan (AAA)	Reduced height and earliness	Gamma 40 Gy	IAEA
	Latundan (AAB)	Large fruit size and reduced height	3 Gy fast neutrons	
Sri Lanka	Embul (AAB)/Embul	Earliness and reduced height	Gamma rays	Sri Lanka

Source: Jain, S.M. and M. Maluszyuynski, 2004, Induced mutations and biotechnology in improving crops. In: In vitro *application in crop improvement*, A. Mujib, M. Cho, S Predieri and S. Banerjee, Eds., 169–202. Science Publishers; Plymouth, UK. With permission.

gamma irradiation. Irradiation of *in vitro* multiple shoots adversely affected multiplication, except for doses of 10–20 Gy, which were observed to significantly enhance multiplication.

10.2.2.1 Shoot Tips

Shoot tips propagated *in vitro* have also been used for chemical mutagenesis. Omar et al. (1989) studied the effects of EMS on the growth and development of excised shoot tips of two banana clones: 'SH-3362' (AA) and 'GN-60A,' a mutant of 'Grande Naine' (AAA). The effects of EMS on fresh weight and on the number of newly initiated adventitious buds of cultured shoot tips were evident after 30 days of incubation. On the basis of different studies, a suitable level of EMS treatment for banana clones appears to be 12.41–37.23 mM for 1 to 3 h. Bhagwat and Duncan (1998a) compared the effect of three chemical mutagens—namely sodium azide (NaN_3), diethyl sulphate (DES), and EMS—at various concentrations on shoot tips of *in vitro* grown cultures of the cultivar 'Highgate' (AAA). On the basis of the number of apices that survived the treatment and the number of regenerated shoots, the authors found that the mutagens differed in their mutation induction efficiency. The highest factor of effectiveness (7.8%) was obtained with NaN_3 resulting in 63.3% explant survival and 58.9% shoot regeneration, while DES gave 65.5% survival and 38.2% shoot regeneration, and EMS gave 5.8%, 80% explant survival, and 31.6% shoot regeneration.

10.2.2.2 Embryogenic Cell Suspensions

Embryogenic cell suspension (ECS) cultures can also be used for the induction of mutations with physical (Kulkarni et al., 2004, 2007) and chemical mutagens. Since ECS is of single-cell origin, the number of regenerated chimeric plants is reduced, allowing rapid generation of homohistonts or nonchimeric plants. Over a million embryogenic cells per ml can be treated with mutagens, directly producing mutant somatic embryos that regenerate to mutant plantlets. *In vitro* direct selection pressure can be applied on mutagen-treated cells for the selection of mutant somatic embryos. The initiation of somatic embryogenic cultures can commence by culturing a wide variety of explants including zygotic embryos, nucellus, parts of seeds and fruits, inflorescences, leaf pieces, stem segments, protoplasts, and microspores. Embryogenic callus is subcultured in a liquid medium on a shaker to develop fine cell suspensions (Kulkarni et al., 2007). Superior quality embryogenic cell suspensions are usually characterized by the presence of clusters of round and densely stained cells, and absence or minimum number of enlarged, elongated, and vacuolated cells and debris (Roux et al., 2004). The embryogenic cells are uniformly spread on agar-solidified culture medium for physical mutagen treatment and subsequent somatic embryo formation. Well-developed somatic embryos are germinated to obtain complete plants (somatic seedlings), which can be acclimatized and transferred to the field. Embryo formation is not synchronous, and therefore large and small embryoids coexist. The uniformity of populations can be improved by culturing on media containing high levels of sucrose and/or low levels of abscisic acid. High levels of sucrose and abscisic acid induce reversible dormancy in somatic embryos and thus might be used to temporarily suspend the growth for synchronous development.

10.2.3 Establishment of a Radiosensitive Curve

One of the first steps in mutagenesis experiments is to determine the appropriate mutagen dose. For physical mutagens, this is generally based on understanding the radiosensitivity of the tissues/cells. Theoretically, the highest frequency of mutations can be expected from a mutagen treatment, killing about 50% of the treated materials (LD_{50}) (Van Harten, 1998), and hence the LD_{50} dose of the mutagen is obtained from a radiosensitive curve (Figure 10.1). The plant material (shoot multiples, embryogenic calli, or cell suspensions) are treated over a wide range of doses, and data are recorded on traits such as fresh and dry weight gain, survival of treated cultures, number of

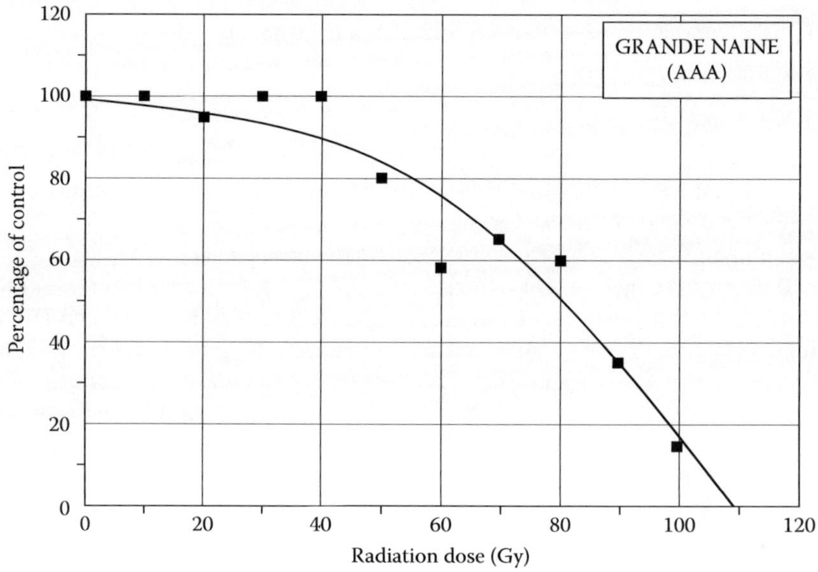

FIGURE 10.1 Radiosensitivity test curve illustrating the effect of increasing dose of gamma rays on survival rate of shoot tips from the cultivar 'Grande Naine' (AAA). Data are expressed in percentage of control (nonirradiated shoot tips). (Reprinted from N. Roux, A. Toloza, J. Doležel, and B. Panis, Usefulness of embryogenic cell suspension cultures for the induction and selection of mutations in *Musa* spp., *Banana improvement: Cellular molecular biology and induced mutations*, Enfield, NH: Science Publishers, 2004, 33–34. With permission.)

complete plantlets regenerated, and LD_{50} values estimated (Jain, 2009; Predieri, 2001). Different genotypes respond differently to mutagenesis of shoot tips, and optimal dose (LD_{50}) has been found to vary considerably. Roux (1997) recommended the following doses for different ploidy levels and banana genomic groups:

- 10–20 Gy of γ rays for diploid clones 'Calcutta 4' (AA) and 'Tani' (BB).
- 30–40 Gy of γ rays for the triploids 'Three Hand Planty' (AAB), 'Grande Naine' (AAA), 'Williams' (AAA).
- 40–50 Gy of γ rays for the triploid 'Cachaco' (ABB).

Less is known about optimal dosages for chemical mutagenesis, but TILLING provides a method to directly assay the density and spectrum of induced nucleotide changes in direct response to mutagen type and dose (see Section 10.4).

10.3 POSTMUTAGENESIS

Postmutagenesis handling of mutagenized populations is critical as *in vitro* mutagenesis of multicellular meristems of *Musa* spp. leads to a high degree of chimerism. In general, mutated cells are difficult to monitor. However, mutations that result in a change in chromosome numbers are an exception. The mutagen treatment of plant organs often leads to chimeras that require dissociation by subsequent subcultures of shoot cultures up to the M_1V_4 generations (Kulkarni et al., 2007). The number of subcultures depends primarily on the genotype, LD_{50} dose, and other factors such as proliferation rate, number of plants to be field evaluated, and so forth. *In vitro* shoots can also be directly rooted in the greenhouse under 70–90% humidity to avoid the additional step of *in vitro* rooting. The number of plants developed is dependent on the greenhouse facilities.

10.3.1 Dissociation of Chimera

In an effort to dissociate chimeras, Roux et al. (2001) assessed three different propagation systems (shoot-tip culture technique, multi-apexing culture technique, and corm-slice culture technique). The average percentage of cytochimera was reduced from 100% to 36% after three subcultures using shoot-tip culture, from 100% to 8% after the same number of subcultures using the multi-apexing technique, and from 100% to 24% when propagating by the corm-slice culture technique. Although none of the systems studied eliminated chimerism completely, the study showed that the possibilities of reducing chimeras depend on the type of shoot produced (axillary or adventitious) and the multiplication rate (number of new shoots produced per subculture). Nevertheless, in all cases, after three subcultures the proportion of chimeras tends to stabilize. This is due to the formation of periclinal chimeras, which are difficult to eliminate using multicellular propagation systems. In a multicellular propagation system, it is expected that more than three to four subcultures will not significantly increase the dissociation of chimeras, and it will not be possible to reach 0% chimera as long as mitotic divisions maintain cell layer identity. Although embryogenic cell suspensions (ECS) can be a useful system for producing mutants in banana, the system has to be fully optimized. Compared to shoot-meristem cultures, which often yield chimeras, ECS allow large cell populations to be used for mutagenesis under controlled conditions, and chimerism can be avoided as somatic embryos are mostly derived from single cells. Roux et al. (2007) demonstrated that colchicine treatment of ECS followed by flow cytometry of regenerated plants produced no mixoploid plants (chimeric). ECS fresh weight gain and regeneration capacity indicated that the optimal irradiation dose ranged from 50 to 75 Gy for 'Williams' (AAA, Cavendish subgroup) and 'Three Hand Planty' (AAB, plantain subgroup). Xu et al. (2008) found that plant regeneration capacity also decreased with an increase in radiation dosage in ECS of 'Beida Aijiao' (*Musa* AAA). At the dose of 60 Gy, ECS did not survive on embryo regeneration medium. It was shown that plant regeneration capacity of a growing cell mass was much higher than that of ECS directly radiated with gamma ray at four radiation dosages, suggesting that cultures of ECS on embryo regeneration medium could reduce the sensitivity of ECS to gamma radiation. Somaclonal variation, probably associated with chromosome instability, can also occur among plantlets produced from ECS and could interfere with this approach for producing mutants in banana. Variable DNA ploidy was detected in ECS and in regenerated plants (Roux et al., 2007). The random nature of mutation warrants the screening of thousands of individuals when taking a forward-genetic approach. The application of banana ECS in mutagenesis research should be able to increase the efficacy of mutation induction and this could revolutionize the isolation and screening of new and useful mutants.

10.3.2 *In Vitro* Selection

Mutagenesis combined with *in vitro* selection is generally regarded as an appropriate tool for enhancing and accelerating induction, selection, and recovery of desirable mutants (Suprasanna et al., 2008). Based upon the information generated on radio sensitivity of different banana cultivars, researchers have attempted to employ irradiated *in vitro* cultures for *in vitro* selection, for example, for aluminum stress (Matsumoto and Yamaguchi, 1990). In another study, Roux and Toloza (2002) applied radiation-induced mutagenesis to select plants resistant to black Sigatoka. Four batches (100 meristems/batch) of the variety 'Grande Naine' were irradiated at 35 Gy and further propagated during four subcultures. The plants were then acclimatized in the greenhouse. The early mass screening method was adopted by using infiltration of juglone, a toxic metabolite of *Mycosphaerella fijiensis*. After screening approximately 4,000 plants, 15 putative mutants showed tolerance (Figure 10.2). Table 10.2 shows various banana mutant varieties/lines have been released/isolated in several countries. In Malaysia, 'Mutiara' (Figure 10.3) and 'Novaria' banana mutant cultivars have been released commercially by United Plantation Bhd.

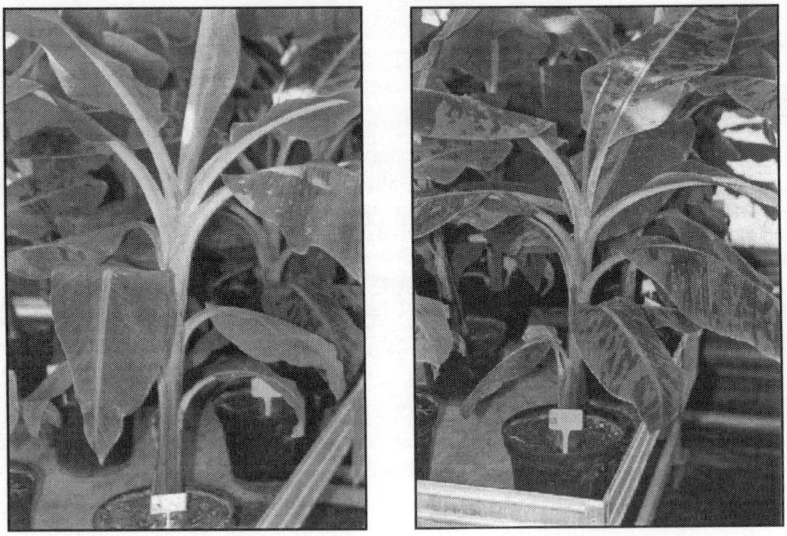

(a) (b)

FIGURE 10.2 Regenerated plants from irradiated shoot tips of the cultivar 'Grande Naine': (A) Susceptible to 25 ppm juglone (5-hydroxy-1,4-naphthoquinone), a toxic metabolite of *Mycosphaerella fijiensis*, the fungus responsible for the black Sigatoka disease. (B) Putative mutant, tolerant to 25 ppm juglone. Note the increased content of anthocyanin. (Reprinted from N. Roux et al. (2003). Mutagenesis and somaclonal variation to develop new resistance to Mycosphaerella leaf spot diseases on *Musa*. In: *Proceedings of 2nd International Workshop on Mycosphaerella Leaf Spot Diseases of Bananas*, San Jose, Costa Rica, 20–23 May 2002: 239–250. With permission.)

FIGURE 10.3 Tolerant 'Mutiara' banana plants surviving in Fusarium wilt-infected hot spot, developed in Malaysia.

10.4 REVERSE-GENETIC STRATEGIES FOR BANANA USING INDUCED MUTATIONS

A reverse-genetic strategy that combines traditional mutagenesis with high-throughput mutation discovery can be considered for banana functional genomics and crop improvement. Reverse-genetic approaches promise enhanced efficiencies over forward-genetic approaches described above. Commonly referred to as TILLING (for targeting induced local lesions in genomes), the reverse-genetic method first developed for *Arabidopsis thaliana* has since been applied to a large number of seed-propagated crops, including maize, rice, soybean, and wheat (McCallum et al., 2000; Till et al., 2003, 2007; Till, Reynolds, et al. 2004; Slade et al., 2005; Cooper et al., 2008b). For TILLING, a mutant population is prepared in advance, typically comprised of several thousand individuals. DNA is collected from each plant and the germplasm is stored. To identify mutations, PCR is performed with gene-specific primers and the resulting amplicons evaluated for induced mutations via enzymatic mismatch cleavage or alternative mutation discovery technologies (Till, Burtner, et al., 2004). Mutant individuals are then recovered from the germplasm bank for further molecular and phenotypic characterization. Since only a small number of plants are expected to harbor deleterious mutations in a target gene, the method can greatly reduce phenotyping efforts compared to forward-genetic screens. It is further advantageous because mutant alleles can be obtained from a population that by themselves do not produce a phenotype but will when combined with others, as for example with duplicated genes or homoeologues in polyploids. The strength of the approach will continually improve as genomic sequences from more species are made available and hypotheses for plant gene function become more refined.

The choice of mutagen, tissue, and optimal dose of treatment are important factors when developing a TILLING population. Many mutagenesis strategies produce chimeric tissues with different cells having different genotypes in the first (M_1) generation. As described above, chimeras need to be dissociated before mutation discovery is performed so that inheritance of mutations can be ensured and phenotypes accurately measured. In seed-propagated plants, chimeras are easily dissociated through sexual crossing. A structured population is typically created whereby M_1 plants are self-fertilized and a single-seed descent strategy is followed with tissue collected for mutation discovery in the M_2 generation. M_3 seed from a self-cross of the M_2 is collected and stored for future phenotypic analysis (see, for example, Till et al., 2003; Till et al., 2007; Caldwell et al., 2004). What has emerged from TILLING studies is a large data set of mutations caused by chemical mutagenesis.

Chemicals such as EMS have been shown to cause single nucleotide changes randomly throughout the genome (Greene et al., 2003). This means that with the right combination of mutation density and population size, multiple mutations in any gene in the genome can be obtained. The resulting point mutations include changes expected to knockout gene function (nonsense mutations and those affecting RNA splicing) as well as missense changes that can have varying consequences on protein function. Thus TILLING can provide knockouts and less severe mutations for more nuanced studies of gene function and for the study of essential genes where a knockout would be lethal. Computational tools have been developed that make predictions on the effect of missense mutations, and many steps of the TILLING procedure can be automated and converted to high-throughput methods, allowing the development of large-scale community mutation screening services (Ng and Henikoff, 2003; Taylor and Greene, 2003; Till et al., 2003; Cooper et al., 2008a). Less is known about the spectrum and density of induced mutations from physical irradiation, such as treatment with gamma rays, but a broader spectrum of changes has been reported, including single nucleotide changes and small indels (Sato et al., 2006).

Many challenges arise when TILLING is considered for banana. Widely consumed triploid varieties are largely sterile, infertile, and/or parthenocarpic, thus requiring vegetative propagation, making seed mutagenesis impractical. Mutation induction using *in vitro* material can therefore be considered. Embryogenic cell suspensions have been prepared in banana, and mutagenesis of cell suspensions is potentially highly efficient because each cell in the suspension accumulates different mutations and each cell can produce a single plantlet, rapidly providing many nonchimeric plants immediately suitable for mutation screening (Strosse et al., 2003). Mutagenesis is also performed using isolated shoot apical meristems (shoot tips) with the resulting adult plant (M_1V_1) being chimeric. As described, chimerism can be reduced through successive rounds of meristem isolation, followed by cutting meristems and allowing plantlets to regenerate (Roux, 2004). Through this process the number of totipotent stem cells in the central zone of the meristem that produce adult plant tissue is reduced, therefore reducing the genotypic complexity of the resulting plantlet. A population of ~1500 EMS mutagenized 'Grande Naine' AAA plantlets subcultured six times (M_1V_6) to remove chimeric sectors and obtain enough plant material was recently prepared. A pilot TILLING screen with this population revealed a spectrum of mutations as expected for EMS and a density of mutations expected for a triploid species (Joanna Jankowicz-Cielslak, Chikelu Mba, and Bradley J. Till, unpublished). The identification of mutations stably inherited in M_1V_6 suggests that induced mutations and reverse genetics can be considered as realistic options for vegetatively propagated species. It also provides genotypic evidence for the dissolution of chimeras at the level of single-nucleotide mutations. A major genetic bottleneck in this strategy, though, is the expectation that most single-nucleotide changes will produce recessive, loss of function alleles, and mutations must be homozygous before phenotypes can be observed. The ratio of recessive to dominant gain of function phenotypes in the M_1 generation, however, may be lower in vegetatively propagated banana compared to other plants because of the lack of meiosis, recombination, and selective pressure to remove deleterious alleles from the population. Thus a situation may exist where two of three alleles have naturally accumulated deleterious mutations, and mutagenesis of the third allele would reveal a phenotype. The frequency of such events is currently unknown. Another perhaps important consideration is that mutagenized and vegetatively propagated plants are unique in that they represent nonchimeric multicellular eukaryotes that have not undergone meiosis or fertilization. Therefore, potentially dominant large structural chromosomal differences that are meiotically lethal would be maintained in the population. Analysis of such changes would require different mutation discovery assays such as array comparative genome hybridization (Bruce et al., 2009; Rios et al., 2008). Such an analysis would produce knowledge of the full spectrum of mutations caused by a particular mutagen and the types of changes are filtered out through meiosis, gametogenesis, and fertilization. Careful controls, however, are needed to separate induced changes from large chromosomal aberrations that might occur due to propagation by tissue culture (Veilleux and Johnson, 1998). It is also worth considering that a particular mutagen and dose may be optimal for producing phenotypes in

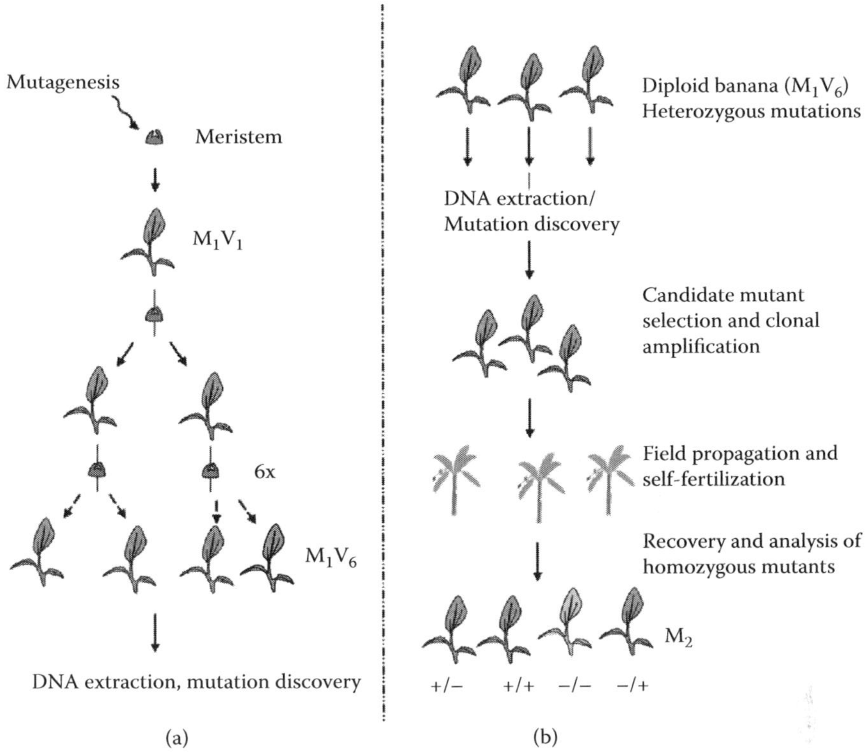

FIGURE 10.4 Meristematic mutagenesis and TILLING strategies for banana. Meristematic tissue is isolated and treated with mutagen (A). The resulting M_1V_1 generation plant is chimeric because different cells in the meristem accumulated different mutations. Plants are made genotypically homogeneous through six rounds of meristem isolation, cutting the meristem in two, and generating two new plants. Tissue is collected for DNA extraction and mutation screening at the M_1V_6 generation. TILLING allows for preselection of candidate mutants for self-fertilization in sexually propagating diploids (B). To overcome low fecundity, clonal amplification can be used prior to sexual crossing, allowing for sufficient seed production for segregation analysis in the M_2 generation.

the M_1 generation, perhaps those conditions favoring large chromosomal aberrations. At some level, however, too many genes are disrupted and reverse-genetic approaches become impractical.

More straightforward is the consideration of reverse genetics in sexually fertile diploid bananas. Here, a strategy can be envisioned whereby vegetative propagation and TILLING are combined to make an efficient system in varieties that are normally recalcitrant to genetic investigation and improvement due to issues of low fecundity and a heavy demand on field resources (Figure 10.4). In normal forward-genetic strategies including mutation breeding, thousands of M_2 or higher lines must be screened for a reasonable chance in finding a deleterious mutation in a gene capable of causing a desired phenotype. Also required is the recovery of sufficient seed from a self-cross to ensure recovery of homozygous mutant alleles. Thus, forward genetics is labor intensive and inefficient for banana. Reverse-genetic strategies provide a method for making genetic studies tractable. Using the strategy described for triploid banana, thousands of nonchimeric diploid plants can be produced and maintained in tissue culture. Each plant will harbor unique heterozygous mutations. TILLING can then be performed to identify deleterious mutations in candidate genes. With knowledge of mutation densities in many diploid populations, the expectation is that screening a population of 3,000 would result in the discovery of several, but fewer than 10, potentially deleterious alleles per target screening region (Greene et al., 2003; Ng and Henikoff, 2003). Field labor could therefore be reduced by as much as three orders of magnitude. Furthermore, preknowledge of the plant with a candidate mutation allows clonal propagation of that plant prior to sexual fertilization. Intercrossing between clonal siblings to produce a suitable amount of M_3 seed to recover homozygous alleles then becomes possible. Mutagenized populations of diploid 'Calcutta 4' AA for TILLING are currently being developed at INIVIT in Cuba as part of an FAO/IAEA Coordinated Research Project, where mutagenesis of cell cultures is also being investigated (J. López, personal communication). With the complete banana genomic sequence expected in the near future, many candidate genes will be available to exploit this approach. For example, 'Calcutta 4' is resistant to *Mycosphaerella fijiensis*, the causative agent of black Sigatoka disease (Mobambo et al., 1997). Identification of knockouts in candidates' resistant genes could allow for the identification of the gene(s) conferring resistance. The 'Calcutta 4' resistance gene(s) could then, for example, be used in *cis*- or transgenic approaches to improve resistance in edible triploid varieties. Another interesting topic is the identification of mutations in genes causing parthenocarpy. In both cases TILLING is advantageous because interesting mutations can be propagated *in vitro* in a heterozygous form in near perpetuity. Finally, it is also worth considering the use of reverse genetics for the improvement of parental material for breeding programs.

10.5 CONCLUSIONS AND PROSPECTS

Considerable research efforts have been made to speed up *in vitro* mutagenesis for banana improvement. Although shoot apical meristems have been used, somatic embryogenic cell suspension cultures are highly suitable for mutation induction in banana genetic improvement strategies. *In vitro* selection systems are at the forefront for the isolation of disease resistant mutants, such as tolerance to black Sigatoka, caused by *M. fijiensis*, a devastating banana disease. So far, putative black Sigatoka–tolerant banana mutants have been isolated (Figure 10.2), with confirmation of disease tolerance still in progress. Such mutants may or may not possess the required resistance, but the mutant germplasm will be useful for genetic studies and in gene discovery. In vegetatively propagated crops like banana, mutation induction should be adopted as a useful genetic improvement strategy that can contribute to banana improvement programs.

The gains in our understanding of basic plant biology and in crop improvement using traditional mutagenesis and forward genetics have been immense. Over 3,000 induced mutant varieties are registered in the FAO/IAEA Mutant Variety and Genetic Stock Database (http://mvgs.iaea.org/). A number of bottlenecks exist in banana, making progress slower than in sexually propagating crops like cereals. Through studies described here, knowledge has been gained with regards to overcoming bottlenecks such as the induction of chimeric plant sectors and limitations of vegetative propagation. The development of new mutagenesis methods promise to increase allelic diversity. Furthermore, projected acquisition of whole genome sequence for banana is expected to provide gene sequences hypothesized to be important for key agronomic traits. By applying this knowledge along with new strategies such as reverse genetics, gains in the efficiency of using induced mutations for functional genomics and breeding can be envisioned. This is timely as changes in climate and an increasing world population are expected to put new pressures on global agricultural production.

REFERENCES

Bhagwat, B. and E.J. Duncan. 1998a. Mutation breeding of banana cv. Highgate (*Musa acuminata*, AAA group) for tolerance to *Fusarium oxysporum* f. sp. *cubense* using chemical mutagens. *Sci. Hort.* 73:11–22.

Bhagwat, B. and E.J. Duncan. 1998b. Mutation breeding of Highgate (*Musa acuminata*, AAA) for tolerance to *Fusarium oxysporum* f. sp. *cubense* using gamma irradiation. *Euphytica* 101:143–150.

Bruce, M., A. Hess, J. Bai, R. Mauleon, M.G. Diaz, N. Sugiyama, et al. 2009. Detection of genomic deletions in rice using oligonucleotide microarrays. *BMC Genomics* 10:129.

Buddenhagen, I.W. 1987. Disease susceptibility and genetics in relation to breeding of bananas and plantains. In: *Banana and plantain breeding strategies*, G.J. Persley and E.A. De Langhe, eds., 95–109. ACIAR Proceedings 21, Canberra, Australia.

Caldwell, D.G., N. McCallum, P. Shaw, G.J. Muehlbauer, D.F. Marshall, et al. 2004. A structured mutant population for forward and reverse genetics in Barley (*Hordeum vulgare* L.). *Plant J.* 40:143–150.

Colbert, T., B.J. Till, R. Tompa, S. Reynolds, M.N. Steine, A.T. Yeung, et al. 2001. High-throughput screening for induced point mutations. *Plant Physiol.* 126:480–484.

Cooper, J.L., B.J. Till, and S. Henikoff. 2008a. Fly-TILL: Reverse genetics using a living point mutation resource. *Fly (Austin)* 2:300–302.

Cooper, J.L., B.J. Till, R.G. Laport, M.C. Darlow, J.M. Kleffner, A. Jamai, et al. 2008b. TILLING to detect induced mutations in soybean. *BMC Plant Biol.* 8:9.

De Guzman, E.V., E.M. Ubalde, and A.G. del Rosario. 1976. Banana and coconut *in vitro* cultures for induced mutation studies, improvement of vegetatively. In: *Propagated plants and tree crops through induced mutations*, 33–54. Vienna: IAEA, TECDOC-194.

Enns, L.C., M.M. Kanaoka, K.U. Torii, L. Comai, K. Okada, and R.E. Cleland. 2005. Two callose synthases, GSL1 and GSL5, play an essential and redundant role in plant and pollen development and in fertility. *Plant Mol. Bio1.* 58:333–349.

Epp, M.D. 1987. Somaclonal variation in bananas: A case study with *Fusarium* wilt. In: *Banana and plantain breeding strategies*, G.J. Persley and E.A. De Langhe, eds., 140–150, ACIAR, Canberra..

Espino, R.C., A.B. Zamora, and R.B. Pimentel. 1986. Mutation breeding on selected Philippine fruit crops. In: *Nuclear techniques and in vitro culture for plant improvement*, 429–433. Vienna: IAEA.

Fortuno, V.J. and C.A. Maldona. 1972. The use of radiation in breeding banana (*Musa sapientum* L.), 485–488. Proceedings of a study group meeting, Buenos Aires, Argentina, IAEA Vienna.

Greene, E.A., C.A. Codomo, N.E. Taylor, J.G. Henikoff, B.J. Till, et al. 2003. Spectrum of chemically induced mutations from a large-scale reverse-genetic screen in Arabidopsis. *Genetics* 164:731–740.

Jain, S.M. 2009. *In vitro* mutagenesis in banana improvement. *Acta Hort.* (In press)

Jain, S.M. and M. Maluszynski. 2004. Induced mutations and biotechnology in improving crops. In: *In vitro application in crop improvement*, A. Mujib, M. Cho, S. Predieri and S. Banerjee, eds. 169–202. Plymouth, UK: Science Publishers, Inc.

Jain, S.M. and R. Swennen. 2004. *Banana improvement: Cellular, molecular biology and induced mutations*, xi–xii. Enfield, NH: Science Publishers, Inc.

Karmarkar, V.M., V.M. Kulkarni, P. Suprasanna, V.A. Bapat, and P.S. Rao. 2001. Radio-sensitivity of *in vivo* and *in vitro* cultures of banana cv Basrai (AAA). *Fruits* 56:67–74.

Kulkarni, V.M., T.R. Ganapathi, P. Suprasanna, V.A. Bapat, and P.S. Rao. 1997. Effect of gamma irradiation on *in vitro* multiple shoot cultures of banana (*Musa species*). *J. Nucl. Agric. Biol.* 26:232–240.

Kulkarni, V.M., T.R. Ganapathi, V.A. Bapat, and P.S. Rao. 2004. Establishment of cell-suspension cultures in banana cv. Grand Naine and evaluation of its sensitivity to gamma-irradiation. *Curr. Sci.* 86:902–904.

Kulkarni, V.M., T.R. Ganapathi, P. Suprasanna, and V.A. Bapat. 2007. *In vitro* mutagenesis in banana (*Musa* spp.) using gamma irradiation. In: *Protocols for micropropagation of woody trees and fruits*, S.M. Jain and H. Haggman, eds., 543–559. The Netherlands: Springer Publishers.

List of new mutant cultivars; *Musa* sp. (banana). 1990. *Mutation Breed. Newslett.* 35:32–41.

Matsumoto, K. and H. Yamaguchi. 1990. Selection of aluminium-tolerant variants from irradiated protocorm-like bodies in banana. *Trop. Agric. (Trinidad)* 67:229–232.

McCallum, C.M., L. Comai, E.A. Greene, and S. Henikoff. 2000. Targeted screening for induced mutations. *Nat. Biotechnol.* 18:455–457.

Menendez, T. 1973. A note on the effect of ethyl-methanesulphonate on *Musa acuminata* seeds. In: *Induced mutations in vegetatively propagated plants*, 85–90. Vienna: IAEA.

Miri, S.M., A. Mousavi, R. Mohammad, A. Naghavi, M. Mirzaii, A.R. Talaei, et al. 2009. Analysis of induced mutants of salinity resistant banana (*Musa acuminata* cv Dwarf Cavendish) using morphological and molecular markers. *Iranian J. Biotech.* 7:86–92.

Mishra, P.J., T.R. Ganapathi, P. Suprasanna, and V.A. Bapat. 2007. Effect of single and recurrent gamma irradiation on *in vitro* shoot cultures of banana. *Int. J. Fruit Sci.* 7:47–57.

Mobambo, K.N., C. Pasberg, F. Gauhl,and K. Zuofa. 1997. Host response to black Sigatoka in Musa germplasm of different ages under natural inoculation conditions. *Crop Prot.* 16:359–363.

Morpurgo, R., H. Brunner, G. Grasso, M. Van Duren, N. Roux, and R. Afza. 1997. Enigma of banana breeding: A challenge for biotechnology. *Agro-Food Ind. Hi-Tech*: 16–21.

Ng, P.C. and S. Henikoff. 2003. SIFT: Predicting amino acid changes that affect protein function. *Nucleic Acids Res.* 31:3812–3814.

Novak, F.J., R. Afza, M. Van Duren, and M.S. Omar. 1990. Mutation induction by gamma irradiation of *in vitro* cultured shoot-tips of banana and plantain (*Musa* cvs). *Trop. Agric. (Trinidad)* 67:21–28.

Omar, M.S., F.J. Novak, and H. Brunner. 1989. *In vitro* action of ethyl-methanesulphonate on banana shoot tips. *Sci. Hort.* 40:283–295.

Pillay, M. and L. Tripathi. 2006. Banana: An overview of breeding and genomics research in *Musa*. In: *Genome mapping and molecular breeding in plants*, Vol. 4, *Fruits and nuts*, C. Kole, ed., 281–301. Heidelberg: Springer-Verlag.

Pillay, M. and L. Tripathi. 2007. Banana breeding. In: *Breeding major food staples*, M.S. Kang and P.M. Priyadarshan, eds., 393–428. New York: Blackwell Science.

Pillay, M., E. Ogundiwin, D.C. Nwakanma, G. Ude, and A. Tenkouano. 2001. Analysis of genetic diversity and relationships in East African banana germplasm. *Theor. Appl. Genet.* 102:965–970.

Ploetz, R.C. 1994. Panama disease: Return of the first banana menace. *Intern. J. Pest Manage.* 40:326–336.

Predieri, S. 2001. Mutation induction and tissue culture in improving fruits. *Plant Cell Tiss. Org. Cult.* 64:185–210.

Reyes-Borja, W.O., I. Sotomayor, D. Garzón, M. Vera, B. Cedeño, A. Castillo, et al. 2007. Alteration of resistance to black Sigatoka (*Mycosphaerella fijiensis* Morelet) in banana by *in vitro* irradiation using carbon ion-beam. *Plant Biotech.* 24:349–353.

Rios, G., M.A. Naranjo, D.J. Iglesias,O. Ruiz-Rivero, M. Geraud, et al. 2008. Characterization of hemizygous deletions in citrus using array-comparative genomic hybridization and microsynteny comparisons with the poplar genome. *BMC Genomics* 9:381.

Roux, N. 1997. Improved methods to increase diversity in *Musa* using mutation and tissue culture techniques. Report of second research coordination meeting of FAO/IAEA/BADC Co-ordinated Research Project, 49–56. Kuala Lumpur, Malaysia.

Roux, N.S. 2004. Mutation induction in *Musa*—Review. In: *Banana improvement: Cellular, molecular biology, and induced mutations*, S.M. Jain and R. Swennen, eds., 23–32. Enfield, NH: Science Publishers.

Roux, N.S. and A. Toloza. 2002. Update in resistance to black Sigatoka through induced mutations. Proceedings of the 15th International Meeting ACORBAT, October 27–November 2, 2002. Cartagena de Indias, Colombia.

Roux N., A. Toloza, J.P. Busogoro, B. Panis, H. Strosse, P. Lepoivre, R. Swennen, and F.J. Zapata-Arias. 2003. Mutagenesis and somaclonal variation to develop new resistance to Mycosphaerella leaf spot diseases on *Musa*. *Proceedings of 2nd International Workshop on Mycosphaerella Leaf Spot Diseases of Bananas*, San Jose, Costa Rica: 239–250.

Roux, N., R. Afza, H. Brunner, R. Morpurgo, and M. van Duren. 1994. Complementary approaches to cross-breeding and mutation breeding for *Musa* improvement, 213–218. Proceedings of the first global conference on International *Musa* Testing Program, FHIA, Honduras.

Roux, N.S., J. Dolezel, R. Swennen, and F.J. Zapata-Arias. 2001. Effectiveness of three micropropagation techniques to dissociate cytochimeras in *Musa* sp. *Plant Cell Tiss. Org. Cult.* 66:189–197.

Roux, N.S, A. Toloza, J. Dolezel, and B. Panis. 2004. Usefulness of embryogenic cell suspension cultures for the induction and selection of mutations in *Musa* spp. In: *Banana improvement: Cellular, molecular biology and induced mutations*, S.M. Jain and R. Swennen, eds., 33–34. Enfield, NH: Science Publishers.

Roux, N.S., A. Toloza, H. Strosse, J.P. Busogoro, and J. Doležel. 2007. Induction and selection of potentially useful mutants in banana. ISHS-International Symposium on Recent Advances in Banana Crop Protection for Sustainable Production and Improved Livelihoods.

Rowe, P.R. 1984. Breeding bananas and plantains. *Plant Breed. Rev.* 2:135–155.

Sato, Y., K. Shirasawa, Y. Takahashi, M. Nishimura, and T. Nishio. 2006. Mutant selection from progeny of gamma-ray-irradiated rice by DNA heteroduplex cleavage using brassica petiole extract. *Breed. Sci.* 56:179–183.

Slade, A.J., S.I. Fuerstenberg, D. Loeffler, M.N. Steine, and D. Facciotti. 2005. A reverse genetic, nontransgenic approach to wheat crop improvement by TILLING. *Nat. Biotechnol.* 23:75–81.

Smith, M.K., S.D. Hamill, D.K. Becker, and J.L. Dale. 2005. *Musa* spp. banana and plantain. In: *Biotechnology of fruit and nut crops*, R.E. Litz, ed., 366–391. Cambridge, MA: CABI Publishing.

Smith, M.K., S.D. Hamill, P.W. Langdon, J.E. Giles, V.J. Doogan, and K.G. Pegg. 2006. Towards the development of a Cavendish banana resistant to race 4 of Fusarium wilt: Gamma irradiation of micropropagated Dwarf Parfitt (*Musa* spp., AAA group, Cavendish subgroup). *Aus. J. Exp. Agri.* 46:107–113.

Stotzky, G., E.A. Cox, R.D. Goos, R.C. Wornick, and A.M. Badger. 1964. Some effects of gamma irradiation on seeds and rhizomes of *Musa*. *Am. J. Bot.* 51:724–729.

Strosse, H., R. Domergue, B. Panis, J.V. Escalant, and F. Cote. 2003. Banana and plantain embryogenic cell suspensions. In: *INIBAP technical guidelines and the international network for the improvement of banana and plantain*, A. Vezina and C. Picq, eds. Montpellier, France: International Plant Genetic Resources Institute.

Suprasanna, P., S. Meenakshi, and V.A. Bapat. 2008 Integrated approaches of mutagenesis and *in vitro* selection for crop improvement. In: *Plant tissue culture, molecular markers and their role in crop productivity*, A. Kumar and N.S. Shekhawat. eds., 73–92. New Delhi, India: IK International Publishers.

Taylor, N.E. and E.A. Greene. 2003. PARSESNP: A tool for the analysis of nucleotide polymorphisms. *Nucleic Acids Res.* 31:3808–3811.

Till, B.J., S.H. Reynolds, E.A. Greene, C.A. Codomo, L.C. Enns, et al. 2003. Large-scale discovery of induced point mutations with high throughput TILLING. *Genome Res.* 13:524–530.

Till, B.J., C. Burtner, L. Comai, and S. Henikoff. 2004. Mismatch cleavage by single-strand specific nucleases. *Nucleic Acids Res.* 32:2632–2641.

Till, B.J., S.H. Reynolds, C. Weil, N. Springer, C. Burtner, et al. 2004. Discovery of induced point mutations in maize genes by TILLING. *BMC Plant Biol.* 4:12.

Till, B.J., J. Cooper, T.H. Tai, P. Colowit, E.A. Greene, et al. 2007. Discovery of chemically induced mutations in rice by TILLING. *BMC Plant Biol.* 7:19.

Van Harten, A.M. 1998. *Mutation breeding: Theory and practical applications.* Cambridge: Cambridge University Press.

Veilleux, R.E., and A.A.T. Johnson. 1998. Somaclonal variation: Molecular analysis, transformation interaction, and utilization. In: *Plant breeding reviews*, J. Janick, ed., 229–268. New York: John Wiley & Sons, Inc.

Vuylsteke, D., R. Ortiz, S. Ferris, and R. Swennen. 1995. 'PITA-9': A black Sigatoka-resistant hybrid from the 'False Horn' plantain gene pool. *HortScience* 30:395–397.

Xu, C.X., H.B. Chen, and J. Luo. 2008. Plant regeneration via embryogenesis after gamma radiation in banana (*Musa* AAA). *Acta Hort.* 773:135–140.

11 Biotechnology in *Musa* Improvement

Leena Tripathi

CONTENTS

11.1 INTRODUCTION

Biotechnology can greatly impact the genetic improvement of vegetatively propagated crops. Bananas and plantains (*Musa* spp.) are staple foods for millions of people in the tropical and subtropical regions of the world. The annual banana production in the world is estimated at 1.3×10^{11} kg, of which less than 15% enters the international commercial market, indicating that the crop is far more important for local or domestic consumption than for export (FAO, 2008). Almost 87% of the bananas grown worldwide are produced by small-scale farmers for consumption and for local markets, leaving only 13% for international trade. Nonetheless, banana is an extremely important export commodity, especially in Latin America and the Caribbean. *Musa* cultivation is affected by many pests and diseases, including black Sigatoka (*Mycosphaerella fijiensis*), Fusarium wilt (*Fusarium oxysporum* f. sp. *cubense*), bacterial wilt (*Xanthomonas campestris* pv *musacearum*), viruses (banana bunchy-top virus, banana streak virus), nematodes, and weevils. Biotechnology could provide solutions to some of the limitations of conventional banana breeding, such as sterility, long generation times, and limited genetic variability. This chapter reviews the progress and challenges in tapping the potential of biotechnology for the genetic improvement of banana.

11.2 RECENT ADVANCES IN BIOTECHNOLOGY

Plant biotechnology, like all aspects of the life sciences, has changed very rapidly over the past 20 years (Flavell, 2003). The new ways of analyzing genes, molecules, and processes constitute, together with established plant breeding systems, new platforms for discovery, design, and synthesis of novel products (Flavell, 2003). Such products include improved sources of food, feed, fiber, energy, chemicals, and drugs. Knowledge of the ways plants develop their architecture; survive abiotic and biotic stresses; utilize fertilizer, water, and CO_2; and make complex secondary metabolites and macromolecules will increase more rapidly in the next few decades (Flavell, 2003). More importantly the genes and alleles that program these attributes will become known rapidly from the combination of genomics, genetics, and plant breeding, particularly for intensively studied species (Flavell, 2003). Due to the sterility of most triploid banana varieties, plant propagation has been achieved traditionally by vegetative multiplication using naturally occurring plant offshoots (Haicur et al., 1998). *In vitro* multiplication of banana is mostly performed through proliferation of vegetative meristems, which has the advantage of producing healthy homogeneous plants in large numbers. Micropropagation of banana is now an industrial process and up to 50 million tissue-cultured plants are produced annually (Teisson and Cote, 1997). The recent development of embryogenic suspension cultures has paved the way for future mass production of banana plants at low cost (Haicur et al., 1998). Suspension cultures can also be used for cryopreservation and for genetic transformation (Haicur et al., 1998). Plant regeneration from protoplasts has been achieved (Ducreux et al., 2001; Xiao et al., 2008). Protoplast fusion is particularly promising for banana improvement, as most cultivars are related, triploid, and sterile. Fusion between haploid plants and selected diploid genotypes could facilitate the production of new triploid bananas (Xiao et al., 2009). The first transgenic banana plants were produced in 1995 through particle bombardment of embryogenic suspension cultures (Sagi et al., 1995).

Agrobacterium-mediated transformation has also been achieved (Hernandez, 1999; Ganapathi et al., 2001; Khanna et al., 2004). Transformation has made it possible to introduce valuable agronomic traits for pest and disease resistance, fruit maturation, and storage into banana (Haicur et al., 1998). Thus, at the onset of the 21st century, improvement of banana and plantain through biotechnology should help ensure food security by stabilizing production levels in sustainable cropping systems geared towards meeting domestic and export market demand (Haicur et al., 1998; Rout et al., 2000).

11.3 APPLICATION OF TISSUE CULTURE FOR BANANA IMPROVEMENT

Besides the production of healthy planting material, tissue culture has been used in crop improvement since the 1940s to create genetic variability and to increase the number of desirable germplasm (http://anrcatalog.ucdavis.edu.). Specifically, *in vitro* techniques for the culture of protoplasts, anthers, microspores, ovules, and embryos have been used to create new genetic variation in the breeding lines, often via haploid production with crop improvement potential (Brown and Thorpe, 1995). The culture of single cells and meristems can be effectively used to eradicate pathogens from planting material and thereby dramatically improve the yield of established cultivars (Badoni and Chauhan, 2010). Tissue-culture techniques, in combination with molecular techniques, have been successfully used to incorporate specific traits through gene transfer.

11.3.1 MICROPROPAGATION

Traditionally, bananas and plantains are propagated vegetatively from suckers or corms. Unlike seeded commercial cultivars of other crop species, production of planting materials through suckers takes a long time and very few suckers are produced per mother plant (Pillay and Tripathi, 2007). *In vitro* propagation has many advantages, such as higher rates of multiplying, pest- and disease-free planting material, and less space required to multiply large numbers of plants (refer to Chapter 15).

11.3.2 Embryogenic Cell Culture

Recently, much progress has been made in the establishment of embryogenic cell culture from banana explants from a variety of cultivars. Novak et al. (1989) established embryogenic cell suspensions (ECS) from somatic tissues such as leaf sheaths and corm sections of the cultivar 'Grand Nain' while Dheda et al. (1991) cultured highly proliferated shoot-tip cultures of 'Bluggoe.' Young male flowers are the most responsive explants for initiating embryogenic cultures of 'Grand Nain' (Escalant et al., 1994), 'Rasthali' (Ganapathi et al., 2001), and the hybrid cultivar 'FHIA-21' (Daniels et al., 2002). Suspension cultures can be regenerated into plantlets through somatic embryogenesis at high frequencies and grown in the field (Dheda et al., 1991; Novak et al., 1989). The plant recovery frequencies were as high as 81.5% (Daniels et al., 2002). Plant regeneration from cell suspension cultures was investigated for its potential in mass propagation and as a tool in genetic transformation using recombinant DNA technology. However, most of the procedures are still laborious and genotype dependent.

11.3.3 Embryo Culture

Embryo culture plays an important role in *Musa* breeding. Embryo culture has been used to rescue hybrid plants from wide crosses, which often fail to produce mature viable seeds (refer to Chapter 15).

11.3.4 Anther Culture

Anther culture is generally used in the production of F_1 hybrid varieties (Drew, 1997). The first step is to develop inbred parental lines by repeated self-pollination. This can be a very slow process in banana, which normally requires more than a year to flower. However, by culturing pollen grains from diploids, haploid plants that contain only one copy of each chromosome can be produced. These plants can be induced to double their chromosome number by a chemical means, resulting in plants that have two identical sets of chromosomes, or are completely inbred (homozygous). This process facilitates the selection of recessive traits. Assani et al. (2003) reported the production of haploid plants of *Musa balbisiana* (BB). Callus was induced from anthers in which the majority of the microspores were at the uninucleate stage. The frequency of callus induction was 77%. About 8% of the anthers developed androgenic embryos and of the 147 plantlets obtained, 41 were haploids ($n = x = 11$), but the frequency of regeneration was low (1.1%). The technique is genotype specific.

11.3.5 Germplasm Conservation

Tissue culture is useful in germplasm conservation of bananas since most cultivars are seedless. The main disadvantage of this method is the propensity for somaclonal variation (Oh et al., 2007). Germplasm collections in fields require extensive use of labor, large space requirements, and are costly. Field-grown plants are exposed to pests and diseases and can be lost. The propagation of various cultivars of banana and plantain by conventional methods is laborious and time consuming. *In vitro* collections offer a safer and cheaper alternative. In addition, *in vitro* plantlets are the materials of choice for the international exchange of germplasm, simplifying quarantine procedures as they are pest- and disease-free, safer and easier to handle than bulky suckers. Shoot-tip culture and disease-indexed cultures for germplasm exchange have been widely adopted for conserving and distributing important banana and plantain collections.

The international *Musa* germplasm collection of INIBAP (International Network for the Improvement of Banana and Plantain) Transit Centre at K.U. Leuven, Belgium, uses *in vitro* culture for germplasm storage. The collection also contains improved materials from breeding programs.

The *in vitro* maintenance of *Musa* germplasm is constrained by the occurrence of somaclonal variation (Vuylsteke et al., 1991). As the frequency of somaclonal variation could be linked to multiplication and growth rates, among other factors, *Musa* germplasm is now stored under slow growth conditions as well as at ultralow temperatures. Shoot-tip cultures maintained at low temperatures (15–18°C) have been used for the slow-growth storage of *Musa* germplasm (Banerjee and De Langhe, 1985).

Cryopreservation is considered to be the only valid alternative for long-term preservation of *Musa* germplasm because ultralow temperatures arrest physical and chemical reactions and eliminate time-related changes. Cryopreservation techniques have been developed for more than 80 different plant species cultivated under various forms including cell suspensions, calluses, apices, and somatic and zygotic embryos (Engelmann, 1997; Panis et al., 1990). Cryopreservation is not feasible in all cultivars since the production of somatic embryos and cell suspensions is cultivar dependent. In the case of banana, slow freezing and vitrification were both totally ineffective, while the encapsulation-dehydration method resulted in a survival rate of 8.1% (Panis et al., 1990). A simple technique was recently developed for cryopreservation of meristem cultures, which involves preculture on high-sucrose medium followed by rapid freezing (Panis et al., 1996). This method was tested on seven banana cultivars of different genomic groups and resulted in viability rates of 12% to 72% depending on the cultivar. This method with improvements can be used as a routine cryopreservation method for banana gene banks.

11.4 GENOMICS FOR BANANA IMPROVEMENT

11.4.1 Genomics for Banana Improvement

Recent breakthroughs in genomics in plants include the whole genome sequencing of many crop plants. The application of genomics in banana and plantain improvement is still in its infancy. The genome size of Musa is 550–600 Mbp (Dolezel et al., 1994). This is larger than the genomes found in species such as rice or *Arabidopsis thaliana* (between 150 and 500 Mbp), but much smaller than the *Triticeae* cereals (5,500 Mbp in barley, 9,000 Mbp in rye, and 17,000 Mbp in wheat) (Heslop-Harrison and Schwarzacher, 2007).

The *Musa* genome display several characteristics that could be exploited for gaining fundamental insights into the genomes of other species (Vilarinhos et al., 2003).

The small size of the haploid genome, the different ploidy levels, and the coexistence of various genome combinations, as well as the combination of parthenocarpy with sterility and vegetative propagation, place the banana as a good model to study gene expression in different chromosomal environments (Musagenomics, 2007). To reinforce banana genomic research, The Global Genomics Consortium has been established (http://www.musagenomics.org).

One of the first tools necessary to develop banana genomics research is a banana bacterial artificial chromosome (BAC) library (Shizuya et al., 1992). Several BAC clone libraries were developed from both A and B genome diploid *Musa* species (Vilarinhos et al., 2003; Safár et al., 2004; Musagenomics, 2007) with a total genomic coverage of more than 30 times. The Musa Genomics Resource Centre was established in the Czech Republic to distribute these resources, and by 2007, had 42 BACs totaling 3·3 Mbp (Aert et al., 2004) and 4·5 Mbp of BAC end sequences (Cheung and Town, 2007). These allow an overview of the *Musa* genome showing that it has 47% guanine-cytosine (GC) content and, together with the development of large numbers of *polymerase chain reaction* (PCR) based markers, a comparison of genes and genomic structures could be made with other species (Heslop-Harrison and Schwarzacher, 2007).

Several thousand expressed sequence tags (ESTs), important for examination of gene expression, responses and differentiation of the plants, and examination of diversity, have been published (Santos et al., 2005) and many more are becoming available.

Comparisons of EST libraries are very valuable for identification of genes that are differentially expressed under stress conditions. Santos et al. (2005) made libraries from plants grown in cold (5°C) and hot (45°C) conditions, and found that about 30% of the genes in their library had been identified in other species as being involved in responses to environmental stress and that there were substantial differences in the expression between the two libraries. Coemans et al. (2005) reported on alternative method for gene-expression profiling: SuperSAGE (super-serial analysis of gene expression), which they suggest will be very useful for transcript profiling and gene discovery.

Like all plant genomes, the *Musa* genome consists of repetitive DNA and single-copy sequences, and understanding the composition and organization of the genome at the large-scale level will be helpful to allow gene isolation and to understand long-term and short-term evolutionary processes (Schmidt and Heslop-Harrison, 1998). Valarik et al. (2002) cloned and characterized many repetitive DNA sequences and located those that were not related to rDNA or retroelements in the centromeric region of the chromosomes.

Retroelements, class I transposable elements or transposons, are abundant in the *Musa* genome (Balint-Kurti et al., 2000), as in other species (Heslop-Harrison et al., 1997). Automated annotation of the BAC libraries shows that more than a third of the open reading frames are related to retroelements (Musagenomics, 2007). The analysis of two BACs by Aert et al. (2004) revealed that one BAC consisted of 45 kb of gene-rich sequence without retroelements, followed by 28 kb containing mostly transposon-like sequences and repetitive DNA. BAC-end sequencing, allowing a survey of the whole genome, showed that 36% of the BESs contained sequences homologous to transposable elements (Cheung and Town, 2007). The evidence suggests that *Musa* has repeat-rich regions in the centromeres and perhaps elsewhere, and there may be gene-rich regions, as suggested in other species (Heslop-Harrison, 1991). Cheung and Town (2007) compared pairs of BAC end sequences from single BACs and found that a small number also mapped to adjacent regions of the rice genome, indicating conserved microsynteny over a larger taxonomic range and showing how part of the *Musa* genome can be anchored to rice. This provides a cost-effective and efficient way to understand *Musa* genes and the genome by informatics and conserved synteny with the model reference species rice and *Arabidopsis thaliana* (Heslop-Harrison and Schwarzacher, 2007).

11.4.2 Molecular Markers in *Musa*

There is rapid progress in the development of molecular markers and gene mapping in crop plants. A number of molecular markers such as *restriction fragment length polymorphism* (RFLP), simple sequence repeat (SSR), *random amplification of polymorphic DNA* (RAPD), and *amplified fragment length polymorphism* (AFLP) were applied in banana and plantain mostly for diversity analysis and genetic mapping (Faure et al., 1993; Ude et al., 2002a, 2002b; Creste et al., 2004; Oreiro et al., 2006). Ferreira et al. (2004) employed RAPD markers to characterize banana diploids (AA) with contrasting levels of black Sigatoka (*Mycosphaerella fijiensis*) and yellow Sigatoka (*M. musicola*) resistance.

A putative RAPD marker for Sigatoka resistance has been identified at the National Research Center for Banana (NRCB), India. The marker has been cloned, sequenced, and converted into a sequence characterized amplified region (SCAR) marker and is being validated using contrasting parents for expression of Sigatoka (*Mycosphaerella musicola*) disease resistance and their progenies. Parallel studies have led to the identification of a putative RAPD marker for nematode resistance (Uma, personal communication). Another RAPD marker has been identified for salt-tolerance clones of cv 'Dwarf Cavendish' obtained through induced mutagenesis (Miri et al., 2009). Attempts are under way at NRCB towards the development of resistant gene analogue-cleaved amplified polymorphic (RGA-CAP) markers, which would facilitate the genetic mapping of candidate resistant gene(s) in *Musa* (Backiyarani et al., 2010).

New markers such as Diversity Arrays Technology or DArT markers (Jaccoud et al., 2001), a microarray technology that can detect and type DNA variation at several hundred of genomic loci in parallel without prior knowledge of sequence information has been applied in banana (Amorim et al., 2009).

11.5 TRANSGENIC TECHNOLOGY FOR BANANA IMPROVEMENT

Transgenic technology has become an important tool for crop improvement. Genetic engineering—that is, the introduction and stable integration of genes into the nuclear genome and their expression in a transgenic plant—offers a better alternative for the genetic improvement of cultivars not amenable to conventional crossbreeding, such as Cavendish bananas and False Horn plantains. Due to lack of cross-fertile wild relatives in many banana-producing areas, as well as the male and female sterility of most edible cultivars and clonal mode of propagation, gene flow is not an issue for this crop, making a transgenic approach even more attractive.

The successful genetic transformation in plants requires the production of normal, fertile plants expressing the newly inserted gene(s). The process of genetic transformation involves several distinct steps: identification of useful gene, the cloning of the gene into a suitable plasmid vector, and delivery of the vector into plant cell, followed by expression and inheritance of the foreign DNA encoding a polypeptide. With the advent of plant biotechnology and the rapid development of gene transfer techniques, the potential to introduce desirable character traits is no longer restricted to those occurring in close relatives. Despite technical difficulties of transforming a monocot species, transformation protocols are available for many banana cultivars (Table 11.1). Development of stable and reproducible transformation and regeneration technologies opened new horizons in banana breeding (Table 11.2).

11.5.1 Transformation Procedures

Transformation protocols are available for most banana cultivars. Genetic transformation using microprojectile bombardment of embryogenic cell suspension is now routine and has been used for direct gene transfer in cooking banana cultivar 'Bluggoe,' plantain 'Three Hand Planty' (Sagi et al., 1995), and Cavendish banana 'Grand Nain' (Becker et al., 2000).

Agrobacterium-mediated transformation offers several advantages over direct gene transfer methodologies, such as the possibility to transfer only one or few copies of DNA fragments carrying

TABLE 11.1
Summary of Transformation of Various Cultivars of Banana and Plantain

Cultivar of Banana/Plantain	Method of Transformation	Reference
Bluggoe	Microprojectile Bombardment	Sagi et al., 1995
Three Hand Planty	Microprojectile Bombardment	Sagi et al., 1995
Cavendish banana cv Grand Nain	Microprojectile Bombardment; *Agrobacterium*	Becker et al., 2000; Khanna et al., 2004; May et al., 1995
Rasthali	*Agrobacterium*	Ganapathi et al., 2001
Lady Finger	*Agrobacterium*	Khanna et al., 2004
Agbagba	*Agrobacterium*	Tripathi et al., 2005
EAHB[a] cv Mpologoma	*Agrobacterium*	Tripathi et al., 2008
EAHB cv Nakitembe	*Agrobacterium*	Tripathi et al., 2008
Gonja Manjaya	*Agrobacterium*	Author's lab

[a] EAHB: East African Highland banana.

TABLE 11.2
Summary of Genetic Modification of Banana for Important Traits

Method	Transgene	Trait	Reference
Microprojectile bombardment	BBTV	Resistance to BBTV	Becker et al. 2000
Agrobacterium	MSI-99	Resistance to fungal diseases	Chakrabarti et al., 2003
Agrobacterium	*hrap*	Resistance to Xanthomonas wilt	Tripathi et al., 2009, 2010
Agrobacterium	*pflp*	Resistance to Xanthomonas wilt	Tripathi et al., 2009
Agrobacterium	Chitinase	Resistance to black Sigatoka disease	Kiggundu, 2007
Agrobacterium	Anti-apotosis	Resistance to Fusarium wilt	Paul, 2009
Agrobacterium	Cystatin	Resistance to nematode	Atkinson et al., 2004

the genes of interest at higher efficiencies with lower cost and the transfer of very large DNA fragments with minimal rearrangement (Shibata and Liu, 2000). Banana was generally regarded as recalcitrant for *Agrobacterium*-mediated transformation, until Hernandez et al. (1999) reported that *A. tumefaciens* was compatible with banana. *Agrobacterium*-mediated transformation of embryogenic cell suspensions of the banana cultivars 'Rasthali,' 'Cavendish,' and 'Ladyfinger' has since been achieved (Ganapathi et al., 2001; Khanna et al., 2004). Banana functional genomics and plant improvement initiatives demand higher transformation frequencies and a standard protocol that can be used to transform all banana genomic groups. Khanna et al. (2004) described centrifugation-assisted *Agrobacterium*-mediated transformation protocol developed using banana cultivars from two economically important genomic groups (AAA and AAB) of cultivated banana.

Although most transformation protocols use cell suspensions, establishing cell suspension is a lengthy process and is cultivar dependent. Protocols have also been established using meristematic tissues from various cultivars of banana (May et al., 1995; Tripathi et al., 2005, 2008). This technique is applicable to a wide range of banana cultivars irrespective of ploidy or genotype (Tripathi et al., 2003, 2005, 2008). This process does not incorporate steps using disorganized cell cultures but uses micropropagation, which has the important advantage that it allows regeneration of homogeneous populations of plants in a short period of time. This procedure offers several potential advantages over the use of embryogenic cell suspensions, as it allows for rapid transformation of banana species, and meristematic tissues have the potential to regenerate plants from many different cultivars, unlike somatic embryogenesis, which is restricted to only a few cultivars. The transformation of meristematic cells may result in chimeric plants when only one or a few cells receive T-DNA. To obtain uniformly transformed plants, two steps of selection and regeneration are performed to avoid regeneration of any nontransformed cells.

11.5.2 Disease Resistance

Recent advances in genetic engineering offer ways to transfer a resistance gene found in any plant into crop varieties without changing other favorable traits. Plant defense genes from other plants and antimicrobial proteins are now a potential source of plant resistance. One of the strategies to control highly destructive fungal diseases like black Sigatoka in banana is the production of transgenic disease-resistant plants based on expression of genes encoding antimicrobial peptides (AMPs). The AMPs usually have a broad-spectrum activity against fungi and bacteria, and most are nontoxic to plant and mammalian cells. Examples of AMPs are cecropins (Boman and Hultmark, 1987), magainin (Bevins and Zasloff, 1990), and plant defensins (Broekaert et al., 1997). The cecropin (Alan and Earle, 2002) and its derivatives (D4E1; Rajasekaran et al., 2001) have been found to inhibit the *in vitro* growth of several important bacterial and fungal pathogens. Transgenic tobacco plants expressing cecropins have increased resistance to *Pseudomonas syringae* pv *tabaci*, the cause of tobacco wildfire (Huang et al., 1997). Similarly, magainin is effective against plant pathogenic fungi

(Kristyanne et al., 1997). Chakrabarti et al. (2003) reported successful expression of this synthetic peptide and enhanced disease resistance in transgenic tobacco and banana.

The AMPs of plant origin may be potential candidates for fungal resistance in banana as they have high *in vitro* activity to *Mycosphaerella fijiensis* and *Fusarium oxysporum f. sp. cubense*, two major fungal pathogens of banana, and are nontoxic to human and banana cells. Several AMPs isolated from radish and dahlias are toxic to both fungal pathogens. A large number of transgenic lines of plantain expressing defensin-type AMPs has been developed at the Catholic University of Leuven (Remy, 2000). Many transformed lines have been generated and screened under screen-house conditions in Belgium for disease resistance, and the most promising lines of transgenic bananas and plantains were evaluated in the greenhouse and under field conditions in Cuba and Costa Rica (Agnet, 2004). The transgenic banana containing antifungal chitinase genes is also being tested for resistance against black Sigatoka in a confined field trial in Uganda.

Plants have their own mechanisms of defense against plant pathogens and include a vast array of proteins and other organic molecules produced prior to infection or during pathogen attack. Pathosystem-specific plant resistance (R) genes have been cloned from several plant species. R genes cloned from resistant varieties can be transferred to susceptible cultivars of the same plant species, making them resistant to pathogens. It is also possible to transfer R genes from one plant species to another.

A series of resistance gene analogues have been isolated from banana, using degenerate PCR primers targeting highly conserved regions in proven plant resistance genes. The disease-resistant genes were isolated from the somaclonal mutant 'CIEN-BTA-03' (resistant to both *M. fijiensis* and *M. musicola*) and the parent 'Williams' (Kahl, 2004). All the resistance genes were fully sequenced, and eight of them were also transcribed in the mutant, its parental genotype, 'Pisang Mas' and a diploid *M. acuminata*. The R gene candidate (RGC-2) from *Musa acuminata* ssp. *malaccensis*, a wild diploid banana segregating for resistance to *Fusarium oxysporum* f. sp. *cubense* (*Foc*) race 4, has been isolated and completely sequenced.

Recently, scientists at the Queensland University of Technology (QUT) inserted this R gene for resistance to Fusarium wilt, or Panama disease, into the banana genome (Dale et al., 2004). The scientists at QUT also introduced the anti-apoptosis gene into the banana genome of two commercially important banana cultivars 'Grand Naine' and 'Lady Finger' for developing resistance to fusarium wilt, or Panama disease (Paul, 2009). The gene stops cells dying when attacked by the disease. These transgenic plants are under evaluation in a confined field in Queensland, Australia (Science Alert, 2008).

Hypersensitive response-assisting protein (HRAP) is a novel plant protein that can intensify the harpinPss-mediated hypersensitive response in plants (Chen et al., 2000). The *pflp* has been shown to delay the hypersensitive response (HR) induced by *Pseudomonas syringae* pv *syringae* in nonhost plants through the release of harpinPss. Transgenic rice carrying the *pflp* gene showed enhanced resistance to *Xanthomonas oryzae* pv *oryzae* (Tang et al., 2001). The *pflp* has also been shown to enhance resistance in transgenic orchids against *E. carotovora* (Liau et al., 2003). The elicitor-induced resistance is not specific against particular pathogens, so it could be very useful strategy for developing broad-spectrum resistance. This is the strategy pursued by IITA, in collaboration with the National Agriculture Research Organization (NARO, Uganda) and the African Agricultural Technology Foundation (AATF). This research aims at "designing" a genetically modified banana that is resistant to the most devastating bacterial disease, banana Xanthomonas wilt (BXW) (Biruma et al., 2007; Tripathi, 2008). Hundreds of transgenic lines with *pflp* or *hrap* genes have been developed using a protocol based on the *Agrobacterium tumefaciens* technology (Tripathi et al., 2009, 2010). These transformed lines of various cultivars have been validated via PCR assay and Southern blot analysis. They have been tested for disease resistance under laboratory conditions. Most promising lines will be evaluated for efficacy against BXW in confined fields. The transgenic lines will also be tested for environmental and food safety in compliance with target country regulations.

The most promising transgenic strategies to control ssDNA viruses like BBTV is to express a defective gene that encodes an essential virus life-cycle activity. For instance, the replication of the virus can be encoded in the replication gene or genes (Rep) whereby the resultant Rep protein may retain the ability to bind to its target viral DNA without the functions of the Rep (Brunetti et al., 2001). The defective Rep protein binds to the invading viral DNA and is thought to outcompete the native viral Rep protein, thus reducing or eliminating virus DNA replication. Lucioli et al. (2003) expressed the first 630 nucleotides of the Rep gene of tomato yellow leaf curl Sardinia virus to generate resistance. The duration of the resistance was related to the ability of the invading virus to switch off transgene expression through post-transcriptional gene silencing (PTGS). Many researchers are trying to develop transgenic plants of *Musa* resistant to BBTV, targeting the PTGS mechanism using mutated or antisense Rep genes.

11.5.3 Pest Resistance

There are several possible approaches for developing transgenic plants with improved weevil and nematode resistance. A variety of genes are available for genetic engineering for pest resistance (Sharma et al., 2000). Among these are proteinase inhibitors, *Bacillus thuringiensis* (*Bt*) toxins, plant lectins, vegetative insecticidal proteins (VIPs), and alpha-amylase inhibitors (AI). Proteinase inhibitors contribute to host-plant resistance against pests and pathogens (Green and Ryan, 1972). They operate by disrupting protein digestion in the insect mid-gut via inhibition of proteinases. The two major proteinase classes in the digestive systems of phytophagus insects are the serine and cysteine proteinases. Coleopteran insects, including the banana weevil, mainly use cysteine proteinases (Murdock et al., 1987) and studies indicate a combination of both serine and cysteine proteinases is useful for insect control (Gerald et al., 1997). These inhibitors have already been used for insect control in transgenic plants (Leple et al., 1995). Presently, cysteine proteinase activity has been identified in the mid-gut of the banana weevil and *in vitro* studies have shown that cysteine proteinases are strongly inhibited by both a purified recombinant rice (oryzacystatin-I [OC-I]) and papaya cystatin (Abe et al., 1987; Kiggundu et al., 2003).

The use of proteinase inhibitors (PIs) as nematode antifeedants is an important element of natural plant defense strategies (Ryan, 1990). This approach offers prospects for novel plant resistance against nematodes and reduces use of nematicides. The potential of PIs for transgenic crop protection is enhanced by a lack of harmful effects when humans consume them in seeds such as rice and cowpea. Cysteine PIs (cystatins) are inhibitors of cysteine proteinases and have been isolated from seeds of a wide range of crop plants, including those of sunflower, cowpea, soybean, maize, and rice (Atkinson et al., 1995). Transgenic expression of PIs provides effective control of both cyst and root-knot nematodes. The cystatins were shown to mediate nematode resistance when expressed in tomato hairy root (Urwin et al., 1995), rice (Vain et al., 1998), pineapple (Urwin et al., 2000), and potato (Urwin et al., 2001). The partial resistance was conferred in a small-scale potato field trial on a susceptible cultivar by expressing cystatins under control of the CaMV35S promoter (Urwin et al., 2001). The enhanced transgenic plant resistance to nematodes has been demonstrated by using dual proteinase inhibitor transgenes (Urwin et al., 1998). The resistance using cystatins has also shown to be effective in banana (Atkinson et al., 2004). In partnership with the University of Leeds, transgenic plants with proteinase inhibitors and repellent genes are being developed by IITA for plantain (Tripathi, 2009).

The expression and biological activity of the *Bt* toxins has been investigated in genetically modified (GM) plants for insect control. *Bt* gene technology is currently the most widely used system for lepidopteran control in commercial GM crops (Sharma et al., 2000). The expression of a selected *Bt* gene for weevil resistance may be a rather long-term strategy since no potential *Bt* gene with high toxic effects against the banana weevil has been identified as yet (Kiggundu et al., 2003). Some *Bt* proteins are also effective against saprophagous nematodes (Borgonie et al., 1996). The Cry5B protein is toxic to wild type *Caenorhabditis elegans*; other *C. elegans* mutants are resistant to Cry5B but susceptible to the Cry6A toxin (Marroquin et al., 2000). The approach using *Cry* genes has potential for plant nematode control (Wei et al., 2003).

Alpha-amylase inhibitors (AI) and chitinase enzymes might also have a future potential for weevil control. Alpha-amylase inhibitors operate by inhibiting the enzyme alpha-amylase, which breaks down starch to glucose in the insect gut (Morton et al., 2000). Transgenic adzuki beans are produced with enhanced resistance to bean bruchids, which are Coleopteran insects like weevils (Ishimoto et al., 1996). Chitinase enzymes are produced as a result of invasion either by fungal pathogens or insects. Transgenic expression of chitinase has shown improved resistance to Lepidopteran insect pests in tobacco (Ding et al., 1998).

11.5.4 Enhanced Micronutrients

Nutritionally enhanced crops could make a significant contribution to the reduction of micronutrient malnutrition in developing countries (Bouis et al., 2002). Biofortification (the development of nutritionally enhanced crops) can be advanced through the application of several biotechnologies in combination. Genomic analysis and genetic linkage mapping are needed to identify the genes responsible for natural variation in nutrient levels of common foods. These genes can then be transferred into familiar cultivars through conventional breeding and marker-assisted selection or, if sufficient natural variation does not occur within a single species, through genetic engineering.

Vitamin and mineral deficiencies are a major cause of child mortality and morbidity every year in the developing world, but this can be easily prevented by adding just a few key nutrients to staple foods. Genetic modification of banana has also been considered as a path towards increasing the value of this crop to health and nutrition in developing countries. As a crop that is widely consumed as a weaning food by children and as a starchy staple by all sectors of the community in some countries, banana has been advocated as a source of carotenoids that can counteract debilitating vitamin A deficiency. Although much of the necessary technology is now available, these applications have yet to advance to the stage of practical evaluation. However, recent works in engineering rice with genes for β-carotene biosynthesis and the development of golden rice (Ye et al., 2000) have shown the feasibility of enriching foods with vitamin A through biotechnology. Recently iron has been enhanced in rice by the introduction of the ferritin gene driven by endosperm-specific promoter (Vasconcelos et al., 2003). Researchers at QUT and NARO are developing biofortified bananas using a number of genes for synthesis of provitamin A or iron, under the control of constitutive or fruit-specific promoters (Dale and Tushemeirewe, 2008). These transgenic bananas are currently being regenerated for field trials in Australia and Uganda.

11.6 CHALLENGES FOR DEVELOPMENT OF TRANSGENIC BANANAS

The production of genetically modified banana plants is routine in many laboratories. The initial technical hurdles have been overcome but new challenges that must be addressed if the technology is to move forward are becoming apparent, from proof of concept to product development, deregulation, and commercial release. The new challenges include intellectual property rights (IPR), an underdeveloped regulatory environment, and a lack of biosafety infrastructures within target countries, besides some technical and sustained funding issues.

Most innovations in biotechnology are developed using the knowledge or technologies generated from previous innovations. Many plant biotechnology products or techniques are "modular" in that they are assembled from a number of previously developed technologies/transgenes, each of which may be subject to a separate patent. The commercialization of many proprietary biotechnology products is typically contingent on other proprietary biotechnology products or processes, and in particular on agreements between IPR holders regarding the relative contributions of different proprietary technologies to the product in question. Many biotechnology products (for example, transgenic seeds or transgene constructs) now have a complex IPR pedigree because a large number of proprietary products or processes are involved in developing the product.

Some public sector agricultural research institutions and universities have become involved in the "defensive" patenting of technologies they develop that may have commercial value or might have some value in the future. Many biotechnology discoveries and enabling technologies (for example, *Agrobacterium* and biolistic transformation methods) generated with public funding in research institutions and agricultural universities are also protected and no longer being treated as "public goods." Such discoveries now primarily flow from the public sector to the private sector and, for use in public research institutes, usually such technologies come under material-transfer agreements (MTAs) that significantly restrict their use, usually for research purposes only, and often include reach-through provisions to capture results of future research.

The IP issues are becoming a major factor, limiting the deployment of transgenic technologies in developing countries, but this problem can be overcome if the correct procedures are followed early in the product delivery program. Institutions with large portfolios of relevant IP are in a better negotiating position to access the IP of others. As IPRs are in greater use by the private sector than the public sector, it is likely that public sector research institutions such as the Consultative Group on International Agricultural Research (CGIAR), the National Agricultural Research Stations, and individual university researchers will not be in a strong negotiating position regarding access to useful proprietary plant biotechnologies. The CGIAR is establishing a biotechnology transfer unit with expertise in IPR law in an effort to strengthen its negotiating position with other IPR holders of useful biotechnologies.

While there are still many nonproprietary biotechnologies available, the cost of access to patented technologies is likely to be a growing issue for many public sector research institutions. It is illustrative that commercially oriented research in many biotechnology companies has to now follow the research route of least cost in terms of royalty payments to other companies for enabling technologies used to develop a commercial product (Mascarenhas, 1998). Unfortunately, no public sector body has yet compiled a directory of which useful plant biotechnologies are freely accessible in the public domain, especially for scientists in developing countries. Conversely, there is a corresponding lack of publicly available studies on what the current patent situation is for key enabling biotechnologies. However, in 1998 the CGIAR Panel on Proprietary Science and Technology conducted a study of proprietary science and technology within the CGIAR system (Cohen et al., 1998). This CGIAR study included an initial review by International Service for National Agricultural Research (ISNAR) of the extent of use of proprietary plant biotechnology tools in each of the International Agricultural Research Centers (IARCs).

The current generation of transgenic bananas and their testing, however, highlights some problems that need to be avoided in future. Some genes of agronomic interest were owned by the industry, and it took much effort by the Catholic University of Leuven before these genes could be used freely for plantain and cooking bananas (Tollens et al., 2004). Therefore, it is urgent that a mechanism be put in place whereby an authority at the global level interacts with industry to negotiate access to protected technologies for developing countries. This is in contrast to the utility patent system that extends protection to the seed and progeny of patented plants so breeders cannot legally use protected varieties as breeding material.

The African Agricultural Technology Foundation (AATF) has been instrumental in facilitating technology transfer negotiations whereby proprietary biotechnologies have been made available to Africa. Recently, IITA has negotiated a royalty-free license from the patent holder Academia Sinica, Taiwan, through the AATF for access to the *pflp* and *hrap* genes for production of commercial banana varieties resistant to bacterial wilt in sub-Saharan Africa. The AATF has signed the licensing agreement with Academia Sinica and granted a sublicense to IITA for developing the improved varieties. Further, IITA has signed tripartite agreement with NARO and AATF for developing disease-resistant transgenic bananas and joint ownership of the developed material.

Given the novelty of commercial application of genetic engineering techniques, the regulatory frameworks around the world are in the process of being formulated and reformulated in response to, for example, consumer reactions to these new products. Existing regulations differ substantially both in scope and stage of implementation, varying from very restrictive regulations in certain industrial countries to nonexistent in certain developing countries.

Many developing countries are beginning to enact regulations related to genetically engineered products. Furthermore, operational field-testing regulations have been implemented in, for example, Argentina, Brazil, Mexico, Chile, Costa Rica, Cuba, India, the Philippines, and Thailand. But still the majority of developing countries currently do not have a regulatory system for genetically modified organisms (GMOs) in place. Developing a regulatory framework may be a costly and time-consuming process involving extensive consultation and effort. Many countries in sub-Saharan Africa are now establishing national biosafety committees and biosafety regulations regarding the use of GMOs. There is also need to harmonize biosafety regulations at the regional level.

Although many developing countries have assembled the regulatory structures required to carry out field trials, the existing legislation is often outdated and the process itself has either never been put into practice or is too slow. This situation obviously presents significant obstacles for those wishing to develop a product within such regions. The programs such as Plant Biosafety Systems (PBS) and Agricultural Biotechnology Support Project II (ABSPII), both funded by the U.S. Agency of International Development, are trying to modernize the regulatory infrastructure and build capacity of regulatory bodies in developing countries. Thus, as a record of safe field trials is established, the process will become progressively simplified and routine.

Desired traits within a transgenic plant must be expressed at the required level under natural cultivation conditions. Presently, very little data is available regarding how transgene expression will be affected when transgenic banana plants are cultivated in the field. Since the aim is to enhance existing farmer-preferred germplasm without altering their desirable traits, it is essential that capacity to produce genetically transformed plants is expanded into the agronomically most important varieties within the major banana-growing regions.

Rather, studies in East Africa suggest that varieties modified for pest or disease resistance will be incorporated into the range of varieties already grown as part of a strategy to reduce risk, provide multiple products, and satisfy varying tastes (http://archives.foodsafety.ksu.edu/agnet-archives.htm). In the meantime, various biotechnologies are already contributing to conventional breeding efforts and are expected to become even more effective in this area as genetic maps and markers are refined. The use of tissue-culture plants is already contributing to the development of novel production systems for smallholder farmers. Indeed, tissue culture is expected to be much more widely used in increasing the productivity and sustainability of such systems as part of a balanced program of deploying biotechnologies cost effectively in developing countries in the future.

11.7 CONCLUSIONS

Bananas are seriously threatened by several diseases and pests. Thus resistance to biotic stresses is an important part of regional or national efforts. Since the major cultivated varieties of banana are sterile and therefore do not set seed, traditional breeding is more difficult than genetic transformation using molecular techniques. Attempts to produce transgenic bananas are proceeding slowly, but public acceptance of these novel plants and their products will depend on sound information and risk assessment. There is major public concern for the transfer of transgenes from transgenic field material to wild species, but the chances for this happening in banana are expected to be negligible in view of the sterility of many cultivars. The scope for further improvement of banana is large; along with other methods of crop improvements, transgenic technology should provide fast and effective methods.

Currently, no transgenic bananas and plantains are commercially available. However, many research institutes, organizations, and universities are concentrating on the development of pest- or disease-resistant varieties and improving the nutritional contents of banana and plantain. The number of experimental transgenic bananas is continuously increasing. Transformation protocols, including tissue culture techniques, suitable transformation constructs with modified promoters driving one or more transgenes, appropriate transformation techniques, the detection of the transgenes, and characterization of their insertion sites are well developed.

Musa genomics can open up new avenues for more efficient breeding of the crop. It is important to investigate the possibilities by which the primary production and other uses of banana can be promoted for the benefit of the growing world's population. Strategies for future genomics research in *Musa* include the development of molecular markers, construction of genetic and physical maps, identification of genes and gene expression, and whole genome sequencing. Sequencing of other plant genomes such as *A. thaliana* and *O. sativa* has provided an enormous amount of data that could reveal unknown features of their genomes. Such data could also be generated for *Musa*. A Global *Musa* Genomics Consortium was established in 2001 with the goal of ensuring the sustainability of banana as a staple food crop by developing an integrated genetic and genomic understanding, allowing targeted breeding, and transforming and efficiently using *Musa* biodiversity. Basically, the consortium aims to apply genomics to the sustainable improvement of banana and plantain. The consortium believes that genomic technologies such as analysis and sequencing of the banana genome, identification of its genes and their expression, recombination, and diversity can be applied for the genetic improvement of the crop (Frison et al., 2004).

There is enormous potential for genetic manipulation of bananas for disease and pest resistance using the existing transformation systems using the genes isolated from the *Musa* genome. The use of appropriate gene constructs may allow the production of nematode, fungus, bacterial, and virus-resistant plants in a significantly shorter period of time than using conventional breeding, especially if several traits can be introduced at the same time. It may also be possible to incorporate other characteristics such as drought tolerance, thus extending the geographic spread of banana and plantain production, and contributing significantly to food security and poverty alleviation in developing countries. Long-term and multiple disease resistance can be achieved by integrating several genes with different targets or modes of action into the plant genome.

REFERENCES

Abe, K., H. Kondo, and S. Arai. 1987. Purification and characterization of a rice cysteine proteinase inhibitor. *Agr. Biol. Chem.* 51:2763–2768.

Aert, R., L. Sági, and G. Volckaert. 2004. Gene content and density in banana (*Musa acuminata*) as revealed by genomic sequencing of BAC clones. *Theor. Appl. Genet.* 109:129–139.

Alan, A.R. and E.D. Earle. 2002. Sensitivity of bacterial and fungal pathogens to the lytic peptides, MSI-99, magainin II, and cecropin B. *Mol. Plant-Microbe Interactions* 15:701–708.

Amorim, E.P., K.O. Cohen, V.B.O. Amorim, J.A. Santos-Serejo, S.O. Silva, A.D. Vilarinhos, et al. 2009. The genetic diversity of carotenoid-rich bananas measured by Diversity Arrays Technology (DArT). *Genet. Mol. Biol.* 32:96–103.

Assani, A., F. Bakry, F. Kerbellec, R. Haïcour, G. Wenzel, and B. Foroughi-Wehr. 2003. Production of haploids from anther culture of banana [*Musa balbisiana* (BB)]. *Plant Cell Rep.* 21:511–516.

Atkinson, H.J., P.E. Urwin, E. Hansen, and M.J. McPherson. 1995. Designs for engineered resistance to root-parasitic nematodes. *Trends Biotechnol.* 13:369–374.

Atkinson, H.J., S. Grimwood, K. Johnston, and J. Green. 2004. Prototype demonstration of transgenic resistance to the nematode *Radopholus similis* conferred on banana by a cystatin. *Transgenic Res.* 13:135–142.

Backiyarani, S., S. Uma, C. Anuradha, M.S. Saraswathi, and G. Arunkumar. 2010. Development of RGA-CAP markers for genetic mapping of candidate resistance Genes(s) in *Musa.* Paper presented at the 18th Plant and Animal Genomes Conference, January 9–13, 2010, San Diego, CA.

Badoni, A. and J.S. Chauhan. 2010. Conventional *vis-à-vis* biotechnological methods of propagation in potato: A review. *Stem Cell* 1(1):1–6.

Balint-Kurti, P.J., S.K. Clendennen, M. Dolezelova, M. Valarik, J. Dolezel, P.R. Beetham, et al. 2000. Identification and chromosomal localization of the monkey retrotransposon in *Musa* sp. *Mol. Gen. Genet*. 263:908–915.

Banerjee, N. and E. De Langhe. 1985. A tissue culture technique for rapid clonal propagation and storage under minimal growth conditions of *Musa* (banana and plantain). *Plant Cell Rep.* 4:351–354.

Becker, D.K., B. Dugdale, M.K. Smith, R.M. Harding, and J. Dale. 2000. Genetic transformation of Cavendish banana (*Musa* spp. AAA group) cv. Grand Nain via microprojectile bombardment. *Plant Cell Rep.* 19:229–234.

Bevins, C.L. and M. Zasloff. 1990. Peptides from frog skin. *Ann. Rev. Biochem.* 59:395–414.

Biruma, M., M. Pillay, L. Tripathi, G. Blomme, S. Abele, M. Mwangi, et al. 2007. Banana Xanthomonas wilt: A review of the disease, management strategies and future research directions. *Afr. J. Biotechnol.* 6:953–962.

Boman, H.G. and D. Hultmark. 1987. Cell free immunity in insects. *Ann. Rev. Microbiol.* 41:103–126.

Borgonie, G., M. Claeys, F. Leyns, G. Arnaut, D. De Waele, and A. Coomans. 1996. Effect of nematicidal *Bacillus thuringiensis* strains on free-living nematodes. *Fund. Appl. Nematol.* 19:391–398.

Bouis, H.E., S. Harris, and D. Lineback. 2002. Biotechnology-derived nutritious foods for developing countries: Needs, opportunities, and barriers. *Food Nutr. Bull.* 23:342–383.

Broekaert, W.F., B.P.A. Cammue, M. De Bolle, K. Thevissen, G. De Samblanx, and R.W. Osborn. 1997. Antimicrobial peptides from plants. *Crit. Rev. Plant Sci.* 16:297–323.

Brown, D.C.W. and A. Thorpe. 1995. Crop improvement through tissue culture. *World J. Microbiol. Biotechnol.* 11:409–415.

Brunetti, A., R. Tavazza, E. Noris, A. Luciolo, G. Accotto, and M.Tavazza. 2001. Transgenically expressed T-Rep of tomato yellow leaf curl Sardinia virus acts as a trans-dominant-negative mutant, inhibiting viral transcription and replication. *J. Virol.* 75:10573–10581.

Chakrabarti, A., T.R. Ganapathi, P.K. Mukherjee, and V.A. Bapat. 2003. MSI-99, a magainin analogue, imparts enhanced disease resistance in transgenic tobacco and banana. *Planta* 216:587–596.

Chen, C.H., H.J. Lin, M.J. Ger, D. Chow, and T.Y. Feng. 2000. The cloning and characterization of a hypersensitive response assisting protein that may be associated with the harpin-mediated hypersensitive response. *Plant Mol. Biol.* 43:429–438.

Cheung, F. and C.D. Town. 2007. A BAC end view of the *Musa acuminata* genome. *BMC Plant Biol.* 2007:7:29.

Choudhury, S.R., S. Roy, P.P. Saha, S.K. Singh, and D.N. Sengupta. 2008. Characterization of differential ripening pattern in association with ethylene biosynthesis in the fruits of five naturally occurring banana cultivars and detection of a GCC-box-specific DNA-binding protein. *Plant Cell Rep.* 27:1235–1249.

Coemans, B.H., H. Matsumura, R. Terauchi, S. Remy, R. Swennen, and L. Sági. 2005. SuperSAGE combined with PCR walking allows global gene expression profiling of banana (*Musa acuminata*), a non-model organism. *Theor. Appl. Genet.* 111:1118–1126.

Cohen, J.I., C. Falconi, J. Komen, and M. Blakeney. 1998. The use of proprietary biotechnology research inputs at selected CGIAR centers. Report of an ISNAR study commissioned by the Panel on Proprietary Science and Technology of the CGIAR. The Hague, Netherlands: ISNAR.

Creste, S., A.T. Neto, R. Vencovsky, S. Oliveira Silva, and A. Figueira. 2004. Genetic diversity of *Musa* diploid and triploid accessions from the Brazilian banana program estimated by microsatellite markers. *Gen. Resour. Crop Evol.* 51:723–733.

Dale, J.L. and W. Tushemeirewe. 2008. Biofortification of banana: A grand challenge in global health pages. Paper presented at the conference on Banana and Plantain in Africa: Harnessing International Partnerships to Increase Research Impact, October 5–9, 2008, Mombasa, Kenya.

Dale, J., B. Dugdale, M. Webb, D. Becker, H. Khanna, S. Peraza-Echeverria, et al. 2004. Strategies for transgenic disease resistance in banana. www.africancrops.net/abstracts2/banana/dale.htm (accessed 24/09/10).

Daniels, D., R.G. Kosky, and M.R. Vega. 2002. Plant regeneration system via somatic embryogenesis in the hybrid cultivar FHIA-21 (*Musa* spp. AAAB group). *In Vitro Cell Dev. Biol.* 38:330–333.

Dheda, D., F. Dumortier, B. Panis, D. Vuylsteke, and E. Langhe. 1991. Plant regeneration in cell suspension cultures of the cooking banana cv. Bluggoe (*Musa* spp. ABB group). *Fruits* 46:125–135.

Ding, X., B. Gopalakrishnan, L.B. Johnson, F.F. White, X. Wang, T.D. Morgan, et al. 1998. Insect resistance of transgenic tobacco expressing an insect chitinase gene. *Transgenic Res.* 7:77–84.

Dolezel, J., M. Doelzelova, and F.J. Novak. 1994. Flow cytometric estimation of nuclear DNA amount in diploid bananas (*Musa acuminata* and *M. balbisiana*). *Biologia Plantarum* 36:351–357.

Drew, R.A. 1997. The application of biotechnology to the conservation and improvement of tropical and subtropical fruit species. www.fao.org/ag/agp/agps/pgr/drew1.htm (accessed 5/4/2010).

Ducreux, M., E. Aguillar, and A. Grapin. 2001. Plant regeneration from protoplasts of dessert banana cv. Grande Naine (*Musa* spp., Cavendish sub-group AAA) via somatic embryogenesis. *Plant Cell Rep.* 20:482–488.

Engelmann, F. 1997. *In vitro* conservation methods. In: *Biotechnology and plant genetic resources: Conservation and use*, J.A. Callow, B.V. Ford-Lloyd, and J.H. Newbury, eds., 119–161. Wallingford: CAB International.

Escalant, J.V., C. Teisson, and F.X. Cote. 1994. Amplified somatic embryogenesis from male flowers of triploid banana and plantain cultivars (*Musa* sp.). *In Vitro Cell Dev. Biol.* 30:181–186.

FAOSTAT Agriculture Data. 2008. http://faostat.fao.org (accessed 24/09/10).

Fauré, S., J.L. Noyer, J.P. Horry, F. Bakry, C. Lanaud, D. González de León. 1993. A molecular marker-based linkage map of diploid bananas (*Musa acuminata*). *Theor. Appl. Genet.* 87:517–526.

Ferreira, C.F., S. Oliveira Silva, N.P.D. Sobrinho, S.C.S. Damascena, F.S. Oliveira Alves and O.P. Paz. 2004. Molecular characterization of Banana (AA) diploids with contrasting levels of Black and Yellow Sigatoka resistance. *American J Applied Sciences* 1:276–278.

Ferreira, C.F., S. Oliveira Silva, N.P.D. Sobrinho, S.C.S. Damascena, F.S. Oliveira Alves and O.P. Paz. 2004. Molecular characterization of Banana (AA) diploids with contrasting levels of Black and Yellow Sigatoka resistance. *American J Applied Sciences* 1:276–278.

Flavell, R.B. 2003. Introduction to plant biotechnology. In: *Handbook of plant biotechnology*, P. Christou and H. Klee, eds., 1–7. London: John Wiley & Sons Ltd.

Frison, E.A., J.V. Escalant, and S. Sharrock. 2004. The Global *Musa* Genomic Consortium: A boost for banana improvement. In: *Banana improvement: Cellular, molecular biology, and induced mutations*, M.S. Jain and R Swennen, eds., 341–350. Plymouth, UK: Science Publishers, Inc.

Ganapathi, T.R., N.S. Higgs, P.J. Balint-Kurti, C.J. Arntzen, G.D. May, and J.M. Van Eck. 2001. Agrobacterium-mediated transformation of the embryogenic cell suspensions of the banana cultivars Rasthali (AAB). *Plant Cell Rep.* 20:157–162.

Gerald, R.R., K.J. Kramer, J.E. Baker, J.F. Kanost, and C.A. Behke. 1997. Proteinase inhibitors and resistance of transgenic plants to insects. In: *Advances in insect control: The role of transgenic plants*, N. Carozi, and M. Koziel, eds., 157–183. London: Taylor and Francis.

Green, T.R. and C.A. Ryan. 1972. Wound-induced proteinase inhibitors in plant leaves: A possible defence mechanism against insects. *Science* 175:776–777.

Haicour R., V. Bui Trang, D. Dhed'a, A. Assani, F. Bakry, and F.X. Cote. 1998. Banana improvement through biotechnology—Ensuring food security in the 21st century (Abstract in English). *Cahiers Agric.* 7:468–475.

Hernandez, J.B.P., S. Remy, V.G. Sauco, R. Swennen, and L. Sagi. 1999. Chemotactic movement and attachment of *Agrobacterium tumefaciens* to banana cells and tissues. *J. Plant Physiol.* 155:245–250.

Heslop-Harrison, J.S. 1991. The molecular cytogenetics of plants. *J. Cell Sci.* 100:15–21.

Heslop-Harrison, J.S. and T. Schwarzacher. 2007. Domestication, genomics and the future for banana. *Ann. Bot.* 100(5):1073–84.

Heslop-Harrison, J.S., A. Brandes, S. Taketa, T. Schmidt, A.V. Vershinin, E.G. Alkhimova, et al. 1997. The chromosomal distributions of Ty1-*copia* group retrotransposable elements in higher plants and their implications for genome evolution. *Genetica* 100:197–204.

Huang, Y., R.O. Nordeen, M. Di, L.D. Owens, and J.H. McBeth. 1997. Expression of an engineered cecropin gene cassette in transgenic tobacco plants confers disease resistance to *Pseudomonas syringae* pv. tabaci. *Phytopathology* 87:494–499.

Ishimoto, M., T. Sato, M.J. Chrispeels, and K. Kitamura. 1996. Bruchid resistance of transgenic azuki bean expressing seed α-amylase inhibitor of common bean. *Entomol. Exp. Appl.* 79:309–315.

Jaccoud, D., K. Peng, D. Feinstein, and A. Kilian. 2001. Diversity Arrays: A solid-state technology for sequence information independent genotyping. *Nucleic Acids Res.* 29:e25.

Kahl, G. 2004. The banana genome in focus: A technical perspective. www.fao.org/docrep/007/ae216e/ae216e0o.htm (accessed 5 April 2010).

Khanna, H., D. Becker, J. Kleidon, and J. Dale. 2004. Centrifugation assisted *Agrobacterium tumefaciens*-mediated transformation (CAAT) of embryogenic cell suspensions of banana (*Musa* spp. Cavendish AAA and Ladyfinger AAB). *Mol. Breed.* 14:239–252.

Kiggundu, A., M. Pillay, A. Viljoen, C. Gold, W. Tushemereirwe, and K. Kunert. 2003. Enhancing banana weevil (*Cosmopolites sordidus*) resistance by plant genetic modification: A perspective. *Afr. J. Biotechnol.* 2:563–569.

Kristyanne, E.S., K.S. Kim, and J.M. Stewart. 1997. Magainin 2 effects on the ultrastructure of five plant pathogens. *Mycologia* 89:353–360.

Leple, J.C., M. Bonadebottino, S. Augustin, G. Pilate, V.D. Letan, A. Delplanque, D. Cornu, and L. Jouanin. 1995. Toxicity to *Chrysomela tremulae* (Coleoptera, Chrysomelidae) of transgenic poplars expressing a cysteine proteinase inhibitor. *Mol Breed* 1:319–328.

Liau, C.H., J.C. Lu, V. Prasad, J.T. Lee, H.H. Hsiao, S.J. You, et al. 2003. The sweet pepper ferredoxin-like protein (pflp) conferred resistance against soft rot disease in Oncidium orchid. *Transgenic Res*. 12:329–336.

Lucioli, A., E. Noris, A. Brunetti, R. Tavazza, V. Ruzza, A. Castilla, et al. 2003. Tomato yellow leaf curl Sardinia virus Rep-derived resistance to homologous and heterologous geminiviruses occurs by different mechanisms and is overcome if virus-mediated transgene silencing is activated. *J. Virol.* 77:6785–6798.

Marroquin, L.D., D. Elyassnia, J.S. Griffits and R.V. Aroian. 2000. *Bacillus thuringiensis* (Bt) toxin susceptibility and isolation of resistance mutants in the nematode *Caenorhabditis elegans*. *Genetics* 155:1693–99.

Mascarenhas, D. 1998. Negotiating the maze of biotech "tool patents." *Nature Biotechnol.* 16:1371–1372.

May, G.D., A. Rownak, H. Mason, A. Wiecko, F.J. Novak, and C.J. Arntzen. 1995. Generation of transgenic banana (*Musa acuminata*) plants via *Agrobacterium*-mediated transformation. *Bio/Technol.* 13:486–492.

Morton, R.L., H.E. Schroeder, K.S. Bateman, M.J. Chrispeels, E. Armstrong, and T.J.V. Higgins. 2000. Bean α-amylase inhibitor-I in transgenic peas (*Pisum sativum*) provided complete protection from pea weevil (*Bruchus piorum*) under field conditions. *Proc. Nat. Acad. Sci.* 97:3820–3825.

Murdock, L.L., G. Brookhart, P.E. Dunn, D.E. Foard, S. Kelley, L. Kitch, et al. 1987. Cysteine digestive proteinases in *Coleoptera*. Comp. *Biochem. Physiol.* 87:783–787.

Musagenomics. 2007. http://www.musagenomics.org (accessed 5 April 2010).

Novak, F.J., R. Afza, M. Van Duran, M. Perea-Dallos, B.V. Conger, and X.L. Tang. 1989. Somatic embryogenesis and plant regeneration in suspension cultures of dessert (AA and AAA) and cooking (ABB) bananas (*Musa* spp.). *Biotechnol.* 7:154–159.

Oh, T.J., M.A. Cullis, K. Kunert, I. Engelborghs, R. Swennen, and C.A. Cullis. 2007. Genomic changes associated with somaclonal variation in banana (*Musa* spp.). *Physiologia Plantarum* 129:766–774.

Oreiro, C.E., O.A. Odunola, Y. Lokko, and I. Ingelbrecht. 2006. Analysis of B-genome derived simple sequence repeat (SSR) markers in *Musa* spp. *African Journal of Biotechnol.* 5:126–128.

Panis, B., J. Whthers, and E.A.L. De Langhe. 1990. Cryopreservation of *Musa* suspension culture and subsequent regeneration of plants. *Cryo-Letters* 11:337–350.

Panis, B., N. Totté, K. Van Nimmen, L. Withers, and R. Swennen. 1996. Cryopreservation of banana (*Musa* spp.) meristem cultures after preculture on sucrose. *Plant Sci.* 121:95–106.

Pillay, M. and L. Tripathi. 2007. Banana breeding. In: *Breeding major food staples*, M.S. Kang and P.M. Priyadarshan, eds., 393–428. New York: Blackwell Science.

Rajasekaran, K., K.D. Stromberg, J.W. Cary, and T.E. Cleveland. 2001. Broad-spectrum antimicrobial activity *in vitro* of the synthetic peptide D4E1. *J. Agr. Food Chem.* 49:2799–2803.

Reimann-Philipp, U., S. Behnke, A. Batschauer, E. Schäfer, and K. Apel. 1989. The effect of light on biosynthesis of leaf-specific thionins in barley (*Hordeum vulgare*). *Eur. J. Biochem.* 182:283–289.

Remy, S. 2000. Genetic transformation of banana (*Musa* sp.) for disease resistance by expression of antimicrobial proteins. PhD diss., Catholic University of Leuven, Belgium

Rout, G.R., S. Samantaray, and P. Das. 2000. Biotechnology of the banana: A review of recent progress. *Plant Biol.* (Stuttgart) 2(5):512–524.

Ryan, C.A. 1990. Proteinase inhibitors in plants: Genes for improving defences against insects and pathogens. *Ann. Rev. Phytopath.* 28:425–449.

Safar, J., J.C. Noa-Carrazana, J. Vrana, J. Bartos, O. Alkhimova, X. Sabau, et al. 2004. Creation of a BAC resource to study the structure and evolution of the banana (*Musa balbisiana*) genome. *Genome* 47:1182–1191.

Sagi, L., B. Panis, S. Remy, H. Schoofs, K. De Smet, R. Swennen, and B. Cammus. 1995. Genetic transformation of banana (*Musa* spp.) via particle bombardment. *Bio/Technol.* 13:481–485.

Santos, C.M.R., N.F. Martins, H.M. Horberg, E.R.P. de Almeida, M.C.F. Coelho, R.C. Togawa, et al. 2005. Analysis of expressed sequence tags from *Musa acuminata* ssp. *burmannicoides*, var. Calcutta 4 (AA) leaves submitted to temperature stresses. *Theor. Appl. Genet.* 110:1517–1522.

Schmidt, T. and J.S. Heslop-Harrison. 1998. Genomes, genes and junk: The large scale organization of plant chromosomes. *Trends Plant Sci.* 3:195–199.

Science Alert. 2008. Disease resistant bananas created. http://www.sciencealert.com.au/news/20083010-18370-3.html (accessed 5 April 2010).

Sharma, H.C., K.K. Sharma, N. Seetharama, and R. Ortiz. 2000. Prospects for using transgenic resistance to insects in crop improvement. *Eur. J. Biotechnol.* 3(2). www.ejbiotechnology.info/content/vo13/issue2/full/3 (accessed 14/09/10).

Shibata, D. and Y.G. Liu. 2000. *Agrobacterium*-mediated plant transformation with large DNA fragments. *Trends Plant Sci.* 5:354–357.

Shizuya, H., B. Birren, U.J. Kim, V. Mancino, T. Slepak, Y. Tachiiri, and M. Simon, et al. 1992. Cloning and stable maintenance of 300-kilobase-pair fragments of human DNA in Escherichia coli using an F-factor-based vector. *Proc. Nat. Acad. Sci.* (USA) 89:8794–8797.

Tang, K., X. Sun, Q. Hu, A. Wu, C.H. Lin, H.J. Lin, R.M. Twyman, et al. 2001. Transgenic rice plants expressing the ferredoxin-like protein (AP1) from sweet pepper show enhanced resistance to *Xanthomonas oryzae* pv. *oryzae Plant. Sci.* 160:1035–1042.

Teisson, C. and F.X. Cote. 1997. Micropropagation of *Musa* (bananas). In: *Biotechnology in agriculture and forestry* 39, Y.P.S. Bajaj, ed., 103–126. Berlin: Springer-Verlag.

Tollens, E., M. Demont, and R. Swennen. 2004. Agrobiotechnology in developing countries North–South partnerships are the key. *Outlook Agric.* 33(4):231–238.

Tripathi, L., J.N. Tripathi, R.T. Oso, J.d'A. Hughes, and P. Keese. 2003. Regeneration and transient gene expression of African *Musa* species with diverse genomic constitution and ploidy levels. *Tropical Agric.* 80:182–187.

Tripathi, L. 2008. The future of African bananas. *IITA R4D Rev.* 1:46–48.

Tripathi, L. 2009. Biotechnology and nematodes. *IITA R4D Rev.* 2:47–48.

Tripathi, L., J.N. Tripathi, and J.d'A. Hughes. 2005. *Agrobacterium*-mediated transformation of plantain cultivar Agbagba (*Musa* spp.) *Afr. J. Biotechnol.* 4:1378–1383.

Tripathi, L., J.N. Tripathi, and W.K. Tushemereirwe. 2008. Rapid and efficient production of transgenic East African Highland banana (*Musa* spp.) using intercalary meristematic tissues. *Afr. J. Biotechnol.* 7:1438–1445.

Tripathi, L., M. Mwangi, S. Abele, V. Aritua, W.K. Tushemereirwe, and R. Bandyopadhyay. 2009. Xanthomonas wilt: A threat to banana production in East and Central Africa. *Plant Dis.* 93:440–451.

Ude, G., M. Pillay, D. Nwakanma, and A. Tenkouano. 2002a. Genetic diversity in *Musa acuminata* Colla and *M. balbisiana* Colla and some of their natural hybrids using AFLP markers. *Theor. Appl. Genet.*104:1246–1252.

Ude, G., M, Pillay, D. Nwakanma, and A. Tenkouano. 2002b. Analysis of genetic diversity and sectional relationships in *Musa* using AFLP markers. *Theor. Appl. Genet.* 104:1239–1245.

Tripathi, L., H. Mwaka, J.N. Tripathi, and W.K. Tushemereirwe. 2010. Expression of sweet pepper *hrap* gene in banana enhances resistance to *Xanthomonas campestris* pv. *musacearum. Mol. Plant Pathol.* http://onlinelibrary.wiley.com/doi/10.1111/fj.1364-3703.2010.00639.x/abstract (accessed 24/09/10) .

Urwin, P.E., H.J. Atkinson, D.A. Waller, and M.J. McPherson. 1995. Engineered oryzacystatin-I expressed in transgenic hairy roots confers resistance to *Globodera pallida. Plant J.* 8:121–131.

Urwin, P.E., M.J. McPherson, and H.J. Atkinson. 1998. Enhanced transgenic plant resistance to nematodes by dual protease inhibitor constructs. *Planta* 204:472–479.

Urwin, P.E., A. Levesley, M.J. McPherson, and H.J. Atkinson. 2000. Transgenic resistance to the nematode *Rotylenchulus reniformis* conferred by *A. thaliana* plants expressing protease inhibitors. *Mol. Breed.* 6:257–264.

Urwin, P.E., K.M. Troth, E.I. Zubko, and H.J. Atkinson. 2001. Effective transgenic resistance to *Globodera pallida* in potato field trials. *Mol. Breed.* 8:95–110.

Vain, P., B. Worland, M.C. Clarke, G. Richard, and M. Beavis. 1998. Expression of an engineered cysteine protease inhibitor (Oryzacystatin-I1D86) for nematode resistance in transgenic rice plants. *Theor. Appl. Genet.* 96:266–271.

Valarik, M., H. Simkova, E. Hribova, J. Safar, M. Dolezelova, and J. Dolezel. 2002. Isolation, characterization and chromosome localization of repetitive DNA sequences in bananas (*Musa* spp.). *Chrom. Res.* 10:89–100.

Vasconcelos, M., K. Datta, N. Oliva, M. Khalekuzzaman, L. Torrizo, S. Krishnan, et al. 2003. Enhanced iron and zinc accumulation in transgenic rice with the ferritin gene. *Plant Sci.* 164:371–378.

Vilarinhos, A.D., P. Piffanelli, P. Lagoda, S. Thibivilliers, X. Sabau, F. Carreel, et al. 2003. Construction and characterization of a bacterial artificial chromosome library of banana (*Musa acuminata* Colla). *Theor. Appl. Genet.* 106:1102–1106.

Vuylsteke, D., R. Swennen, and E. De Langhe. 1991. Somaclonal variation in plantains (*Musa* spp, AAB group) derived from shoot-tip culture. *Fruits* 46:429–439.

Wang, G.L., W.Y. Song, D.L. Ruan, S. Sideris, and P.C. Ronald. 1996. The cloned gene, *Xa* 21, confers resistance to multiple *Xanthomonas oryzae* pv. *Oryzae* isolates in transgenic plants. *Mol. Plant-Microbe Interact.* 9:850–855.

Wei, J.Z., K. Hale, L. Cara, E. Platzer, C. Wong, S.C. Fang, et al. 2003. *Bacillus thuringiensis* crystal proteins that target nematodes. *Proc. Nat. Acad. Sci. (USA)* 10:2760–2765.

Xiao, W., X. Huang, Y.R. Wei, J.T. Zhao, X.M. Dai, X.L. Huang. 2008. Plant regeneration from protoplast culture of *Musa* AAB Silk cv. Guoshanxiang. *Act. Hortic. Sin.* 35(6):873–878, in Chinese with English abstract.

Xiao, W., X. Huang, Q. Gong, X-M. Dai, J-T. Zhao, Y-R. Wei, et al. 2009. Somatic hybrids obtained by asymmetric protoplast fusion between *Musa* Silk cv. Guoshanxiang (AAB) and *Musa acuminate* cv. Mas (AA). *Plant Cell Tiss. Organ Cult.* 97:313–321.

Ye, X., S. Al-Babili, A. Kloti, J. Zhang, P. Lucca, P. Beyer, et al. 2000. Engineering the provitamin A (beta-carotene) biosynthetic pathway into (carotenoid-free) rice *endosperm. Science* 287:303–305.

12 Genotype by Environment Interaction and *Musa* Improvement

Rodomiro Ortiz and Abdou Tenkouano

CONTENTS

12.1 INTRODUCTION

This chapter provides the state of knowledge on genotype by environment interactions (or GxE hereafter) in *Musa* and components of phenotypic stability for some traits (especially bunch weight and host-plant resistance to black Sigatoka) in this crop. This chapter reviews how to manage GxE to have efficient selection schemes and multilocation testing prior to release of improved cultivars.

The clonal phenotype, which corresponds to a specific genotype, can vary from year to year in the same location and or from location to location within an agro-ecozone in the same year (Ortiz and Ekanayake, 2000). This phenomenon, which affects genotype ranking in different environments, is known as genotype by environment interaction.

Plantain and banana breeding programs aim to identify genotypes that have both a high and a stable yield in a range of environments across a targeted a region (Vuylsteke et al., 1997). In the presence of a significant GxE, both the stratification of environments according to their agro-climatological similarities and the determination of stability parameters for genotypes across environments are important tools for the management of this interaction.

Several techniques have been developed to determine the most stable genotype in a set of replicated trials across years and locations or combinations of both environments (Ortiz and Ekanayake, 2000). Postdictive models are, however, not useful in the identification of which genotypes and environments contribute to the GxE interaction. Moreover, a breeder may be interested to identify which genotypes are adapted to specific environments or to predict their performance in a specific location. The additive main effects and multiplicative interaction (AMMI) model was developed to provide answers for such questions (Gauch, 1992). AMMI uses the analysis of variance (ANOVA) to study the main effects of genotypes and environments and principal component analysis (PCA) for the residual multiplicative interaction. In this regard, Ortiz (1996) assessed the potential of AMMI analysis for field assessment of *Musa* genotypes to banana streak virus (BSV) infection in West Africa. The AMMI1 model revealed that an increase in clonal susceptibility resulted in a more unstable response to BSV, whereas the AMMI2 model showed that seasonal rather than locational diversity accounted for most of the interaction patterns. Such results suggested a low level of BSV strain differentiation in this sub-Saharan Africa region.

12.2 PHENOTYPIC STABILITY IN BANANA AND PLANTAIN

Plantain and banana breeding programs need to assess thoroughly their materials through multilocation testing (Ortiz, 1998). A sequence of multisite trials will lead to selecting promising clones as potential new cultivars in the targeted agro-ecozone.

12.2.1 Multienvironment Testing

The first GxE analysis of *Musa* multilocation trials in West Africa showed that this interaction affected all traits except fruit circumference (De Cauwer et al., 1995). Moreover, the genotype by location was significant for bunch weight, number of hands, number of fruits, and fruit weight, whereas most of the traits were not affected by genotype by cycle interaction in the humid and degraded forest of Nigeria. These observations suggest that multilocation trials may be more efficient than single-site trials over several years in the humid forest of West Africa.

12.2.2 Stability Analysis

Further stability analysis of bunch weight and yield potential, based on the phenotypic coefficient of variation (PCV), allowed the identification of high- and stable-yielding tetraploid plantain-banana hybrids for West African humid forests (Ortiz et al., 1997). Selections from the cross ['Obino l'Ewai' × 'Calcutta 4'] were among the highest yielders in this set of multilocation trials, which confirmed the potential of this cross to generate diverse but stable high-yielding hybrids for further advanced testing and subsequent cultivar release(s) in the "plantain belt" of West Africa. Likewise, de Cauwer and Ortiz (1999) indicated that the cooking banana cultivar 'Cardaba' had the most stable yield potential in the West African humid and degraded forest as revealed by the AMMI biplots of the GxE analysis.

Stability and AMMI analyses provided a means for identifying *Musa* cultivars and experimental hybrids with stable bunch weight across environments in sub-Saharan Africa (Ortiz, 1998). These analyses also facilitated the interpretation of multilocation testing results, which suggested that the yield potential of a *Musa* clone should be assessed in the ratoon cycle because plantain and banana are perennial crops. Furthermore, Tenkouano et al. (2002) showed that there was a higher genetic expression for most yield components during the second crop cycle in all the environments, which confirms that the selection should be better in ratoon than in the first crop cycle.

Selection of high-yielding clones with specific adaptation should be in environments showing the respective stress to select for. Baiyeri et al. (2000b) suggest that farm gate and postharvest processing traits should be taken into account when selecting for broad or specific adaptation. Selection indices considering both groups of traits could allow *Musa* breeders to perform multitrait selection and enhance breeding efficiency. Further research showed that simultaneous use of different stability statistics may protect *Musa* breeders from wrongly identifying presumably stable genotypes (Baiyeri et al., 1999a). For example, *Musa* genotypes selected by AMMI and the regression coefficient were also classified as stable by PCV and Kang's statistic for simultaneous selection for high and stable yield.

12.3 EFFECT OF GENOTYPE BY ENVIRONMENT INTERACTIONS ON *MUSA* BREEDING

With reduced budgets allocated for agricultural research, site rationalization had become an important issue to consider when carrying out multilocation testing of promising selections. Experimental results from West Africa showed that multilocation testing is more profitable than single-site evaluation over several years in a *Musa* breeding station. Ortiz and de Cauwer (1998) also suggested, based on correlated responses across environments for yield potential, that a selection site in the humid forest could be dropped when selections in one site may be well adapted to the other location in the same agro-ecozone. Their results further indicated that the relatively poor performance of most *Musa* genotypes in dry environments reinforced the importance of early testing across a wide range of environments. In this way, selections with broad or specific adaptation may be identified for further release to targeted farmers.

12.3.1 Genotype by Environment Interactions and Trait Heritability and Repeatability

The "easy to breed" traits are those that can be scored easily and have a constant phenotypic expression in all environments: high heritability and low environmental influence. Traits with low (or nonsignificant) GxE, although they may be affected by the environment, are more important for assessing their potential across environments.

Components of variance are estimated to calculate broad-sense heritability (the ratio between the genetic and phenotypic variances) and repeatability (or the ratio between the genetic and the GxE plus the environment variances). Table 12.1 lists heritability and repeatability estimates for growth, bunch, and fruit traits in *Musa* germplasm. Plant height and bunch weight appear to be significantly influenced by the GxE and thus showed lower heritability than traits such as pseudostem girth, number of fruits, and fruit sizes, which are highly heritable and repeatable across environment, and thereby likely to be more responsive to selection than the former.

12.3.2 Indirect and Multitrait Selection, Selection Index, and Ideotype Breeding

Indirect selection may be useful for selection of traits that are difficult to score or that show low heritability. Intraclass correlations (Ortiz, 1997b) suggest that selection for large fruit may lead to heavy fruits in plantain-banana hybrids, thereby resulting in plants bearing heavy bunches. However, significant correlations between plant height and bunch weight in polyploidy hybrids indicate that selection of dwarf cultivars with heavy bunches may be difficult. Further research by Tenkouano et al. (2002) showed that associations between bunch weight and yield components were higher than between bunch weight and phenological traits. In the former cases, the genetic correlation was similar to the phenotypic correlation, indicating that GxE effect on the relationship between bunch weight and yield components was neutral. This result confirms that yield components could serve as indirect selection criteria for bunch weight in *Musa*.

Breeding efforts for genetic improvement of banana and plantain have gradually shifted from individual trait selection to simultaneous improvement of several traits following an ideotype concept in

TABLE 12.1
Broad-Sense Heritability (H^2) and Repeatability (R) Estimates for Growth, Bunch, and Fruit Traits in Triploid *Musa* Germplasm after Trials across Crop and Ratoon Cycles in a Single West African Humid Forest Location

	Plantain (75) and Banana (17) Cultivars[a]		Plantain (75) and Banana (18) Cultivars[b]		Plantain (52) and Banana (51) Cultivars[c]	
Trait	**H^2**	**R**	**H^2**	**R**	**H^2**	**R**
Days to flowering	–	–	0.80	1.50	–	–
Plant height	0.84	2.16	0.91	1.52	0.75	1.47
Plant girth	0.72	1.29	0.89	1.26	0.86	3.08
Leaf number	–	–	0.86	2.30	–	–
Leaf length/width ratio at flowering	–	–	0.69	1.06	–	–
Tallest sucker height at flowering	0.61	0.45	0.89	2.36	0.82	2.31
Tallest sucker height at harvest	0.75	1.06	0.78	1.51	0.82	2.20
Days to harvest	–	–	0.80	1.43	–	–
Days to fruit filling	–	–	0.76	1.58	–	–
Bunch weight	0.42	0.87	0.66	0.98	0.68	1.05
Fruit weight	0.86	2.18	0.89	4.11	–	–
Hand number	0.89	4.13	0.93	6.61	0.78	1.70
Fruit number	0.96	13.06	0.94	8.40	0.80	1.99
Fruit per hand	0.94 0.96	7.86 12.83	–	–	0.82 0.92	2.26 5.49
Fruit length	–	–	0.95	9.67	0.91 0.93	4.87 6.42
Fruit girth	–	–	0.82	2.14	0.86 0.90	3.48 4.24

[a] Baiyeri, K.P. and R. Ortiz, 2000, Agronomic evaluation of plantain and other triploid banana in Africa, *Acta Hort.*, 540, 125–135.

[b] Ortiz, R., D. Vuylsteke, R.S.B. Ferris, J.U. Okoro, A.N. Guessan, O.B. Hemeng, et al., 1997, Developing new plantain varieties for Africa, *Plant Var. Seeds*, 10, 39–57.

[c] Ortiz, R., 1997, Morphological variation in *Musa* germplasm, *Genet. Res. Crop Evol.*, 44, 393–404.

Musa. There are, however, a few common pathways determining yield potential among plantain landraces, making it difficult to define a common ideotype for plantain breeding (Ortiz and Langie, 1997). This finding suggests that plantains may possess different genes controlling similar pathways, or different traits contributing to yield potential. Likewise, defined ideotypes may differ for each landrace, according to the production system. Plantain breeders should therefore consider those relationships that are affected by both genotype and production system when selecting for improved hybrid germplasm.

12.3.3 Genotype by Crop Management Interactions

Careful management of organic matter is essential to achieve sustained perennial productivity of plantain under large-scale field production conditions. Agro-forestry systems such as alley cropping and the management of regrowth of natural bush fallow species in plantain fields have been investigated for their capacity to maintain productivity over long periods of cultivation without degradation of the resource base (Ortiz and Langie, 1997). Alley cropping generally leads to better performance in plantains than sole cropping (Baiyeri et al., 1999b, 2000b, 2004).

The performance of improved genotypes relative to cropping systems should be also assessed because on-site cultural practices used in breeding trials may differ considerably from typical farmer practice and could influence genotype selection as well as the size of testing plots (Ortiz, 1995). For example, early selection of sucker for the ratoon crop and other crop management options that will enhance healthy growth of the plant crop will sustain high yield in *Musa* genotypes (Baiyeri et al.,

2000). Baiyeri et al. (2004) warn on generalizing *Musa* cultivar recommendation across cropping systems because they found a significant genotype by cropping cycle interaction for yield and post-harvest traits in multilocation trials in Nigeria (Baiyeri et al., 1999b).

12.3.4 Black Sigatoka Host-Plant Resistance Interactions with the Environment

The aim of any resistance breeding program is to develop superior high-yielding genotypes with durable resistance (Craenen and Ortiz, 2003). Host-plant resistance breeding needs therefore methods that are reliable for discriminating between resistant and susceptible genotypes across target environments and strains of the pathogen. Moreover, epidemiological studies aimed at the identification of different pathotypes with variable degrees of virulence on well-known standard cultivars (or differentials) help in monitoring the appearance of more virulent strains and to establish a resistance breeding strategy able to maximize the probability of generating durable resistance.

Although field-screening methods are relatively simple, they are time consuming and affected by environmental factors such as weather and soil, which may affect symptom expression. Hence, it is usual to include genotypes of known host response as checks (from susceptible to highly resistant) and to gather a large dataset to validate results. Theory from population genetics indicates that the smaller the selection intensity(s), the lower the response to selection; that is, increases in frequency of favorable alleles will be at maximum when there are no escapes during screening (or maximum selection efficiency). Host-plant resistance breeding becomes inefficient (and it may be worthless) with increasing escape rates due to the screening method.

Genotype by environment interaction influences black Sigatoka reaction in plantain and banana, which explains the low repeatability of host-plant resistance traits to this fungal pathogen (Craenen and Ortiz, 1997). GxE was, however, not important when the analyses were based on the marker genotype bs_1, which is regarded as the major gene in the host response (slow disease development) to black Sigatoka (Ortiz and Vuylsteke, 1994). This finding confirms that Mendelian genetic markers are seldom affected by the environment.

Genotype by location effects are more important that the nonsignificant genotype by year interaction (Ortiz et al., 1997). These differences between locations may be attributed to the bias of different surveyors. Nonetheless, results from multilocation trials in sub-Saharan Africa showed that all partially resistant hybrids had a homeostatic resistant host response to black Sigatoka. Their slow disease development may lead to durable resistance to black Sigatoka because such host-plant response slows the progress of an epidemic without inhibiting its initiation.

12.3.5 Genotype by Environment Interactions on Root and Reproductive Traits

12.3.5.1 Roots

Root phenotypic plasticity refers to the ability of roots to adapt their structure to changes in the environment. Soil temperature, moisture level, partial pressure of carbon dioxide and oxygen, and nutrient availability are among the most important edaphic factors influencing root development. Likewise, air temperature, day length, light intensity, and partial pressure of carbon dioxide affect the provision of nutrients and growth regulators from the shoot to the root system and the shoot–root ratio of a plant. Results from trials in two contrasting Nigerian agro-ecozones showed that location significantly affects most aerial growth, corm, and root traits (except for the leaf number), total cord root length of the mother plant and mat, and average cord root diameter of the mother plant (Blomme et al., 2006). The genotype by location interaction was also significant for all above traits.

12.3.5.2 Reproductive Traits

The environment influences seed set in female-fertile plantain cultivars (Jenny et al., 1993). Ortiz and Vuylsteke (1995) indicated that seed production in plantain, as well as embryo and seed germination success of plantain-banana hybrids, followed a seasonal variation pattern in which production of desired tetraploids hybrids (after triploid-diploid crosses) are highest when hand pollinations are done under high solar radiation and low relative humidity. Such weather factors seem to be the most convenient for relatively high production of $2n$ eggs as compared to haploid gametes. There was also a seasonal variation in $2n$ pollen production in some diploid banana cultivars that was positively correlated with solar radiation (Ortiz, 1997c), which confirms that high solar radiation could enhance $2n$ gametes production in *Musa*.

Environmental conditions may also influence the quantity and quality of pollen produced in fertile *Musa* clones (Dumpe and Ortiz, 1996). Ortiz et al. (1998) indicated that solar radiation was also significantly associated with pollen stainability in *Musa*. Their results further confirm that this weather factor as well as temperature, total pan evaporation, rainfall, and minimum relative humidity correlate significantly with $2n$ pollen production. Hence, *Musa* breeders should identify the best time of the year to maximize both fertility and $2n$ gamete production to enhance the synthesis of polyploid *Musa* hybrids through sexual polyploidization.

12.4 GENOTYPE X ENVIRONMENT INTERACTIONS ON REACHING THE END USER

Plant breeding programs aim to develop and deploy improved cultivars that consistently display distinct phenotypic superiority in cultivation or utilization when compared to existing cultivars across their cropping range in farmers' fields. The fundamental determinants of the performance of a cultivar (P) are the biological potential embedded in the genetic makeup of the cultivar (G), the quality of the environmental context in which it is grown (E), and the responsiveness of the cultivar to changes in the environment (GxE) so that $P = G + E + \text{GxE}$.

Typically, superior genotypes are identified and selected in experimental fields that only marginally represent the range of target environments, but the breeders' reward depends on the suitability of such superior cultivars to the biophysical and socioeconomical circumstances of the farmers. GxE interactions may cause discrepancies between expected and observed performance of bananas and plantains both spatially and temporally as has been observed in Nigeria (Baiyeri, 1998), Uganda (Baiyeri et al., 2008), and outside Africa (Orjeda, 2000).

Ideally, therefore, each cultivar should be targeted to the environment that maximizes the expression of its biological potential. Causes of spatial variation include differences in climate (rainfall pattern and temperature), soil quality (biophysical characteristics), and cultural practices. The same factors can change over time and explain temporal variations. Understandably, plant breeders cannot replicate each target environment at their breeding stations even if the target environments were not subject to stochastic variations. Nor is it logistically possible for breeders to carry out validation tests of selected lines in the whole range of targeted environments. However, it is conceivable that data on long-term climatic characteristics, soil quality, and other factors can be used to define broad target zones within which representative sites could be chosen for additional evaluation of most promising lines selected in research stations.

Perhaps equally or more important than the biophysical context, it is the cultural and economic context of the farming communities that dictates cultivar choice for adoption (Tenkouano et al., 2009).

12.4.1 MARKET POTENTIAL: BEYOND THE FARM GATE

The concept of market potential (Bressani, 1976) has been adapted to banana and plantain by Baiyeri et al. (1999b), who used it to discuss the likelihood of adoption of a new cultivar by a farming community as a function of the value of the products derived from the harvested yield. The utilizable fraction of the yield is a function of pulp weight of the harvested fruit, dry matter content of the pulp, and shelf life of the fruit. Dry matter content determines the biological value of the crop, that is, the harvested fraction that is available for processing into food products (Sanchez et al., 1968). Ripening patterns—that is, shelf life—dictate processing options and the array of food products that can be derived from the crop, and this may determine the economic value of the crop (Dadzie and Orchard, 1996; Ferris et. al., 1996).

Ripening is associated with progressive hydrolysis of starch into simple sugars, which induces changes in pulp firmness and peel color (Stover and Simmonds, 1987). Three ripening stages, easily distinguished by the color of the peel and the consistency of the fruit, can be broadly defined. The first stage encompasses the period from complete greenness (unripe fruits) to yellowing of the fruit tip. The fruit is very firm and can be subjected to transport over relatively long distances and processed into dried/solid products, such as flour, that can be subsequently used in various preparations. The second stage goes from the end of the first stage to about 60% yellowing of the fruit. This is usually the preferred stage of consumption of dessert bananas with sufficient hydrolysis of starch into sugars while retaining a soft/firm consistency. Some solid food products can be derived at this stage, but options for dried products become limited or uneconomical while those for liquid products, such as soft drinks, become appealing.

The third stage begins with the end of the second stage, with increased browning of the peel and increased conversion of the sugars from hydrolyzed starch into alcohol, through fermentation. Virtually no solid product can be derived at this stage without addition of solidifiers from other crops for traditional preparations, but options for fermentation and distillation become attractive. In some West African countries, the market price of ripe fruits is higher than that of the green fruit (Ahenkora et al., 1996), but the increased number of food processing options with green fruits compared to ripe fruits would normally make fruits in earlier stages of ripening more marketable than those in later stages.

Therefore, the duration of each stage for a cultivar, not just the total shelf life, is an important determinant of what the cultivar can be processed into, hence its market potential. Taking this further, it is known that cultural preferences exist for traditional food preparations. Therefore, it is likely that a given food preparation option may have higher social value in some environments but not in others. Consequently, a high-yielding cultivar that has a good segmentation of its shelf life so that it can be used for a given food preparation preferred by populations in a particular area should be targeted to that area for release. On the breeding side, knowledge of the biological characteristics associated with a given food preparation can be used to understand the underlying genetic causes of these characteristics and use this information for developing new cultivars that would fit the particular food preparation, a process that can be termed "end-usage breeding."

The extension of this concept to industrial products is straightforward, with a broadening of the target clients from traditional farmers producing for their own consumption or local village markets to specialty crop growers feeding industrial processors. Thus, a market potential index can be estimated from biological yield components, the number of days in each ripening stage, and the value of corresponding food products across target environments, using the following formula:

$$M = Y\,\Sigma_{\mathrm{ij}} s_{\mathrm{i}} a_{\mathrm{jk(i)}}$$

where M is the estimated market potential index of the cultivar, Y is the dry pulp yield of the cultivar, s_{i} is the number of days spent in the ith ripening stage by the cultivar, and $a_{\mathrm{jk(i)}}$ is a subjective index value attributed to the jth product in the kth target environment (Figure 12.1).

12.4.2 Defining Best Bets for Dissemination

Before attempting to introduce new cultivars to farmers, it is important to establish their likely performance when grown under smallholder-managed environments. This can be done by bringing farmers' voices into the cultivar development process, notably through farmer-participatory variety evaluation, whereby breeders and farmers jointly conduct and evaluate trials in order to identify

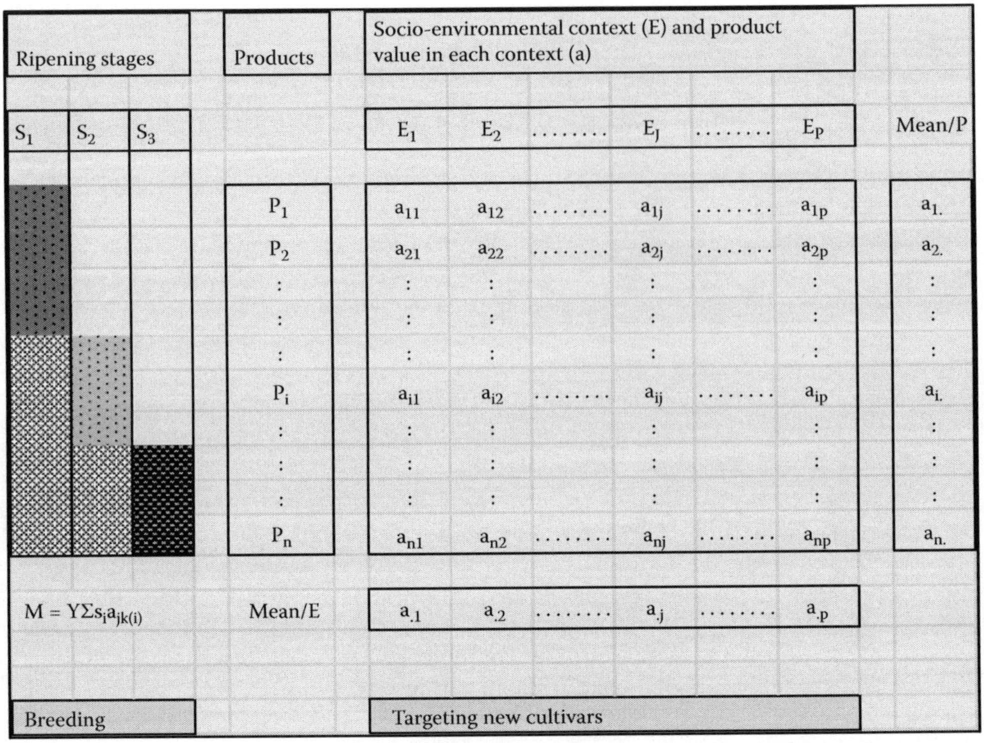

FIGURE 12.1 Illustration of market potential as a breeding and targeting tool: Harvested fruits undergo a ripening process made of distinct stages with genetically governed duration (*S*). There is a range of processed food products (*P*) obtainable from each cultivar and the value of such products (*a*) depends on the cultural and economic context (*E*). Thus, besides the yield (*Y*), the adoption potential of a cultivar depends on the relative value components of shelf life and the value of the processed products. A new cultivar that is good for a processed product that has high value in a given environmental context should be targeted to that environment. Likewise, knowledge of the biological characteristics associated with a given food preparation can be used to develop niche-specific cultivars.

those cultivars most attractive to the farmers, which are often location specific. This allows farmers to cast their cultivar choices without exposing the household to any risk (Ceccarelli and Grando, 2004; Sperling et al., 2001) and facilitates the subsequent deployment of the chosen cultivars on a large scale in matching regions, constituting a cost-efficient means of enhancing crop productivity and farmers' incomes. Regardless of the method chosen for capturing farmers' voices, cultivars can be classified into four broad categories, depending on the extent of congruence between agronomic performance and farmers' assessments (Figure 12.2).

12.4.3 Nondisruptive Dissemination

Resource-poor farmers in many regions of the world, particularly in Africa, have adopted cropping strategies based on intraspecific (cultivar mixtures) and interspecific (various forms of intercropping) to maximize land and labor use efficiency and to minimize risks of crop failure. Cultivar mixtures are common in subsistence farming systems, offering growers diversity of diet, stability of income, and reduced losses to pests (Smithson and Lenne, 1996; Selatsa et al., 2009). Thus, the chances for adoption of new cultivars from breeding programs would partly depend on the extent that traditional cropping and consumption practices are not drastically disrupted.

One of the lessons learned over the years by the breeding programs of the International Institute of Tropical Agriculture (IITA) was that as much as farmers seemed attracted to new cultivars,

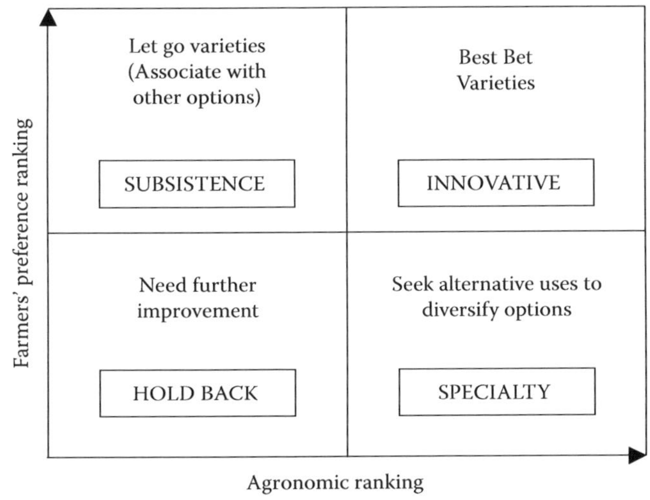

FIGURE 12.2 Schematic representation of the outcome of farmer participation in cultivar selection. High congruence between agronomic performance ranking and farmers' preference ranking indicates that the cultivars are best bets for further evaluation under farmers' conditions (top right quadrant) or should be held back and subjected to further improvement (bottom left quadrant); in the lack of congruence between agronomic performance and farmers' preference, one group of cultivars can be promoted if they lend themselves to economically attractive ways of processing unfamiliar to the grower, provided the new processing options are included in the dissemination process (bottom right quadrant). This group can then easily move into the group of best bets. The last group of cultivars are those not performing well agronomically but are nonetheless attractive to farmers (top left quadrant), which the farmers should be allowed to have.

owing notably to their higher yields under disease pressure where the traditional cultivars succumb, they also insisted on keeping their traditional cultivars. This suggested that the disease-resistant cultivars should be introduced into the farmers' cropping system through association with their own landraces and other crops.

Nondisruptive dissemination of new cultivars through mixtures promotes plant diversity. This has the potential of increasing the productivity and stability of the mixture, but it depends on the success of each individual in the mixture composition, which in turns depends on the extent of competition among individuals in the mixture. The performance of plants in communities departs significantly from their performance in isolation, a discrepancy that is usually attributed to negative or positive competitive interactions, but the relative importance of competition and facilitation may vary inversely along gradients of abiotic stress (Callaway et al., 2002). Diversity is considered to be a stabilizing factor for an ecosystem that operates like insurance, allowing for greater resilience of the ecosystem in the face of stressful conditions (Kahmen et al., 2005).

There may be different options for cultivar mixtures, but the simplest approach could be one that achieves equal representation of new and old cultivars. Binary mixtures can achieve a 1:1 old-to-new cultivar ratio or 50% landrace substitution level. This approach was used for large-scale on-farm testing and dissemination of improved cultivars in Nigeria, using a checkerboard design whereby each hybrid plant was surrounded by four plants of a landrace and vice versa (Tenkouano and Swennen, 2004; Tenkouano et al., 2009). The mixture was effective in reducing black Sigatoka effects on the landrace, whereby the index of nonspotted leaves (INSL) increased from 43.2% to 51–56% when the landrace was mixed with the hybrids, and the bunch weight of the landrace increased from 4.9 kg when grown solely to 7.1–8.1 kg when grown in mixture with the hybrids (Table 12.2). Similar results were observed in Cameroon, with the mixture increasing the INSL of the local variety 'Essong' from 24.3% when grown solely to 30.8–37.7% when mixed with hybrids.

TABLE 12.2
Improved Hybrids Act as "Bio-Pesticides" to Enhance Landrace Performance and Preserve Diversity (50% Substitution) in Cultivar Mixtures in Nigeria and Cameroon

Clone	Treatment	Bunch Weight (kg)	Index of Nonspotted Leaves (%)
		Nigeria	
Agbagba	Agbagba sole	4.9 ± 0.2	43.2 ± 3.1
Agbagba	Agbagba + BITA3	8.1 ± 0.6	55.7 ± 2.1
Agbagba	Agbagba + PITA14	7.1 ± 0.4	51.4 ± 3.6
Agbagba	Agbagba + PITA17	7.3 ± 0.3	52.8 ± 2.7
BITA3	Agbagba + BITA3	16.7 ± 1.1	73.8 ± 2.8
PITA14	Agbagba + PITA14	12.7 ± 0.6	89.7 ± 2.0
PITA17	Agbagba + PITA17	12.7 ± 0.7	75.4 ± 3.0
		Cameroon	
Essong	Essong sole	9.6 ± 1.9	24.3 ± 0.5
Essong	Essong + BITA3	9.1 ± 2.6	30.8 ± 3.7
Essong	Essong + PITA14	11.2 ± 0.8	37.7 ± 3.6
Essong	Essong + PITA21	10.6 ± 3.0	39.6 ± 7.5
BITA3	Essong + BITA3	11.5 ± 1.4	68.7 ± 1.2
PITA14	Essong + PITA14	9.0 ± 1.2	72.0 ± 0.6
PITA21	Essong + PITA21	6.6 ± 0.7	66.6 ± 1.8

Note: The cultivars 'Agbagba' and 'Essong' are among the most preferred plantains in Nigeria and Cameroon, respectively. They were grown in mixture with the improved hybrids 'BITA3,' 'PITA14,' and 'PITA17' (Nigeria) or 'PITA21' (Cameroon).

Likewise, increased bunch weight of the local cultivar was observed with mixtures, although not statistically different from the bunch weight obtained under sole cropping (Table 12.2).

It is noteworthy that although the performance of the landrace is significantly enhanced, this is still less than that of the hybrids. Thus, cultivar mixtures appear to enhance the performance of the landraces without altering that of the hybrids. This mechanism not only helped to increase the yield of the landraces (preferred for their culinary properties), but it also preserved genetic diversity while exposing farmers to new, high-yielding hybrids. Hence, cultivar mixtures can be considered a nondisruptive mechanism for introducing new hybrids into farmers' fields, where the landraces can satisfy traditional consumption needs while hybrids can be processed for commercial purposes (there is a wide variety of options).

The epidemiological factors that are responsible for the apparent improvement of the performance of the landraces have not been fully investigated, but there are several theoretical possibilities, such as inoculum trapping by resistant cultivars resulting in reduced disease spread to plants of the susceptible landraces. Neighbors can beneficially change the situation of a plant, most commonly when the neighbors contribute to alleviation of environmental conditions that are unfavorable to the survival, growth, and reproduction of the plant (Bertness and Hacker, 1994). Thus, the mixture approach should be based on a careful analysis of the blending ability of the cultivars proposed for dissemination. The blending ability can be experimentally calculated using formulas analogous to those developed for analysis of genetic combining ability (Gallandt et al., 2001; Springer et al., 2001).

Regardless of the theoretical basis of the beneficial effects of mixtures, mixtures may be a better option than chemical control. Chemical control is done either with protectant or systemic fungicides. Protectant fungicides are applied topically to the leaves and are effective pre-infectionally but they have adverse effects on leaf physiology depending on the amount of oil emulsions and weather conditions. Systemic fungicides are effective post-infectionally but have been reported to cause development of resistance without loss of fitness in *M. fijiensis* populations (Marin et al., 2003).

Growing susceptible cultivars is not recommended, unless there are environmentally acceptable means of controlling the biological organisms to which the cultivars are susceptible. Biological principles dictate that the causal organisms thrive to maintain a balance between their growth that is detrimental to the host and the need to circumvent extinction of the host, which is synonymous of extinction for the organisms unless alternate hosts are available. Similarly, exclusive deployment of cultivars that are highly resistant could impose extreme pressure on the biological threat, forcing it to mutate to more virulent forms in order to ensure its survival.

Under such circumstances, it is advisable to develop a strategy that maintains an acceptable balance between threatened and threatening organisms. Thus, growers of resistant cultivars should provide a biological refuge consisting of susceptible cultivars to the threatening organisms, similar to the practice for deployment of transgenic plants with *Bt* resistance to insects (Gopalaswamy et al., 2003). The cultivar mixture approach to the dissemination of new cultivars of banana and plantain adheres to this principle.

REFERENCES

Ahenkora, K., M.A. Kyei, E.K. Marfo, and B. Banful. 1996. Nutritional composition of False Horn 'Apantu pa' plantain during ripening and processing. *Afr. Crop Sci. J.* 4(2):243–247.

Baiyeri, K.P. 1998. Evaluation of growth, yield and yield components of 36 *Musa* genotypes under four different environments. PhD diss., University of Nigeria, Nsukka, Nigeria.

Baiyeri, K.P. and R. Ortiz. 2000. Agronomic evaluation of plantain and other triploid banana in Africa. *Acta Hort.* 540:125–135.

Baiyeri, K.P., B.N. Mbah, and A. Tenkouano. 1999a. Comparing yield stability of *Musa* genotypes in Nigeria using four statistical methods. *J. Trop. Forest Res.* 15:53–67.

Baiyeri, K.P., A. Tenkouano, B.N. Mbah, and J.S.C. Mbagwa. 1999b. Genetic and cropping system effects on yield and postharvest characteristics of *Musa* species in Southeastern Nigeria. *Afr. Crop Sci. J.* 7:1–8.

Baiyeri, K.P., B.N. Mbah, and A. Tenkouano. 2000a. Relationships between phenological and yield traits of the plant crop and first ratoon crop of *Musa* genotypes as affected by ploidy level and genomic group. *Agro-Science* 1:113–121.

Baiyeri, K.P., B.N. Mbah, and A. Tenkouano. 2000b. Yield components of triploid and tetraploid *Musa* genotypes in Nigeria. *HortScience* 35:1338–1343.

Baiyeri, K.P., A. Tenkouano, B.N. Mbah, and J.S.C. Mbagwa. 2004. Phenological and yield evaluation of *Musa* genotypes under alley and sole cropping in Southeastern Nigeria. *Trop. Subtrop. AgroEcosyst.* 4:137–144.

Baiyeri, K. P., M. Pillay, and A. Tenkouano. 2008. Phenotypic relationships between growth, yield and black leaf streak disease responses of *Musa* genotypes. *Journal of Crop Improvement* 21(1): 41–54.

Bertness, M.D. and S.D. Hacker. 1994. Physical stress and positive associations among marsh plants. *Am. Nat.* 144:363–372.

Blomme, G., R. Swennen, R. Ortiz, and A. Tenkouano. 2006. Root system and shoot growth of banana (*Musa* spp.) in two agro-ecological zones in Nigeria. *InfoMusa* 15:18–23.

Bressani, R. 1976. Productivity and improved nutritional value of basic food crops. In: *Improving nutrient quality of cereals* II. Report of second workshop on breeding and fortification, H.L. Wilcke, ed., 265–287. Boulder, CO, 12–17 September, Washington, DC.

Callaway, R.M., R.W. Brooker, P. Choler, Z. Kikvidze, C.J. Lortie, R. Michalet, et al. 2002. Positive interactions among alpine plants increase with stress. *Nature* 417:844–848.

Ceccarelli, S. and S. Grando. 2004. Decentralized-participatory plant breeding. In: *Proceedings of the international congress "In the Wake of the Double Helix: From Green Revolution to the Gene Revolution,"* R. Tuberosa, R.L. Phillips, and M. Gale, eds., 27–31. May, 2003, Bologna, Italy.

Craenen, K. and R. Ortiz. 1997. Effect of the bs_1 gene in plantain-banana hybrids on response to black Sigatoka. *Theor. Appl. Genet.* 95:497–505.

Craenen, K. and R. Ortiz. 2003. Genetic improvement for a sustainable management of the resistance. In: *Mycosphaerella leaf spot diseases of bananas: Present status and outlook*, L. Jacome, P. Lepoivre, D. Marin, R. Ortiz, R. Romero, and J.V. Escalant, eds., 181–198. Montpellier, France: INIBAP.

Dadzie, B.K. and J.E. Orchard. 1996. *Post-harvest criteria and methods for routine screening of banana/plantain hybrids*. Montpellier, France: INIBAP.

De Cauwer, I. and R. Ortiz. 1998. Analysis of the genotype-by-environment interaction in *Musa* trials. *Exp. Agric.* 34:177–188.

De Cauwer, I., R. Ortiz, and D. Vuylsteke. 1995. Genotype-by-environment interaction and phenotypic stability of *Musa* germplasm in West and Central Africa. *Afr. Crop Sci.* 3:425–432.

Dumpe, B.B. and R. Ortiz. 1996. Apparent male fertility in *Musa* germplasm. *HortScience* 31:1019–1022.

Ferris, R.S.B., T. Adeniji, U. Chukwu, Y.O. Akalumhe, D. Vuylsteke, and R. Ortiz. 1996. Post-harvest quality of plantains and cooking bananas. In: *Plantain and banana: Production and research in West and Central Africa*, R. Ortiz and M.O. Okoroda, eds., 5–22. Proceedings of a regional workshop held at IITA High Rainfall Station, Onne, Rivers State, Nigeria: IITA.

Gallandt, E.R., S.M. Difing, P.E. Reisenauer, and E. Donaldson. 2001. Diallel analysis of cultivar mixtures in winter wheat. *Crop Sci.* 41:792–796.

Gauch, H.G. 1992. *Statistical analysis of regional yield trials: AMMI analysis of factorial designs*. Amsterdam: Elsevier.

Gopalaswamy, S.V.S., G.V. Subbaratnam, and H.C. Sharma. 2003. Development of resistance in insects to transgenic plants with *Bacillus thuringiensis* genes: Current status and management strategies. *Resistant Pest Manage. Newslet.* 12(2):16–26.

Jenny, C., E. Auboiron, D. Vuylsteke, and R. Ortiz. 1993. Influence of genotype and environment seed set in plantains. *MusAfrica* 3:3.

Kahmen, A., J. Perner, and N. Buchmann. 2005. Diversity-dependent productivity in semi-natural grasslands following climate perturbations. *Functional Ecol.* 19:594–601.

Marin, D.H., R.A. Romero, M. Guzmán, and T.B. Sutton. 2003. Black Sigatoka: An increasing threat to banana cultivation. *Plant Dis.* 87:208–222.

Orjeda, G. 2000. International Musa testing programme: A worldwide effort of *Musa* scientific community. In: *Banana—Improvement, production and utilization*, H.P. Singh and K.L. Chadha, eds., Conference on challenges for banana production and utilization in 21st century, 24–25 September, 1996, Trichy, India.

Ortiz, R. 1995. Plot techniques for assessment of bunch weight in banana trials under two systems of crop management. *Agron. J.* 87:63–69.

Ortiz, R. 1996. The potential of AMMI analysis for field assessment of *Musa* genotypes to virus infection. *HortScience* 31:829–832.

Ortiz, R. 1997a. Morphological variation in *Musa* germplasm. *Genet. Res. Crop Evol.* 44:393–404.

Ortiz, R. 1997b. Genetic and phenotypic correlations in plantain-banana euploid hybrids. *Plant Breed.* 116:487–491.

Ortiz, R. 1997c. Occurrence and inheritance of 2*n* pollen in *Musa*. *Ann. Bot.* 79:449–453.

Ortiz, R.1998. AMMI and stability analyses of bunch mass in multilocational testing of *Musa* germplasm in sub-Saharan Africa. *J. Am. Soc. Hort. Sci.* 123:623–627.

Ortiz, R. and I. De Cauwer. 1998. Genotype-by-environment interaction and testing environments for plantain and banana (*Musa* spp. L.) breeding in West Africa. *Tropicultura* 16–17:97–102.

Ortiz, R. and I.J. Ekanayake. 2000. Assessment of genotype x environment interaction and role of physiological analyses for crop breeding. In: *Genotype by environment interaction analysis of IITA mandate crops in sub-Saharan Africa*, I.J. Ekanayake and R. Ortiz, eds., 10–31. Ibadan, Nigeria: International Institute of Tropical Agriculture.

Ortiz, R. and H. Langie. 1997. Path analysis and ideotypes for plantain breeding. *Agron. J.* 89:988–994.

Ortiz, R. and D. Vuylsteke. 1994. Inheritance of black Sigatoka resistance in plantain-banana (*Musa* spp.) hybrids. *Theor. Appl. Genet.* 89:146–152.

Ortiz, R. and D. Vuylsteke. 1995. Factors influencing seed set in triploid *Musa* spp. L. *Ann. Bot.*75:151–155.

Ortiz, R. and D. Vuylsteke. 1998. Quantitative variation and phenotypic correlations in banana and plantain. *Scientia Hort.* 72:239–253.

Ortiz, R. and D. Vuylsteke. 1998. 'PITA 14': A black Sigatoka resistant tetraploid plantain hybrid with virus tolerance. *HortScience* 33:360–361.

Ortiz, R., D. Vuylsteke, R.S.B. Ferris, J.U. Okoro, A.N. Guessan, O.B. Hemeng, et al. 1997. Developing new plantain varieties for Africa. *Plant Var. Seeds* 10:39–57.

Ortiz, R., F. Ulburghs, and J.U. Okoro. 1998. Seasonal variation of apparent male fertility and 2*n* pollen in plantain and banana. *HortScience* 33:146–148.

Sánchez Nieva, F., I. Hernández, G. Colom Covas, R. Guadalupe Luna, N. Diaz, and C.B. Vinas. 1968. A comparative study of some characteristics of two plantain cultivars which affect yields and product quality *J. Agric. Univ. Puerto Rico* 52(4):323–338.

Selatsa, A.A., A. Tenkouano, E. Njukwe, R.N. Iroume, and P.J. Bramel. 2009. Morphological diversity of plantain (*Musa* sp. L. AAB group) in Cameroon: Relationships to farmer's cultural practices. *Afr. J. Plant Sci. Biotechnol.* 3(1):51–58.

Smithson, S. B. and J. M. Lenné. 1996. Varietal mixtures: A viable strategy for subsistence agriculture. *Ann. Appl. Biol.* 1288:127–158.

Sperling, L., J. Ashby, M.E. Smith, W. Weltzien, and S. McGuire. 2001. A framework for analyzing participatory plant breeding approaches and results. *Euphytica* 122:439–450.

Springer, T.L., G.E. Aiken, and R.W. McNew. 2001. Combining ability of binary mixtures of native, warm-season grasses and legumes. *Crop Sci.* 41:818–823.

Stover R. H. and N. W. Simmonds. 1987. *Bananas*, 3rd edition, Longman, London. 468 pp.

Tenkouano, A. and R.L. Swennen. 2004. Progress in breeding and delivering improved plantain and banana to African farmers. *Chronica Horticulturae* 44:9–15.

Tenkouano, A., K.P. Baiyeri, and R. Ortiz. 2002. Phenotypic and genetic correlations in *Musa* populations in Nigeria. *Afr. Crop Sci.* J. 10:121–132.

Tenkouano, A., B.O. Faturoti, and K.P Baiyeri. 2009. On farm evaluation of *Musa* hybrids in Nigeria. *Tree Forestry Sci. Biotechnol.* Special Issue 1. Bananas, plantains and ensete (accepted February 2009).

Vuylsteke, D., R. Ortiz, R.S.B. Ferris, and J.H. Crouch. 1997. Plantain improvement. *Plant Breed. Rev.* 14:267–320.

13 Quality Improvement of Cultivated *Musa*

Edson Perito Amorim, Sebastião de Oliveira e Silva, Vanusia Batista de Oliveira Amorim, and Michael Pillay

CONTENTS

13.1 INTRODUCTION

The world's production of banana and plantain is estimated at about 125 million tons per year, placing the crop first in the world's ranking of fruit production, followed by grape, orange, mango, and pineapple (FAO, 2010). Bananas are grown in more than 130 countries in the tropical and subtropical zones in five continents (Jones, 2000). The world's largest producers are India, Uganda, the Philippines, China, and Brazil. Besides its importance as a food crop, banana cultivation has a strong socioeconomic significance, since it is the only source of income for many small-scale growers (Mattos et al., 2010; Bakry et al., 2009).

Banana is a fruit that is rich in natural antioxidants such as vitamin C and vitamin E (Someya et al., 2002; Amorim et al., 2009a). The antioxidant properties of many fruits are associated with flavonoids and β-carotenes. Flavonoids are found in the pulp and peel of banana while β-carotenes are present only in the pulp of some banana varieties (Vijayakumar et al., 2008). β-carotenes are precursors of vitamin A. In low-income regions of the world, such as parts of Asia, Africa, and Latin America, high levels of vitamin A deficiency lead to serious health problems, especially in children (Bloem et al., 2005). Vitamin A is associated with cell differentiation, vision, bone growth, reproduction, and integration of the immune system and is considered to play a role in the prevention of cancer, cardiovascular disease, cataracts, and macular diseases as well as neurologic, inflammatory, and immune disorders (Arora et al., 2008a, b). Micronutrient deficiencies of iron and zinc also result in serious health problems such as mental and physical retardation, reduced resistance to infections, and hypogonadism (Whittaker, 1998). Despite various efforts to improve the vitamin and micronutrients intake of people the past 50 years, over 2 billion people, about one third of the world's population, suffer from vitamin A, Zn, and/or Fe deficiencies (FAO, 1997).

To maintain a healthy status, it is believed that the human body requires more than 20 minerals and about 40 nutrients, especially vitamins and essential amino acids that can be easily obtained from a healthy diet. However, the vast majority of the world's population has little or no access to sufficient quantity, quality, and variety of foods, resulting in deficiencies in minerals and essential nutrients (Pfeiffer and McClafferty, 2007). García-Alonso et al. (2004) suggested that there is a direct relation between diets rich in fruits and vegetables and a reduced risk of cardiovascular diseases and some types of cancer. In this context, the biofortification of banana by conventional breeding combined with the use of biotechnological tools has the potential to increase the concentrations of micronutrients (Fe, Zn) and vitamin A in new cultivars, with the aim of improving the health status of populations in both rural and urban areas of developing countries. Bananas are already used in special diets where ease of digestibility, low fat, minerals, and vitamins are required. These special diets are used for babies, the elderly, and patients with stomach problems, gout, and arthritis (Nakasone and Paull, 1999).

This chapter deals with some aspects of banana as a food crop, emphasizing the nutritional value and breeding strategies for the development of biofortified cultivars. Findings of different research groups working on the identification, measurement, and use of genotypes with functional properties are presented.

13.2 NUTRITIONAL VALUE

The nutritional value of any food crop depends on several variables such as growth stage, climatic conditions, soil quality, and of particular importance, the genotype (Mozafar, 1994; Lee and Kader, 2000). Therefore assessments of the same micronutrients in a crop can differ significantly, depending on growth conditions and method of evaluation. This observation is also valid for banana.

13.2.1 Nutrition in Banana

The pulp of ripe banana consists essentially of sugar and is therefore easily digestible. The fruit is composed of approximately 70% water, 27% carbohydrate, 0.3% fat, and 1.2% protein (Robinson, 1996). In addition, each gram contains approximately one calorie of energy. Twelve vitamins are found in the fruit, which is considered a good source of the vitamins A, B_1, B_2, and C (Robinson, 1996). Banana and plantain differ in moisture content, with water content of banana averaging 75%, while plantains contain around 65% of water (www.answers.com/topic/banana-and-plantain). Therefore the conversion of starch to sugars is quicker in bananas. Plantains are generally cooked before being eaten, although they are eaten as a fresh fruit when ripe. The total lipid content is low both in banana and plantain (<0.2%) and the energy the fruits provide is therefore not related to fat. Similarly, the concentrations of amino acids are usually low (http://www.nal.usda.gov).

There are no significant amounts of antinutritional compounds in banana (Adeniji et al., 2007). This contrasts sharply with other staple crops such as cassava that contains cyanogenic glycosides (Koch et al., 1992; White et al., 1994) and potato that has high concentrations of glycoalkaloids (Hellenas et al., 1995), which are toxins. Banana contains high concentrations of serotonin and dopamine, two neurotransmitters that act directly in controlling the release of some hormones and in regulating the circadian rhythms of sleep and appetite (Adao et al., 2005).

Recently, banana has been labeled as an ideal vehicle for the development of vaccines, in view of the easy digestion, pleasant taste, and great acceptability of banana by children. Studies are underway to develop different vaccines in banana (Sala et al., 2003; Arntzen et al., 2004; Sunil Kumar et al., 2005).

The quantity of different nutrients in bananas is presented in Table 13.1 and includes protein, sugar, minerals, vitamins, lipids, and amino acids, obtained from the U.S. Department of Agriculture (UDA) based on the evaluation of banana samples, particularly of the subgroup Cavendish (AAA).

13.2.2 Identification of Genotypes with Functional Food Attributes

Studies on the identification and use of banana genotypes with functional properties have been conducted around the world by different research institutions, including the following: Embrapa Cassava and Tropical Fruits (Brazil), Catholic University of Leuven (Belgium), Bioversity (France), University of Queensland (Australia), Tohoku University (Japan), University of Kerala (India), Agricultural Research Service (United States), and the International Institute of Tropical Agriculture (IITA), among others. This chapter presents the research results of the identification, assessments, and quantification of functional compounds in different banana genotypes. Particular attention was given to the presence of the following compounds: carotenoids and their fractions, vitamin C, total polyphenols, flavonoids, and antioxidant activity.

Sharrock and Lutsy (1999) assessed the contents of carbohydrates, protein, fats, vitamins, and minerals in banana and plantain, and compared the nutritional value of banana with those of sweet potato, potato, cassava, and apple. The vitamin C content of banana and plantain was similar to that found in other species (18.4 mg 100 g^{-1}). Vitamin A content was higher in plantain (1127 UI/IU) than in banana (81 IU) and exceeded that found in potato, cassava, and apple.

Kanazawa and Sakakibara (2000) estimated the dopamine content, a strong natural antioxidant in banana (subgroup Cavendish). The amount varied from 80 to 560 mg 100 g^{-1} in the peel and from 2.5 to 10 mg 100 g^{-1} in the flesh. The antioxidant activity of dopamine was higher than of butylated hydroxyanisole (BHA), butylated hydroxytoluene (BHT), flavonoids, catechin, and glutamine, and was similar to that of other antioxidants, such as gallocatechin gallate and vitamin C. According to the authors, dopamine is easily absorbed by the body and plays an important role in the control of Parkinson's disease. Someya et al. (2002) also found higher levels of the antioxidant gallocatechin in the peel (158 mg/100 g dry wt) than in the pulp (29.6 mg/100 g dry wt) in Cavendish banana. Together, these results suggest that bananas should be considered as a good source of antioxidants.

Englberger et al. (2003a, 2003b, 2003c) identified banana genotypes, commonly known as the Fe'i bananas (section *Australimusa*), with much higher concentrations of β-carotene than in Cavendish cultivars, which account for approximately 41% of the world's total banana production. There was a wide variability for β-carotene in the Fe'i bananas, with concentrations ranging from 56 μg 100 g^{-1} to 6360 μg 100 g^{-1} (Table 13.2). The genotypes from the islands of Micronesia (*Karat* and *Uht en Yap*, *Musa troglodytarum*) contained 275 times more carotenoids than Cavendish. These genotypes, with upward-growing bunches and fruits with variable and dark coloration, are promising sources of alleles for improving the nutritional quality of other bananas in breeding programs. Adeniji et al. (2006) determined micronutrients in plantain and banana hybrids in Nigeria ('PITA 14,' 'PITA 17,' 'BITA 3,' 'FHIA 17,' 'FHIA 23,' and 'Agbagba,' used as a control). The cultivars 'Agbagba' and 'FHIA 17' had high Fe (36.5 μg g^{-1} and 16.13 μg g^{-1}) and zinc contents (12.5 μg g^{-1} and 12.49 μg g^{-1}), respectively. The carotenoid contents of the cultivars were similar (mean of 3.6 μg g^{-1}) with the highest values recorded in 'PITA 17' (4.80 μg g^{-1}). The authors are of the opinion that it is possible to develop new banana hybrids with high yield and high micronutrient concentrations. Such varieties may be potentially useful in reducing health problems related to deficiency of vitamin A and/or other micronutrients.

Wall (2006) determined the mineral composition and vitamins C and A content in the banana varieties 'Prata,' 'Santa Catarina,' and 'Williams' (Cavendish) and in papaya cultivars in Hawaii. The variety 'Prata' contained 1.5 times more carotenoids than 'Williams' and the mean vitamin C content was 12.7 mg 100 g^{-1} for 'Prata' and 4.45 mg 100 g^{-1} for 'Williams.' In addition, the P, Ca, Mg, Mn, and Zn contents of 'Prata' were higher than of Williams, indicating a better nutritional quality of the former.

TABLE 13.1
Nutritional Values for Banana and Plantain (per 100 g Raw Edible Portion)

Nutrient	Units	Value per 100 g	
		Banana	Plantain
Water	G	74.91	65.28
Energy	Kcal	89	122
Energy	kJ	371	510
Protein	G	1.09	1.30
Total lipid (fat)	G	0.33	0.37
Ash	G	0.82	1.17
Carbohydrate, by difference	G	22.84	31.89
Fiber, total dietary	G	2.6	2.3
Sugars, total	G	12.23	15.00
Sucrose	G	2.39	–
Glucose (dextrose)	G	4.98	–
Fructose	G	4.85	–
Lactose	G	0.00	–
Maltose	G	0.01	–
Galactose	G	0.00	–
Starch	G	5.38	–
Minerals			
Calcium, Ca	Mg	5	3
Iron, Fe	Mg	0.26	0.60
Magnesium, Mg	Mg	27	37
Phosphorus, P	Mg	22	34
Potassium, K	Mg	358	499
Sodium, Na	Mg	1	4
Zinc, Zn	Mg	0.15	0.14
Copper, Cu	Mg	0.078	0.081
Manganese, Mn	Mg	0.270	–
Fluoride, F	mcg	2.2	–
Selenium, Se	mcg	1.0	1.5
Vitamins			
Vitamin C, total ascorbic acid	Mg	8.7	18.4
Thiamin	Mg	0.031	0.052
Riboflavin	Mg	0.073	0.054
Niacin	Mg	0.665	0.686
Pantothenic acid	Mg	0.334	0.260
Vitamin B-6	Mg	0.367	0.299
Folate, total	mcg	20	22
Folic acid	mcg	0	0
Folate, food	mcg	20	22
Folate, DFE	mcg_DFE	20	22
Choline, total	Mg	9.8	13.5
Betaine	Mg	0.1	–
Vitamin B-12	mcg	0.00	56
Vitamin B-12, added	mcg	0.00	0
Vitamin A, RAE	mcg_RAE	3	56
Retinol	mcg	0	0

(continued)

TABLE 13.1 (CONTINUED)
Nutritional Values for Banana and Plantain (per 100 g Raw Edible Portion)

Nutrient	Units	Value per 100 g	
		Banana	Plantain
Carotene, beta	mcg	26	457
Carotene, alpha	mcg	25	438
Cryptoxanthin, beta	mcg	0	0
Vitamin A	IU	64	1127
Lycopene	mcg	0	0
Lutein + zeaxanthin	mcg	22	30
Vitamin E (alpha-tocopherol)	Mg	0.10	0.14
Vitamin E, added	Mg	0.00	0.00
Tocopherol, beta	Mg	0.00	–
Tocopherol, gamma	Mg	0.02	–
Tocopherol, delta	Mg	0.01	–
Vitamin D (D2 + D3)	mcg	0.0	0.0
Vitamin D	IU	0	0
Vitamin K (phylloquinone)	mcg	0.5	0.7
	Lipids		
Fatty acids, total saturated	G	0.112	0.143
4:0	G	0.000	0.000
6:0	G	0.000	0.000
8:0	G	0.000	0.000
10:0	G	0.001	0.001
12:0	G	0.002	0.002
14:0	G	0.002	0.002
16:0	G	0.102	0.096
18:0	G	0.005	0.005
Fatty acids, total monounsaturated	G	0.032	0.032
16:1 undifferentiated	G	0.010	0.009
18:1 undifferentiated	G	0.022	0.021
20:1	G	0.000	0.000
22:1 undifferentiated	G	0.000	0.000
Fatty acids, total polyunsaturated	G	0.073	0.069
18:2 undifferentiated	G	0.046	0.043
18:3 undifferentiated	G	0.027	0.025
18:4	G	0.000	0.000
20:4 undifferentiated	G	0.000	0.000
20:5 n-3 (EPA)	G	0.000	0.000
22:5 n-3 (DPA)	G	0.000	0.000
22:6 n-3 (DHA)	G	0.000	0.000
Cholesterol	Mg	0	0
Phytosterols	Mg	16	–
	Amino Acids		
Tryptophan	G	0.009	0.015
Threonine	G	0.028	0.034
Isoleucine	G	0.028	0.036
Leucine	G	0.068	0.059
Lysine	G	0.050	0.060

(continued)

TABLE 13.1 (CONTINUED)
Nutritional Values for Banana and Plantain (per 100 g Raw Edible Portion)

Nutrient	Units	Value per 100 g	
		Banana	Plantain
Methionine	G	0.008	0.017
Cystine	G	0.009	0.020
Phenylalanine	G	0.049	0.044
Tyrosine	G	0.009	0.032
Valine	G	0.047	0.046
Arginine	G	0.049	0.108
Histidine	G	0.077	0.064
Alanine	G	0.040	0.051
Aspartic acid	G	0.124	0.108
Glutamic acid	G	0.152	0.116
Glycine	G	0.038	0.045
Proline	G	0.028	0.050
Serine	G	0.040	0.041
Other			
Alcohol, ethyl	G	0.0	0.0
Caffeine	Mg	0	0
Theobromine	Mg	0	0

Source: USDA National Nutrient Database for Standard Reference, Release 22, 2009, http://www.nal.usda.gov. With permission.

Davey et al. (2007, 2009) reported that the genetic variation for carotenoids is strongly related to differences in fruit maturity and that orange-colored varieties (plantains of the AAB group) usually had higher carotenoid concentrations than triploid AAA fruits. The authors found variability for carotenoid content within fruit, within hand, and within plant, as well as between plants. It was suggested that sampling methods for correct carotenoid analyses should take two fruits in the central region of the bunch.

Arora et al. (2008a, b) quantified the content of carotenoids, β-carotene, and antioxidant enzymes in some banana genotypes in India. 'Red Banana' was found to have the highest carotenoid concentrations (241.91 ug 100 g^{-1} in the peel and 117.2 ug 100 g^{-1} in the flesh), followed by 'Karpooravalli' (1786.0 μg g^{-1} dry wt in peel and 544.85 μg g^{-1} dry wt in flesh). The catalase enzyme activity in the peel of these varieties ranged from 5.66 to 35.57 nmol min^{-1} mg^{-1} protein. Both the peel and flesh of these varieties are rich sources of bioactive compounds such as carotenoids, enzymes, and antioxidant carbohydrates.

Davey et al. (2009) assessed the genetic variability for provitamin A carotenoids (pVACs), lutein, and micronutrients (Fe and Zn) in 171 different banana cultivars, including diploid (genome AA and AB), triploid (genome AAA, AAB, and ABB), and tetraploid accessions of unknown ploidy. The results indicated a substantial variation in the pVAC concentration between genotypes, genomic constitution, and ploidy, although the frequency of genotypes with high pVAC content was low. The cultivars with highest pVAC values were 'Bantol Red' (unknown ploidy), 'Pusit' (unknown ploidy), 'Iholena Lele' (AAB), 'Henderneyargh' (AAS), 'Katimor' (AAB), and 'Chek Porng Mean' (unknown ploidy), with values of 1500–2800 μg t-BCE 100 g^{-1} dry wt.

González-Montelongo et al. (2010) quantified the antioxidant activity in banana and observed that the peel contains a large amount of dopamine and L-dopamine, two catecholamines with considerable antioxidant activity.

TABLE 13.2
Carotenoid Content of Selected Cultivars of Ripe Micronesian Banana (μg 100 g^{-1} Edible Portion)

Sci. Name	Local Name	Source	Color of Raw Flesh	β-Carotenoids
Mt	*Uht en yap*	Pohnpei	Orange	6360
Mt	*Uht en yap*	Pohnpei	Orange	5860
Ms	*Usr wac*	Kosrae	Orange	2082
Ms	*Uht ipali*	Pohnpei	Orange	1181
Mt	*Uht karat*	Pohnpei	Yellow-orange	918
Mt	*Uht karat*	Pohnpei	Yellow-orange	578
Ms	*Usr wac essie*	Kosrae	Yellow-orange	686
Ms	*Usr wac essie*	Kosrae	Yellow-orange	309
Ms	*Usr taiwang*	Kosrae	Yellow	662
Ms	*Usr taiwang*	Kosrae	Yellow	571
Ms	*Usr kuria*	Kosrae	Yellow	653
Ms	*Usr kuria*	Kosrae	Yellow	218
Ms	*Usr macao*	Kosrae	Yellow-orange	589
Ms	*Uht akatan*	Pohnpei	Yellow	515
Ms	*Uht akatan*	Pohnpei	Yellow	227
Ms	*Usr in yeir*	Kosrae	Yellow	421
Ms	*Usr in yeir*	Kosrae	Yellow	360
Ms	*Uht enkerinis*	Pohnpei	Yellow	310
Ms	*Marechg*	Chuuk	Yellow	189
Ms	*Usr fiji/uhten fijih*	Fiji	White	56

Note: Ms: *Musa* spp.; Mt: *Musa troglodytarum*.

Source: Englberger, L., J. Schierle, G.C. Marks, and M.H. Fitzgerald, 2003, Micronesian banana, taro, and other foods: Newly recognized sources of provitamin A and others carotenoids, *J. Food Composition Anal.*, 16, 3–19. With permission.

Prebreeding initiated at Embrapa to identify accessions with functional properties has identified 80 accessions that have been characterized for the presence of carotenoids, flavonoids, polyphenols, vitamin C, and antioxidant activity. The active germplasm bank (AGB) of banana maintained by Embrapa currently contains 276 accessions of different genomic groups (Figure 13.1). Amorim et al. (2007) estimated the amount of functional foods of 11 genotypes. The mean content of total polyphenols was 46.68 mg 100 g^{-1}, and ranged from 21.58 mg 100 g^{-1} in the triploid 'Ambei' to 120.97 mg 100 g^{-1} for the diploid 'Khai.' For vitamin C, the mean content was 38.46 mg 100 g^{-1}, ranging from 20.76 mg 100 g^{-1} for 'Ambei' to 54.20 mg 100 g^{-1} for diploid 'Lidi.' The mean total carotenoid content was 4.39 μg g^{-1} and the highest mean was observed for diploid 'Khai' with 9.02 μg g^{-1} (Figure 13.2).

Cohen et al. (2009a, b) quantified the flavonoid and polyphenol concentrations in 26 diploid banana accessions at Embrapa (Table 13.3). The mean total polyphenol content was 37.06 mg 100 g^{-1}, ranging from 12.84 mg 100 g^{-1} for 'Thu Mambee' to 120.97 mg 100 g^{-1} for 'Khai.' The mean flavonoid content was 2.58 mg 100 g^{-1}, ranging from 1.29 mg 100 g^{-1} for 'NBA 14' to 4.68 mg 100 g^{-1} for 'Pipit' (Figure 13.3). Cohen et al. (2009a, b) also determined the antioxidant activity of three banana accessions rich in vitamin C, carotenoids, and polyphenols (Table 13.4). The antioxidant activity of tetraploid 'Teparod' (ABBB) was considerably higher (119.19 μM trolox g^{-1}) than those of 'Khai' (6.32 μM trolox g^{-1}) and 'F_3P_4' (6.44 μM trolox g^{-1}). Mattos et al. (2010) characterized 26 banana accessions of the Embrapa AGB for agronomic, physical, and physical-chemical characteristics and by microsatellite molecular markers. Wide variability was observed for most agronomic traits, particularly for fruit number and bunch weight. In the physical and physical-chemical assessments

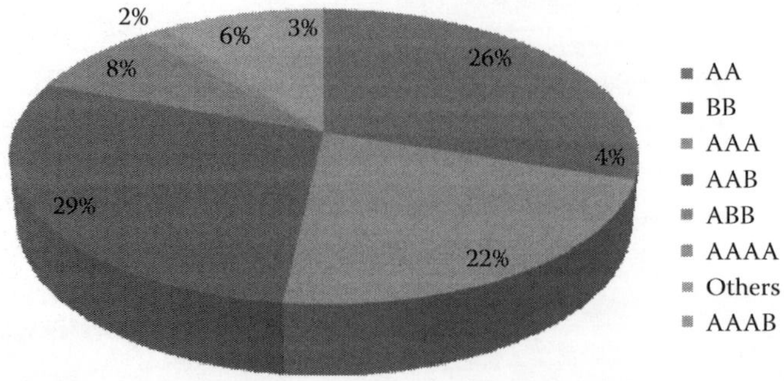

FIGURE 13.1 Frequency of banana genomic groups in the germplasm collection of Embrapa Cassava and Tropical Fruits, Cruz das Almas, Bahia, Brazil, 2010.

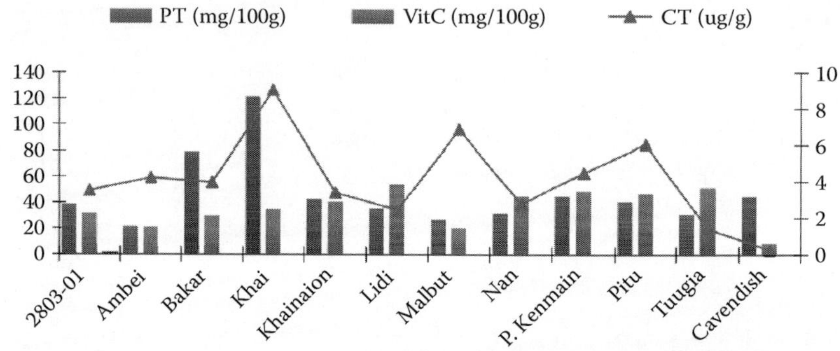

FIGURE 13.2 Polyphenols (PT), vitamin C (Vit C), and carotenoids in 12 banana genotypes, including diploids (AA) and triploids (AAA) of the germplasm collection of Embrapa Cassava and Tropical Fruits, Cruz das Almas, Bahia, Brazil, 2010.

TABLE 13.3
Banana Genotypes of Germplasm Collection of the Embrapa Cassava and Tropical Fruits and Their Respective Origins (Cruz das Almas, Bahia, Brazil, 2010)

Accessions	Ploidy	Origin	Accessions	Ploidy	Origin
2803–01	AA	Brazil	Mambee Thu	AA	New Guinea
Berlin	AA	Indonesia	Modok Gier	AA	New Guinea
F3P4	AA	Ecuador	NBA 14	AA	New Guinea
Jambi	AA	Indonesia	Ouro	AA	Brazil
Jaran	AA	Indonesia	P. Kenmain	AA	New Guinea
Jari Buaya	AA	Honduras	Pa Phatalung	AA	Thailand
Khai	AA	Thailand	Pipit	AA	Indonesia
Khai Nai On	AA	Thailand	P. Nangka	AA	Thailand
Khi Maeo	AA	Thailand	Pitu	AA	Indonesia
Lidi	AA	Honduras	SN2	AA	New Guinea
M-48	AA	Ecuador	Sowmuk	AA	New Guinea
M-61	AA	Ecuador	Tongat	AA	Honduras
Malbut	AA	New Guinea	Tuugia	AA	Hawaii

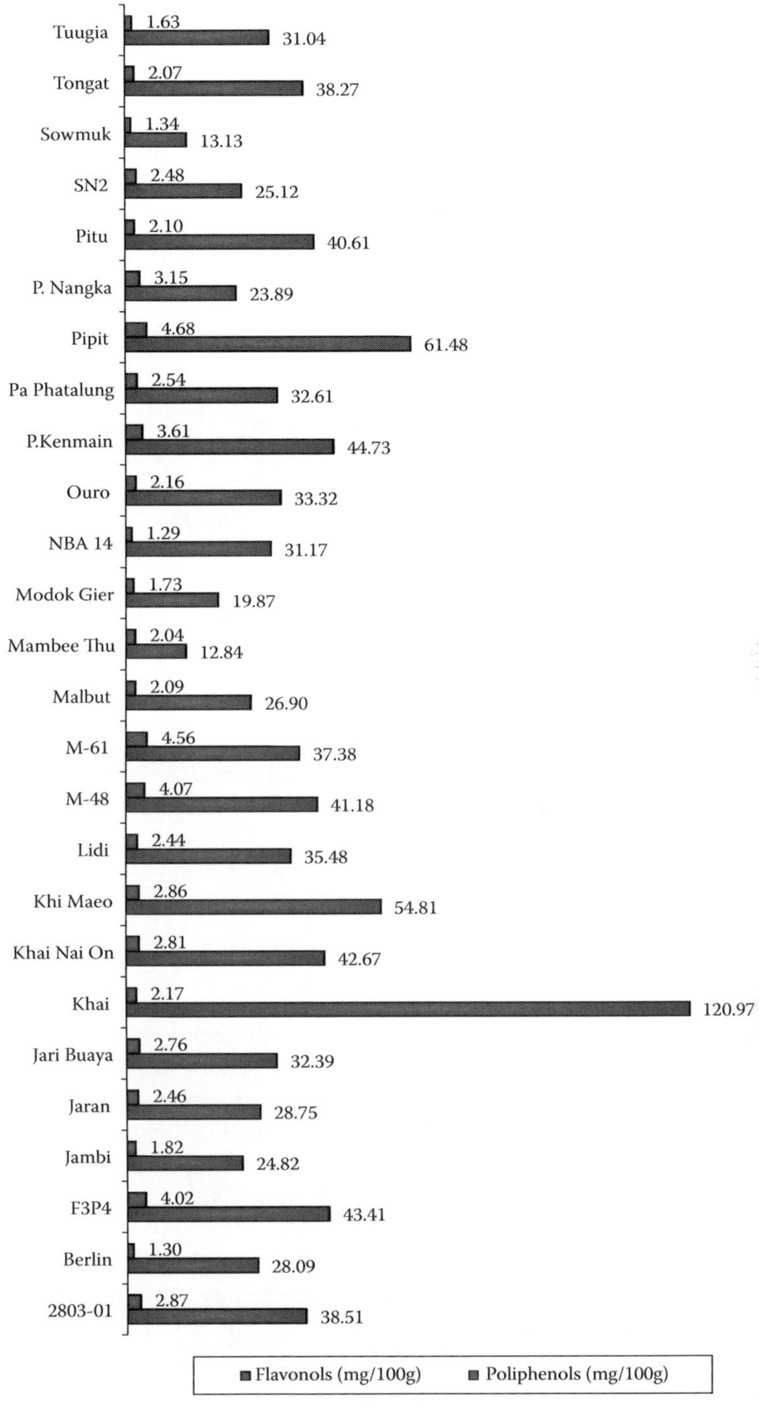

FIGURE 13.3 Flavonols and polyphenols in 26 diploids of the Embrapa Cassava and Tropical Fruits and their respective origins, Cruz das Almas, Bahia, Brazil, 2010.

TABLE 13.4
Functional Compounds in Banana Diploids of the Embrapa Cassava and Tropical Fruits and Their Respective Origins (Cruz das Almas, Bahia, Brazil, 2010)

Genotype	Ploidy	Origin	Vit. C	CT	PET	FLAV	AAT
Teparod	ABBB	Indonesia	76.82a	1.44b	257.80a	6.63a	119.19
Khai	AA	Thailand	34.51b	9.02a	120.97b	2.17b	6.32
F3P4	AA	Equador	17.85c	2.52b	43.41c	4.02a	6.44

Note: Vit. C: vitamin C (mg 100 g^{-1}); CT: carotenoids (μg g^{-1}); PET: polyphenols (mg 100 g^{-1}); FLAV: flavonols (mg 100 g^{-1}); AAT: antioxidant activities (μM trolox g^{-1}). Averages followed by the same letter in the columns belong to the same group by the Tukey test at 5% probability.

of the fruits, accessions with high concentrations of carotenoids (diploid 'Jaran'), polyphenols (triploid 'Caipira' and tetraploid 'Teparod') and vitamin C (diploid 'Tuu gia' and an unknown triploid AAA) were identified (Table 13.5). Thirteen microsatellite primers were used to assess diversity detecting a mean of 7.23 alleles. Using the mean genetic divergence as cut-off point, three groups were formed: G1 with the diploids 'Jaran,' '028003-01,' and 'M-48'; G2 containing the diploids 'Malbut' and 'Ido 110'; and G3 with 21 tri-and tetraploid accessions, including 'Tuu gia' a diploid. The triploids with the B genome, 'Thap Maeo,' 'Walha,' 'Pacha Nadan,' and 'Champa Madras' were grouped in G2 (Figure 13.4). Segregating populations for carotenoids and other functional compounds are being developed at Embrapa with the aim of analyzing the genetic control of these characteristics and identifying markers linked to the genes responsible for metabolic pathways.

13.3 BREEDING OBJECTIVES FOR QUALITY IMPROVEMENT

The goal of conventional banana breeding is to develop hybrids with high yield, short stature, short cycle, and resistance to abiotic (water stress, cold, and so forth) and biotic factors (such as pests, yellow and black Sigatoka, Panama disease). The specific features of breeding for biofortification are presented and discussed in this section.

13.3.1 Biofortification

Biofortification is a new approach that combines conventional breeding with modern tools of biotechnology with the aim of increasing the micronutrient concentrations in staple crops (Zimmermann and Hurrell, 2002). Biofortification represents a great opportunity to improve the health status of low-income populations in rural and urban areas of developing countries (Bouis et al., 2003). According to the World Health Organization (WHO), the micronutrients considered critical to human health are iron (Fe), zinc (Zn), and vitamin A. To minimize the deficiency of these elements, biofortification needs to become a multidisciplinary endeavor with public health being one of the targets of agricultural research.

Research aimed at biofortification has been carried out primarily with cassava (*Manihot esculenta*), rice (*Oryza sativa*), maize (*Zea mays*), sweet potato (*Ipomoea batatas*), and common bean (*Phaseolus vulgaris*) (Welch, 2003). These crops were designated as "Phase I crops" for the HarvestPlus Challenge Program coordinated by the Consultative Group for International Agricultural Research (CGIAR), with the participation of more than 70 scientists from 46 research institutions worldwide, including Embrapa. Other crops designated as "Phase II crops" included banana (*Musa* spp.), barley (*Hordeum vulgare*), cowpea (*Vigna unguiculata*), lentil (*Lens culinaris*), pearl millet (*Pennisetum glaucum*), pea (*Cajanus cajan*), potato (*Solanum tuberosum*), sorghum (*Sorghum bicolor*), and yam (*Dioscorea spp.*).

TABLE 13.5
Average of the 10 Agronomical and Physical-Chemical Characteristics Evaluated in 26 Banana Accessions (Cruz das Almas, Bahia, Brazil, 2010)

Accessions	Ploidy	ALP	DPC	NPC	NFR	PSC	SIF	CTN	FLA	PLF	VIT
Jaran	AA	2.83b	15.73d	8.00b	148.00a	3.20d	2.35a	8.23a	8.23a	28.76m	17.61k
2803–01	AA	1.76d	9.75f	5.00c	67.00b	3.30d	0.71e	3.53e	3.53e	38.51h	31.52f
Malbut	AA	2.55c	15.00d	6.00c	64.00b	2.93d	2.12a	6.88b	6.88b	26.90n	20.42j
Idu-110	AA	2.33c	9.67f	7.00b	87.00b	3.30d	0.71e	2.86f	2.86f	40.96g	20.10j
Tuugia	AA	2.53c	12.67e	6.00c	59.00b	2.09d	0.71e	1.41g	1.41g	31.051	51.10c
M-48	AA	2.75b	14.67d	6.00c	84.00b	4.57d	0.71e	3.52e	3.52e	41.18g	9.03n
Pipit	AAA	2.21c	12.00e	5.00c	92.00b	3.80d	1.14d	2.95f	2.95f	61.48e	15.721
Caru Roxo	AAA	3.33a	21.00b	5.00c	64.00b	7.00c	1.58c	5.91c	5.91c	33.32j	24.63i
Wasolay	AAA	2.64c	13.33d	5.00c	51.00b	3.03d	0.88e	3.15e	3.15e	17.51p	14.721
Markatooa	AAA	2.45c	17.50c	5.00c	63.00b	4.75d	1.73c	2.29f	2.29f	16.23q	14.411
Bakar	AAA	3.33a	18.00c	6.00c	79.00b	9.80c	1.22d	3.99e	3.99e	79.14c	29.43g
AAA Desc.	AAA	2.90b	16.50d	6.00c	57.00b	6.20c	1.55c	2.45f	2.45f	35.48i	54.20b
Nam	AAA	2.23c	16.50d	6.00c	87.00b	4.33d	1.87b	2.77f	2.77f	31.86k	44.67d
Towoolle	AAA	2.20c	14.50d	4.00c	42.00b	2.85d	1.40c	2.33f	2.33f	12.84s	10.87m
Caipira	AAA	2.46c	17.17c	7.00b	132.00a	9.67c	0.71e	1.05g	1.05g	146.31b	11.48m
Thap Maeo	AAB	3.43a	20.60b	10.00a	158.00a	15.03b	0.71e	3.78e	3.78e	15.71q	37.21e
Walha	AAB	1.44d	14.67b	4.00c	30.00b	1.71d	1.90b	2.52f	2.52f	43.41f	17.85k
P. Nadan	AAB	3.47a	18.00c	7.00b	88.00b	8.47c	1.58c	5.83c	5.83c	64.90d	26.85h
C. Madras	ABB	3.49a	21.50b	7.00b	94.00b	12.90b	0.71e	3.34e	3.34e	27.41n	12.45m
Ambrosia	AAAA	3.54a	24.42a	9.00a	154.00a	21.26a	0.71e	1.39g	1.39g	27.52n	11.60m
Calipso	AAAA	3.15a	24.56a	8.00b	138.00a	18.62a	0.71e	1.40g	1.40g	27.12n	9.49n
Tropical	AAAB	2.76b	20.40b	6.00c	92.00b	9.96c	0.71e	0.98g	0.98g	14.83r	14.681
Maravilha	AAAB	2.66c	20.60b	5.00c	53.00b	6.74c	1.58c	1.89g	1.89g	16.03q	9.66n
Porp	AAAB	2.85b	20.25b	5.00c	51.00b	5.95c	1.22d	2.34f	2.34f	14.77r	13.651
O. da Mata	AAAB	3.22a	20.67b	5.00c	78.00b	6.75c	1.29c	4.70d	4.70d	24.56o	19.45j
Teparod	ABBB	2.93b	18.00c	6.00c	37.00b	3.87d	0.71e	1.44g	1.44g	257.80a	76.83a

(continued)

TABLE 13.5 (CONTINUED)
Average of the 10 Agronomical and Physical-Chemical Characteristics Evaluated in 26 Banana Accessions (Cruz das Almas, Bahia, Brazil, 2010)

Accessions	Ploidy	ALP	DPC	NPC	NFR	PSC	SIF	CTN	FLA	PLF	VIT
F (Trat.)		12.30*	22.05*	4.59*	5.97*	14.93*	22.80*	40.79*	40.79*	34877.64*	1161.36*
CV (%)		9.77	9.12	21.53	32.76	33.77	17.07	12.72	12.72	0.86	2.90
Mean		2.79	17.76	6.00	83.00	7.78	1.08	3.19	3.19	45.31	23.82

Note: PH: Plant height (cm); PD: pseudostem diameter (cm); NS: number of suckers; NH: number of hands; NF: number of fruits, BW: bunch weight (kg); YSF: yellow Sigatoka during flowering; CTN: carotenoids (μg.g^{-1}); FLA: flavonoids (mg 100 g^{-1}); PLF: polyphenols (mg 100 g^{-1}); VIT: vitamin C (mg 100 g^{-1}). * significant at 5%; ^{ns}: not significant. Averages followed by the same letter in the columns belong to the same group by the Scott and Knott (1974) test at 5% probability.

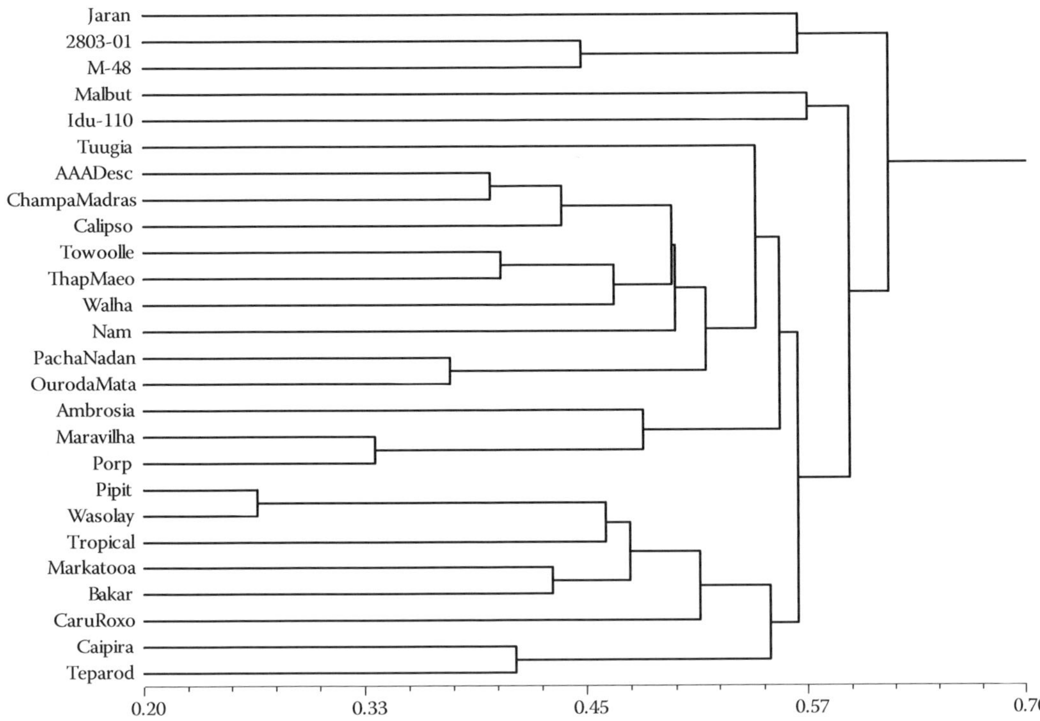

FIGURE 13.4 Genetic diversity between six banana accessions from the Germplasm Bank at Embrapa Cassava and Tropical Fruits integrating agronomical, physical, and physical-chemical data from fruits and molecular data using the Gower (Gower, 1971) algorithm. Cruz das Almas, Bahia, Brazil, 2010.

The success of biofortification depends, in part, on target concentrations of micronutrients in crops that should be set by nutritionists who understand the complexity associated with the expression of these elements and who can quantify their impact on human health. Information on the retention time of nutrients after storage, appropriate processing methods and forms of food preparation, as well as the daily requirements need to be taken into consideration when breeding new biofortified cultivars (Nestel et al., 2006).

Another aspect to be considered is the appearance of the food, in the form of grains, tubers, or fruits. Foods that are rich in provitamin A tend to have an orange color (Krinsky, 1998; Russell, 1998). Food color plays an important role for the consumer, who often chooses a particular food due to its appearance. In general, housewives prefer, for example, white wheat, maize, and cassava flour as well as white or cream-colored banana fruit. On the other hand, micronutrient-rich biofortified cultivars, except in the case of carotenoids, are visually not different from traditional varieties. This may be an obstacle, since it is not possible to identify the biofortified cultivar visually. Before the adoption of a biofortified product, farmers and consumers must be convinced of the benefits of this new variety. Therefore, studies that quantify the benefits of consuming this modified food are required for marketing programs about the benefits of biofortified foods.

13.3.2 Crop Improvement

In conventional breeding, the target characteristics are of direct benefits to farmers, and include pest resistance, drought tolerance, high productivity, and so forth. Breeding for improved nutritional status, however, may in many cases not benefit the growers directly. Consequently this breeding objective has been largely ignored by breeders when selecting promising genotypes.

Breeding for improved nutritional qualities has primarily focused on the exploration and quantification of natural variability for different micronutrients (Bouis et al., 2003). At the same time (or during subsequent selections), agronomic evaluations should be performed. If there is sufficient variability, breeders can exploit the additive effects, heterosis, and transgressive segregation to increase micronutrient concentrations.

The genetic variability can be used in: (1) identification of parents for hybridization, (2) genetic studies, (3) development of molecular markers, and (4) direct use of the genotypes. The next step involves the development and evaluation of new biofortified cultivars/hybrids, based on genetic studies and development of markers for assisted selection. The quantification of environmental effects (genotype by environment [GxE] interaction) on the expression of micronutrient contents must be performed under field conditions in different environments and years.

13.3.3 Genetics

Knowledge of the heritability and expected genetic gain ($G_s = i\sigma_p h^2$) where G_s is the genetic gain from selection, i is a constant based on selection intensity, σ_p is the standard deviation of the phenotypic variance, and h^2 is the heritability is crucial for the choice of adequate breeding methods and appropriate strategies to quantify the GxE interaction. Research results of different crops have indicated that Fe and Zn concentrations are controlled by two to five genes, and that heritability is low (Philip and Maloo, 1996; Maloo et al., 1998; Long et al., 2004; Cichy et al., 2005). Moreover, these two micronutrients are strongly correlated, which facilitates simultaneous improvement of these traits.

Provitamin A is apparently controlled by few genes (~2), and heritability is high (Egesel et al., 2003; Gruneberg et al., 2005). For both minerals (Fe and Zn) and provitamin A, the additive gene effects and general combining ability are high. Transgressive segregation is also found for provitamin A.

13.3.4 GxE Interaction

The expression of a particular trait and the extent of GxE interaction in different environments determine the method of selection, breeding, and evaluation, and can influence results in terms of heritability, genetic variation, and even the genetic gain.

Research results indicate that the expression of provitamin A is relatively stable, independent of the environment (Egesel et al., 2003; Menkir and Maziya-Dixon, 2004).

The expression of Zn and Fe is more variable due to environmental variations. In the case of Fe, the environmental influence on the trait expression is particularly high.

Micronutrient-rich germplasm should be selected using more stringent criteria, due to the variation observed for these elements between plants of the same genotype. The use of controls, a greater number of replications, and statistical designs that minimize the effects of soil variation (microenvironments) are therefore recommended. Large plots are more suitable to sample the genetic variation and allow a comparison of results between different experiments. In addition, results of experiments installed at different weather stations may also vary, and these temporal and spatial variations hamper the selection of germplasm for higher micronutrient contents.

13.3.5 Breeding for High Yield and Micronutrient Density

Breeding for characteristics that are not productivity related often results in a low progress rate for yield traits. When resources are not limiting, breeding can substantially increase both σ_p and selection intensity i ($G_s = i\sigma_p h^2$) and provide genetic gain for productivity without affecting the selection of micronutrient-rich genotypes.

Another strategy to increase the genetic gain over breeding generations is to breed and test micronutrient-rich genotypes in controlled environments (greenhouse) to simulate the climate and soil conditions for which these new genotypes will be recommended. By this practice the heritability and correlation between selection and environmental variation can be increased and a greater number of plants can be selected in little space and in less time. In addition, molecular markers can be used for assisted selection. Promising results of this strategy have been reported in the literature (Guzmán-Maldonado et al., 2003, Wong et al., 2004).

13.3.6 Strategies for Agronomic Superiority

The increase in productivity of a crop may be due to: (1) genetic gain in yield potential, (2) genetic gain in tolerance or resistance to biotic and abiotic factors, (3) genetic gains through management practices, and (4) synergistic effects of all the above factors. Indirect selection for tolerance to biotic and abiotic factors has led to yield gains in banana. The development of hybrids resistant to black Sigatoka and Panama disease in the breeding program of Embrapa Cassava and Tropical Fruits can be cited as an example (Amorim et al., 2009b).

Pleiotropic effects associated with high micronutrient contents and yield performance have been reported in the literature. Cakmak et al. (1990) identified a strong relationship between Zn content in wheat seed and seedling vigor, resulting in greater uniformity of the plantation and, consequently, higher productivity. Higher economic returns through tiered pricing for higher-quality fruits that address specific market niches can motivate the adoption of biofortified banana cultivars by farmers. Additionally, federal programs that encourage the consumption of biofortified banana cultivars in public schools have led to an increased demand for banana.

13.4 CONCLUSIONS

Biofortification of banana, especially in terms of provitamin A, Fe, and Zn is possible and can potentially minimize health problems caused by the deficiency of micronutrients in the world's low-income populations. The accessibility of the fruit and the facts that it is a staple food of millions of people, especially in Africa, suggests that banana can play a fundamental role as a health-promoting food. There is enough genetic variability for micronutrients, making genetic improvement and biofortification of banana possible.

REFERENCES

Adão, R.C. and M. Beatriz A. Glória. 2005. Bioactive amines and carbohydrate changes during ripening of 'Prata' banana (*Musa acuminata* × *M. balbisiana*). *Food Chem.* 90:705–711

Adeniji, T.A., L.O. Sanni, L.S. Barimalaa, and A.D. Hart. 2006. Determination of micronutrients and colour variability among new plantain and banana hybrids flour. *World J. Chem.* 1:23–27.

Adeniji, T.A., L.O. Sanni, L.S. Barimalaa, and A.D. Hart. 2007. Nutritional composition of five new Nigerian Musa hybrid cultivars: Implications for adoption in human nutrition. *Fruits* 62:135–142.

Amorim, E.P., K.O. Cohen, N.S. Paes, J.A. Santos-Serejo, and S.O. Silva. 2007. Compostos funcionais em genótipos de banana. Comunicado Técnico. *Embrapa Mandioca e Fruticultura* 1:1–4.

Amorim, E.P., K.O. Cohen, V.B.O. Amorim, J.A. Santos-Serejo, S.O. Silva, A.D. Vilarinhos, D.C. Monte, N.S. Paes, and R.V. Reis. 2009a. The genetic diversity of carotenoid-rich bananas measured by Diversity Arrays Technology (DArT). *Genet. Molecul. Biol.* 32:96–103.

Amorim, E.P., L.S. Lessa, C.A.S. Ledo, V.B.O. Amorim, R.V. Reis, J.A. Santos-Serejo, and S.O. Silva. 2009b. Caracterização agronômica e molecular de genótipos diplóides melhorados de bananeira. *Revista Brasileira de Fruticultura* 31:154–161.

Arntzen, C., H. Mason, R. Mahoney, and D. Kirk. 2004. Oral vaccines derived from transgenic plants. In: *Vaccines: Preventing disease and protecting health*, C. de Quadros, ed., 256–262. Scientific Publication No. 596. Washington, DC: Pan American Health Organization.

Arora, A., D. Choudary, G. Agarwal, and V.P. Singh. 2008a. Compositional variation in β-carotene content, carbohydrate and antioxidant enzymes in selected banana cultivars. *Int. J. Food Sci. Technol.* 43:1913–1921.

Arora, A., D. Choudhary, G. Agarwal, and P.S. Virender. 2008b. Compositional variation in β-carotene content, carbohydrate and antioxidant enzymes in selected banana cultivars. *Int. J. Food Sci. Technol.* 43:1913–1921.

Bakry, F., F. Carreel, C. Jenny, and J.P. Horry. 2009. Genetic improvement of banana. In: *Breeding plantations tree crops*, S.M. Jain and P.M. Priyadarshan, eds., 3–50. New York: Springer Science+Business Media.

Bloem, M.W., S. de Pee, and I. Darnton-Hill. 2005. Micronutrient deficiencies and maternal thinness: First link in the chain of nutritional and health events in economic crises. In: *Primary and secondary nutrition*, 2nd ed., A. Bendich and R.J. Deckelbaum, eds., 357–373. Totowa, NJ: Humana Press.

Bouis, H.E., S. Harris, and D. Lineback. 2003. Biotechnology-derived nutritious foods for developing countries: Needs, opportunities, and barriers. *Food Nutr. Bull.* 23:342–383.

Cakmak, I., M. Kalayc, H. Ekiz, H.J. Braun, Y. Kılınc, and Y. Yılmaz. 1990. Zinc deficiency as a practical problem in plant and human nutrition in Turkey: A NATO science-for-stability project. *Field Crops Res.* 60:175–188.

Cichy, K.A., K.F. Forster, K.F. Grafton, and G.L. Hosfield. 2005. Inheritance of seed zinc accumulation in navy bean. *Crop Sci.* 45:864–870.

Cohen, K.O., E.P. Amorim, N.S. Paes, S.O. Silva, J.A. Santos-Serejo, A.M. Costa, H.N. Souza, and M.V. Baiocchi. 2009a. Teores de flavonóides e polifenóis totais em genótipos diplóides de bananeira. Comunicado Técnico. *Embrapa Mandioca e Fruticultura* 1:1–4.

Cohen, K.O., E.P. Amorim, N.S. Paes, S.O. Silva, M.V. Baiocchi, and H.N. Souza. 2009b. Vitamina C, carotenóides, compostos fenólicos e atividade antioxidante em genótipos de banana. Comunicado Técnico. *Embrapa Mandioca e Fruticultura* 2:1–4.

Davey, M.W., W. Stals, G. Ngoh-Newilah, K. Tomekpe, C. Lutsy, R. Markham, R. Swennen, and J. Keulemans. 2007. Sampling strategies and variability in fruit pulp micronutrient contents of West and Central African bananas and plantains (*Musa* sp.). *J. Agric. Food Chem.* 55:2633–2644.

Davey, M.W., I. Van den Berg, R. Markham, R. Swennen, and J. Keulemans. 2009. Genetic variability in *Musa* fruit provitamin A carotenoids, lutein and mineral micronutrient contents. *Food Chem.* 115:806–813.

Egesel, C.O., J.C. Wong, R.J. Lambert, and T.R. Rocheford. 2003. Combining ability of maize inbreds for carotenoids and tocophenols. *Crop Sci.* 43:818–823.

Englberger, L., I. Darnton-Hill, T. Coyne, M.H. Fitzgerald, and G.C. Marks. 2003a. Carotenoid-rich bananas: A potential food source for alleviating vitamin A deficiency. *Food Nutr. Bull.* 24:303–318.

Englberger, L., J. Schierle, G.C. Marks, and M.H. Fitzgerald. 2003b. Micronesian banana, taro, and other foods: Newly recognized sources of provitamin A and other carotenoids. *J. Food Composition Anal.* 16:3–19.

Englberger, L., W. Aalbersberg, P. Ravi, E. Bonnin, G.C. Marks, M.H. Fitzgerald, and J. Elymore. 2003c. Further analyses on Micronesian banana, taro, breadfruit and other foods for provitamin A carotenoids and minerals. *J. Food Composition Anal.* 16:219–236.

Food and Agriculture Organization of the United Nations (FAO). 1997. *Preventing micronutrient malnutrition: A guide to food-based approaches.* Washington, DC: International Life Science Institute.

Food and Agriculture Organization of the United Nations (FAO). 2010. www.faostat.fao.org/site/340/default.aspx (accessed 20 January 2010).

García-Alonso, M., S. Pascual-Teresa, C. Santos-Buelga, and J.C. Rivas-Gonzalo. 2004. Evaluation of the antioxidant properties of fruits. *Food Chem.* 84:13–18.

González-Montelong, R., M.G. Lobo, and M. González. 2010. Antioxidant activity in banana extracts: Testing extraction conditions and related bioactive compounds. *Food Chem.* 119:1030–1039.

Gower, J.C. 1971. A general coefficient of similarity and some of its properties. *Biometrics* 27:857–871.

Gruneberg, W.J., K. Manrique, D. Zhang, and M. Hermann. 2005. Genotype x environment interactions for a diverse set of sweet-potato clones evaluated across varying ecogeographic conditions in Peru. *Crop Sci.* 45:2160–2171.

Guzmán-Maldonado, S.H., O. Martínez, J.A. Acosta-Gallegos, F. Guevara-Lara, and O. Paredes-López. 2003. Putative quantitative trait loci for physical and chemical components of common bean. *Crop Sci.* 43:1029–1035.

Hellenas, K.E., C. Branzell, H. Johnsson and P. Slanina. 1995. High levels of glycoalkaloids in the established Swedish potato variety *Magnum bonum*. *J. Sci. Food Agric.* 68:249–255.

Jones, D.R. 2000. Introduction to banana, abaca and enset. In: *Diseases of banana, abaca and enset*, D.R. Jones, ed., 1–36. Wallingford, UK: CABI Publishing,.

Kanazawa, K. and H. Sakakibara. 2000. High content of dopamine, a strong antioxidant, in Cavendish banana. *J. Agric. Food Chem.* 48:844–848.

Koch, B., V. Nielsen, B. Halkier, C. Olsen, and B. Moller. 1992. The biosynthesis of cyanogenic gylcosides in seedlings of cassava (*Manihot esculenta* Crantz). *Arch. Biochem. Biophys.* 292:141–150.

Krinsky, N.I. 1998. The antioxidant and biological properties of the carotenoids. *Ann. NY Acad. Sci.* 854:443–447.

Lee, S.K. and A.A. Kader. 2000. Preharvest and postharvest factors influencing vitamin C content of horticultural crops. *Postharvest Biol. Technol.* 20:207–220.

Long, J.K., M. Banziger, and M.E. Smith. 2004. Diallel analysis of grain iron and zinc density in Southern African-adapted maize inbreds. *Crop Sci.* 44:2019–2026.

Maloo, S.R., J.S. Salanki, and S.P. Sharma. 1998. Genotypic variability for quality traits in finger millet (*Eleusine coracana* (L.) Gaertn). *Int. Sorghum Millets Newslett.* 39:126–128.

Mattos, L.A., E.P. Amorim, V.B.O. Amorim, K.O. Cohen, and S.O. Silva. 2010. Agronomical and molecular characterization of banana accessions. *Pesq. Agropec. Bras.* (in press).

Menkir, A. and B. Maziya-Dixon. 2004. Influence of genotype and environment on β-carotene content of tropical yellow-endosperm maize genotypes. *Maydica* 49:313–318.

Mozafar, A. 1994. *Plant vitamins: Agronomic, physiological and nutritional aspects*. Boca Raton, FL: CRC Press.

Nakasone, H.Y. and R.E. Paull. 1999. *Tropical fruits: Banana*. Wallingford, UK: CABI Publishing.

Nestel, P., H.E. Bouis, J.V. Meenakshi, and W. Pfeifer. 2006. Biofortification of staple food crops. *J. Nutr.* 136:1064–1067.

Pfeiffer, W.H. and B. McClafferty. 2007. Harvestplus: Breeding crops for better nutrition. *Crop Sci.* 47(S3):S38–S115.

Philip, J. and S.R. Maloo. 1996. An evaluation of *Setaria italica* for seed iron content. *Int. Sorghum Millets Newslett.* 37:82–83.

Robinson, J.C. 1996. *Bananas and plantains*. Wallingford: CABI Publishing.

Russell, R.M.1998. Physiological and clinical significance of carotenoids. *Int. J. Vitam. Nutrient Resour.* 68:349–353.

Sala, F., M.M. Rigano, A. Barbante, B. Basso, A.M. Walmsley, and S. Castiglione. 2003. Vaccine antigen production in transgenic plants: Strategies, gene constructs and perspectives. *Vaccine* 21:803–808.

Scott, A. J. and M.A. Knott. 1974. A cluster analysis method for grouping means in the analysis of variance. *Biometrics* 30: 507–512.

Sharrock, S. and C. Lutsy. 1999. Nutritive value of banana. In: *INIBAP annual report*. Montpellier, France: INIBAP.

Someya, S., Y. Yumiko, and K. Okubob. 2002. Antioxidant compounds from bananas (*Musa Cavendish*). *Food Chem.* 79:351–354.

Sunil Kumar, G.B., T.R. Ganapathi, C.J. Revathi, L. Srinivas, and V.A. Bapat. 2005. Expression of hepatitis B surface antigen in transgenic banana plants. *Planta* 222:484–493.

Vijayakumar, S., G. Presannakumar, and N.R. Vijayalakshmi. 2008. Antioxidant activity of banana flavonoids. *Fitoterapica* 79:279–282.

Wall, M.M. 2006. Ascorbic acid, vitamin A, and mineral composition of banana (*Musa* sp.) and papaya (*Carica papaya*) cultivars grown in Hawaii. *J. Food Composition Anal.* 19:434–445.

Welch, R.M. 2003. Farming for nutritious foods: Agricultural technologies for improved human health. Paper presented at the IFA-FAO Agriculture Conference on Global Food Security and the Role of Sustainable Fertilization, Rome, Italy, 26–28 March 2003.

White, W., J. McMahon, and R. Sayre. 1994. Regulation of cyanogenesis in cassava. *Acta. Hort.* 375:69–77.

Whittaker, P. 1998. Iron and zinc interactions in humans. *Am. J. Clin. Nutr.* 68:442S–446S.

Wong, J.C., R.J. Lambert, E.T. Wurtzel, and T.R. Rocheford. 2004. QTL and candidate genes phytoene synthase and *β*-carotene desaturase associated with the accumulation of carotenoids in maize. *Theor. Appl. Genet* 108:349–359.

Zimmermann, M.B. and R.F. Hurrell. 2002. Improving iron, zinc and vitamin A nutrition through plant biotechnology. *Curr. Opin. Biotechnol.* 13:142–145

14 Postharvest Processed Products from Banana

Cherukatu Kalathil Narayana and Michael Pillay

CONTENTS

14.1 INTRODUCTION

Bananas and plantains are staple food crops grown in over 130 countries in the tropics and subtropics. They are considered to be one of the most important sources of energy in the diet of millions of people in Africa, the Caribbean, Latin America, Asia, and the Pacific (Jones, 2000). Most of the bananas produced in many countries such as India, Uganda, Brazil, and China are consumed locally, with only a small percentage being exported (Pillay and Tripathi, 2007). Bananas and plantains are consumed in various ways in different countries. Traditionally dessert bananas are consumed raw in the ripe stage while plantains and cooking bananas are cooked as part of the staple diet, or for processing into longer lasting products such as flour (Wainwright and Burdon, 1991; Dadzie, 1995). The need for processed banana products has not been high in many countries since fresh bananas are readily available throughout the year (Sole, 1996, 2005). The development of processed products from bananas has been slow compared with that of other crops. Processing of bananas has been stimulated by the high volumes of bananas that are lost after harvest, especially those that are rejected for the export market in the main banana-growing countries of South America. The climatic conditions of the tropics that favor banana production also play a major role in increasing postharvest losses. The optimum storage temperature for green bananas and plantains is 13–14°C, at which temperature ripe bananas can be stored for 1 to 2 weeks (Chia and Huggins, 2003). Storage facilities are not available in many banana-producing countries to guarantee a longer shelf life (Wills et al., 1989). The exact quantity of postharvest banana losses in different countries is not known but in general losses of fruits and vegetables in the tropics are reported to be about 40–50% (Mejia, 2003). One report estimated that 40% of bananas produced in Brazil are lost after harvest (Agrianual, 2003). Postharvest banana losses vary from country to country according to market chains and modes of consumption (Tchango Tchango et al., 1999). The marketing chain in many developing countries is not well organized, and farmers incur losses in many ways, including: the bulky nature of bananas makes transportation and handling costs from the rural to urban areas very high and unaffordable for many small-scale farmers, small bunches are considered below market size and are rejected, the shelf life of both green and ripe banana is limited, and there is a lack of appropriate technology, as well as insufficient or scarce access to information (Pekke et al., 2004; Mukhtar, 2009). The shelf life of green banana is shortened further by black Sigatoka (*Mycosphaerella fijiensis*), a widespread disease of the crop (Chillet et al., 2008).

Dehydration is one of the oldest methods used to preserve agricultural products (Adams, 2004). Drying prevents or reduces the growth of microorganisms responsible for the decay of food. Advances in dehydration techniques and development of novel drying methods have in recent years enabled the preparation of a wide range of dehydrated products and convenience foods from fruits and vegetables meeting the quality, stability, and functional requirements coupled with economy (Jayaraman and Das Gupta, 1992).

Dehydration is traditionally used in West Africa (Dadzie and Wainwright, 1995) and to a limited extent in East Africa (Aked and Kyamuhagire, 1996) to preserve unripe banana. In addition to drying and preparation of banana flour, banana is now being processed into several products that are being consumed and traded throughout the world. For example, banana "figs" produced by sun drying of ripe bananas are one of the oldest processed food products. Other popular processed banana products are dehydrated flakes and banana powder. Brazil, Honduras, and Mexico were the main countries exporting dried, desiccated, or evaporated bananas (Sole, 1996). Banana puree or mashed bananas are at present the banana product with the highest volume processed in the world. The Philippines have emerged as one of the important countries exporting large quantities of processed banana products with an exported volume of 16,964 tons of banana chips/crackers valued at US$18.7 million in 1998 (hvcc.da.gov.ph/pdf/banana_phil_prodn_market.pdf, accessed 14 May 2010). The Philippines also exported 1,474 tons of banana catsup/ketchup valued at US$1.3 million in the same year. The biggest importers are the United States, capturing almost half of the total export volume, followed by Canada, Saudi Arabia, and the United Arab Emirates. Over 2.2 tons of banana flour, meal, and powder valued at US$14,771 were exported solely to Japan in 1998 (Narayana and Sathiamoorthy, 2002). Banana is favorable for industrial processing due to its rich content of soluble solids, minerals, and low acidity (Carvalho et al., 2009a, b). The major processed banana products are discussed fully in Sole (2005). Two products from banana that appear to be ubiquitous in many banana-producing countries are banana flour and banana chips.

14.2 BANANA/PLANTAIN FLOUR

One strategy to increase the utilization of banana and plantain includes the production of flour from ripe and unripe fruit and to incorporate the flour into various innovative products such as slowly digestible cookies (Aparicio-Saguilan et al., 2007), high-fiber bread (Juarez-Garcia et al., 2006), and edible films (Sothornvit and Pitak, 2007). Many methods to prepare flour from unripe banana and plantain have been reported (Oyesile, 1987; Ukhum and Ukpebor 1991; Rodriguez-Ambriz et al., 2008). The flour from unripe banana possesses thickening and cooking properties that are almost identical to that of starch (Suntharalingam and Ravidran, 1993). As a technology to preserve banana for later use, the production of banana flour is widespread in Africa and in countries such as

Bolivia. Large-scale production of banana flour is carried out by peeling and slicing green fruits and exposure to sulfur dioxide gas or dipping in sulfurous acid, and then drying to a moisture content of 8% in a countercurrent tunnel dryer for 7–8 hours with an inlet temperature of 75°C and outlet temperature of 45°C, and finally milling (Sole, 2005). The full three-quarter-maturity stage of fruits contains the correct proportions of starch and sugar that produces good quality flour. If immature fruits are used, the resultant flour tastes bitter and astringent due to a high tannin content. It was shown that the different dehydration methods used to produce flour significantly affect its proximate composition and physical characteristics (Pacheco-Delahaye et al., 2008).

In the traditional method, the banana/plantain is peeled, cut into small pieces, and sun dried for a few days. Nowadays, new miniature solar dryers that are more hygienic have been developed in many low-income countries. The dried pulp is then ground into a powder in a mortar and pestle. Flour produced in this manner is usually brown due to the effect of enzymes such as polyphenol oxidase and peroxidase or other nonenzymatic reactions (Cano et al., 1997). It is known that bananas undergo rapid browning as a result of tissue disruption and exposure to oxygen during peeling and slicing operations (Cano et al., 1997). Enzymatic browning of banana can cause undesirable changes in quality during handling, processing, and storing of the fruit. Various methods have been used to try and reduce the browning of processed banana, including: the addition of sodium bisulphate (Tonaki et al., 1973; Garcia et al., 1985), blanching to inactivate the enzymes (Ngalani, 1989; Cano et al., 1990; Giami, 1991), and mild heat treatment with addition of sodium bisulphate, citric acid, and potassium sorbate. Cano et al. (1997) showed that microwave and steam blanching significantly reduced the enzymes that cause browning in banana, with blanching being more effective. However, blanching had a significant negative effect on the proximate analysis, mineral content, and pasting properties of whole flour prepared from plantain and banana hybrids (Adeniji and Tenkouano, 2008). Steaming has been used to reduce enzymatic discoloration before processing (Hanson, 1976; Suntharalingam and Ravidran, 1993). Steaming made peeling easier, reduced discoloration, improved rehydration, and reduced cooking loss in the East African Highland banana (Muyonga, 2000). Color changes in foods during the drying process are affected by a number of variables such as temperature, moisture content, degree of ripeness, wet-bulb temperature, and air flow (Baini and Langrish, 2009). Bananas were dried in a kiln at dry-bulb temperature of 50–100°C. The bananas were dried continuously and intermittently for 72 h. For continuous drying it was found that browning increased and reached equilibrium. The rate of browning was found to decrease with drying time and moisture content (Baini and Langrish, 2009). The rate of browning for overripe and ripe bananas was higher than those of unripe banana, suggesting that the sugar content may affect browning.

Peeling of banana for flour production is difficult and time consuming. Haslinda et al. (2009) prepared flour from both peeled and unpeeled banana. They observed that flour prepared from unpeeled banana ('Awak,' AAB) had enhanced nutrition values with higher concentrations of minerals, dietary fiber, and total phenolics. This is supported by the work of Izonfuo and Omuaru (1988), who reported that plantain peel is richer in minerals such as potassium, calcium, magnesium, phosphorus, copper, and iron when compared with the pulp and that the concentrations of potassium, calcium and iron increase as the fruit ripens. In addition, flour fortified with peel showed higher antioxidant activity and had better pasting properties than flour without peel. Choo (2007) showed that noodles containing 30% banana ('Awak') flour and oat β-glucan showed very high antioxidant properties and very low glycemic index (GI) and high rate of carbohydrate digestibility. These noodles compared well with control samples in a sensory test. Banana peel and pulp flour was also found to be useful for controlling starch hydrolysis when used in the preparation of yellow noodles (Ramli et al., 2009). Acid treatment of unripe banana flour produced a fiber-rich product that may be important for the development of food and medical products (Aguirez-Cruz et al., 2008).

Banana flour has usually been obtained from unripe banana and plantain. Flour from ripe fruits has the potential to offer new products for industrial and domestic uses (Abbas et al., 2009). Flour from ripe banana containing a small quantity of sugar can be incorporated into food products requiring solubility, sweetness, and high energy content.

Studies have shown that ripe banana flours from different cultivars ('Cavendish,' AAA, and *Musa acuminata* 'Pisang Berangan,' AAA, can be differentiated from each other (Abbas et al., 2009).

Banana flour is used as an adjuvant in several food preparations and baby food formulations. Banana is an important weaning food in Uganda (Bukusuba et al., 2008). Soyamusa, a baby food made from plantain flour (60%), full-fat soybean flour (32%), sucrose (8%), and fortified with 0.15% of multivitamins and 0.85% calcium carbonate, has been made and used in Nigeria (Ogazi et al., 1991; Ogazi, 1996). Banana flour can be blended with other cereal flours for making chapattis (Indian flat bread), biscuits, instant porridge, bread, *papadams*, and extruded foods. Pekke et al. (2004) reported that the inclusion of flours from millet, cassava, and soybean improved the acceptability of banana flour products. The acceptability of porridge made from a combination of banana-millet-soybean flour in the ratio 7:2:1 was rated highly in Uganda. Bukusuba et al. (2008) showed that pregelatinization and extrusion cooking can significantly raise the energy content of banana with significant impact on its viscosity and reduction of its bulkiness. This has implications for weaning foods in Uganda since high rates of malnutrition were reported in banana-consuming areas in the country. In western Nigeria, plantain flour is mixed with an appropriate quantity of boiling water to prepare thick dough that is eaten with vegetable soup (Ogazi, 1998).

Banana flour has been used as an additive in the preparation of bread and whole maize meal. It was shown that 15% plantain flour substitution could be adopted in the bread-making processes, without affecting the quality of the bread that is made exclusively with wheat flour (Olaoye et al., 2006). Similarly it was found that different formulations of banana flour, especially from Cavendish, when added to whole-maize meal imparted improvements in formulated products in terms of mouth feel and binding properties with a significant difference in the color of the base product (Daramola and Osanyinlusi, 2006).

In the 1980s dietary fiber (DF) was identified as an important component of a healthy diet. The interest in foods rich in DF has increased in recent decades and there is a large market for fiber-rich products and ingredients (Juarez-Garcia et al., 2006). Unripe banana represents a good source of indigestible carbohydrate, due to the starch content of the pulp and high cellulose, hemicelluloses, and lignin levels. Bread made with banana flour had significantly higher quantities of resistant starch, dietary fiber, and indigestible fractions than control bread made primarily from wheat flour. Bakery products with banana flour have a low glycemic index and could be used as a dietary aid for people with low caloric requirements (Juarez-Garcia et al., 2006).

Banana flour is produced for export in Ecuador, Colombia, Canada, and Switzerland, generally in small volumes. Confoco S.A. Trobana in Ecuador produces green banana flour for export (Sole, 1996). Commercial banana flour production is not yet common in Asia (Abbas et al., 2009), although this industry is gaining popularity in major banana-growing countries in Africa (Emaga et al., 2008).

14.3 BANANA CHIPS/CRISPS

The production of banana and plantain chips is a widespread activity in many banana-growing countries. Chips are the most popular snack in many fast food outlets in Bangladesh (Molla et al., 2009) and are the most popular plantain product in Nigeria (Onyejegbu and Olorunda, 1995). Chips are an important ingredient in breakfast cereals. The chips are either fried in oil or dried to be processed into flour. Chips can also be coated with sweeteners and then fried. Chips can be preserved for a long time (months, possibly years) with adequate packaging and storage facilities. Several studies have been conducted to determine the best *Musa* cultivars for chip making (Adeniji and Tenkouano, 2007; Molla et al., 2009).

In India, the most popular processed banana product is 'Nendran' chips, which has grown from a household art to a well-established cottage industry in the states of Kerala and Tamil Nadu. In India, banana chips or crisps are made primarily from the cultivar 'Nendran' (AAB) while 'Saba' (ABB) is mainly used in the Philippines. The Philippines also export large quantities of banana chips to the United States and Canada. In Africa, plantains are used for chip production in West Africa while other varieties are used in East Africa. Banana chips/crisps are made by deep-frying raw banana slices of 1.75 to 2.0 mm thickness in a suitable cooking medium and salting them. Coconut oil is the most preferred medium in India, while cottonseed oil or corn oil is used in other countries. Many variations of banana chips exist. Chips are coated with cane sugar and refried to produce sugar-coated banana chips. Another variation is that chips are coated with a mixture of banana puree and sugar before frying the second time. The stage of maturity and use of antioxidants play a very important role in the quality and storability of banana chips.

Plantains are commonly used to prepare chips in Nigeria and Cameroon (Onyejegbu and Olorunda, 1995). Chips made from plantains absorb less frying oil than those produced from cooking and dessert banana (Lemaire et al., 1997). Hydrocolloids such as alginate, carboxyl methyl cellulose, and pectin were shown to decrease the amount of oil absorption in banana chips (Singthong and Thongkaew, 2009). Plantains and certain cooking banana also have the advantage of not turning brown during chip preparation and antioxidizing treatments are not necessary (Lemaire et al., 1997). Badia (1985) reported that the degree of ripeness of plantain is important in chip preparation and that browning occurs if the sugar content is higher than 1% in the fruits. Varieties used for chip preparation have to be selected carefully since genetic differences exist among cultivars in pigment composition and/or pulp browning potential (Tourjee et al., 1998). This was also shown for plantain and plantain-derived hybrids in Nigeria (Adeniji and Tenkouano, 2007). Studies on the effects of systems for contact of banana slices (solid phase) and palm oil during deep frying of banana chips showed that shaking homogenizes the temperature of the oil and enhances mass energy transfer. Exchange of matter and heat took place during the first 3 minutes of cooking. Diaz et al. (1996) observed that chip quality can be improved if the frying temperature is maintained at 167°C. At this temperature, water loss was less than 5 g/100 g of raw material and gain in lipids was less than 16 g/100 g. The optimum cooking time was 200 s with a minimum ratio of energy consumption to evaporation of water.

Suvittawat and Babpraserth (1996) tried to identify the best cultivar for banana chip production in Thailand using nine banana cultivars with different genome compositions. The most preferred cultivar was the tetraploid 'Khai' (unknown genomes). At room temperature (25–30°C) and supermarket temperature (20–25°C) the chips could be stored for up to 1 or 3 months, respectively. Shere et al. (1993) studied the effect of maturity stage on the quality of fried banana chips by using the cultivar 'Basrai,' AAA, Cavendish subgroup. A maturity stage of 100 days after flowering was not ideal for chip preparation because of the high sugar content that caramelized at high temperatures. Bananas harvested after 85 days from flowering produced better chips.

The shelf life of banana and plantain chips depends to a large extent on the type of packaging. Visual appearance, especially color, is the major quality criterion for determining the commercial quality and cost of banana chips (Anand et al., 1982). Chips must be packed in moisture-proof bags to prevent loss of crispness (Stover and Simmonds, 1987). Chips packed in polyethylene bags become rancid with time, due to oxidation and changes in color and taste of the product (Ogazi, 1996). Addition of antioxidants like propylene glycol, propyl gallate, and citric acid has been recommended for reducing the rancidity (Badia, 1985; Ogazi, 1996; Narayana and Mustaffa, 2000). The inclusion of different absorbents inside the polyethylene bag can prolong the shelf life of banana chips by 3–5 days (Goswami and Barua, 1996). These researchers found that inclusion of the ethylene absorbent ($KMnO_4$) together with the carbon dioxide absorber [$Ca(OH)_2$] showed the best results not only in terms of shelf life but in physicochemical properties and incidence of fungal rots. The viability of such a technical method may not be useful to small-scale producers of banana chips.

In Nigeria, cellophane has been recommended as the packing material for plantain chips (Ogazi, 1996). A number of packaging techniques have been outlined by González-Aguilar et al. (2010). Some of these techniques may be suitable for banana products and need further investigation.

Banana chips could play an important role in intervention programs to combat micronutrient deficiencies by virtue of their iron, zinc, and total carotenoid content (Adeniji and Tenkouano, 2007). Chips are a ready-to-eat food and are a favorite among children and adults alike. It was already shown that chips made from some banana varieties could contribute substantially to the recommended daily allowance (RDA) of retinol, iron, and zinc of both children and adults (Adeniji and Tenkouano, 2007). Unpublished data (Fungo and Pillay) showed that banana accessions from Papua New Guinea have very high levels of ß-carotene with values ranging from 205 μg/100 g to 2594 μg/100 g. It would be interesting to produce chips from such varieties and assess the effect of processing on their micronutrients.

14.4 BANANA PUREE/PULP

Banana puree is obtained by pulping peeled, ripe bananas and then preserving the pulp by one of three methods: aseptic canning, acidification followed by normal canning, or quick-freezing (www.vuatkerala.org, accessed 22 April 2010).

Most of the world's banana puree is processed by aseptic canning. Peeled, ripe fruits are conveyed to a pump that forces them through a plate with 8 mm holes, then on to a homogenizer, followed by a centrifugal deaerator and into a receiving tank with 100 kPa vacuum, where removal of air helps in preventing discoloration. The sterilized puree is then packed aseptically into steam-sterilized cans, which are closed in a steam atmosphere (Sole, 1996, 2005). Pretreatments of plantain by blanching in hot water, steam, microwave treatment, or calcium chloride modifies the texture and the organoleptic characters. Calcium chloride and microwave pretreatments hardened frozen banana pulp. Microwave pretreatment seemed to be the most suitable for frozen banana (Giami, 1991). Garcia et al. (1987) conducted a study to reduce the time required to lower the pH of banana puree by lactic acid fermentation. The best production conditions were found to be 37°C and 10% (w/w) of inoculum, under which conditions the pH decreased from 4.8 to 4.3 in approximately 8 hours.

14.5 BANANA POWDER

Banana powder is made from fully ripe banana pulp. Passage through a colloidal mill converts the pulp into a fine paste. The color of the final product is improved by adding approximately 1–2% potassium metabisulfite solution. In addition, the pulp is spray or drum dried and the solids are recovered. In the latter method, the moisture content of the pulp is reduced to 8–12% and then it is further decreased to 2% by drying at 60°C in a tunnel or cabinet dryer. The dry powder is highly hygroscopic and is packed in bag-in-box packages. The bags are made from laminated material with moisture barrier layers. This product has a high market value as it is used in the confectionary industry, making of ice creams, and weaning and baby foods. The edible pulp of the cultivar 'Bhimkol,' ABB, was shown to be a nutritious baby food (Barthakur and Arnold, 1990). The fruit was found to be very rich in macro- and micronutrients, and 100% of a 6-month-old infant's RDA for K, Mn, Zn, and Se could be satisfied with 100 g of fresh pulp. It appears that this cultivar may be useful for the production of banana powder.

14.6 BANANA FIGS AND DEHYDRATED BANANA

Fruit softening is characteristic of fruit ripening, and mature preclimacteric bananas lose their firmness rapidly, even after 4 days of storage (Yang et al., 2008).

The production of banana figs represents another way in which the shelf life of ripe banana can be prolonged. Banana figs are dried or dehydrated banana with a sticky fig-like consistency and very sweet taste. Banana figs were initially produced in the tropics by sun drying, but hot air circulation in tunnel or cabinet dryers is now used. The product is made from fully ripe fruits. Varieties with a high total soluble solids (TSS) content and a sugar content of about 19.5% are highly suitable for making figs. Dried banana is an important food product in Ethiopia (Osman and Asrat, 1985). The dried bananas are produced by treating batches of sliced banana for 5 minutes in a boiling 50% sugar solution and then drying them in the sun in glass-covered black boxes.

In Brazil, 'Nanica,' AAA, Cavendish subgroup, bananas are dried to produce *banana-passa*. *Banana-passa* is obtained by the natural or artificial drying of ripe bananas until they reach a moisture content of 20–25% wet weight (Cano-Chauca et al., 2004). *Banana-passa* is a commercial product of Brazil and is exported in limited quantities to the United States, Germany, France, and the United Kingdom. The commercialized product is often rejected due to microbiological contamination and physical degradation of the product. Therefore various drying procedures were studied to produce the best *banana-passa* (Nogueira and Park, 1992; Cano-Chauca et al., 2004; Leite et al., 2007). The finding of Nogueira and Park (1992) suggested a temperature of 60°C or 70°C and airflow of 1.5 m/s. The best drying time to reach a final moisture content of 23–25% depends on temperature, drying speed, and relative humidity (Cano-Chauca et al., 2004). Large-scale production of *banana-passa* is likely to become a profitable enterprise since it requires low initial investment. The export market for this product is largely unexplored.

In India, osmotic dehydration of 'Karpuravalli,' ABB, Pisang Awak, banana in 60–70% sugar syrup for 12 hours followed by drying in a hot air oven at 50°C produced high-quality figs with excellent color and taste (Narayana and Sathiamoorthy, 2002). Wrapping the figs in cellophane paper and packaging in polymeric containers gave an extended storage life of 6 months under ambient conditions.

In a study in Africa, undertaken to develop an organoleptically acceptable and shelf-stable dry plantain product using an osmotic drying technique, Ukkuru and George (1996) showed that an ideal plantain product can be obtained when plantains with sugar concentration of 70°Br were immersed in an osmotic solution for 60 min and heated to a temperature of 60°C.

Moisture content is an important quality attribute that directly influences the storability of fruits and vegetables (Romano et al., 2008). It was found that ripe banana halves dehydrated to a moisture content of about 30%, followed by bleaching and/or sulfating, are quite astringent (Ramirez-Martinez et al., 1977). Therefore accurate methods for determining the moisture content of dried banana are essential. In this regard, Romano et al. (2008) developed an approach for monitoring the moisture content of banana with laser-light backscattering imaging. Significant relationships were obtained between changes in backscattering area and moisture content especially at temperatures of approximately 53°C. This technology will be useful in rapidly evaluating the moisture content of dried banana and preventing spoilage of dried banana products.

Ecuador is one of the main producers of banana figs, but substantial amounts are now produced in India, Brazil, Philippines, Costa Rica, and other countries (Sole, 2005). The main importers of the product are the European countries, especially France and Germany.

14.7 BANANA FLAKES

Banana flakes are made by drying puree in large chrome-plated drum dryers. The puree, extracted in a similar way to that of banana puree, is fed directly to the dryers. The water evaporates as the drum turns, forming a film of dried material on the drum surface. Carefully adjusted knives remove the dry film as a continuous sheet, which passes to an air-conditioned room where it is broken into flakes. It is then sifted and packaged in bag-in-box containers (Sole, 1996). Confoco S.A. is the world's largest supplier of banana flakes, produced in their factories in Ecuador (Sole, 2005).

14.8 BANANA JAM

Several varieties of banana are suitable for preparing jam. Banana jam is prepared by cooking ripe fruit pulp with an equal quantity of sugar and with pectin and acid in right proportions until the product sets well. This product has a good commercial value and is used in several mixed-fruit jams (Narayana and Mustaffa, 2000).

14.9 BANANA BEVERAGES

14.9.1 BANANA JUICE

Banana has a widely appreciated flavor and aroma and will be able to compete favorably in the fruit juice market either as banana juices or mixed with other juices (Viquez et al., 1981). Generally juices are extracted from fruits by crushing or grinding. Banana pulp is too pectinaceous to yield juice by simple pressing or centrifugation (Viquez et al., 1981). This method produces a sticky, lumpy mass with no juice (Surendranathan et al., 2003). Therefore, various methods have been utilized to extract banana juices. Earlier methods for juice extraction involved the use of calcium oxide and sulfuric acid. Through pectolytic enzyme clarification a clear transparent juice to an extent of 75–80% of pulp weight could be obtained with a TSS content of 23–26°Br depending on the cultivar (Narayana and Sathiamoorthy, 2002). The clarified juice could be dispensed as a "ready-to-serve" (RTS) beverage by adjusting the TSS to 18–20°Br and acidity to 0.3% with sugar, citric acid, and water. After chilling it forms an excellent thirst-quenching and nutritious drink (Narayana et al., 2003). The RTS can also be carbonated using carbon dioxide gas at an appropriate level. Sole (1989) presents, in his patent, a method to recover concentrated clarified banana juice that will probably be used more as a "filler" ingredient in several other fruit juice blends rather than being served as a sole beverage. In fact, the first such product launched by Chiquita Brands International Inc. was an orange-banana juice. However, it has grown to become an "anytime drink," slightly eroding the market for carbonated beverages (Sole, 1996, 2005).

Koffi et al. (1991) reported that two different combinations of pectinase, cellulase, and hemicellulase were more effective in reducing viscosity and improving the filterability of both green and ripe banana purees when producing banana juice. The addition of potassium metabisulfate at 100 mg/L was more effective in producing a light-colored juice with a stable color. By adding cellulase (0.06%) and pectinase (0.05%) at 45°C for 2 hours, a 73% clear juice yield can be obtained. On the other hand, amylase was not effective in obtaining juice from banana (Pheantaveerat and Anprung, 1993; Sim and Bates, 1994).

In East Africa, banana juice is extracted with traditional technology. The banana pulp is mixed with grasses, especially *Imperata cylindrica*, and worked manually until the juice appears. The extracted juice is a clear, creamy liquid with a strong banana taste and aroma with a sugar content of 18–28% soluble solids (Aked and Kyamuhagive, 1996). This method is laborious and time consuming, and mechanical methods will be required if banana juice has to be marketed as a profitable fresh drink. However, the banana juice is often fermented to prepare different types of "banana beer," which is widely consumed in many countries in Africa (Stover and Simmonds, 1987).

14.9.2 BANANA BEER AND WINE

Fermenting banana juice for making banana beer and wine is an age-old traditional practice. In Central and East Africa, the juice from the ripe "beer banana" fruits are consumed either fresh or fermented to a low alcohol content having a short shelf-life (Sharrock, 1996). These 'Mbide' varieties (AAA) are used principally for making banana beer, which is a typical feature of the banana-based highland cropping systems in Burundi, Kenya, Rwanda, Tanzania, Uganda, and Zaire (now Democratic Republic of Congo). Making banana beer in East Africa consists of four main stages: (1) forced maturation of green banana, (2) pressing and mixing, (3) recovery of filtered juice and fermentation, and (4) racking, filtering, and cooling (Davies, 1995).

Various researchers have reported on methods for producing banana wine from different banana varieties in different continents (Adeniji, 1995; Narayana et al., 2002a; Akubor et al., 2003; Carreno and Aristizabal, 2003). The method of Narayana et al. (2002a) involved mashing ripe pulp with about 10% water and after adjusting the TSS to 26°Br, the pulp was sterilized and inoculated with wine yeast, *Saccharomyces cerevisiae* var. *ellipsoideus*, and incubated at 24–26°C for 7 days with intermittent aeration. After a week, the fermented juice is filtered and kept under anaerobic conditions with a water seal for secondary fermentation. After a further 2 weeks the secondary fermentation is terminated. The wine is racked or centrifuged, bottled, and pasteurized at 50°C for 20 minutes. The pasteurized wine is aged, giving the wine a characteristic flavor and aroma. Akingbala et al. (1994) reported that banana and mango juice can be fermented to produce wine using *Saccharomyces cerevisiae*. Pasteurization of the banana-based alcoholic beverages increases the ester and alcohol content.

Banana has been used as an adjunct in the production of regular beer (Carvalho et al., 2009b). The wort of regular beer was adjusted with different concentrations of banana juice. The results showed that ethanol production increased by approximately 0.4 g/g ethanol yield, suggesting that banana could be used in brewing methods for the development of new products. This provides a new opportunity for banana growers and a different outlet for excess banana.

14.9.3 Banana Alcohol

In Uganda and Sudan, banana beer is distilled to produce banana alcohol, or "Waragi," that is sold commercially. Homemade banana alcohol, although considered illegal, is widely produced in many countries in Africa and many small-scale rural farmers depend on this for their income. Commercial or medicinal alcohol from banana has also been produced in several countries. Iizuka et al. (1985) reported a 100% transformation with glucoamylase and yeast at 70°C; pectic enzymes were added to reduce the viscosity of the crushed material. Large quantities of Cavendish (AAA) bananas in the Philippines are rejected for export and are used for preparation of alcohol. Juice extracted from the ripe fruits was found to contain approximately 126 g /L total sugars. A technique for preparation of the inoculum, immobilization of yeast, and fermentation of juice to alcohol has been described by Del Rosario and Pamatong (1985). Tewari et al. (1986) produced alcohol from banana peels by subjecting the material to saccharification treatments using either sulfuric acid, steam under pressure, or cellulase. The fermentation was carried out with *Saccharomyces cerevisiae* var. *ellipsoideus*. The alcohol yield was approximately 150 ml/kg of peel. Commercial alcohol was produced at the National Research Centre for Banana, Tiruchirapalli, from banana peel of ripe banana fruits through saccharification followed by fermentation and distillation. The yield of alcohol was around 200 ml/kg peel (Sree Prabha et al., 2002).

14.9.4 Banana Vinegar

Vinegar is produced in Panama from banana wastes by subjecting it to alcoholic and acetic fermentation (Ortega and Blanco, 1986). A simple technique for processing rejects or overripe 'Robusta' (AAA), Cavendish subgroup, banana into vinegar was developed by Suresh and Ethiraj (1991). The pulp was crushed with an equal volume of water, and pectinase was added at 40 mg/kg. This method produced a clear juice that was seeded with a strain of *Saccharomyces cerevisiae* var. *ellipsoideus* to initiate fermentation. A good quality alcoholic base for producing vinegar containing 5–6% acetic acid was obtained. Simmonds (1966) reported that vinegar has been prepared by fermenting a mash of banana pulp and peel.

14.10 BANANA SAUCE

In the Philippines, bananas are used to produce ketchup on a commercial scale. Banana sauce resembles tomato ketchup in appearance but not in flavor. A recipe for the preparation of banana sauce was developed at the National Research Centre for Banana, Tiruchirapalli, where the 'Monthan'

(ABB) cultivar was blanched, peeled, mashed, and cooked with spices and vinegar to yield banana sauce with favorable acceptability. The product is stable for almost a year (Narayana et al., 2002b).

14.11 OTHER BANANA-BASED PRODUCTS

The technology for several other value-added products based on banana flour such as biscuits, *papads*, health drink, baby food, sweet chutney, and cakes have been developed at the National Research Center for Banana, Tiruchirapalli, India (Narayana and Sathiamoorthy, 2002; Narayana et al., 2002c; Ramajayam et al., 2003). Dried, ground, and roasted green bananas are also used as a coffee substitute in some places. Starch can be extracted from the pseudostems of banana and plantain. Banana starch has been used for producing glue used in manufacturing of cartons for export of fresh banana (Sharrock, 1996). Maini and Sethi (2000) reported that 5% edible starch can be obtained from 20–25 tons of banana pseudostem from 1,000 plants. Dried banana and banana/cassava pancakes called *Kabalagala* are consumed in Uganda (Aked and Kyamuhagive, 1996). Ogazi (1996) reported that in Nigeria unripe plantain peel and pulp were pounded to obtain a meal and eaten with any of the popular soups. Various delicious banana products (savory and sweet dishes) were prepared by Manimegalai et al. (1999), the *Annual Report of the National Research Centre* (2002), and the *Annual Report of NATP* (2002) using raw and ripe plantain, pseudostems, and flowers from banana.

Green pseudostems, corm, and sun-dried ripe banana peels provide feed for cattle and sheep. Ripe banana rejects, supplemented with protein, vitamins, and minerals are also used for fattening hogs. Good silage can be made from equal parts of chopped green bananas and grass or from chopped green banana mixed with 1.5% molasses. In the Philippines meal made from dehydrated rejected bananas could form up to 14% of total broiler rations without any adverse effects. Peeled leaf sheaths are used fresh or after drying as packaging material for flowers, betel leaves, fruits, and other items. The leaf sheaths are stripped into shreds, dried, and used for tying packages and making garlands in Southern India.

14.11.1 BANANA FIBER

Bananas and plantains are frequently used as a source of fiber. Banana fiber is extracted from dried petioles and leaf sheaths that make up the pseudostem. Dependent on the cultivar and method of extraction, a fresh banana plant yields about 0.6–1.0% fiber (Uma et al., 2002). Banana fiber is gaining popularity because it can be used in the pulp and paper industry for manufacturing of certain papers, particularly where great strength is required. The paper is used for, among other things, making tea bags and bank notes. The fiber has numerous other uses, including textile manufacture; making rope, string, and thread; and for production of handicrafts such as baskets, toys, table mats, wall hangings, and lamp shades (Uma et al., 2002). Mechanical devices for efficient extraction of fiber from banana pseudostem have been developed by the Central Institute for Research on Cotton Technology, Mumbai (Nachene and Uma, 2003). Although fiber can be extracted from all cultivars, variability for fiber yield exists according to genomic groups.

The various uses of banana fiber are explained by Uma et al. (2005) and can be summarized as follows: (1) as a natural absorbent for spillages including oil; (2) as a base material for bioremediation and recycling; (3) as a natural water purifier; (4) as a base material for the pulp and paper industry; (5) as a medium for mushroom production; (6) for handicrafts and textiles; (7) for printing currency; and (8) as a cottage industry for handicrafts and source of income.

14.11.2 BANANA STARCH

The production of starch from rejected bananas in the major banana-producing countries could be an important source of income and employment (Zhang et al., 2005). Rejected bananas are apparently dumped in rivers in countries such as Costa Rica, causing serious pollution problems. Bananas are rich in starch and when green the pulp contains about 70–80% starch on a dry weight basis (Guilbot and Mercier, 1985; Waliszewski et al., 2003). The starch content of banana is comparable with that found in the endosperm of corn and the flesh of white potato. There is growing interest in processing banana starch because of its importance for food and other industrial purposes (Waliszewski et al., 2003). A number of studies have been conducted on the isolation and characterization of starch from different banana varieties (Lii et al., 1982; Bello-Perez et al., 1999, 2000a, 2000b; Nunez-Santiago et al., 2004). These studies showed that banana starch has potential in both its digestion and functional properties to be used in processed foods and become a commercially viable starch product. These studies also suggested that banana starch has low stability to freezing treatments, and its use in frozen products is not recommended (Bello-Perez et al., 1999). Other studies (Siriwong et al., 2003) showed that the stability of banana starch paste over a range of pH makes it suitable for food applications that require high amylase and high retrogradation after processing. Banana starch has potential applications in food systems that require high-temperature processing, such as jellies, sausages, and bakery and canned products (Aurore et al., 2009). Chemical modifications of banana starch can produce improvements to its chemical properties. For example, phosphorylated and hydroxypropylated banana starches showed improvement in clarity and phosphated starches have shown the best freeze–thaw ability. Debranching of native banana starch from the cultivar 'Nandigobe' (EA-AAA) and retrogradation under different storage temperatures and starch concentrations allowed the production of a high-quality resistant starch with prebiotic properties with health applications (Lehmann et al., 2002).

Further details about banana starch are available in a number of reviews, including one by Zhang et al. (2005).

14.12 CONCLUSION

Postharvest losses of most crops, including banana, are relatively high especially in developing countries. Advances in biotechnology are now being applied to food commodities to increase the production of new products. The development of processing industries will see an increase in the production of processed products (chips, flour, dried pulp, jam, and beverages) from banana. Although there are many innovative laboratory studies of processed banana products, the industrial application of these products is still not exploited (Aurore et al., 2009). But there is marked increase in the number of studies that are attempting to utilize banana in various forms. In addition to exploring various ways of reducing the large postharvest losses incurred in banana production, these studies are developing new products from banana. The discovery that banana flour has a low glycemic index has the potential of increasing the utilization of banana flour, especially for controlling diabetes. The fact that alcohol can be produced from banana also suggests that banana could be used for the production of biofuel.

The large genetic diversity of banana is an asset for any program attempting to explore the various uses of this genus. Better use of bananas and plantains could be achieved by investigating their suitability for different types of processing (Aurore et al., 2009). Bananas represent a great potential raw material for food and nonfood processing industries.

REFERENCES

Abbas, F.M.A., R. Saifullah, and M.E. Azhar. 2009. Assessment of physical properties of ripe banana flour prepared from two varieties Cavendish and Dream banana. *Int. Food Res. J.* 16:183–189.

Adams, K.L. 2004. Food dehydration options: Value-added technical note, ATTRA publication IP-147. National Sustainable Agriculture Information Service.

Adeniji, T.A. 1995. Recipe for plantain/banana wine. *MusAfrica* 8:23–24.

Adeniji, T.A. and A. Tenkouano. 2007. Effect of processing on micronutrient content of chips produced from some plantain and banana hybrids. *Fruits* 62:345–352.

Adeniji, T.A. and A. Tenkouano. 2008. Effect of processing on the proximate, mineral, and pasting properties of whole flour made from some plantain and banana hybrids pulp and peel mixture. *J. Trop. Agric. Food Environ. Ext.* 7:99–105.

Agrianual. 2003. *Anuario da agricultura Brasileira*. Sao Paulo: FNP Consultoria & Comercio.

Aguirez-Cruz, A., A. Alvarez-Castillo, H. Yee-Madeira and L.A. Bello-Perez. 2008. Production of fiber-rich powder by the acid treatment of unripe banana flour. *J. Appl. Polym. Sci.* 109:382–397.

Aked, J. and W. Kyamuhagive. 1996. Post harvest aspects of highland bananas in Uganda. *Trop. Sci.* 36:54–64.

Akingbala, J.O., G.B. Oguntimein, B.A. Olunlade, and J.O. Aina. 1994. Effects of pasteurization and packaging on properties of wine from overripe mango and banana juices. *Trop. Sci.* 34:345–352.

Akubor, P.I., S.O. Obio, K.A. Nwadomere, and E. Obiomah. 2003. Production and quality evaluation of banana wine. *Plant Foods Human Nutr.* 58:1–6

Anand, J.C., S.B. Maini, and B. Diwan. 1982. A simple way to preserve potato for chips making *J. Food Sci. Technol.* 19:267.

Annual report of National Research Centre for Banana. 2002. Tiruchirapalli, Tamil Nadu: National Research Council for Banana.

Annual report of NATP. 2002. Standardization of processes for product development, value addition and waste utilization in bananas and plantains. Tiruchirapalli, Tamil Nadu: National Research Council for Banana.

Aparicio-Saguilan, A., S.G. Sayago-Ayerdi, A. Vargas-Torres, J. Tovar, T.E. Ascencio-Otero, and L.A. Bello-Perez. 2007. Slowly digestible cookies prepared from resistant starch-rich lintnerized banana starch. *J. Food Composition Anal.* 20:175–181.

Aurore, G., B. Parfait, and L. Fahrasmane. 2009. Bananas, raw materials for making processed food products. *Trends Food Sci. Technol.* 20:78–91.

Badia, I.A. 1985. Processing plantain chips in Honduras. Proceedings of the third meeting of IARPB, Abidjan, Côte d'Ivoire, May 27–31, 1985.

Baini, R. and T.A.G. Langrish. 2009. Assessment of colour development in dried bananas—Measurements and implications for modelling. *J. Food Eng.* 93:177–182.

Barthakur, N.N. and N.P. Arnold 1990. Chemical evaluation of *Musa* "Bhimkol" as a baby food. *J. Sci. Food Agric.* 53:497–504.

Bello-Perez, L.A., E. Agama-Acevedo, L. Sanchez-Hernandez, and O. Paredes-Lopez. 1999. Isolation and partial characterization of banana starches. *J. Agric. Food Chem.* 47:854–857.

Bello-Perez, L.A., R. Romero-Manilal, and O. Paredes-Lopez. 2000a. Preparation and properties of physically modified banana starch prepared by alcoholic-alkaline treatment. *Starch* 52:154–159.

Bello-Perez, L.A., E. Agama-Acevedo, S.G. Sáyago-Ayerdi, and E. Moreno-Damian. 2000b. Some structural, physicochemical and functional studies of banana starches isolated from two varieties growing in Guerrero, México, *Starch* 52:68–73.

Bukusuba, J., F.I. Muranga, and P. Nampala. 2008. Effect of processing technique on energy density and viscosity of cooking banana: Implication for weaning foods in Uganda. *Int. J. Food Sci. Technol.* 43:1424–1429.

Cano, P., M.A. Marín, and C. Fúster. 1990. Effects of some thermal treatments on polyphenol oxidase and peroxidase activities of banana (Musa cavendishii, var enana). *J. Sci. Food Agric.* 51:223–231.

Cano, M.P., B de Ancos, M.G. Lobo, and M. Santos. 1997. Improvement of frozen banana (*Musa cavendishii*, cv Enana) color by blanching: Relationship between browning, phenols and polyphenol oxidases and peroxidase activities. *Z. Lebensm Unters Forsch A.* 204:60–65.

Cano-Chauca, M., A.M. Ramos, P.C. Stringheta, and Jose A.M. Pereira. 2004. Drying curves and water activity evaluation of dried banana. *Drying* C: 2013–2020.

Carreno, A.C. and M.L. Aristizabal. 2003. Post-harvest use of plantain "Dominico harton" to make wine. *InfoMusa* 12:2–4.

Carvalho, G.B.M., D.P. Silva, C.V Bento, J.C. Santos, H.J.I. Filho, A.A. Vincente, et al. 2009a. Total soluble solids from banana: Evaluation and optimization of extraction parameters. *Appl. Biochem. Biotechnol.* 153:34–43.

Carvalho, G.B.M., D.P. Silva, J.C. Santos, A.A. Vincente, J.A. Teixeira, M. das Gracas, et al. 2009b. Banana as adjunct in beer production: Adaptability and performance of fermentative parameters. *Appl. Biochem. Biotechnol.* 155:356–365.

Chia, C.L. and C.A. Huggins. 2003. Bananas, community fact sheet BA-3(A) Fruit, Hawaii Cooperative Extension Service, CTAHR, University of Hawaii.

Chillet, M., C. Abadie, O. Hubert, Y. Chilin-Charles, and L. de Lapeyre de Bellaire. 2008. Sigatoka disease reduces the greenlife of bananas. *Crop Prot.* 28:41–45.

Choo, C.L. 2007. Utilization of matured green banana (*Musa paradisiaca* var Awak) flour and oat–beta glucan as fibre ingredients in noodle, master's thesis, Universiti Sains Malaysia.

Dadzie, B.K.K. 1995. Cooking qualities of black Sigatoka resistant hybrids. *InfoMusa* 4:7–9.

Dadzie, B.K.K. and H. Wainwright. 1995. Plantain utilization in Ghana. *Trop. Sci.* 35:405–410.

Daramola, B. and S.A. Osanyinlusi. 2006. Production, characterization and application of banana (*Musa* spp.) flour in whole maize. *Afr. J. Biotechnol.* 5:992–995.

Davies, G.E. 1995. Farmer knowledge in East-African highlands: A note on the preparation of banana based alcoholic beverages. *MusAfrica* 3(6):8–11.

Del Rosario, E.J. and F.V. Pamatong. 1985. Continuous flow fermentation of banana fruit pulp sugar into ethanol by carrageenan-immobilised yeast. *Biotechnol. Lett.* 7:819–820.

Diaz, A., A. Totte, F. Giroux, M. Reynes, and Q.L. Raoult Wack. 1996. Deep fat frying of plantain (*Musa paradisiaca*. L). Pt. 1, Characterization of control parameters. *Food Sci. Technol.* 29:489–497.

Emaga, T.H., C. Robert, S.N. Ronkart, B. Wathelet, and M. Paquot. 2008. Dietary fiber components and pectin chemical features of peels during ripening in banana and plantain varieties. *Bioresource Technol.* 99:4346–4354.

Garcia, R., M.C. de Arriola, E. de Porres, and C. Rolz. 1985. Process for banana puree preservation at rural level. *Lebensm. Wiss. Technol.* 18:323–327.

Garcia, R., L.Villatoro, and H. Orantes. 1987. Reduction of the time required to lower the pH of banana puree by lactic fermentation. Proceedings of Latin American symposium of biotechnology for the production of biomass and waste treatments, ICAITI, Guatemala (GTM).

Giami, S.Y. 1991. Effects of pretreatments on the texture and ascorbic acid content of frozen plantain pulp (*Musa paradisiaca*. L.). *J. Sci. Food Agric.* 55:661–666.

González-Aguilar, G.A., J.F. Ayala-Zavala, G.I. Olivas, L.A. de la Rosa, and E. Álvarez-Parrilla. 2010. Preserving quality of fresh-cut products using safe technologies. *J. Verbraucherschutz Lebensmittelsicherheit* 5:65–72.

Goswami, J. and P.C. Barua. 1996. Extension of shelf life of banana by the help of some chemical absorbents. Paper presented at the international conference on Technological Advancement in Banana/Plantain Production and Processing, August 20–24, 1996, Mannuthy, Kerala, India.

Guilbot, A. and C. Mercier. 1985. Starch. In: *The polysaccharides*, O. Aspinall, ed., 209–282. New York: Academic Press.

Hanson, L.P. 1976. *Commercial processing of fruits*. Park Ridge, NJ: Noyes Data Corporation.

Haslinda, W.H., L.H. Cheng, L.C. Chong, and A.A. Nooraziah. 2009. Chemical composition and physicochemical properties of green banana (*Musa acuminata* x *balbisiana* Colla cv Awak) flour. *Int. J. Food Sci. Nutr.* 60:232–239.

Iizuka, M., K. Uenakai, O. Svendsby, and T. Yamamoto. 1985. Alcohol fermentation of green bananas. *J. Fermentation Technol.* 63:475–477.

Izonfuo, W.A.L. and V.O.T. Omuaru. 1988. Effect of ripening on the chemical composition of plantain peels and Pulps (*Musa paradisiaca*). *J. Sci. Food Agric.* 45:333–336.

Jayaraman, K.S. and D.K. Das Gupta. 1992. Dehydration of fruits and vegetables: Recent developments in principles and techniques. *Drying Technol.* 10:1–50.

Jones, D.R. 2000. Introduction to banana, abaca and enset. In: *Diseases of banana, abaca and enset*, D.R. Jones, ed., 1–36. Wallingford, UK: CABI Publishing.

Juarez-Garcia, E., E. Agama-Acevedo, S.G. Sayago-Ayerdi, S.L. Rodriguez-Ambriz, and L.A. Bello-Perez. 2006. Composition, digestibility and application in breadmaking of banana flour. *Plant Foods Human Nutr.* 61:131–137.

Koffi, E.K., C.A. Sims, and R.P. Bates. 1991. Viscocity reduction and prevention of browning in the preparation of clarified banana juice. *J. Food Qual.* 14:209–218.

Lehmann, U., G. Jacobasch, and D. Schmiedl. 2002. Characterization of resistant starch type III from banana (*Musa acuminata*). *J. Agric. Food Chem.* 50:5236–5240.

Leite, J.B., M.C. Mancini, and S.V. Borges. 2007. Effect of drying temperature on the quality of dried bananas cv *prata* and *d'agua*. *Food Sci. Technol.* 40:319–323.

Lemaire, H., M. Reynes, J.A. Ngalani, J. Tchango Tchango, and A. Guillaumont. 1997. Aptitude à la friture de cultivars de plantains et bananes à cuire. *Fruits* 52:273–282.

Lii, C.-Y., S.-M. Chang, and Y.-L. Young. 1982. Investigation of the physical and chemical properties of banana starches. *J. Food Sc.* 47:1493–1497.

Maini, S.B. and V. Sethi. 2000. Utilization of fruits and vegetables processing waste. In: *Postharvest technology of fruits and vegetables*, L.R. Verma and V.K. Joshi, eds., 1006–1018. New Delhi, India: Indus Publishing Co.

Manimegalai, G., K. Bairavi, K. Jothilakshmi, and S.R. Subbalakshmi. 1999. Delicious banana product. Project report, Department of Home Science. Madurai: Home Science College and Research Institute.

Mejia, D.J. 2003. Preservation of fruits and vegetables by combined methods technologies. *Int. Trop. Fruits Network* 1–2.

Molla, M.M.,T.A.A. Nasrin, and M. Nazrul-Islam. 2009. Study on the suitability of banana varieties in relation to preparation of chips. *J. Agric. Rural Dev.* 7:81–86.

Mukhtar, F.B. 2009. Effect of storage temperature on post harvest deterioration of banana and plantain (*Musa* spp.). *Int. J. Phys. App. Sci.* 3:28–38.

Muyonga, J.H. 2000. Production and evaluation of precooked dehydrated unripe banana slices. *Afr. Crop Sci. J.* 8:1021–1027.

Nachene, R.P. and S. Uma. 2003. Processing of banana pseudostem for extraction of fibres. Paper presented at sixth Agricultural Science Congress, February 13–15, 2003, Bhopal, India.

Narayana, C.K. and M.M. Mustaffa. 2000. Annual report. Tiruchirapalli, Tamil Nadu: National Research Centre for Banana.

Narayana, C.K. and S. Sathiamoorthy. 2002. Value addition and product diversification in banana. Souvenir published on occasion of Global Conference on Banana and Plantain. India: Improvement in Production and Utilization of Banana.

Narayana, C.K., S. Sathiamoorthy, A. Evelin Mary, and D. Ramajayam. 2002a. Banana wine: A novel alcoholic beverage. Paper presented at Agro India, September–October 2002.

Narayana, C.K., S. Sathiamoorthy, and D. Ramajayam. 2002b. Preliminary studies on preparation of banana sauce. Paper presented at Global Conference on Banana and Plantains, October 28–31, 2002, Bangalore, India.

Narayana, C.K., S. Sathiamoorthy, and D. Ramajayam. 2002c. Evaluation of Monthan (cooking) bananas for making pickles and its quality changes during storage. Paper presented at the Global Conference on Banana and Plantains, October 28–31, 2002, Bangalore, India.

Narayana, C.K., S. Sathiamoorthy, A. Evelin Mary, and D. Ramajayam. 2003. Studies on quality changes of ready-to-serve banana juice during storage. Paper presented at sixth Agricultural Science Congress, February 13–15, 2003, Bhopal, India.

Ngalani, J.A. 1989. Valorisation du plantain: étude de l'obtention, de la caractérisation et de l'utilisation des farines—Conservation en frais. PhD diss., Université des Sciences et Techniques du Languedoc, France.

Nogueira, R.I. and K.J. Park. 1992. Drying parameters to obtain "Banana-Passa." In: *Drying*, A.S. Mujumdar, ed., 874–883. New York: Elsevier Science Publisher.

Nunez-Santiago, M.C., L.A. Bello-Perez, and A. Tecante. 2004. Swelling-solubility characteristics, granule size distribution and rheological behavior of banana (*Musa paradisiaca*) starch. *Carbohydr. Polym.* 56:65–75.

Ogazi, P.O. 1996. *Plantain: Production, processing and utilization.* Aku-Okigwe, Okigwe, Imo State, Nigeria: Paman and Associates Limited.

Ogazi, P.O. 1998. Plantain storage and processing. Proceeding of a Post-Harvest Conference, 2 November–1 December 1998, Accra, Ghana.

Ogazi, P.O., O. Smith-Kayode, H.M. Solomon, and S.A.O. Adeyemi. 1991. Packaging considerations and shelf life study of a new plantain based weaning food in Nigeria. Proceedings of the eighth World Congress of Food Science and Technology, 29 September–4 October 1991, Toronto, Canada.

Olaoye, O.A., A.A. Onilude, and O.A. Idowu. 2006. Quality characteristics of bread produced from composite flours of wheat, plantain and soybeans. *Afr. J. Biotechnol.* 5:1102–1106.

Onyejegbu, C.A. and A.O. Olorunda. 1995. Effects of raw materials, processing conditions and packaging on the quality of plantain chips. *J. Sci. Food Agric.* 68:279–283.

Ortega, M.M. and C. Blanco. 1986. Vinegar from banana wastes in Panama. *Informe Mensual* 10:47–52.

Osman, A.I. and W. Asrat. 1985. Sun drying of banana. *Ethiopian Nutr. Inst.* 1–18.

Oyesile, O.O. 1987. Comparative studies on cooking banana and plantain products, master's thesis, Department of Food Technology, University of Ibadan, Nigeria.

Pacheco-Delahaye, E., R. Maldonado, E. Perez, and M. Schroeder. 2008. Production and characterization of unripe plantain (*Musa paradisiaca* L) flours. *INCI* (Caracas) 33(4).

Pekke, M., K. Nowakunda, W.K. Tushemereirwe, C. Nankinga, J. Namaganda, and R. Kalyebara. 2004. Farmer evaluation of banana based products. *Afr. Crop Sci. J.* 12:27–31.

Pheantaveerat, A. and P. Anprung. 1993. Effect of pectinases, cellulases and amylases on production of banana juice. *Food* 23:188–196.

Pillay, M. and L. Tripathi. 2007. Banana breeding. In: *Breeding major food staples*, M.S. Kang and P.M. Priyadarshan, eds., 393–428. New York: Blackwell Science.

Ramajayam, D., C.K. Narayana, S. Sathiamoorthy, and A. Evelin Mary. 2003. Physico-chemical and quality changes of banana sweet chutney during storage. Paper presented at sixth Agricultural Science Congress, February 13–15, 2003, Bhopal, India.

Ramirez-Martinez, J.R., A. Levi, H. Padua, and A. Bakal. 1977. Astringency in an intermediate moisture banana product. *J. Food Sci.* 42:1201–1203.

Ramli, S., A.F M. Alkarkhi, Y.S. Yong, L. Min-Tze, and A.M. Easa. 2009. Effect of banana pulp and peel flour on physicochemical properties and in vitro digestibility of yellow alkaline noodles. *Int. J. Food Sci. Nutr.* 60:326–340.

Rodriguez-Ambriz, S.L., J.J. Islas-Hernandez, E. Agama-Acevedo, J. Tovar, and L.A. Bello-Perez. 2008. Characterization of a fiber-rich powder prepared by liquefaction of unripe banana flour. *Food Chem.* 107:1515–1521.

Romano, G., L. Baranyai, K. Gottschalk, and M. Zude. 2008. An approach for monitoring the moisture content changes of drying banana slices with laser light backscattering imaging. *Food Bioprocess Technol.* 1:410–414.

Sharrock, S. 1996. Uses of *Musa*. INIBAP annual report 1996, 42–44. Montpellier, France: INIBAP.

Shere, D.M., S.L. Kotapalle, and D.N. Kulkarni. 1993. Effect of maturity stage on qualities of fried banana chips. *J. Maharashtra Agric. Univ.* 18:338.

Sim, C.A. and R.P. Bates. 1994. Challenges to processing tropical fruit juices: Banana as an example. *Proc. Fla. St. Hort. Soc.* 107:315–319.

Simmonds, N. 1966. *Bananas*, 2nd ed. *Tropical agriculture*. London: Longman.

Singthong, J. and C. Thongkaew. 2009. Using hydrocolloids to decrease oil absorption in banana chips. *Food Sci. Technol.* 42:1199–1203.

Siriwong, W., V. Tulyathan, and Y. Waiprib. 2003. Isolation and physicochemical characterization of starches from different banana varieties. *J. Food Biochem.* 27:471–484.

Sole, P. 1989. October 17, 1989. US Patent 4,874,617.

Sole, P. 1996. Bananas (processed). In: *Processing fruits: Science and technology.* Vol. 2, *Major processed products*, L.P. Somogyi, D.M. Barret, and Y.H. Hui, eds., 361–385. Lancaster, PA: Technomic Publishing Co. Inc.

Sole, P. 2005. Bananas (Processed). In: *Processing fruits: Science and technology.* Vol. 2. *Major processed products*, D.M. Barrett, L.P. Somogyi, and H.S. Ramaswamy, eds., 658–678. Boca Raton, FL: CRC Press.

Sothornvit, S. and N. Pitak. 2007. Oxygen permeability and mechanical properties of banana films. *Food Res. Int.* 40:365–370.

Sree Prabha, S., C.K. Narayana, and S. Sathiamoorthy. 2002. Studies on production of ethanol from Poovan and Rasthali banana peel. Paper presented at Global Conference on Banana and Plantain, October 28–31, 2002. Bangalore, India.

Stover, R.H. and N.W. Simmonds. 1987. Utilization of fruit and plant byproducts. In: *Bananas*, 3rd ed. London: Longman Scientific and Technical

Suntharalingam, S. and G. Ravindran. 1993. Physical and biochemical properties of green banana flour. *Plant Foods Human Nutr.* 43:19–27.

Surendranathan, K.K., N.K. Ramaswamy, and R.K. Mitra. 2003. In: *Biotechnological strategies in agro-processing*, S.S. Marwaha and J.K. Arora, eds., 432–437. New Delhi: Asiatech Publishers Inc.

Suresh, E.R. and S. Ethiraj. 1991. Utilization of overripe bananas for vinegar production. *Trop. Sci.* 31:317–320.

Suvittawat, K. and C. Babpraserth. 1996. Banana cultivars for banana chip production. Paper presented at International Symposium on Technological Advancements in Banana/Plantain Production and Processing, August 20–24, 1996, Mannuthy, Kerala, India.

Tchango Tchango, T., A. Biko, R. Achard, J.V. Escalant, and J.A. Ngalani. 1999. Centre de Recherches Regionales sur Bananiers et Plantains, http://www.fao.org/inpho/content/compend/text/ch14.htm (accessed 30 April 2010).

Tewari, H.K., S.S.Marwaha, and K. Rupal. 1986. Ethanol from banana peels. *Agric. Wastes* 16:135–146.

Thomas, A.S. 1940. Bananas III. Uganda banana varieties and their uses. In: *Agriculture in Uganda*, J.D. Tothik, ed., 116–120. Kampala: Uganda Department of Agriculture.

Tonaki, K.I., J.E. Brekke, H.A. Frank, and C.G. Cavaletto. 1973. Banana puree processing. Research report n. 202, Hawaii Agricultural Experimental Station, College of Tropical Agriculture, University of Hawaii.

Tourjee, E.R., D.M. Barrett, M.V. Romero, and T.M. Gradziel. 1998. Measuring flesh color variability among processing clingstone peach genotypes differing in carotenoid composition. *J. Am. Soc. Hort. Sci.* 123:433–437.

Ukhum, M.E. and I.E. Ukpebor. 1991. Production of instant plantain flour, sensory evaluation and physic-chemical changes during storage. *Food Chem.* 42:287–299.

Ukkuru, P.M. and D. George. 1996. Organoleptic and qualitative evaluation of osmotically dried plantain products (*Musa*, AAB group, Palayankodan). Paper presented at International Symposium on Technological Advancement in Banana/Plantain Production and Processing, August 20–24, 1996, Mannuthy, Kerala, India.

Uma, S., S. Kalpana, and S. Sathiamoorthy. 2002. Annual Report of the ICAR-funded project, Physico-Chemical and Structural Characteristics of Banana Pseudostem Fibre. Tiruchirapalli, India: National Research Centre for Banana.

Uma, S., S. Kalpana, S. Sathiamoorthy, and V. Kumar. 2005. Evaluation of commercial cultivars of banana (*Musa* spp.) for their suitability for the fibre industry. *Plant Genet. Resourc. Newslett.* 142:29–35.

Viquez, F., C. Laetreto, and R.D. Cooke. 1981. A study of the production of clarified banana juice using pectinolytic enzymes. *J. Food Technol.* 16:115–125.

Wainwright, H. and J.N. Burdon.1991. Problems and prospects for improving the post-harvest technology of cooking bananas. *Postharvest News Inf.* 2:249–253.

Waliszewski, K.N., M.A. Aparicio, L.A Bello, and J.A. Monroy. 2003. Changes of banana starch by chemical and physical modification. *Carbohydr. Polym.* 52:237–242.

Wills, R.B.H., W.B. McGlasson, D. Graham, T.H. Lee, and E.G. Hall. 1989. *Postharvest. An introduction to the physiology and handling of fruits and vegetables*. New York: Van Nostrand Reinhold.

Yang, S., X. Su, KN. Prasad, B. Yang, G. Cheng, Y. Cheng, et al. 2008. Oxidation and peroxidation of postharvest banana fruit during ripening. *Pak. J. Bot.* 40:2023–2029.

Zhang, P., R.L. Whistler, J.N. BeMiller, and B.R. Hamaker. 2005. Banana starch: Production, physicochemical properties, and digestibility: A review. *Carbohydr. Polym.* 59:443–458.

15 Propagation Methods in *Musa*

Michael Pillay, Christopher A. Cullis, David Talengera, and Leena Tripathi

CONTENTS

15.1 INTRODUCTION

Bananas (*Musa* spp.) are propagated vegetatively using suckers or corms. The rate of multiplication of suckers is slow and variety dependent. Suckers may be infected with diseases and pests and are transferred to new fields. With the exception of some wild diploids, bananas do not set seeds. Although the seeds from such diploids have very high germination rates, seeds obtained from crossing programs have very low rates of germination (Pillay and Tripathi, 2007). Embryo culture is widely used to increase germination rates in breeding programs. The potential of *in vitro* propagation, via shoot-tip cultures, has gained worldwide popularity for rapid mass multiplication of planting materials. Since the first report of banana *in vitro* clonal propagation in the 1960s, the tissue-culture technology in banana has undergone significant improvement and is now used widely in banana production worldwide. Small-scale farmers are, however, unable to afford the cost of tissue-culture-derived planting material. Low-cost multiplication techniques have been developed to rapidly produce planting materials. Tissue culture is universally associated with somaclonal variation. The genome of banana has also been shown to vary during tissue culture. Variation has been observed at the cytogenetic level as well as in specific DNA sequences. As with other plants, these changes in DNA sequence are limited to a subset of the genome and may also include the activation of some transposable element families. A series of genomic scanning techniques has been applied to characterize the DNA variation, and two of these techniques have identified molecular markers that could be useful in identifying one of the common off-types in banana, namely the dwarf phenotype. This chapter examines embryo culture, *in vitro* multiplication, low-cost multiplication, and somaclonal variation in *Musa*.

15.2 EMBRYO CULTURE IN *MUSA*

15.2.1 Background

Embryo culture was probably the first tissue-culture technique to be applied to plant improvement (Drew, 1997). Embryo culture has been used to rescue hybrid plants from wide crosses, which often fail to produce mature viable seeds (Hu and Wang, 1986). In these cases the immature embryonic tissue is removed from the developing seeds and cultured in the laboratory to produce the hybrid plants. Immature embryos represent the most regenerable tissue for many species, including banana. Embryo culture enables the breeder to access a much wider range of genes that can be used for genetic improvement of crop plants. Wide crosses and embryo culture have been valuable tools, especially for the transfer of disease resistance genes from wild relatives into crop plants (Drew, 1997).

Embryo culture was the first application of a tissue-culture technique in banana nearly 40 years ago (Cox et al., 1960). The principal applications of embryo culture are rescuing embryos after interspecific hybridization, clonal propagation of recalcitrant species, and overcoming seed dormancy and seed sterility. Interspecific crosses are often attempted to transfer desired traits such as disease resistance, stress tolerance, or high yield from wild species into important crop species (Liang et al., 1978). However, embryo abortion often occurs due to breakdown of the endosperm or embryo-endosperm incompatibility (Hu and Wang, 1986; Elmsweller and Uhring, 1962). *In vitro* culture of hybrid embryos often successfully bypasses postzygotic incompatibilities (Drew, 1997).

Seed set in triploid banana cultivars is very low due to high levels of sterility (Ortiz and Vuylsteke, 1995). This makes crossbreeding of plantain and banana difficult. Nonetheless, several triploid plantain and banana cultivars produce seeds after hand pollination with diploid parents (Ssebuliba et al., 2006). Axenic *in vitro* germination of hybrid seed has been a cornerstone of banana breeding programs. Indeed, embryo culture increases rates of seed germination by a factor of 10 or more (Vuylsteke et al., 1990). At the International Institute of Tropical Agriculture (IITA), *in vitro* embryo culture is being used to create interspecific mapping populations and to rescue hybrid materials from breeding programs.

In vitro technique of banana embryo culture involves surface sterilizing of banana seeds, aseptically cracking them to extract the embryos, and culturing the embryos on appropriate nutrient medium and growth conditions to induce germination. Since the early 1970s (Menendez and Shepherd, 1975), embryo culture has been relevant in banana conventional breeding work where poor fertility due to low seed-set rates has been aggravated by poor seed germination. Seed germination as low as 1% in plantains (Swennen et al., 1992) and 1.4% in East African Highland bananas (AAA-EA) (Talengera et al., 1996) has been recorded from direct seed sowing. However, embryo culture has improved embryo germination to 22% in East African Highland bananas (EAHB, AAA-EA) (Ssebuliba et al., 2006) and up to 60% in plantains AAB and 'Popoulou' (ABB) (Bakry and Horry, 1992).

15.2.2 Factors Affecting Success in Embryo Culture

In vitro banana embryo germination rates vary considerably and are ascribed to several factors.

15.2.2.1 Embryo Culture Media Composition

Embryo culture media are based on the low nutrient formulations of Knudson (1950) or MS (Murashige and Skoog, 1962) at half strength (Afele and De Langhe, 1991; Vuylsteke et al., 1990) and the mineral rich salts of MS at full strength (Bakry and Horry, 1992; Vuylsteke and Swennen, 1992; Ssebuliba et al., 2005) and without complex natural supplements. Modifications of the culture media have been employed to improve embryo germination. Addition of growth regulators such as BAP (benzylaminopurine) at 1–2 mg/L and IAA (indoleacetic acid) at 0.4–1 mg/L to the medium (Bakry and Horry, 1992; Vuylsteke and Swennen, 1992; Jenny et al., 1994) gave varied germination rates and adverse responses. Callus formation was observed on East African Highland banana embryos when the above-growth regulators were used in the medium. Rescue of immature embryos from young fruits has been suggested but the low seed fertility in bananas limits this operation. Instead, white to cream mushroom-shaped banana embryos are routinely obtained from seeds extracted from mature and ripened fruits. Such embryos are mature and do not require complicated nutrient medium (Pierik, 1987).

15.2.2.2 Parental and Seasonal Effects

In vitro embryo germination rates vary considerably from cross to cross, genomic composition, and ploidy of parents. In addition, factors that influence fertility and seed quality positively affect embryo germination. Generally, $2x \times 4x$ crosses gave higher germination rates than $3x \times 2x$ crosses. Besides, the presence and proportion of the B genome in the parents is associated with higher seed quality and embryo germination. Seasonal effects on embryo germination have been reported, with high germination rates obtained in seeds extracted from bunches pollinated during the wet seasons (Vuylsteke et al., 1990). Embryo germination is strongly associated with the seed quality and number of extracted embryos (Bakry and Horry, 1992; Ssebuliba et al., 2005).

15.2.2.3 Effects of Seed Quality

Aborted and empty seeds with soft coats and those lacking endosperm are frequently obtained from crosses and must be discarded. These seeds can be identified by squeezing them between the thumb and finger or by a water gravity test in which they would float. Caution is needed with seeds derided from crosses involving EAHB as they do not have a compact endosperm such as the seeds derived from BB, AAB, and ABB genotypes, and the majority will float. Seed abnormality has been attributed to failure of fertilization, failure of embryo sac development, and zygote development (Shepherd, 1954; Vuylsteke et al., 1990). The proportion of such imperfect seeds vary with the parents used in the crossing scheme and ranges between 10 and 70% in 'Popoulou' (ABB) and plantains (AAB) (Vuylsteke et al., 1990; Bakry and Horry, 1992) and 41% in AAA–EA pollinated bunches (Ssebuliba et al., 2005).

15.2.2.4 Seed Treatments

Soaking the seeds for 5 days in water doubled the embryo germination in *Musa balbisiana* (Afele and De Langhe, 1991) compared to nontreated seeds. Three days of seed soak in plantain diploids raised germination rates remarkably from 2 and 30% (untreated) to 45 and 75% (treated) (Talengera, unpublished). However, seed soaking increases contamination with fungi, yeasts, and bacteria that can be controlled by a harsh disinfectant such as 1% (w/v) silver nitrate instead of sodium hypochlorite (0.75% w/v). The effectiveness of sterilization can be improved further by sterilizing the seeds for 15 minutes before soaking and prior to embryo extraction. Disinfection is made easier by wrapping the seeds in cotton gauze and immersing them in the disinfectant.

15.2.2.5 Culture Physical Environment

Banana embryos are cultured at 26–32°C, reflecting the tropical nature of the crop. Although continuous light was used by Afele and De Langhe (1991), embryos are generally incubated in darkness. Upon germination, the seedlings are exposed to light for further growth, as is required in normal *in vitro* micropropagation work.

15.2.2.6 Culturing Techniques

Germination of cultured embryos of hybrids occurs within 1 to 2 weeks compared to the 6-week period required in *Musa balbisiana* (Afele and De Langhe, 1991). Late germination usually produces normal seedlings. After germination, seedlings are cultured for 2 months before they are ready for potting. Where several plants of a seedling are required, embryo culture has been combined with shoot-tip culture technique (Vuylsteke, 1998). This involves cutting back the seedling near the base and culturing them onto medium supplemented with cytokinins such as BAP. To reduce the potential for somaclonal variation due to high cytokinin levels, the BAP concentration can be scaled down to 2.5 mg/L.

For *in vitro* rooting, microshoots are separated and cultured onto root-inducing media supplemented with 0.225 mg/L BAP and 0.186 mg/L NAA (naphthalene acetic acid) (Vuylsteke, 1998). However, phytohormone-free medium can be used to root the shoots (Talengera et al., 1994). Rooted shoots are potted for a month or two acclimatized in shade before the plantlets are ready for field establishment. Rouging out of off types arising from aneuploidy is required at this stage.

15.3 MICROPROPAGATION

Micropropagation has been defined as *in vitro* regeneration of plants from cells or protoplasts, tissues, or organs (Beversdorp, 1990) on specially formulated nutrient media. Under the correct conditions, an entire plant can be regenerated from a single cell. Plant-tissue culture has been in existence for more than 30 years. Tissue culture is seen as an important technology for developing countries for the production of disease-free, high-quality planting material and the rapid production of many uniform plants. Micropropagation is the production of multiple copies of a single plant using tissue-culture techniques. Micropropagation increases the number of planting materials to facilitate distribution and large-scale planting. In this way, thousands of copies of a plant can be produced in a short time. Micropropagated plants are observed to establish more quickly, grow more vigorously, have a shorter and more uniform production cycle, and produce higher yields than conventional propagules. Often, the tissue used is the meristem from which new leaves and stems are produced. Each growing tip on a plant can be excised and grown into a complete new plant.

Micropropagation of plants has many advantages over conventional methods of vegetative propagation, which suffer from several limitations (Nehra and Kartha, 1994). Tissue-culture systems allow propagating plant material with high multiplication rates in an aseptic environment. For example, an *in situ* banana plant can produce about 10 shoots per year whereas *in vitro* meristem culture can produce about 125–144 shoots within 8 weeks. Through repeated subculturing of proliferating shoots, an open-ended system can be maintained (Tripathi et al., 2003). Propagation from existing meristems yields plants that are genetically identical with the donor plants (Hu and Wang, 1983).

Tissue-culture propagation is capable of producing disease-free planting materials. Several major diseases of bananas are known to be transmitted through vegetative planting materials. In addition, some pests, such as nematodes and weevils, and some pathogens, such as those causing *Fusarium* wilt, moko, and bacterial wilt, are also transmitted through the soil and root tissues. Micropropagation of banana can eliminate virus diseases. The use of tissue-culture and disease-free planting materials has rehabilitated the banana industry in many countries (Molina, 2002). In Taiwan, the epidemic of Panama wilt, *Fusarium* race 4, on the popular export cultivar Cavendish had been successfully managed with the massive use of tissue culture, coupled with selection and use of somaclonal-variation-derived resistant varieties. In India and China, the use of tissue-culture planting materials is a very effective control tactic in reducing the incidence of banana bunchy top virus (BBTV). In the commercial export banana production in the Philippines, tissue-culture-planted farms are known to have fewer problems with nematodes. Fewer nematicide treatments are required than those of the traditional sucker-planted farms and perennially maintained plantations.

The micropropagation technique for cultivated *Musa* is now well established (Cronauer and Krikorian, 1984; Banerjee and De Langhe, 1985; Vuylsteke and De Langhe, 1985). Use of this technique may be limited by the risk of somaclonal variation, a widespread occurrence in some *in vitro* cultures (Vuylsteke et al., 1988). Shoot-tip culture is simple, easy, and applicable to a wide range of *Musa* genotypes (Vuylsteke, 1989). An efficient regeneration protocol, which seems to be independent of ploidy level and genomic background, was developed for *Musa* species using apical meristems (Tripathi et al., 2003). The selected species represented major groups of *Musa*, including fertile diploid bananas (AA and BB genomes), the sterile triploid plantains (AAB), Cavendish bananas (AAA), and tetraploid hybrids (AAAA and AAAB).

Propagation of banana through encapsulated shoot tips has also been reported (Ganapathi et al., 1992). An efficient, simple, and rapid regeneration system was also established for bananas using sections of corm containing intercalary meristematic tissues as explants (Tripathi and Tripathi, 2008). Several workers have reported that cellular differentiation and organogenesis in tissue culture are controlled by concentrations of cytokinin and auxin (George, 1993). George (1993) suggested that organogenesis in monocotyledonous plants was promoted on media supplemented with cytokinin/weak auxin combinations. The effects of cytokinin/auxin interactions on *in vitro* shoot proliferation of bananas were investigated (Arinaitwe et al., 1999). Higher proliferation rates were observed in cytokinin/auxin combinations in which a weak auxin IBA (indole-3-butyric acid) was included.

Micropropagation has played a key role in plantain and banana improvement programs worldwide (Vuylsteke et al., 1997). Planting material derived from micropropagation performs equal to or superior to conventional material (Smith and Drew, 1990; Vuylsteke, 1998). Micropropagated banana and plantain establish faster, grow more vigorously, are taller, have a shorter and more uniform production cycle, and yield higher than conventional propagules (Vuylsteke and Ortiz, 1996). Bananas propagated from apical meristem in Kenya have been shown to have increased vigor and suffer lower yield loss from weevils, nematodes, and fungal diseases (ISAAA, 2006). Micropropagation was considered an appropriate option to provide sufficient quality and quantity of such materials. With proper management and field hygiene, yield losses caused by pests and diseases at the farm level have been reduced substantially. Tissue-culture technology has made it possible for farmers to have large quantities of superior, clean planting materials that are early maturing, with bigger bunch weights and higher annual yield per unit of land. Moreover, uniformity in orchard establishment and simultaneous plantation development can make marketing easier to coordinate with the possibility of transforming banana growing from merely subsistence to a commercial enterprise. Tissue-culture banana production is more remunerative as an enterprise than traditional banana production.

15.4 LOW-COST MULTIPLICATION METHODS FOR BANANA

Cultivated banana and plantain are triploid and seedless. Most cultivars are female sterile and only produce seeds during artificial hybridization. Seeds are only used during the breeding process for the genetic improvement of banana. On the contrary, wild diploids are fertile and produce copious amount of seeds that are capable of germination after dispersal. Bananas are clonally propagated. Their survival in nature and geographical dispersal is not possible without intervention by humans. Large plantations purchase new planting material from tissue-culture laboratories. The high cost of tissue-culture planting material (averaging about $US1.00 per plant) is unaffordable by most small-scale farmers in most countries. Bananas in general have poor suckering ability (Robinson,

1996). The difficulty in obtaining healthy planting material is one of the most important constraints for large-scale banana production or expansion of existing plantations in many countries. Several types of planting materials that vary in their degree of suitability include: the maiden sucker, water sucker, sword suckers, butt, peeper, and bits that can be used for the establishment of new banana plantations (Ndubizu and Obiefuna, 1982; Baiyeri and Ndubizu, 1994). The major concern of vegetative planting material of this nature is its tendency to perpetuate pests (weevils, nematodes), wilts (Fusarium, Xanthomonas), and viral diseases (banana bunchy top, BSV) from existing to newly established fields. The movement of unhealthy planting material is probably the biggest factor in transmitting diseases and pests not only from one country to another but from one continent to another.

An alternative to tissue culture for producing large numbers of planting material in a relatively short time will benefit farmers who consider the cost of tissue-cultured plants prohibitive. In addition, most farmers in the developing countries where banana is grown are not trained in handling tissue-culture plantlets. For example, a plant arising from tissue culture produces a large number of suckers in the very early growth stage due to the residual effect of the growth hormones. These suckers have to be rouged out carefully without damaging the mother plant. The underground banana stem (rhizome) has numerous axillary buds at the base of the leaf sheaths which provide a source of a large number of plantlets if they are allowed to develop (Barker, 1959). One of the first experiments in the vegetative multiplication of bananas involved splitting of the rhizome and planting the pieces under nursery conditions (La multiplication, 1955). Navarre (1957) reported that over 180 plants were obtained from one rhizome by initiating callus formation. The technique involved digging, cleaning, and injuring of the rhizome to initiate callus formation. The callus tissue gave rise to new plants that were removed and rooted. This method was also adopted by Hamilton (1965) who obtained over 150 plants from a single rhizome in 5 to 7 months (Hamilton, 1965). Barker (1959) used the stripping technique in which the outer leaf sheaths of the pseudostem were removed to expose the axillary buds. Soil was then mounded around the buds to initiate sucker development. The process was repeated when a sucker reached 1 m in height. This method was capable of producing about 20 plants from a single parent plant. In a series of experiments, Loor (1978, cited in Menendez and Loor, 1979) tried to improve vegetative multiplication of bananas. One experiment identified the most suitable kind of corm to use while another experiment defined the type of wound that produced the best results. A third experiment looked at the action of hormones and cytokinins in stimulating the growth of adventitious buds.

In vitro banana cell culture research was first studied in Jamaica by Cook (Menendez and Loor, 1979). Later *in vitro* work was carried out in Honduras (Berg and Bustamante, 1974), Taiwan (Su-Shien and Shii, 1974; Ma and Shii, 1972, 1974), and the Philippines (de Guzman et al., 1976). Micropropagation using meristem/tissue culture can rapidly multiply and produce disease-free planting materials. Modern methods, especially micropropagation using meristem/tissue culture, are very efficient in rapid multiplication and production of healthy, vigorous, and disease-free planting materials. The method, however, requires sophisticated techniques, skill, and care to handle (Vuylsteke and Talengera, 1998). Tissue culture as a method of generating planting materials is being established in many developing countries. Though effective and fast, tissue culture (*in vitro* multiplication) is not an option for the majority of traditional banana producers.

A number of robust, low-cost means of multiplying bananas are now available (Pillay and Tripathi, 2007). They include false decapitation (Figure 15.1a, b), complete decapitation (Figure 15.2a, b), excised corm (Figure 15.3a, b), split corm (Figure 15.4a, b), and a bud manipulation technique (Figure 15.5a–i). These techniques, known as macropropagation, are easy affordable alternatives for tissue culture for large-scale sucker production at the farm level. Macropropagation methods can generate from 16 to 50 or more plantlets from one sword-sucker corm utilizing sawdust as a plantlet initiation medium. Other media such as rice hulls could also be used (Baiyeri and Aba, 2005).

FIGURE 15.1 False decapitation method. (a) The actively growing region (meristem) of the plant is destroyed by cutting a small hole (arrow in picture) in the pseudostem (the trunk) with a sharp knife. New plantlets will appear around the plant in about 1 month. (b) The suckers are detached carefully once they reach a height of 20 to 30 cm and have three to four leaves and are then planted directly to the field. (Reprinted from Pillay, M. and L. Tripathi, 2007, *Breeding major food staples*, M.S. Kang and P.M. Priyadarshan, eds., 393–428. New York. With permission from John Wiley.)

False decapitation destroys the actively growing region (meristem) of the mother plant by carefully cutting a hole through the pseudostem base with a sharp instrument such as a knife (Figure 15.1a). New plantlets arise from axillary buds at the base of the plant (Figure 15.1b). Suckers with a height of 30–40 cm can be detached carefully with a sharp instrument and planted directly in the field. Complete decapitation differs from the earlier method in that the pseudostem is completely cut down

(a)

(b)

FIGURE 15.2 Complete decapitation method: (a) The pseudostem is completely cut down. The meristem is killed and the corm left to sprout (within a month). Young plants are detached and planted directly. (b) New suckers formed from mat. (Reprinted from Pillay, M. and L. Tripathi, 2007, *Breeding major food staples*, M.S. Kang and P.M. Priyadarshan, eds., 393–428. New York. With permission from John Wiley.)

and the meristematic tissue is removed or destroyed (Figure 15.2a). The destruction of the apical meristem stimulates the growth of the axillary buds (Figure 15.2b), which are detached and planted in a field. Best results with both false and complete decapitation occurs when there is no competing ratoon sucker in the same mat. A sucker in the same mat retards the growth of the axillary buds, perhaps by exerting apical dominance. In the excised corm technique, the large visible buds are

(a)

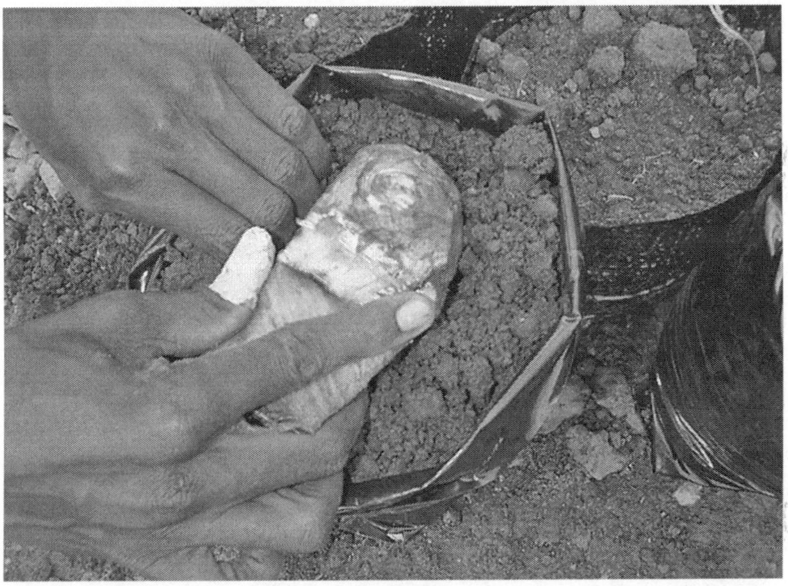

(b)

FIGURE 15.3 Excised corm method: (a) Buds are cut out in mini sets (about 100 g each). (b) Each bud is grown in a separate polyethylene bag and transferred to the field when three to four leaves have been attained. (Reprinted from Pillay, M. and L. Tripathi, 2007, *Breeding major food staples*, M.S. Kang and P.M. Priyadarshan, eds., 393–428. New York. With permission from John Wiley.)

carefully removed together with some of the surrounding corm tissue and sown in suitable-sized polythene bags (Figure 15.3a, b). Enough water is applied to keep the soil moist. The split-corm technique, as the name suggests, involves splitting of the corm into two or more fragments. The pieces are planted, preferably in sterile soil with the axillary buds facing the soil (Figure 15.4a, b). Sprouted plantlets are detached and potted before transplanting to the field. The technique that has the highest potential of producing the largest number of suckers is the bud manipulation technique. This is a modification of the technique of Barker (1959). Medium-sized corms are obtained from healthy mats (Figure 15.5a). The sucker is pared with a sharp knife to remove roots, old dead tissue,

(a)

(b)

FIGURE 15.4 Slit corm method. (a) The corm is dug out of the soil, pared, and split into two or more fragments depending on its size. (b) Plant split corm face down in polyethylene bag. Detach sprouted plantlets and pot in appropriate substrate. (Reprinted from Pillay, M. and L. Tripathi, 2007, *Breeding major food staples*, M.S. Kang and P.M. Priyadarshan, eds., 393–428. New York. With permission from John Wiley.)

and pseudostem remains (Figure 15.5b). Removal of the outer corm tissue gets rid of superficial weevil and nematode eggs. The pseudostem is then cut about 20 cm above the collar (Figure 15.5c). The leaf sheaths are removed individually in order to expose the axillary buds (Figure 15.5d). The apical meristem is destroyed. A well-prepared corm in the bud manipulation technique is represented in Figure 15.5e. The bud is carefully placed in sawdust or an appropriate medium in a plastic

(a)

(b)

(c)

(d)

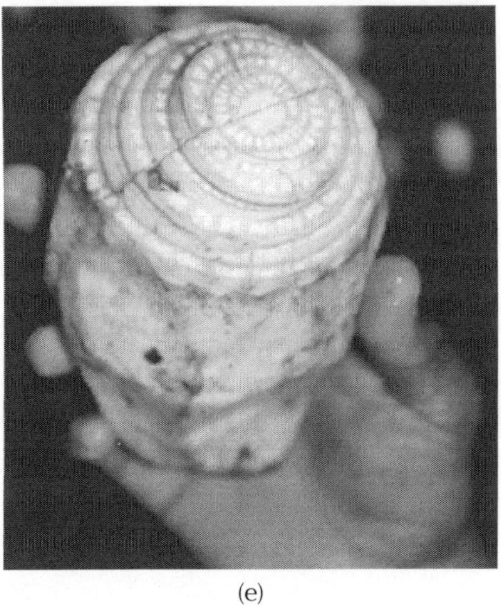

(e)

FIGURE 15.5 Bud manipulation method. (a) Medium-sized sucker obtained from healthy mats. (b) Sucker is pared. (c) The pseudostem is cut about 20 cm above the collar. (d) The sheaths are removed individually in order to expose the buds at the v end of each sheath. (e) The innermost section (meristematic doom) is cut to create a small depression. The corm is placed in sawdust in the propagation chamber. *(continued)*

(f)

(g)

(h)

(i)

FIGURE 15.5 (Continued) Bud manipulation method. (f) Corms in a propagation chamber, regular watering is required. Presence of condensed water droplets on the inside of the plastic material indicates sufficient humidity in the chamber. First sprouting commences from the second week of operation. (g) Manipulation of primary bud sprouts. Sawdust is removed to expose the bulbous shoot bases. The shoots are cut off just above the collar and meristem excised to induce production of secondary bud sprouts. They are recovered with sawdust. Secondary bud sprouts are observed 2 to 4 weeks after primary bud manipulation. (h) A clump of plantlets regenerated from bud sprouts. A range of 16 to 30 plantlets is obtainable between 10 to 18 weeks. At this stage plantlets are detached and put in plastic materials with nutrient-rich porous topsoil. (i) Plantlets just detached from the mother corm ready to be taken to the root initiation chamber from where they are weaned after developing small whitish roots. (Reprinted from Pillay, M. and L. Tripathi, 2007, *Breeding major food staples*, M.S. Kang and P.M. Priyadarshan, eds., 393–428. New York. With permission from John Wiley.)

growth chamber as shown in Figure 15.5f. The axillary buds begin to develop after approximately 3 weeks (Figure 15.5g). This technique is capable of producing 16–30 plantlets in 10–18 weeks (Figure 15.5h). The plantlets are detached, rooted in soil in a low-light chamber, and then hardened under direct sunlight (Figure 15.5i).

Although macropropagation is genotype specific, it certainly provides a low-cost method of producing new planting material. The method has been used to rapidly multiply and supply farmers with planting material in emergency situations such as the recent threat of Xanthomonas wilt in eastern Africa.

15.5 MOLECULAR CHARACTERIZATION OF SOMACLONAL VARIATION IN *MUSA*

15.5.1 Background

The term "somaclonal variation" was coined to describe the variation observed in plants when they were taken through one or more cycles of culture and regeneration (Larkin and Scowcroft, 1981). This observation was unexpected, as the process of tissue culture and regeneration was not originally considered to be mutagenic. The characteristics of the somaclonal variants have led to varying hypotheses concerning their origin. The appearance of a very limited range of aberrant phenotypes is inconsistent with the notion that the mutagenic effect is simply a random process within the genome. Across many plant species, flower and leaf abnormalities are very common. Therefore there may be a common mechanism that targets a specific compartment of the genome for variation while cells are being propagated through tissue culture. Additionally, the observation that some of the somaclonal variants can revert during growth has led to the suggestion that the process is "epigenetic" (Kaeppler et al., 2000); that is, it is not based in DNA sequence variation but in modifications such as cytosine methylation that can be reversed.

In banana, the set of phenotypic aberrations, known as off-types, is limited but can arise at a variable frequency, which can be as high as 90% (Cote et al., 1993; Israeli et al., 1995; Reuveni et al., 1996). Among the most common variants are 'Dwarf,' 'Giant,' 'Massada,' and chlorotic variegation. The 'Massada' variant is mosaic variegation on the leaves while the remaining names are descriptive of the phenotype. This limited set of variants coupled with the high rates of appearance indicate that a search for molecular markers associated with this phenomenon would be successful.

15.5.2 Molecular Techniques for Detecting Genomic Changes in Somaclonal Variants

Since the somaclonal variants appear to be heritable, and some somaclonal variants are also meiotically transmissible, genome scanning techniques have been applied to banana to identify DNA variations associated with somaclonal variation. Cytogenetics and flow cytometry methods (Roux et al., 2004; Gimenez et al., 2001), random amplified polymorphic DNA (RAPD) (Damasco et al., 1996; Walther et al., 1997; Sheidai et al., 2008), the selective amplification of microsatellite polymorphic loci (SAMPL) (Gimenez et al., 2005), and representation difference analysis (RDA) (Oh et al., 2007) have all been used to characterize normal and off-type bananas. In addition, there is evidence that retrotransposons are also activated during culture (Khayat et al., 2004).

15.5.2.1 Cytogenetics and Flow Cytometry

Genetic instability in the DNA content of embryogenic cell suspension cultures of triploid clones has been identified using flow cytometry and chromosome counting (Roux et al., 2004). They demonstrated that the ploidy level of *Musa* embryogenic cell suspension cultures can be determined by flow cytometry. The ploidy level of such cells can vary over a short time in culture but the contribution of such ploidy alterations is debatable since the abnormal ploidy levels coincide with lowered regeneration ability. However, a somaclonal variant of banana, 'CIEN BTA-03,' resistant to yellow Sigatoka disease (Gimenez et al., 2001), was shown to be a tetraploid clone through cytogenetics and flow cytometry.

15.5.2.2 Random Amplified Polymorphic DNA (RAPD)

A number of studies have used RAPD comparisons between normal and off-type bananas (Sahijram et al., 2003; Lakshmanan et al., 2007; Sheidai et al., 2008). These studies have screened large numbers of RAPD primers to assess variation among parental identified off-types from micropropagation. In one study, 66 primers were used in an initial screening of approximately equal numbers of normal and dwarf plants (Damasco et al., 1996). Nineteen of the 66 primers identified polymorphisms between normal and dwarf plants, and one primer, OPJ-04 (5′-CCGAACACGG-3′), could be used to identify a specific polymorphism between normal and dwarf plants. The polymorphism was in the form of an approximately 1.6 kb band, which was consistently present in all normal but absent in all dwarf plants. This RAPD band was converted into a sequence characterized amplified region (SCAR) marker for use in detection of dwarf plants in regenerated bananas (Ramage et al., 2004). Since the test is one of a band present in the DNA from normal plants and not from dwarf off-types, a control for all the amplifications needed to be included in order to distinguish between the absence of a PCR product due to the plant being a dwarf or simply a problem with the PCR amplification. None of the other polymorphisms that were observed between normal and dwarf phenotypes were specifically associated with the dwarf phenotype. In another study, Sheidai et al. (2008) used 30 RAPD primers to study somaclonal variation among the parental as well as regenerated plants. A total of 289 bands were scored, of which 147 were polymorphic. There were no associations with any of the polymorphisms and particular phenotypes. In contrast to these two studies, Lakshmanan et al. (2007) screened a large number of random primers but found no changes in the banding patterns of tissue-cultured plants and those of the mother plant. Since Sheidai et al. (2008) found that the level of variation was associated with the culture conditions, this last set of data (Lakshmanan et al., 2007) supports previous studies that the appearance of genomic variation is dependent on both the variety and the culture conditions.

15.5.2.3 Selective Amplification of Microsatellite Polymorphic Loci

The selective amplification of microsatellite polymorphic loci (SAMPL) technique is a modification of the amplified fragment length polymorphism (AFLP) procedure (Vos et al., 1995). In the case of SAMPL, the primers are compound microsatellite sequences combined with AFLP oligonucleotides. High numbers of monomorphic bands were amplified in all cultivars and somaclones but some markers were present in only a single cultivar or somaclone (Giminez et al., 2005).

15.5.2.4 Representation Difference Analysis (RDA)

RDA is a technique whereby all the fragments that are in common between two genomes are removed, leaving only those that are different between the genomes (Lisitsyn et al., 1993). These subtractions were performed using the tall (off-type) and dwarf forms of the cultivar 'Curare enano' (Oh et al., 2007). The series of different products that were isolated were sequenced. Using the available sequence, primers were designed and used in a PCR-based system to monitor changes in genomic integrity of *in vitro*–produced banana plants. A subset of the primers identified dwarf from normal plants. The most informative primer pair amplified a band in the normal or giant phenotypes, but failed to amplify a band in the dwarf phenotypes. This is identical to the SCAR marker described earlier, although there was no similarity between the two sequences. A sequence analysis over an extended region containing the marker identified numerous other small DNA variations between the phenotypes that are consistent with the notion that specific genomic regions are targeted when banana cells are propagated through tissue culture.

An extensive analysis of this characterized region also identified some differences in the methylation status of the DNAs from normal and dwarf phenotypes, with the level of methylation being greater in the dwarf than the normal phenotype. More studies on the levels of methylation covering a greater part of the banana genome are necessary to determine the role of methylation in somaclonal variation in banana. Are the methylation differences a cause or a consequence of the mechanisms that are responsible for the genomic restructuring?

15.5.2.5 Transposable Element Activation

Retrotransposons have been identified as being activated by stresses in plants (Grandbastien, 1998), including the stresses imposed by tissue culture. Following tissue-culture cycles, retrotransposon activation was observed in banana (Khayat et al., 2004). As with many of the other observations concerning the genomic changes associated with passage through tissue culture, there were no directly identifiable changes in the retrotransposon patterns with a particular phenotype. The activation of previously quiescent retrotransposons by stresses is one hallmark of some epigenetic changes where the instability is mediated by modification of the heterochromatin.

15.6 CONCLUSIONS

Despite efforts to improve the rate of embryo rescue in banana, the current *in vitro* technique does not allow total germination of all embryos. Failure to achieve total germination even after ridding the embryo of a suspected inhibitor from the endosperm and seed coat suggests the presence of embryo-related dormancy. There is need to explore dormancy-breaking treatments on the excised embryos. Attempts to improve embryo germination must consider other factors such as seasonal effects on seed fertility. Genomic composition and ploidy of the hybrids should also be considered to improve the rate of embryo germination.

The main advantage of tissue-culture technology lies in the production of high-quality and uniform planting material that can be multiplied on a year-round basis under disease-free conditions anywhere irrespective of the season and weather. However, the technology is capital, labor, and energy intensive. Although labor is cheap in many developing countries, the resources of trained personnel and equipment are often not readily available. In addition, energy, particularly electricity, and clean water are costly. In many African countries there are difficulties in adoption of tissue-culture material because of the high cost of the material in comparison to conventional planting material. Hence, it is necessary to have low-cost options for weaning, hardening of micropropagated plants, and finally growing them in the field. Low-cost tissue-culture technology is the adoption of practices and use of equipment to reduce the unit cost of the micropropagule and plant production. Low-cost options should lower the cost of production without compromising the quality of the micropropagules and plants. Alternative low-cost macropropagation techniques for producing large numbers of planting material in a relatively short time will benefit farmers who consider the cost of tissue-cultured plants prohibitive.

The data for the genomic changes in banana in response to passage through tissue culture is consistent with that observed for other plants and includes cytological abnormalities, frequent qualitative and quantitative phenotypic mutation, sequence change, and gene activation and silencing. Clearly there are a number of different mechanisms in play during the occurrence of somaclonal variation in banana. The changes in ploidy level as well as possible chromosome rearrangements are directly observable. The physical changes within the genome that have been identified through the application of a series of molecular screens of the genome include polymorphisms identified through RAPD and RFLP analysis, the characterization of difference products by RDA, and the activation of transposable elements. Finally, some variation in the methylation status of regions of the genome has also been identified. The primary events that are responsible for the appearance of the variants have not yet been definitively identified. However, two different methods of analysis have both identified changes that identify dwarf from normal banana plants. In both cases, a

DNA fragment is absent from the dwarf but is present in the normal plants. The observation that a reproducible genomic event is associated with one of the phenotypes is again consistent with all the other plant data that limits the genomic variation in response to tissue culture to a small subset of the genome. The evidence for the involvement of epigenetic alterations in banana somaclonal variation is minimal at present, but future research will determine the relative importance of epigenetic versus sequence or chromosome variation in the appearance of somaclones. In some respects the appearance of somaclonal variation resembles tumorigenesis in animals. Here again, the genome is subject to many different restructuring processes and it has been difficult to directly assign the root event responsible for the tumor progression. In the case of somaclonal variation, the same problem exists: Which of the many variations that have been observed are the specific causes of the altered phenotype and which are simply associated with the restructuring that accompanies the stresses imposed by the growth environment of cells in culture?

REFERENCES

Afele, J.C. and E. De Langhe. 1991. Increasing *in vitro* germination of *Musa balbisiana* seed. *Plant Cell Tissue Organ Cult.* 27:33–36.

Arinaitwe, G., P.R. Rubaihayo, and M.J.S. Magambo. 1999. Effects of auxin/cytokinin combinations on shoot proliferation in banana cultivars. *Afr. Crop Sci. J.* 7:605–611.

Baiyeri, K.P. and S.C. Aba. 2005. Response of *Musa* species to macro-propagation. I: Genetic and initiation media effects on number, quality, and survival of plantlets at prenursery and early nursery stages. *Afr. J. Biotechnol.* 4:223–228.

Baiyeri, K.P. and T.O.C. Ndubizu. 1994. Variability in growth and field establishment of Falsehorn plantain suckers raised by six cultural methods. *MusAfrica* 4:1–3.

Bakry, F. and J.P. Horry. 1992. Tetraploid hybrids from interploid 3x/2x crosses in cooking bananas. *Fruits* 47(6):641–647.

Banerjee, N. and E. De Langhe. 1985. A tissue culture technique for rapid clonal propagation and storage under minimal growth conditions of *Musa* (banana and plantain). *Plant Cell Rep.* 4:351–354.

Barker, W.G. 1959. A system of maximum multiplication of the banana plant. *Trop. Agric.* (Trinidad) 36:275–284.

Berg, L.A. and M. Bustamante. 1974. Heat treatment and meristem culture for production of virus free bananas. *Phytopathology* 64:320–322.

Beversdorp, W.D. 1990. Micropropagation in crop species. In: *Progress in plant cellular and molecular biology*, H.J.J. Nijkamp, L.H.W. Van Der Plas, and J. Van Aartrijk, eds., 3–12. Dordrecht: Kluwer Academic Publishers.

Cote, F.X., J.A. Sandoval, P.H. Marie, and E. Auboiron. 1993. Variations in micropropagated bananas and plantains: Literature survey. *Fruits* 48:11–18.

Cox, E.A., G. Stotzky, and R.D. Goos. 1960. *In vitro* culture of *Musa balbisiana* Colla embryos. *Nature* 185:403–404.

Cronauer, S.S. and A.D. Krikorian. 1984. Rapid multiplication of bananas and plantains by *in vitro* shoot tip culture. *HortScience* 19:234–235.

Damasco, O.P., G.C. Graham, R.J. Henry, S.W. Adkins, M.K. Smith, and I.D. Godwin. 1996. Random amplified polymorphic DNA (RAPD) detection of dwarf off-types in micropropagated Cavendish (Musa spp. AAA) bananas. *Plant Cell Rep.* 16:118–123.

De Guzman, E., E. Ubalde, and A.G. Rosario. 1976. *Banana and coconut in vitro cultures for induced mutation studies: Improvement of vegetatively propagated plants and tree crops through induced mutations*, 33–54. Vienna, Austria: International Atomic Energy Agency.

Drew, R.A. 1997. The application of biotechnology to the conservation and improvement of tropical and subtropical fruit species. www.fao.org/ag/agp/agps/pgr/drew1.htm (accessed 5 April 2010).

Elmsweller, S.L. and J. Uhring. 1962. Endosperm-embryo incompatibility in *Lilium* species hybrids. *Adv. Hort. Sci. Their Appl.* 2:360–367.

Ganapathi, T.R., P. Suprasanna, V.A. Bapat, and P.S. Rao. 1992. Propagation of banana through encapsulated shoot tips. *Plant Cell Rep.* 11:571–575.

George, E.F. 1993. *Plant propagation by tissue culture.* Part 1, *The technology*. London: Exegetics LTD.

Giménez, C., A. Garcia, N.X. De Enrech, and I. Blanca. 2001. Somaclonal variation in banana: Cytogenetic and molecular characterization of the somaclonal variant CIEN BTA-03 in vitro. *Cell. Dev. Biol. Plant* 37:217–222.

Giménez, C., G. Palacios, M. Colmenares, and G. Kahl. 2005. SAMPL: A technique for somaclonal variation fingerprinting in *Musa*. *Plant Mol. Biol. Rep.* 23:263–269.

Grandbastien, M. 1998. Activation of plant retrotransposons under stress conditions. *Trends Plant Sci.* 3:181–187.

Hamilton, K.S. 1965. Reproduction of banana from adventitious buds. *Trop. Agric.* (Trinidad) 42:69–73.

Hu, C.Y. and P.J. Wang. 1983. Meristem, shoot-tip and bud culture. In: *Handbook of plant cell culture*, Vol. 1, *Techniques and applications*, D.A. Evans, W.R. Sharp, and P.V. Ammirato, eds., 177–277. New York: Macmillan Publishing Co.

Hu, C.Y. and P. Wang. 1986. Embryo culture: Technique and applications. In: *Handbook of plant cell culture*, Vol. 4, D.A. Evans, W.R. Sharp and P.V. Ammirato, eds., 43–96. New York: Macmillan.

ISAAA. 2006. Tissue culture and micropropagation. http://www.isaaa.org/Kc/inforesources/publications/biotechinagriculture/default.html#Tissue_Culture_and_Micropropagation_.htm (accessed 27 July 2010).

Israeli, Y., E. Lahav, and O. Reuveni. 1995. In vitro culture of bananas. In: *Bananas and plantains*, S. Gowen, ed., 147–178. London: Chapman and Hall.

Jenny, C., E. Auboiron, and A. Baveraggi. 1994. Breeding plantain-type hybrids at CRBP. In: *The improvement of testing of Musa: A global partnership*, D. Jones, ed., Proceedings of the First Global Conference of the International *Musa* Testing Program held in FHIA, Honduras, April 27–30, 1994, Montpellier, France: INIBAP.

Kaeppler, S.M., H.F. Kaeppler, and Y. Rhee. 2000. Epigenetic aspects of somaclonal variation in plants. *Plant Mol. Biol.* 43:179–188.

Khayat, E., A. Duvdevani, E. Lahav, and B.A. Ballesteros. 2004. Somaclonal variation in banana (*Musa acuminata* cv. Grande Naine). Genetic mechanism, frequency and application as a tool for clonal selection. In: *Banana improvement: Cellular, molecular biology and induced mutations*, S.M. Jain and R. Swennen, eds., 97–109. Enfield, NH: Science Publishers.

Knudson, L. 1950. Germination of seeds of vanilla. *Am. J. Bot.* 37:241–247.

La multiplication vegetative du bananier. 1955.. *Bull. Inf. I.N.E.A.C.* 4:49–52.

Lakshmanan, V., S.R. Venkataramareddy, and B. Neelwarne. 2007. Molecular analysis of genetic stability in long-term micropropagated shoots of banana using RAPD and ISSR markers. *Electron. J. Biotechnol.* doi: 10.2225/vol10-issue5-fulltext-12.

Larkin, P.J. and W.R. Scowcroft. 1981. Somaclonal variation—A novel source of variability from cell culture for plant improvement. *Theor. Appl. Genet.* 60:197–214.

Liang, C., S. Sun, T. Liu, and J. Chiang. 1978. Studies on interspecific hybridization in cotton. *Sci. Sinica* 21:545–556.

Lisitsyn, N., N. Lisitsyn, and M. Wigler 1993. Cloning the differences between two complex genomes. *Science* 259:946–951.

Ma, S.S. and C.T. Shii. 1972. In vitro formation of adventitious buds in banana shoot apex following decapitation. *J. Chin. Soc. Hort. Sci.* 18:135–142 (in Chinese with English summary).

Ma, S.S. and C.T.Shii. 1974. Growing banana plantlets from adventitious buds. *J. Chin. Soc. Hort. Sci.* 20:6–12 (in Chinese with English summary).

Menendez, T. and F.H. Loor. 1979. Recent advances in vegetative propagation and their application to banana breeding. In: *Reunion da ACORBAT* 4:211–222. Anais, Panamá: UPEB.

Menendez, T. and K. Shepherd, 1975. Breeding new bananas. *World Crops* May–June, 104–112.

Molina, A.B. 2002. Tissue culture in the banana industry. International Training Course of Biotechnology for Seed and Seedling Production, 2–6 December, Philippines.

Murashige, T. and F. Skoog, 1962. A revised medium for rapid growth and bioassays with tobacco tissue culture. *Physiol. Plant.*15:473–497.

Navarre, E. 1957. Multiplication des musa a feuilles rouges. *Revue Horticole* (Francia) no. 29:712.

Ndubizu, T.O.C. and J.C. Obiefuna. 1982. Upgrading inferior plantain propagation material through dry-season nursery. *Sci. Hort.* 18:31–37.

Nehra, N.S. and K.K. Kartha. 1994. Meristem and shoot tip culture: Requirements and applications. In: *Plant cell and tissue culture*, I.K. Vasil, and T. Thorpe, eds., 37–70. Norwell, MA: Kluwer Academic.

Oh, T.J., M.A. Cullis, K. Kunert, I. Engelborghs, R. Swennen, and C.A. Cullis. 2007. Genomic changes associated with somaclonal variation in banana (*Musa* spp). *Physiologia Plantarum* 129:766–774.

Ortiz, R. and D. Vuylsteke. 1995. Factors influencing seed set in triploid *Musa* spp. L. and production of euploid hybrids. *Ann. Bot.* 75:151–155.

Pierik, R.L.M. 1987. *In vitro culture of higher plants*. Dordrecht, Netherlands: Martinus Nijhoff Publishers.

Pillay, M. and L. Tripathi. 2007. Banana breeding. In: *Breeding major food staples*, M.S. Kang and P.M. Priyadarshan, eds., 393–428. New York: Blackwell Science

Ramage, C.M., A.M. Borda, S.D. Hamill, and M.K. Smith. 2004. A simplified PCR test for early detection of dwarf off-types in micropropagated Cavendish bananas (*Musa* spp. AAA). *Sci. Hort.* 103:145–151.

Reuveni, O., Y. Israeli, and E. Lahav. 1996. Somaclonal variation in *Musa* (bananas and plantains). In: *Biotechnology in agriculture and forestry—Somaclonal variation in crop improvement II*. Vol. 36, Y.P.S. Bajaj, ed., 174–196. Heidelberg: Springer.

Robinson, J.C. 1996. *Bananas and plantains*. Crop Production Science in Horticulture 5. Wallingford, UK: CAB International.

Roux, N.S., H. Strosse, A. Toloza, B. Panis, and J. Dolozel. 2004. Detecting ploidy level instability of banana embryogenic cell suspension cultures by flow cytometry. In: *Banana improvement: Cellular, molecular biology and induced mutations*, S.M. Jain and R. Swennen, eds., 251–261. Enfield, NH: Science Publishers.

Sahijram, L., J.R. Soneji, and K.T. Bollamma. 2003. Analyzing somaclonal variation in micropropagated bananas (*Musa* spp.) *In Vitro Cell Dev. Biol. Plant.* 39:551–556.

Sheidai, M., H. Aminpoor, Z. Noormohammadi, and F. Farahani. 2008. RAPD analysis of somaclonal variation in banana (*Musa acuminata* L.) cultivar Valery. *Acta Biol. Szeged* 52:307–311.

Shepherd, K. 1954. Seed fertility of the Gros Michel banana in Jamaica. *J. Hort. Sc.* 29(1):1–11.

Smith, M.K. and R.A. Drew. 1990. Current applications of tissue culture in plant propagation and improvement. *Aus. J. Plant Physiol.* 17:267–289.

Ssebuliba, R., D. Talengera, D. Makumbi, P. Namanya, A. Tenkuoano, W. Tushemereirwe, and M. Pillay. 2005. Reproductive efficiency and breeding potential of East African Highland (*Musa* AAA-EA) bananas. *Field Crops Res.* 95:250–255.

Su-shien, M. and C. Shii. 1974. Growing banana plantlets from adventitious buds. *J. Chin. Hort. Sci.* 20:1–7.

Swennen, R., D. Vuylsteke, and S.K. Hahn. 1992. The use of simple biotechnological tools to facilitate plantain breeding. In: *Biotechnology: Enhancing research on tropical crops in Africa*, G. Thottapilly, L.M. Monti, D.R. Mohan Raj, and A.W. Moore, eds., 69–74. Ibadan, Nigeria: CTA/IITA.

Talengera, D., M.J.G. Magambo, and P.R. Rubaihayo. 1994. Testing for a suitable culture medium for micropropagation of the East African highland bananas. *Afr. Crop Sci. J.* 2:17–21.

Talengera, D., D. Vuylsteke, and E. Karamura. 1996. *In vitro* germination of Ugandan banana hybrids. *MusaAfrica* 10:14.

Tripathi, L. and J.N. Tripathi. 2008. High frequency shoot regeneration of various cultivars of banana (*Musa* sp.). *J. Crop Improv*. 22:171–180.

Tripathi, L., J.N. Tripathi, R.T. Oso, J. d'A. Hughes, and P. Keese. 2003. Regeneration and transient gene expression of African *Musa* species with diverse genomic constitution and ploidy levels. *Trop. Agric.* 80:182–187.

Vos, P., R. Hogers, M. Bleeker, M. Reijans, T. Vandelee, M. Hornes, et al. 1995. AFLP: A new technique for DNA fingerprinting. *Nucl. Acids Res.* 23:4407–4414.

Vuylsteke, D. 1989. Shoot-tip culture for the propagation, conservation, and exchange of *Musa* germplasm. *Practical manuals for handling crop germplasm in vitro 2*. Rome, Italy: International Board for Plant Genetic Resources.

Vuylsteke, D. 1998. Shoot-tip culture for the propagation, conservation and distribution of *Musa* germplasm. Ibadan, Nigeria: International Institute of Tropical Agriculture.

Vuylsteke, D. and E. De Langhe. 1985. Feasibility of *in vitro* propagation of bananas and plantains. *Trop. Agr.* 62:323–328.

Vuylsteke, D. and R. Ortiz. 1996. Field performance of conventional vs. *in vitro* propagules of plantain (*Musa* spp., AAB group). *HortScience* 31:862–865.

Vuylsteke, D. and R. Swennen. 1992. Biotechnological approaches to plantain and banana improvement at IITA. In: *Biotechnology: Enhancing research on tropical crops in Africa*, G. Thottapilly, L.M. Monti, D.R. Mohan Raj, and A.W. Moore, eds., 142–150. Ibadan, Nigeria: CTA/IITA.

Vuylsteke, D. and D. Talengera. 1998. *Postflask management of micropropagated bananas and plantains*. Ibadan, Nigeria: IITA.

Vuylsteke, D., R. Swennen, G.F. Wilson, and E. De Langhe. 1988. Phenotypic variation among in vitro propagated plantain (*Musa* sp. cultivar AAB). *Sci. Hort.* 36:79–88.

Vuylsteke, D., R. Swennen, and E. De Langhe. 1990. Tissue culture technology for the improvement of African plantains. In *Sigatoka leaf spot diseases of bananas*, R.A. Fullerton and R.H. Stover, eds., 316–337. Proceedings of an international workshop. Montpellier, France: INIBAP.

Vuylsteke, D., R. Ortiz, R.S.B. Ferris, and J.H. Crouch. 1997. Plantain improvement. *Plant Breed.Rev.* 14:267–320.

Walther, R., A. Ilan, A. Lere, A. Duvdevani, and E. Khayat. 1997. Analysis of somaclonal variation in tissue cultured banana plants (*Musa* AAA cv. 'Grand Nain'). *Acta Hort.* 447:379–383.

16 Hybrid Distribution to Farmers
Adoption and Challenges

Abdou Tenkouano, Michael Pillay, and Ousmane Coulibaly

CONTENTS

16.1 INTRODUCTION

The adoption of new banana cultivars by large-scale producers is a relatively straightforward process when the commercial incentives for adoption are high or when there are no other alternatives. This was evidenced by the replacement of 'Gros Michel' by the Cavendish varieties. Therefore, this chapter will consider hybrid distribution to smallholder farmers who are responsible for the bulk of banana and plantain production in developing countries, usually under complex cultural and technological circumstances.

Productivity gains achievable from improved agricultural technologies have not been fully exploited by farmers in developing countries. There is a big gap between potential and achieved yields on farmers' fields, mainly due to various factors such as poor extension services, institutional and cultural constraints, and farmers' firm adoption of traditional practices and their limited financial ability and willingness to increase input levels in farming (Ghatak and Ingersent, 1984; Ali and Byerlee, 1991; Xu and Jeffrey, 1998; Pingali and Heisey, 1999; Kalirajan and Shand, 2001; Alene and Manyong, 2006a, 2006b). Although measures have been taken to achieve a high rate of adoption of new cultivars, little or no emphasis has been given to the adoption of other types of complementary information, which taken together represent the "best practice" technological package (Kalirajan, 1991; Pingali and Heisey, 1999; Kalirajan and Shand, 2001; Alene and Manyong, 2006a, 2006b). These general principles strongly apply in the case of banana and plantains.

One of the alternative measures taken to achieve a high adoption rate of new cultivars—especially in areas where formal seed systems do not exist, are ineffective, and/or are inefficient—has been the farmer-to-farmer seed diffusion. The basic premise behind such farmer-led technology diffusion is the rich theory of diffusion of innovation, which is described as "the process by which an innovation is communicated through certain channels over time among the members of a social system" (Rogers, 1995). The most striking feature of the diffusion theory is that, for most members of a social system, the innovation decision depends heavily on the innovation decisions of the other members of the system. Farmer-to-farmer technology transfer thus builds upon farmers' traditional seed-transfer methods and is based on the observation that farmers prefer that their fellow farmers are the primary sources of information even when they have alternative sources (Feder and Slade, 1985; Rogers, 1995). This must be considered when designing technology dissemination programs, particularly for improved seeds (Grisley, 1994; Kormawa et al., 2004).

However, when farmers have access to modern inputs such as improved seeds, they often lack the knowledge and agronomic and crop-management technologies that are crucial for bridging the yield gap (Pingali and Heisey, 1999). This is often the case with farmer-to-farmer technology diffusion, because crop management information is either not integrated into the farmer-to-farmer seed transfer scheme or is not effectively communicated.

Promoting improved seeds without the complementary management technologies through farmer-to-farmer transfer means that potential yields remain unexploited. For example, the yield achieved on farmers' fields depends not only on the amount of fertilizer applied but also on when and how it is applied—not only on whether improved cultivars are adopted but also on whether the recommended cropping patterns and systems are followed. In this situation, farmer-to-farmer seed transfer will promote partial adoption and results in underutilization of yield potential of new cultivars by adopters.

It is thus important to conceive seed transfer programs within the context of a technology package, particularly for bananas that have a crop cycle of 12–18 months from planting to harvest. This chapter focuses on the direct components of seed systems, including access to improved cultivars and quality planting materials, which are discussed with emphasis on technical and institutional issues.

16.2 ACCESS TO IMPROVED CULTIVARS AND QUALITY PLANTING MATERIALS

We have discussed the concept of breeding and targeting new cultivars in Chapter 9 and Chapter 12. In essence, we indicated that the prospects for adoption of new cultivars increase as the new cultivars are (1) adapted to biophysical and socioeconomic characteristics of the target area, (2) chosen with inclusive participation of the farmers in the selection process, and (3) introduced in a nondisruptive approach that preserves the existing cultivars. Once these prerequisites are met, it is important to devise a user-friendly approach for supply of planting materials.

16.2.1 Seed Systems: Promotion of Rapid Plant Propagation Techniques

A major constraint to expansion of banana and plantain cultivation is the scarcity of healthy planting materials. Farmers usually depend on natural regeneration of plants for the supply of planting materials. This is a very slow process that often produces small numbers of planting materials that are usually contaminated by various soil-borne pathogens such as nematodes. Transplanting of the contaminated materials often spreads nematode problems and shortens the lifetime of plantations to only one or two cycles of cultivation, beyond which most plants fall and become unproductive. In prelude to the wide-scale distribution of hybrids, the International Institute of Tropical Agriculture (IITA) has been looking at alternative means of producing planting materials, since it is unlikely that the average farmer will develop the capacity to multiply plants by micropropagation (tissue culture). The alternative methods (low-cost multiplication) can be classified into two categories: field techniques based on complete or partial decapitation and detached corm techniques practiced away from the field (Tenkouano et al., 2006). These techniques are discussed in detail in Chapter 15.

These approaches for multiplying planting material have been adopted by farmers on a commercial scale not only in Cameroon and Nigeria but also in Ghana. The method was widely used across eastern (Uganda, Tanzania) and southern (Malawi, Mozambique) Africa for mass-testing and delivery of new hybrids. This low-cost method was also used to rapidly multiply material for distribution to farmers in East Africa due to the ravages of Xanthomonas wilt (Crop Crisis Control Project, 2007).

16.2.2 Farmer-to-Farmer Diffusion of Hybrids in Nigeria

Breeding programs at the International Institute of Tropical Agriculture (IITA), the Fundación Hondureña de Investigación Agrícola (FHIA) in Honduras, and the Centre Africain de Recherches sur Bananiers et Plantains (CARBAP) in Cameroon have developed disease-resistant hybrids with appropriate agronomic characteristics. The development of these disease-resistant and high-yielding plantain and banana hybrids is a significant scientific achievement in recent history in terms of its potential contribution to food security and poverty reduction in Africa. The major varieties include 'PITA-14,' 'PITA-17,' 'PITA-21,' 'PITA-22,' 'PITA-23,' 'PITA-24,' 'PITA-25,' 'PITA-26,' 'BITA-3,' 'BITA-7,' 'FHIA-17,' 'FHIA-18,' 'FHIA-20,' 'FHIA-21,' 'FHIA-23,' 'FHIA-25,' and 'CRBP-39' (Tenkouano and Swennen, 2004).

16.2.2.1 Delivery Package

Multilocation on-farm farmer-participatory trials were set up to evaluate these improved hybrids for the widespread distribution of those materials fitting the ecological and culinary preparation needs in different countries across Africa. Several hybrids were identified as promising in many countries, for example, 'PITA-2' and 'PITA-5' in Ghana, 'PITA-17' and 'PITA-14' in Nigeria, 'BITA-3' in Ghana and Uganda. It must be noted that farmers' preferences for distributed hybrids vary in different countries. Out of four cultivars—'BITA- 3,' 'FHIA-17,' 'FHIA- 21,' and 'CRBP-39'—farmers in Cameroon preferred 'FHIA-21' followed by 'FHIA-17' (M. Pillay and G. Blomme, unpublished data). In Mozambique, farmers preferred the hybrid 'SH-3640' and landrace 'Grande Naine' and showed less preference for 'FHIA-17,' 'FHIA-21,' and 'FHIA-23' (Uazire et al., 2008).

Variety dissemination projects should be designed to have significant economic impact in the medium term. In Nigeria, for example, on-farm demonstrations were launched in 2000 to promote an improved plantain and banana package consisting of: (1) planting materials of hybrids, (2) pests and diseases control methods, (3) agronomic practices, and (4) postharvest utilization. Between 2000 and 2001, over 5,000 farmers obtained planting materials through farmer-to-farmer diffusion of planting materials, and this had grown to over 20,000 farmers by 2005. Similarly, over 4,000 farmers were targeted in Tanzania, Mozambique, Malawi, and Zambia between 2003 and 2005. While the growing number of beneficiary farmers over the years certainly shows the effective dissemination of hybrid planting materials, little is known about the dissemination of the remaining components of the improved hybrid delivery package.

Achieving the productivity gains from hybrids also depends critically on the adoption of the recommended associated technologies. Given that the cropping pattern component of the technology package is provided in the form of information, its dissemination depends not only on the effectiveness of the farmer-to-farmer communication but also on the quality of the technical advice provided during major cropping operation.

16.2.2.2 Delivery Process

Instrumental in the implementation of large-scale variety dissemination is the quality of collaboration among multiple stakeholders, from the farmers to extension institutions, to community-based organizations, including faith-based and self-help groups (Tenkouano et al., 2010). A careful choice of partners is essential in implementing large-scale delivery for evaluation of improved hybrids of banana and plantain to the smallholder farmers. This is because the target regions are usually very large and cut across ecological and linguistic/cultural landscapes, each with unique socioeconomic features. For example, the Nigerian plantain belt (NPB) is a region lying between 3° E and 9° E of longitude and south of 8° N latitude, with a northern boundary running approximately parallel to the coast of the Gulf of Guinea. This region encompasses the territories of 11 states, namely, Abia, Akwa-Ibom, Bayelsa, Cross River, Delta, Edo, Imo, Ogun, Ondo, Oyo, and Rivers. Each state has an agricultural extension network coordinated by erstwhile World Bank–sponsored Agricultural Development Programmes (ADP). Other field operatives of banana and plantain extension include the National Horticulture Research Institute (NiHort) and the Plantain and Banana Development Programme (PBDP), agencies of the federal government of Nigeria. A large number of voluntary service organizations (VSOs) that operate as nongovernmental organizations (NGOs), sometimes as community-immersed or community-based organizations (CBOs), and the network of schools constitute other key components of the landscape.

Four ADPs were chosen by their peers and other stakeholders to serve as relay centers that would be repositories of the disseminated hybrids. Trials carried out at the relay centers were considered to be reference trials where hybrids were grown under two cropping regimes: recommended best agronomic practices and standard farmer practice. Additionally each of these relay centers would also serve as a source of planting material for further distribution of hybrids selected by the growers. Therefore, IITA established model plant-propagation units and trained the staff of the relay centers for operating the propagation facility as a business. The relay centers also served as schools for demonstration of post-harvest processing options that were components of the delivery package. With their triple function, the relay centers were also known as Plantain Resource and Training Centers (PRTC), designed to be self-sustaining. Among the criteria used by peers to designate which ADP would be a PRTC were geographical position (central enough to cater for several ADPs with minimum distance) and logistical endowment (including willingness to cofinance the setting up of the facilities and training events for their own personnel as well as beneficiaries of the dissemination process).

A joint exploratory survey involving the ADPs, NiHORT, PBDP, and IITA was carried out to identify candidate farmers for the study. Thus, shortlists of five farmer households for each of five farming communities per state were provided by the extension agencies and site-screened by IITA on the basis of the suitability of their farms (size and accessibility, biophysical condition) and their personal credentials (experience with plantain cultivation, leadership status in the community).

One farmer household per community was retained, giving a sample of 55 contact farmer households (CFH) for the study. The household approach ensured that collective knowledge remains within the family and not solely in the hands of one person. Thus each CFH served as the social experimental unit, linking to the communities where they are sited as well as those of neighboring villages. The entire set of hybrids in the reference trials was also planted at each CFH (mother trials) to simulate performance under farmers' conditions.

Each CFH was aggregated with one PRTC in such a way that hybrids selected by the communities served by the linked CFHs would be the primary targets for mass propagation for further distribution to new growers in waves of five new growers per crop cycle for each CFH. At this stage, only those hybrids chosen by farmers from the mother trials would undergo further distribution to establish baby trials, usually consisting of subsets of the chosen hybrids.

Each farmer receiving planting materials would have to agree to provide two suckers for each one received so that these could be distributed to more farmers, ensuring a steady farmer-to-farmer diffusion of the best performing hybrids. Also, each receiving farmer would be subjected to training in simple field multiplication and sanitation techniques as described earlier.

Finally, regular progress review and planning meetings involving all the stakeholders were held at least twice a year to exchange experiences and document practices that should be encouraged (do's) and those that should not be recommended (don'ts) with respect to both the delivery package and the delivery process. Table 16.1 illustrates the outcome of one such meeting. These meetings

TABLE 16.1
Highlights of Project Review and Planning Meeting in Nigeria

Proposed Interventions	Specific Tasks
1. Improved hybrids	
1.1.Phase I (Old): Hybrids (first batch introduced in 2000)	Select best bets for large-scale distribution.
	Develop a plan for rapid multiplication of best bets.
	Develop a plan for maintenance and sourcing of best bets.
1.2.Phase II (New): Varieties (second batch introduced in 2005)	Set up large demonstration plots at the four designated PRTCs and at selected sites with proven capacity and commitment to handle large demonstrations. Plots will follow a classical research design, allowing for data collection and statistical analysis as part of a hybrid release scheme.
	Organize field days for public awareness and participatory evaluation of the hybrids.
	Select best bets and undertake rapid multiplication for distribution to contact farmers.
2. Integrated crop management	Superimpose to new hybrids a set of recommended cultural practices to sustain yield. Hybrid demonstrations plots will be divided into two subplots, a well-managed subplot [Do's] and a poorly managed subplot [Don'ts].
	Set up additional demonstration plots on pest management (focus on nematodes) to prolong plantation lifespan.
	Make provision for monitoring and preemptive management of pests (nematodes) and pathogen (Sigatoka, emerging threats) population shifts.
3. Training (capacity building)	At least two intensive training sessions on plantain management and sucker multiplication/sanitation will be encouraged at each PRTC with a well-coordinated calendar. The training is to be decentralized so that the ADP can handle it with backstopping from IITA based on needs.
	Training curricula, supported by pictorial booklets, will be disaggregated into component modules that can be handled separately or concomitantly based on prospective trainees' interests.
	Posters will be developed to capture in a glance key messages and technology descriptions (hybrid characteristics, multiplication techniques, postharvest processing).
4. Governance (resource center)	Ideally, each PRTC should have a governing or advisory council that will oversee the operations of the PRTC. The council should be chaired by the chief executive officer of the host ADP and include one representative from each of the states in the operating zones, one representative each of stakeholders in production and processing, one representative of NIHORT, one representative of PBDP, with IITA participating as an observer.
	Address policy issues with respect to seed certification systems, postharvest quality control (processing, packaging, mycotoxins).
5. Impact assessment of phase I	To be carried out by an independent body in order to assess the delivery model (partnerships, capacity building) and technologies (hybrids, rapid multiplication, postharvest) of HDP-I.

ensure the collective appropriation of project interventions by all parties involved. With a greater understanding of the directions the project is heading to and the way this is done, there is a greater commitment to contributing resources and implementing interventions by all parties.

16.2.2.3 Preliminary Assessment of Impact

Aitchedji et al. (unpublished) attempted to assess preliminary changes associated with the dissemination project. They concluded that farmers' capacity to choose and use planting materials and related production techniques has been improved by training programs over the 2000–2004 period of the dissemination project. The curricula for training included: information on hybrids, crop management practices, pests and diseases management technologies, sucker multiplication, and postharvest and food processing techniques. Also farmer's awareness has been increased with field days and exchange of information on hybrids and associated techniques. All participants reported that the capacity-building curricula were adapted to their needs and opportunities. More than half of the farmers were aware of and had adopted disease-resistant plantain and banana hybrids at least once by 2005.

Most of the adopters grew both local cultivars and hybrids in sole cropping and/or intercropping. The main reasons for adoption evoked by farmers were (1) good/high yield, (2) good taste and high quality, (3) planting material (suckers) availability and accessibility, (4) good sale (high market demand), (5) resistant to pest and disease (black Sigatoka), and (6) early maturity of plant and fruit. The main reason reported for mixing hybrids and local cultivars is to control black Sigatoka and to strengthen resistance of local varieties. Pioneer farmers were key sources of information and suppliers of suckers for other farmers in rural areas.

Farmers' awareness and knowledge of disease-resistant hybrids through participation in on-farm trials and/or field days and demonstration plots are important factors in adoption. In addition, farmers who participated in the project's training programs adopted the hybrids because of their participation and benefits gained from their attendance. The project's collaboration with ADPs for organization of annual training programs on different aspects of plantain and banana contributed to the dissemination of hybrids. The attendance to project training programs on plantain and banana contributed to the adoption of plantain and banana hybrids by small-scale farmers.

The Nigeria experience has since inspired similar large-scale delivery of improved hybrids across Africa, notably in Cameroon, Ghana, Rwanda, and Uganda. It would be informative to document the economic and social benefits of these initiatives.

16.3 UNDERSTANDING AND MANAGING DIVERSITY

Farmers usually grow a number of cultivars that they maintain perennially through successive crops derived from suckers (shoots). Suckers constitute the planting material of choice for farmers, which they acquire on an *ad hoc* basis from neighbors or rural markets, using local names as guides for purchasing the suckers.

It is difficult to distinguish among cultivars on the basis of vegetative characteristics alone, particularly in juvenile plants. Even when planting materials are obtained from mature plants, there is still a high probability of cultivar mixtures, because homonyms might actually be different cultivars. This may result in intrafield diversity that may be unintended (since homonymous suckers may not be identical cultivars) or deliberate (because intrafield diversity may be viewed as risk management strategy) as outlined by Smithson and Lenne (1996) and Kahmen et al. (2005).

Managing diversity in farmers' fields requires a good understanding of the pattern of variation among popular cultivars, particularly in juvenile plants. This in turn determines to what extent plant breeding may succeed in releasing new cultivars into existing production schemes, as indicated in Chapter 9. Selatsa et al. (2009) provide a comprehensive discussion of this matter.

16.4 BANANA TRADE PROSPECTS FOR AFRICA

16.4.1 Trends in Commercial Banana Production in Sub-Saharan Africa

This section is largely drawn from the comprehensive analysis of Arias et al. (2003) of the United Nations Food and Agriculture Organization (FAO). Since the 1960s, banana production in Africa has increased at about 2.2% per annum, while the area harvested has grown at a slightly reduced pace of 1.7%. In the same period land productivity has increased slowly, at less than 1% per annum. However, yield has not been the same in all countries.

Export-oriented banana producers use a relatively high supply of external inputs. Côte d'Ivoire and Cameroon are the most important banana-exporting countries, and their markets are in Europe and, to a lesser extent, neighboring Burkina Faso, Mali, and Senegal. Banana exports from Côte d'Ivoire and Cameroon, which in 2001 accounted for 98% of all banana exports from Africa, have increased in the period between 1987 and 2000 at a sustained rate of 10% per annum. This contrasts sharply with the steady decline in banana exports in the previous period (1960–1986) when exports decreased by 2% per annum.

In Cameroon, banana exports had declined steadily since the early 1960s from some 140,000 in 1961 to little more than 20,000 in 1987. Exports recovered during the 1970s with the support of the Organisation Camerounaise de la Banane (OCB) that helped farmers to fight Panama disease.

However, Sigatoka and weather-related problems affected production during the 1980s and exports declined to a record low in 1987. In 1987 foreign companies began to take a leading role in the development of banana for export. In that year Del Monte engaged in a joint venture with the Cameroon Development Corporation (CDC). Today CDC and Del Monte plant some 2,100 ha of bananas: CDC provides land and labor, while Del Monte provides credit and technical assistance, and markets the produce. CDC is currently the largest employer after the state, and the government allows the company tax exemptions for the importation of banana production inputs.

In 1990, the Organisation Camerounaise de la Banane (OCB) was purchased by the Compagnie Fruitière, the retail company in charge of selling OCB produce in Europe. At present, Compagnie Fruitière is controlled by Dole. Both Dole and Del Monte made large investments into irrigation, fruit-handling facilities, and sanitation equipment. Today bananas in Cameroon provide direct employment to some 10,000 people in rural areas and exports in 2002 reached almost 260,000, having grown at a rate of 10% per annum since 1988. Of these, an estimated 215,000 went to the European Union and the rest to Eastern Europe, North Africa, and neighboring African countries.

Bananas in Côte d'Ivoire are, together with pineapples, the most important exported fruits. Bananas for export are produced by farms of different sizes and under different production techniques, from small farms using low-input technologies to large and input-intensive plantations. However, smaller farms, which are usually located in poor soils with steep slopes and limited water availability, are finding it increasingly difficult to compete and are slowly disappearing. The increase in banana production in recent years (5.4% per annum in the period 1987–2001) was accompanied by an increase in the scale of production, which in turn was caused by the closing down of smaller farms.

Increases in production were accompanied by an expansion of exports in a pattern that resembles that of Cameroon. Following a period of steady decline from the mid 1970s to the mid 1980s, the period between 1987 and 2000 saw a sustained growth in exports of 9% per annum, from 82,000 in 1987 to 240,000 t in the year 2000. The increase coincides with a stronger participation of multinational companies and the dissolution by the government of the Cooperative de Producteurs pour la Commercialization des Fruits et Légumes de la Côte d'Ivoire (COFRUITEL) in 1985, an organization that had since 1978 grouped banana producers and was in charge of marketing.

Nowadays banana production for export consists of an integrated system whereby a few large operators specify the production technology to be applied and market the produce internationally. The government is no longer involved in production, and its role is limited to monitoring the phytosanitary aspects of the exported fruit. Almost 80% of all exports go to the European Union, and the rest to North Africa and the Middle East. The bulk of bananas is produced in 65 plantations covering an area of approximately 5,500 ha and provide employment to about 20,000 people.

Producers are grouped in the Organisation Centrale des Producteurs Exportateurs d'Ananas et de Banane (OCAB), created in 1991. The largest export companies are the Société pour le Développement de la Culture de la Banane (SCB), a subsidiary of Dole with a share of about 50% of all banana exports; Banador, a subsidiary of Chiquita with a share of 25%; and Canavèse, with a share of 10%.

Currently, major efforts are underway to restore the banana export potential of Guinea—targeting countries of the Middle East—with support from the Common Fund for Commodities (CFC) of the

World Bank and from private investors from the Arab Peninsula. It is noteworthy that land areas cultivated with banana and plantain have increased by about 50% over the past 12 years, with a largely untapped potential in the major production zones of Guinée Maritime (coastal Guinea) and Guinée Forestière (bordering Sierra Leone, Liberia, and Côte d'Ivoire).

Finally, some export-oriented banana production occurs in the Volta region of Ghana, with a specific emphasis on fair-trade to target customers in the European Union.

16.4.2 Strategic Implications

Most of banana production is carried out by small-scale farmers in most of the countries in sub-Saharan Africa, including those that have developed an export-oriented production. However, the small-scale farmers are hardly part of the export chain. Rather, they supply domestic or regional markets.

Export-oriented production is carried out by a small number of entrepreneurs operating large-scale farms with high-input technology and in close partnership with transnational trade companies. Therefore, it is advised that the development of a fresh produce banana-based economy should follow a dual approach, focusing on:

- Promoting the development of large-scale farms with significant investment support. This could be done through the promulgation of conducive policies that attract foreign investors or incite domestic investors towards banana production and commercialization.
- Promoting the organization of small-scale farmers in cooperative-like institutions. These institutions will benefit from investment support for access to improved production technologies and serve as brokers for access to international export markets.

It is advised that the domestic and regional markets should also be targeted, particularly in view of the lower transportation costs, and the availability of numerous value-adding postharvest processing options that could (1) rescue from wastage fruits that are rejected for international export and (2) subtend the development of an agro-processing industry.

While international trade opportunities should focus on Cavendish varieties (for example, 'Dwarf Cavendish,' 'Giant Cavendish,' 'Williams,' and 'Grande Naine'), there is also ample scope for non-Cavendish varieties of the plantain and cooking banana types, particularly for domestic and regional markets. Only about 10% of world banana production is traded at the international level. The share of regional trade is not well documented, but it is known that the consumption of banana and plantain products is well entrenched in the dietary habits of the people, and demand for the products is high even in countries lacking the climatic conditions for the production of these crops.

While this section used examples from West and Central Africa, the trends are the same and the implications similar for other regions of sub-Saharan Africa where the major players currently are South Africa, Kenya, and Uganda, but other countries like the Democratic Republic of Congo, Ethiopia, and Mauritius are also potential strongholds for banana production and trade.

16.5 INTERNATIONAL DISTRIBUTION OF HYBRIDS

During the past few decades of plantain and banana research, progress has been made in the area of host-plant resistance to black Sigatoka, through the development of improved hybrids and populations with Sigatoka resistance (Tenkouano and Swennen, 2004). The development of these disease-resistant and high-yielding plantain and banana hybrids constitutes a most significant scientific achievement in recent history in terms of the impact they could have on the alleviation of hunger in Africa. What is now needed is to introduce these hybrids into farmers' fields in sizeable quantities.

There are only a handful of active breeding programs across the world, yet bananas are grown in more than 130 countries. Therefore, international cooperation to facilitate access to improved varieties has been a commendable achievement of the banana research community through the International Network for Improvement of Banana and Plantain (INIBAP; see Chapter 9). Thus, an international transit center (ITC) was set up in a non-banana-producing country (Belgium) to receive, sanitize (when required), and multiply for distribution genetic stocks from all over the world.

Virtually all agencies engaged in genetic improvement through breeding, mutation, and selection have deposited specimen of their products at the ITC (Table 16.2). Lacking in this repository are products from the breeding programs operating in India. Also lacking are cooking banana varieties recently derived from East African Highland bananas by the breeding program jointly operated by IITA and the National Agricultural Research Organization (NARO) in Uganda and new dessert banana varieties developed by French Agricultural Research Centre for International Development (CIRAD) for commercial production in the French Caribbean islands.

There has been no restriction in distributing the available accessions or breeding products across the world for production or research purposes as long as normal precautionary measures to ensure that no biological threats (notably, viruses) are accidentally disseminated through infected plant materials. Through the various mechanisms of INIBAP (now Commodities for Livelihoods, Bioversity), such as the International *Musa* Testing Program (IMTP), new varieties were channeled to an impressive list of recipients across the word (Table 16.3).

16.6 CONCLUSION

The success of plant breeding programs is measured by the extent to which the breeding products are adopted and used by the growers, profitably and durably. Beyond the genetic products, perhaps the more challenging task for the breeders is to understand and help establish the complex battery of institutional and transactional measures that create a conducive delivery framework.

Access to improved hybrids is extremely important. Therefore, the willingness of breeding programs to place their products under custody of a reputable international conduit that helps to maintain the intellectual property of the contributing programs is commendable. It is expected that the influx of improved hybrids into the ITC and efflux from the ITC would be asymmetrical, given the small number of active breeding programs and the large number of institutions needing improved hybrids for research or commercial use. Intellectual property issues may also be responsible for the asymmetrical influx/efflux noted in some cases where countries drawing from the ITC have not yet contributed to the ITC repository, despite having active breeding programs with tangible products. Once the asymmetry is removed and improved hybrids meet the biological requirements for safe international exchange, producers throughout the world would have increased access to a pool of hybrids that would meet their production needs and those of the markets they serve.

TABLE 16.2
Accessions and Breeding Stocks Available for International Distribution at the International Transit Center (ITC) of the International Network for Improvement of Banana and Plantain (INIBAP)

Contributing Breeding Programme	Accession or Breeding Stock Designation
Banana board/Imperial College of Tropical Agriculture (ICTA)	I.C. 2 (ITC0019), Bodles Altafort (ITC0366), 2390 (ITC0367), 2390-2 (ITC0553), Calypso (ITC0555), TU8 (ITC0581), Buccaneer (ITC1248), B7925 (ITC1320)
Centre Africain de Recherches sur les Bananiers et Plantains (CARBAP)	CRBP 01 (ITC1337), CRBP 14 (ITC1338), CRBP 15 (ITC1339), CRBP 37 (ITC1342), CRBP 39 (ITC1344)
Centre de Coopération Internationale en Recherche Agronomique pour le Développement (CIRAD)	IRFA 904 (ITC1266), IRFA 905 (ITC1267), IRFA 908 (ITC1268)
Empresa Brasileira de Pesquisa Agropecuaria (EMBRAPA)	Diploide EMBRAPA 205 (ITC1193), Tetraploide EMBRAPA 401 (ITC1194), PC12-05 (ITC1260), PA03-22 (ITC1261), PV03-44 (ITC1262), JV03-15 (ITC1263), PA 12.03 (ITC1302), PV 42-53 (ITC1310), PV 42-81 (ITC1312), PV 42-320 (ITC1313), JV 42-41 (ITC1314)
Fundación Hondureña de Investigación Agricólà (FHIA)	SH-3142 (ITC0425), Balbisiana tetraploide (ITC0454), FHIA-21 (ITC0503), FHIA-01 (ITC0504), FHIA-02 (ITC0505), FHIA-03 (ITC0506), FHIA-04 (ITC0968), FHIA-17 (ITC1264), FHIA-23 (ITC1265), SH-3640 (ITC1307), FHIA-18 (ITC1319), SH-3751 (ITC1324), FHIA-21 (ITC1332), FHIA-18 (ITC1412), FHIA-25 (ITC1418)
International Atomic Energy Agency (IAEA)	GN 60A (ITC1328), Novaria (ITC1329)
International Institute of Tropical Agriculture (IITA)	PITA-1 (ITC1081), TMPx 2481 (ITC1093), PITA-4 (ITC1094), PITA-9 (ITC1110), TMBx 612-74 (ITC1119), PITA-5 (ITC1141), PITA-7 (ITC1195), PITA-11 (ITC1196), PITA-6 (ITC1198), TMPx 4744-1 (ITC1199), PITA-3 (ITC1200), TMPx 5706-1 (ITC1201), PITA-12 (ITC1203), TMPx 7356-1 (ITC1204), PITA-10 (ITC1205), T6 (ITC1247), PITA-8 (ITC1272), TMP2x 1297-3 (ITC1278), TMP2x 1297-3 (ITC1292), TMPx 4479-1 (ITC1293), PITA-14 (ITC1294), BITA-2 (ITC1296), BITA-3 (ITC1297), TMP2x 1297-3 (ITC1415), TM2x 2829-62 (ITC1416), PITA-16 (ITC1417), TMB2x 9128-3 (ITC1437)
Instituto Nacional de Investigaciones de Viandas Tropicales (INIVIT)	IBP 5–61 (ITC1478), IBP 5-B (ITC1479), IBP 12 (ITC1480), SH-3436-9 (ITC1283), SH-3436-6 (ITC1284), SH-3436–9 (ITC1318)
Taiwan Banana Research Institute (TBRI)	'GCTCV-215' (ITC1271), 'GCTCV-119' (ITC1282), 'GCTCV-215' (ITC1436), 'GCTCV-106' (ITC1442), 'GCTCV-247' (ITC1443)

Note: INIBAP is now a program of Bioversity International. Accessions codes are followed by ITC reference numbers in parentheses.

Source: Data courtesy of Ines Vandenhouwe (Bioversity International, ITC manager).

TABLE 16.3
Distribution of Improved Varieties of Bananas from the International Transit Center (ITC) of the International Network for Improvement of Banana and Plantain (INIBAP)

Breeding Program	Accession Number	Recipient Countries
Banana Board/ICTA	ITC0019	Belgium, Burundi, DR Congo, Germany, Guadeloupe, India, Nigeria, Tanzania, Uganda, Zimbabwe
	ITC0366	Costa Rica, Ethiopia, Guadeloupe, Reunion, United Kingdom
	ITC0367	India
	ITC0553	Costa Rica, DR Congo, Guadeloupe, India, Marshall Islands, Philippines, United States
	ITC0555	Burundi, Costa Rica, India, Philippines, Reunion, United Kingdom
	ITC0581	Kenya, United Kingdom, Zimbabwe
	ITC1248	Costa Rica, Guadeloupe, India, Reunion, United Kingdom
	ITC1320	Bangladesh, Ecuador, India
CARBAP	ITC1342	Cameroon, China, Costa Rica
	ITC1344	Belgium, Burundi, Cambodia, Cameroon, China, Colombia, Costa Rica, Côte d'Ivoire, Cuba, DR Congo, Dominican Republic, Ecuador, France, French Polynesia, Ghana, Guyana, Haiti, Honduras, India, Indonesia, Jamaica, Liberia, Malaysia, Marshall Islands, Martinique, Mauritius, Mexico, Myanmar, Nicaragua, Nigeria, Pakistan, Panama, Papua New Guinea, Peru, Philippines, Puerto Rico, South Africa, Sri Lanka, Taiwan, Tanzania, Thailand, Uganda, United States, Venezuela
CIRAD	ITC1267	India
	ITC1268	France, Guadeloupe, Vietnam
EMBRAPA	ITC1194	Guadeloupe, Vietnam
	ITC1260	Guadeloupe, India, United States
	ITC1261	Belgium, Brazil, Burundi, Cameroon, Colombia, Costa Rica, Cuba, Fiji Islands, Guadeloupe, Honduras, India, Indonesia, Malaysia, Mexico, Nigeria, Philippines, Puerto Rico, Saint Lucia, South Africa, Spain, Taiwan, Thailand, Tonga, Uganda, United States
	ITC1262	Belgium, Brazil, Burundi, Cameroon, Colombia, Costa Rica, Cuba, Dominican Republic, Guadeloupe, Honduras, India, Indonesia, Malaysia, Mexico, Nigeria, Philippines, Puerto Rico, Saint Lucia, South Africa, Spain, Taiwan, Thailand, Tonga, Uganda, United States
	ITC1263	United States
	ITC1302	Australia, Cameroon, Costa Rica, Fiji Islands, Mexico, Puerto Rico, Spain, United States
	ITC1310	Australia, Bangladesh, Cameroon, Costa Rica, Guadeloupe, Malaysia, Rwanda, Spain, Uganda, United States
	ITC1312	Australia, Bangladesh, Cameroon, Costa Rica, Guadeloupe, Puerto Rico, Spain
	ITC1313	Australia, Bangladesh, Cameroon, Costa Rica, DR Congo, Guadeloupe, India, Puerto Rico, Spain, United States
	ITC1314	Australia, Bangladesh, Cameroon, Costa Rica, Dominican Republic, Guadeloupe, Spain, United States
FHIA	ITC0425	Belgium, Guadeloupe, Venezuela
	ITC0454	Germany, India, Philippines
	ITC0503	Burundi, United Kingdom

(continued)

TABLE 16.3 (CONTINUED)
Distribution of Improved Varieties of Bananas from the International Transit Center (ITC) of the International Network for Improvement of Banana and Plantain (INIBAP)

Breeding Program	Accession Number	Recipient Countries
	ITC0504	Australia, Bangladesh, Barbados, Belgium, Bolivia, Brazil, Burundi, Cambodia, Cameroon, China, Colombia, Comoros, Congo, Republic of, Costa Rica, Cuba, Cyprus, DR Congo, Dominican Republic, Ecuador, Eritrea, Fiji Islands, France, French Polynesia, Gabon, Gambia, Germany
	ITC0505	American Samoa, Australia, Bangladesh, Barbados, Belgium, Bolivia, Brazil, Burundi, Cambodia, Cameroon, China, Colombia, Comoros, Congo, Republic of, Costa Rica, DR Congo, Dominican Republic, Ecuador, Eritrea, Fiji Islands, France, French Polynesia, Gabon, Gambia, Germany, Guadeloupe, Guinea, Haiti, Honduras, India, Indonesia, Ireland, Jamaica, Kenya, Malawi
	ITC0506	American Samoa, Australia, Bangladesh, Barbados, Belgium, Bolivia, Brazil, Burundi, Cambodia, Cameroon, China, Colombia, Comoros, Congo, Republic of, Costa Rica, Cuba, DR Congo, Dominican Republic, Ecuador, Eritrea, Fiji Islands, France, French Polynesia, Gabon, Gambia, Germany, Ghana, Guadeloupe, Guinea, Guyana, Haiti, Honduras, India, Indonesia
	ITC0968	Burundi, Cameroon, Colombia, Costa Rica, Honduras, Nigeria, United Kingdom, United States
	ITC1264	American Samoa, Australia, Bangladesh, Belgium, Bolivia, Brazil, Burundi, Cambodia, Cameroon, China, Colombia, Comoros, Costa Rica, Côte d'Ivoire, Cuba, DR Congo, Dominican Republic, Ecuador, Fiji Islands, France, Gambia, Ghana, Guadeloupe, Honduras, India, Indonesia, Jamaica, Jordan, Malawi, Malaysia, Mauritius, Mexico, Myanmar, Nicaragua, Nigeria, Oman, Pakistan, Palau, Papua New Guinea, Peru, Philippines, Reunion, Seychelles, South Africa
	ITC1265	American Samoa, Bangladesh, Bolivia, Brazil, Burundi, Cambodia, Cameroon, China, Colombia, Comoros, Costa Rica, Côte d'Ivoire, Cuba, DR Congo, Dominican Republic, Ecuador, Eritrea, France, French Polynesia, Gambia, Germany, Ghana, Guadeloupe, Honduras, India, Indonesia, Ireland, Jamaica, Jordan, Madagascar, Malawi, Malaysia, Mauritius
	ITC1307	American Samoa, Belgium, Burundi, Cambodia, Cameroon, China, Colombia, Costa Rica, DR Congo, Dominican Republic, Ecuador, Fiji Islands, France, Guadeloupe, India, Indonesia, Jamaica, Malawi, Malaysia, Mauritius, Myanmar, Nicaragua, Oman, Pakistan, Papua New Guinea, Peru, Philippines, Puerto Rico, Reunion, South Africa, Spain, Sri Lanka, Suriname, Taiwan, Tanzania, Thailand, Uganda, United Kingdom
	ITC1319	American Samoa, Bangladesh, Belgium, Bolivia, Burundi, Cambodia, Cameroon, China, Colombia, Costa Rica, Côte d'Ivoire, Cuba, DR Congo, Dominican Republic, Ecuador, Fiji Islands, France, French Polynesia, Gambia, Germany, Ghana, Guadeloupe, Haiti, India, Indonesia, Malawi, Malaysia, Martinique, Mauritius, Myanmar, Nicaragua, Nigeria, Oman, Pakistan, Panama, Papua New Guinea, Peru, Philippines, Puerto Rico, Reunion, Samoa, South Africa, Spain, Sri Lanka, Taiwan, Tanzania, Thailand, Trinidad and Tobago, Uganda, Venezuela, Vietnam
	ITC1324	Cameroon, France

(continued)

TABLE 16.3 (CONTINUED)
Distribution of Improved Varieties of Bananas from the International Transit Center (ITC) of the International Network for Improvement of Banana and Plantain (INIBAP)

Breeding Program	Accession Number	Recipient Countries
	ITC1332	American Samoa, Australia, Bangladesh, Belgium, Bolivia, Burundi, Cambodia, Cameroon, China, Colombia, Costa Rica, Côte d'Ivoire, DR Congo, Dominican Republic, Ecuador, Fiji Islands, France, Gambia, Ghana, Haiti, Honduras, India, Indonesia, Israel, Jamaica, Malawi, Malaysia, Mauritius, Mexico, Myanmar, Nicaragua, Nigeria, Oman, Pakistan, Papua New Guinea, Peru, Philippines, Puerto Rico, Reunion, South Africa, Sri Lanka, Suriname, Taiwan, Tanzania, Thailand, Trinidad and Tobago, Uganda, United Kingdom, Venezuela
	ITC1350	Australia
	ITC1412	China, Comoros, DR Congo, Ecuador, India, Mauritius, Mexico
	ITC1418	American Samoa, Bangladesh, Belgium, Burundi, Cambodia, Cameroon, China, Colombia, Comoros, Costa Rica, Côte d'Ivoire, Cuba, DR Congo, Ecuador, Fiji Islands, Gambia, Germany, Ghana, Guyana, India, Indonesia, Jamaica, Jordan, Malawi, Malaysia, Mauritius, Mexico, Myanmar, Nicaragua, Oman, Pakistan, Papua New Guinea, Peru, Philippines, Puerto Rico, Reunion, South Africa, Sri Lanka, Suriname, Taiwan, Tanzania, Thailand, Uganda, Venezuela, Vietnam
IAEA	ITC1328	Austria, Colombia, Costa Rica, Israel, United States
IITA	ITC1082	United Kingdom, United States
	ITC1094	United States
	ITC1141	United States
	ITC1200	United States
	ITC1202	United States
	ITC1203	United States
	ITC1205	Belgium, Brazil, Cameroon, Colombia, Costa Rica, Cuba, DR Congo, Dominican Republic, Ecuador, Germany, Ghana, Honduras, India, Israel, Jamaica, Mauritius, Nicaragua, Nigeria, Puerto Rico, United States
	ITC1247	Burundi, Costa Rica, Ecuador, Fiji Islands, Guadeloupe, India, Papua New Guinea, Reunion, Uganda, United Kingdom
	ITC1272	Belgium, Brazil, Cameroon, Colombia, Costa Rica, Cuba, DR Congo, Dominican Republic, Ecuador, Ghana, Honduras, India, Jamaica, Nicaragua, Nigeria, Uganda, United Kingdom
	ITC1278	India, Jamaica, Malaysia, Mauritius, Uganda, Venezuela
	ITC1293	Belgium, Brazil, Cameroon, Colombia, Costa Rica, Cuba, DR Congo, Dominican Republic, Ecuador, Germany, Ghana, Honduras, India, Israel, Jamaica, Mauritius, Nicaragua, Nigeria, Puerto Rico, Uganda, United States
	ITC1294	Belgium, Brazil, Cameroon, Colombia, Costa Rica, Cuba, DR Congo, Dominican Republic, Ecuador, Ghana, Honduras, India, Jamaica, Mauritius, Nicaragua, Nigeria, Puerto Rico, Uganda, United Kingdom, United States
	ITC1296	Australia, Belgium, Brazil, Burundi, Cambodia, Cameroon, China, Costa Rica, Cuba, DR Congo, Dominican Republic, Ecuador, Ghana, Guadeloupe, India, Indonesia, Jamaica, Madagascar, Malawi, Malaysia, Marshall Islands, Mauritius, Mexico, Myanmar, Nicaragua, Nigeria, Pakistan, Papua New Guinea, Philippines, Puerto Rico, Puerto Rico, South Africa, Spain, Sri Lanka, Taiwan, Tanzania, Thailand, Uganda, United Kingdom, United States

(continued)

TABLE 16.3 (CONTINUED)
Distribution of Improved Varieties of Bananas from the International Transit Center (ITC) of the International Network for Improvement of Banana and Plantain (INIBAP)

Breeding Program	Accession Number	Recipient Countries
	ITC1297	Australia, Belgium, Brazil, Burundi, Cambodia, Cameroon, China, Colombia, Comoros, Costa Rica, Côte d'Ivoire, Cuba, DR Congo, Dominican Republic, Ecuador, Ghana, Guadeloupe, India, Indonesia, Jamaica, Madagascar, Malawi, Malaysia, Marshall Islands, Mauritius, Mexico, Myanmar, Nicaragua, Nigeria, Pakistan, Panama, Papua New Guinea, Philippines, Puerto Rico, South Africa, Spain, Sri Lanka, Taiwan, Tanzania, Thailand, Uganda, United States
	ITC1415	India
	ITC1416	Dominican Republic, India, Uganda, Venezuela
	ITC1417	Australia, Colombia, Costa Rica, DR Congo, Dominican Republic, Ecuador, Honduras, India, Madagascar, Oman, Pakistan, Peru, Philippines, Puerto Rico, South Africa, Uganda, Venezuela, Vietnam
	ITC1437	China, Dominican Republic, India, Malaysia, South Africa, Venezuela, Vietnam
INIVIT	ITC1283	Austria, Bangladesh, Burundi, Cambodia, Cameroon, China, Colombia, Costa Rica, Cuba, DR Congo, Dominican Republic, Ecuador, France, Guadeloupe, Honduras, India, Indonesia, Jamaica, Malaysia, Myanmar, Nigeria, Oman, Pakistan, Papua New Guinea, Peru, Philippines, Saint Lucia, South Africa, Spain, Sri Lanka, Taiwan, Tanzania, Thailand, Tonga, Uganda, Zimbabwe
	ITC1284	Mexico
	ITC1318	Burundi, Venezuela
TBRI	ITC1271	Bangladesh, Belgium, Brazil, Burundi, Cameroon, Costa Rica, Cuba, Dominican Republic, Ecuador, France, Guadeloupe, Honduras, India, Indonesia, Indonesia, Israel, Madagascar, Malaysia, Mauritius, Mexico, Pakistan, Philippines, South Africa, Spain, Taiwan, Thailand, Uganda
	ITC1282	Bangladesh, Bangladesh, Belgium, Brazil, Burundi, Cambodia, Cameroon, China, Costa Rica, Cuba, Dominican Republic, Ecuador, France, Guadeloupe, Honduras, India, Indonesia, Israel, Madagascar, Malaysia, Mauritius, Mexico, Myanmar, Pakistan, Papua New Guinea, Philippines, Puerto Rico, South Africa, Spain, Sri Lanka, Taiwan, Thailand, Uganda
	ITC1436	Australia
	ITC1442	Australia, Bangladesh, Belgium, Cambodia, China, Colombia, India, Indonesia, Malaysia, Mauritius, Pakistan, Papua New Guinea, Philippines, Thailand, Vietnam
	ITC1443	Australia, Bangladesh, Belgium, Cambodia, China, India, Indonesia, Malaysia, Mauritius, Myanmar, Oman, Pakistan, Papua New Guinea, Philippines, Thailand, Vietnam

Source: Data courtesy of Ines Vandenhouwe (Bioversity International, ITC manager).

REFERENCES

Alene, A.D. and V.M. Manyong. 2006a. Endogenous technology adoption and household food security: The case of improved cowpea varieties in northern Nigeria. *Q. J. Int. Agric.* 45:211–230.

Alene, A.D. and V.M. Manyong. 2006b. Testing farmers' cropping decisions and varietal adoption behavior: The case of cowpea producers in northern Nigeria. *J. Agric. Food Econ.* 1:1–15.

Ali, M. and D. Byerlee. 1991. Economic efficiency of small farmers in a changing world: A survey of recent evidence. *J. Int. Dev.* 3(1):1–27.

Arias, P., C. Dankers, P. Liu, and P. Pilkauskas. 2003. *The world banana economy, 1985–2002*. FAO Commodity Studies. ISBN: 9251050570.

Crop Crisis Control Project (C3P). 2007. Final report submitted to USAID on October 15, 2007. http://c3project.iita.org/Doc/Final%20report%20c3P%20small.pdf (accessed 29 May 2010).

Feder, G. and R. Slade. 1985. The role of public policy in the diffusion of improved agricultural technology. *Am. J. Agric. Econ.* 67(2):4233–4238

Ghatak, S. and K.A. Ingersent. 1984. *Agriculture and economic development*. Baltimore, MD: Johns Hopkins University Press.

Grisley, W. 1994. La révolution du haricot, le K_{20} de Sam Muas: Triomphe africain. Et maintenant la même chose avec le maïs. Cérès, *FAO Rev.* 149:27–30.

Kahmen, A., J. Perner, and N. Buchmann. 2005. Diversity-dependent productivity in semi-natural grasslands following climate perturbations. *Functional Ecol.* 19:594–601.

Kalirajan, K.P. 1991. An analysis of production efficiency differentials in the Philippines. *Appl. Econ.* 23(4A):631–638.

Kalirajan, K.P. and R.T. Shand. 2001. Technology and farm performance: Paths of productive efficiencies over time. *Agric. Econ.* 24(3):297–306.

Kormawa, P., S. Keya, and A. Toure. 2004. Rice research and production in Africa. *Agra Informa* (July 2004 (www.Agra-net.com).

Pingali, P.L. and P.W. Heisey. 1999. *Cereal productivity in developing countries*. CIMMYT Economics paper 99–03. Mexico DE: CIMMYT.

Rogers, E M. 1995. *Diffusion of innovation*, 4th ed. New York: The Free Press.

Selatsa, A.A, A. Tenkouano, E. Njukwe, R.N. Iroume, and P.J. Bramel. 2009. Morphological diversity of plantain (*Musa* sp. L. AAB group) in Cameron: Relationships to farmer's cultural practices. *Afr. J. Plant Sci. Biotechnol.* 3(1):51–58.

Smithson, S.B. and J.M. Lenne. 1996. Varietal mixtures: A viable strategy for subsistence agriculture. *Ann. Appl. Biol.* 1288:127–158.

Tenkouano, A. and R.L. Swennen. 2004. Progress in breeding and delivering improved plantain and banana to African farmers. *Chronica Horticulturae* 44:9–15.

Tenkouano, A., S. Hauser, D. Coyne, and O. Coulibaly. 2006. Clean planting materials and management practices for sustained production of banana and plantain in Africa. *Chronica Horticulturae* 46(2):14–18

Tenkouano, A., B.O. Faturoti, and K.P Baiyeri. 2010. On farm evaluation of *Musa* hybrids in Nigeria. *Tree For. Sci. Biotechnol.* Special Issue 1. Bananas, Plantains and Ensete (in press).

Uazire, A.T., C.M. Ribeiro, C. Ruth Bila Mussane, M. Pillay, G. Blomme, C. Fraser, C. Staver, and E. Karamura. 2008. Preliminary evaluation of improved banana varieties in Mozambique. *Afr. Crop. Sci. J.* 16:17–25.

Xu, X. and S.R. Jeffrey. 1998. Efficiency and technical progress in traditional and modern agriculture: Evidence from rice production in China. *Agric. Econ.* 18(2):157–165.

17 Molecular Breeding of Other Vegetatively Propagated Crops

Lessons for Banana

Michael Pillay, Abdou Tenkouano, and Rodomiro Ortiz

CONTENTS

17.1 INTRODUCTION

Since the beginning of agriculture about 10,000 years ago, there was a concomitant increase in food supply with growth in human population. Increases in food production were ascribed to availability of new agricultural land for greater food production as well as increased yields of crops due to breeding. In the past 50 years, increase in food production has matched and outstripped increases in population growth (Chrispeels and Sadava, 2003). There is uncertainty whether this trend will continue. The human population is expected to reach 7 billion soon. At the same time agricultural production is growing at a slower rate of about 1.8% annually (Altman, 1999). The largest increases in population growth occur in poorer countries where many vegetatively propagated crops play an important role in maintaining food security and providing income. Countries that fall in this category are primarily within the tropical and subtropical regions of Africa, Asia, and Latin America. Some of the major vegetatively propagated staple foods in these countries include banana, cassava, potato, sweet potato, and yam. Currently the production of improved cultivars that are nutritionally acceptable to consumers, with resistance or tolerance to biotic and abiotic stresses, and reduced post-harvest losses, have been met largely through conventional breeding. The pressure of an increasing population and consequent increase in demand for food on the one hand and the depletion of arable land on the other have placed new emphases on conventional plant breeding. However, conventional breeding—especially of vegetatively propagated crops—poses many challenges.

The rapid development of molecular biology techniques and their application to plant breeding have resulted in significant genetic gains in agricultural crops, some of which have already entered the market (James and Krattiger, 1996). Conventional breeding of new improved cultivars heavily depends on choosing the right plant parents. Breeders typically make a large number of "parent crosses" of which only a select few eventually result in marketable cultivars. Following through on each cross is a long process, typically taking 12 to 15 years, and there are no assurances of success. Molecular breeding is simply a form of conventional breeding that allows breeders to have a better understanding of the parents that go into crosses. It is a specialized area of plant science that relies on collecting genetic information about plants. Molecular breeding may or may not involve genetic engineering or the development of genetically modified (GM) crops. Two of the basic tools of molecular breeding are the development of molecular markers that are linked to traits and the construction of molecular maps.

There have been many attempts at boosting banana breeding efficiency through the use of molecular genetics, but very little progress has been made to date in using molecular tools for selection of individuals with useful traits of economic importance. Molecular tools have proven very valuable in genetic and genomic diversity analyses, but breeding applications have been relatively scarce, with the exception of inferring parent–offspring relationships or ascertaining the dihybrid nature of progenies from crosses. That molecular breeding has not become routine for banana breeding is likely due to the complex biology of the species, including polyploidy, heterzygosity, multiple genomes, and vegetative propagation. In this regard, a study of what has been achieved in other species with similar biological characteristics could prove useful for banana breeders of the future.

This chapter examines some aspects of molecular breeding in three major vegetatively propagated crops: potato, cassava, and sugarcane. It is not the authors' intention to provide a detailed analysis of all aspects of molecular breeding of these crops but to examine some key aspects in which molecular breeding can make useful contributions in the development of new improved cultivars.

17.2 ECONOMIC IMPORTANCE OF THE CROPS

17.2.1 Potato

The potato (*Solanum tuberosum* L.) is the third major world food crop, exceeded only by wheat and rice in terms of world production for human consumption (Gebhardt and Valkonen, 2001). The high yield and low cost of cultivation makes potatoes an important part of the diet in many cultures. Potato production was recorded as 309 million tons in 2007 (http://www.scri.ac.uk/news/pgsc, accessed 8 April 2010). It is estimated that by 2020 more than 2 billion people will depend on potato for food, feed, or income. Potatoes are adapted to grow in a wide range of agro-ecologies and altitudes and are grown in over 149 countries from latitudes 65° N to 50° S and at altitudes from sea level to 4,000 m (Hijmans, 2001). Potatoes have been adapted to a wide range of end uses. As well as being a staple food, the potato is grown as a vegetable for table use, is processed into French fries and crisps (chips), and is used for dried products and starch production. New uses can be anticipated for the future such as designer starches (Davies, 1998) and as bioreactors for biopharmaceuticals (Sonnewald et al., 2003).

17.2.2 Cassava

Cassava (*Manihot esculenta* Crantz.) is one of the major sources of carbohydrates for more than 600 million in sub-Saharan Africa, Latin America, and Asia. It is the fourth most important tropical crop worldwide with a production of 162 million metric tons per year. The per capita consumption of cassava in Africa is approximately 80 kg/capita and is double that number in Central America. Cassava is drought tolerant, grows well in poor and degraded soils, and the roots can persist in the soil for 1 or 2 years without decay. This secures rural farmers a carbohydrate source in years with adverse growth conditions when other crops fail and famine would otherwise prevail (Jorgensen et al., 2005). Although cassava is cultivated mainly for its roots, the leaves are also consumed by many African cultures and are regarded as an excellent source of protein and vitamins (Ikoigbo, 1980). Cassava also has important industrial applications such as the production of animal feed, starch, and more recently ethanol. Dependence on the crop will not diminish as production is projected to increase 60% by the year 2020 in response to needs of the growing population that depend on cassava (Scott et al., 2000).

17.2.3 Sugarcane

Sugarcane (*Saccharum* spp.) is an important vegetatively propagated crop that is cultivated for its sugar-rich stalks and contributes about 75% of the world's sucrose supply (McIntyre et al., 2005). The mature stem can accumulate 12 to 16% of its fresh weight and approximately 50% of its dry weight as sucrose (Bull and Glasziou, 1963). Sugarcane is cultivated in more than 20 million hectares in tropical and subtropical regions of the world, producing up to 1.3 billion metric tons of crushable stems (Menossi et al., 2008). Sugarcane has many other uses besides the production of sugars. Sugarcane production is receiving increased attention for the production of ethanol as an important source of renewable energy. Sugarcane bagasse is already being used as a fuel source for sugar mills or for the production of animal feed. Bagasse is also being investigated for use in the production of paper, as a dietary fiber in bread, and as a wood substitute in the production of composite wood and in the synthesis of carbon fibers (Menossi et al., 2008). Today, sugarcane has many industrial uses and is one of the most widely used and cheapest domestic products (Jenkins, 1966). Molasses, a by-product of sugarcane production, is used as a fertilizer for cane soils, stock feed, alcoholic beverages, vinegar, cosmetics and pharmaceuticals, cleaning preparations and solvents, and coatings. It is quite possible that further uses of sugarcane will be developed in the future, but even now it can be seen that sugarcane is a very important and useful plant crop worldwide. The development of sugarcane as a biological system for large-scale production of value-added products is also gaining momentum (Lakshmanan et al., 2005) and is discussed later in this chapter.

17.3 BREEDING CHALLENGES

Genetic improvement of vegetatively propagated crops through conventional breeding is faced with several biological constraints. We examine some of these difficulties below for each of the selected crops.

17.3.1 Potato

Potatoes are polyploid with ploidy ranging from diploid ($2n = 2x = 24$) to hexaploid ($2n = 6x = 72$). Most diploid species are self-incompatible outbreeders, and most of the tetraploids ($2n= 4x = 48$) are self compatible and display tetrasomic inheritance, and the hexaploids are mostly self-compatible allopolyploids that display disomic inheritance (Hawkes, 1990). Crossability groups occur, each defined by an endosperm balance number, although interspecific pollen–pistil incompatibility and nuclear-cytoplasmic male sterility can occur (Camadro et al., 2004). The high levels of heterozygosity and severe inbreeding depression in parental clones require very large populations for selection of individual clones with good agronomic traits. New potato cultivars are exceptionally cumbersome to develop. Selection for desirable characters is time consuming and it may take up to 20 years to develop a new cultivar. The genes for resistance to pests, diseases, frost, and some quality traits such as starch and protein content are introgressed from related wild *Solanum* species. Wild species also carry many undesirable traits, such as low yield. The process of getting rid of all the unwanted traits requires several backcrosses in the breeding process. Cross-pollination for introgression of traits is only possible with very closely related species that do not contain many of the desired traits. The multigenic nature and low heritability of some traits also slows down the breeding process. Potato breeding is also problematic due to its narrow genetic diversity as a result of clonal propagation (Bradshaw and Mackay, 1994). Genetic improvement through conventional breeding is limited, since cultivated potatoes are largely tetraploid and are generally vegetatively propagated.

17.3.2 Cassava

Today's cassava cultivars are regarded as segmental allopolyploids with $2n = 36$ chromosomes (Magoon et al., 1969). Cassava breeding is based mainly on mass phenotypic recurrent selection. Parental lines are selected mainly on their *per se* performance for crossing programs. The high degree of heterozygosity in cassava makes identification of ideal parental material difficult. Synchronized flowering of selected parents is problematic and when considerable difference in flowering times exists, delayed planting or pruning of the earliest flowering genotypes must be practiced. Normally, few seeds are obtained (an average of 1 to 1.5), and acquiring large numbers of seeds is a labor-intensive and tedious process. It takes about 5–6 years from the time of seed set to the regional trial stage in cassava breeding (Ceballos et al., 2004). Very little knowledge exists on the genetics of important agronomic traits in cassava, and precise genetic control is known for relatively few traits. Single-gene control has been demonstrated for leaf lobe width, root surface color, albinism, stem collenchyma color, stem growth habit, root flesh pigmentation, and male sterility (Hershey and Ocampo, 1989). There are no confirmed examples of physiological specialization on a gene-for-gene basis for pests or pathogens. The inheritance patterns of a broad range of agronomically important traits have been studied. Results to date indicate that nearly all these traits are under multigenic control, with a high proportion of additive genetic effects (Iglesias and Hershey, 1994). Genetic variability within cassava for some traits, such as resistance to postharvest deterioration and virus resistance, is lacking or limited.

17.3.3 Sugarcane

Sugarcane cultivars are both highly aneuploid and polyploid (McIntyre et al., 2005). They typically contain more than 100 chromosomes. Modern cultivated sugarcane (*Saccharum* spp.) is a hybrid complex originating from crosses between *S. officinarum* ($2n = 80$, $x = 10$) and *S. spontaneum* ($2n = 40–124$, $x = 8$) and in some lineages *S. sinensis* or *S. barberi* (D'Hont and Layssac, 1998). Most cultivated sugarcane cultivars harbor two genomes with about 80% *S. officinarum* and 20% *S. spontaneum* (D'Hont et al., 1996). Sugarcane chromosomes are assigned to one of eight homology groups (HG) (Aitken et al., 2005). Each HG is estimated to contain between 8 and 14 homo(eo)logous chromosomes. Little is known about the allelic complexity of homo(eo)logous genes on chromosomes within a HG (McIntyre et al., 2006). The complexity and size of the sugarcane genome is a major limitation in genetic improvement. Traditional breeding is time consuming and breeding a new sugarcane cultivar can take more than 15 years from hybridization to commercial cultivation of the selected cultivar (Arvinth et al., 2009). The breeder has to rely on the initial variation created during hybridization, and no amount of selection can produce a good cultivar out of a poor cross (Alwala et al., 2006). The choice of parents to use in crossing is one of the most crucial decisions the breeder has to make. Enhanced breeding for sucrose accumulation is considered to have reached a plateau. The employment of new technologies to assist in the association of traits with genetic markers and genetic maps can aid in achieving further yield increases in breeding programs.

17.4 PRODUCTION CONSTRAINTS

17.4.1 Potato

The major diseases infesting the potato crop are late blight, considered to be the most serious potato disease worldwide, and bacterial wilt (Johnson and Veilleux, 2003). Potato late blight occurs almost everywhere potatoes are grown. Bacterial wilt or brown rot, caused by *Ralstonia* [*Pseudomonas*] *solanacearum*, limits potato production worldwide and especially in Asia, Africa, and Latin America, where it causes severe crop losses in tropical, subtropical, and warm-temperate regions. The major insect pests include potato tuber moth, Colorado potato beetle, and potato leaf miner fly. The potato tuber moth is the most damaging pest of potatoes in fields and stores in warm, dry areas of the world, such as North Africa and the Middle East, Mexico, Central America, and the inter-Andean valleys of South America. A wide variety of viruses also affect potato production. For example, a major devastating viral disease is the potato leaf roll virus (PLRV), which causes yield losses of up to 90% in infected crops. Nematodes especially of the genus *Globodera* cause major damage to potatoes.

17.4.2 Cassava

The production constraints of cassava have been detailed in many publications (Jenning and Iglesias, 2002; Kawano, 2003; Ceballos et al., 2004) and will not be dealt with in detail. In summary the main biotic constraints include viral, bacterial, fungal, and a variety of arthropod pests. There are a variety of abiotic factors that limit cassava productivity such as drought, low fertility, and acidic or alkaline soils. Productivity of cassava is very low and there is a huge gap in cassava production between the ideal (80 tons/ha) and what the average farmer harvests, around 8–12 tons /ha (Taylor et al., 2004).

17.4.3 Sugarcane

Disease and pests cause considerable economic losses to sugar industries worldwide. Sugarcane is susceptible to many viral, bacterial, fungal, and phytoplasma diseases, and there are a number of diseases of unknown etiology (Rott et al., 2000). The major viral pathogens include sugarcane mosaic virus (SCMV), Fiji disease virus (FDV), sorghum mosaic virus (SrMV), sugarcane streak virus (SSV), and sugarcane yellow leaf virus (SCYLV) (Lakshmanan et al., 2005). Viral diseases cause severe stunting, ratoon stunting, chlorotic streak, Fiji disease, and Sereh disease (Purseglove, 1979). There are about 160 fungal and 8 bacterial pathogens of sugarcane (Rott et al., 2000). Major fungal diseases include red rot (*Colletotrichum falcatum*), root rot (*Pythium graminicolum*.), pineapple disease (*Thielaviopsis parodoxa*), downy mildew (*Sclerospora sacchari*), and smut (*Ustilago scitaminea*), while the main bacterial diseases include gumming disease, *Xanthomonas vasculorum*, in which yellowish stripes occur at the leaf tips and the vascular bundles exude a yellowish gum when cut, and leaf scald, *Xanthomonas albilineans*, in which yellow stripes occur on the leaf blade, many side shoots are produced, and the vascular bundles of the stalk are red (Purseglove, 1979). Crop loss due to insect pests is estimated to be at 10 to 20% of sugarcane and is a limiting factor in sugarcane production (Avasthy, 1983). The most destructive insects of sugarcane are stem borers and larvae of several genera of moths. The larvae burrow into the stem and on emergence cause loss of sucrose and weakened stems. Other pests include termites and white grubs. Rats are also a problem in many areas since they eat the cane and introduce pathogens.

17.5 BREEDING OBJECTIVES

The breeding objectives for most crop plants are similar, with the primary focus being on the continuous improvement in productivity. Breeding for yield is a major target followed by breeding for resistance to diseases and pests that have an impact on yield. In addition to yield, disease and pest resistance, and nutritive and organoleptic qualities, other considerations will have to be added to breeding parameters. These parameters, according to Sinha and Swaminathan (1984), include

- Increasing the nutritive value of crops by breeding for enhanced proteins, vitamins, and other essential minerals
- Improving the efficiency of food production for each unit of cultural and solar energy invested
- Producing stable crops over wide environments with resistance to weeds, pests, and pathogens
- Improving and identifying plants as sources of biomass and renewable energy
- Breeding genotypes for optimum response to high, low, and zero inputs
- Breeding efficient plant types for crop-livestock production systems
- Breeding for suitability of crops in drought- and flood-prone areas

Plant breeding aims to improve crop performance or quality and to create new cultivars. However, new cultivars must preserve the quality characteristics and meet consumer demands. The process of developing a new cultivar can take up to 14 years in some crops from the first cross to the cultivar coming to the marketplace. Traditional plant breeding cannot meet all the food requirements of human beings. Therefore breeding through plant biotechnology is a necessity. The effective merging of classical breeding with modern plant biotechnology and the novel tools it provides are the gateway for the continuous increase in agricultural productivity and human survival (Altman, 1999). Some specific breeding objectives for each of the crops discussed in this chapter are outlined below.

17.5.1 Potato

Breeding efforts for potato are faced with the classical questions "What to breed?" and "How to breed?" (Bonnel, 2008). As scientific knowledge of the potato increased, the two questions have become related. The most important characteristics in potato breeding are summarized by Bonnel (2008) and include:

- Yield, tuber number and tuber size
- Tuber appearance
- Complex quality traits, e.g., low reducing sugar, no oxidation or bruising
- Resistance against nematode and insect pests
- Resistance against pathogens causing late blight, early blight, and scabs, viruses, and *Erwinia* spp.
- Resistance against abiotic stresses such as water and heat stress
- Specific traits for niche markets
- Adaptation to specific environments

Bonnel also lists the issues to the question "How to breed potatoes?" (Bonnel, 2008).

17.5.2 Cassava

The breeding objectives for cassava are determined by the ultimate use of the crop. For subsistence use, production stability and cooking quality or starch characteristics are the main traits. For industrial purposes such as starch production and animal feed, productivity is the main target. Some morphological traits such as peel color of the roots, the leaf petiole, or the shoot are associated with good cooking quality. Other root-quality traits include reduced cyanogenic glucosides, early bulking capacity, higher protein content, and reduced postharvest physiological deterioration (Ceballos et al., 2004). Breeding for increased yield, multiple pest and disease resistance, desirable agronomic and consumer preference traits such as early vigor in plant growth (for high foliage yield for leaf vegetable), appropriate plant architecture, and early bulking of storage roots, combined with high dry matter content, low cyanide content, and other favorable traits, such as easy peeling, have been the main breeding objectives. Recently, breeding for improved micronutrient content has been emphasized. Improvement of this crop has been primarily oriented towards relatively few traits, most prominently to yield and host-plant resistance to pathogens and pests. The former is a complex trait with many individual components, each involving numerous biochemical pathways and progress has been most pronounced in those cases where objectives were relatively simple. Resistance to pathogens is often determined by fewer genes, allowing for qualitative progress in a breeding program. Advances in yield potential have been substantial but mainly in more favorable environments. These gains are often less prominent when translated to farmers' conditions, but the products of steady improvement are beginning to have substantial impact (Hershey and Jennings, 1992). Improvements in cassava in terms of adaptation, resistance, productivity, and other traits have not been exhausted.

17.5.3 Sugarcane

The goal of sugarcane breeding is to produce an economic yield of sugar sustained over several ratoons. Sugarcane yields the highest number of calories per unit area of any plant (Heiser, 1981), producing up to 10 tons of sucrose per hectare in Barbados. The highest yields of 22 tons of sucrose per hectare occur in Hawaii, but the crop takes two or more years to mature there. Recovery of raw sugar from cane varies from 11 to 13% (Purseglove, 1979). Increasing the yield of sucrose can be achieved by either increasing biomass at the same concentration of sucrose or increasing the sucrose concentration. The yield of sucrose has increased over the years mainly due to increase in biomass. However, increasing the sucrose content is considered to have reached a plateau via conventional breeding due to the limits in the gene pool (Mariotti, 2002). The employment of new technologies to assist in the association of traits with genetic markers and genetic maps can aid in achieving further yield increases by breeding (Dillon et al., 2007). Sugarcanes are highly polyploid, wind-pollinated outbreeders. They are clonally propagated, highly heterozygous, and intolerant to inbreeding. New cultivars are sought from the first generation progeny of crosses between clones. The five species of interest to cane breeders are: (1) *S. officinarum* ($2n = 80$) has good sugar quality and low fiber, although it is susceptible to most of the main diseases, except gumming disease and smut; (2) *S. spontaneum* ($2n = 40–128$) is a source of resistance to many diseases, including "Sereh," mosaic, gumming, red rot, and downy mildew; (3) *S. barberi* ($2n = 82–124$) is considered the most important breeding cane and is immune to gumming and mosaic, and resistant to downy mildew, but susceptible to smut and red rot; (4) *S. sinense* ($2n = 82–124$) is difficult to breed but has given rise to some useful breeding lines; and (5) *S. robustum* ($2n = 60–194$) has been used to some extent in breeding lines (Wrigley, 1982). Breeding and selection of cane is not a simple process since viable seeds are seldom produced. Breeding occurs at the experiment stations, which are able to provide the proper conditions and techniques required.

17.6 CONVENTIONAL VERSUS MOLECULAR BREEDING

Conventional breeding efforts have attempted to address many of the constraints facing vegetatively propagated crops, but progress has been slow because most of the crops have complex genetic systems that makes it difficult to breed efficiently. Ploidy manipulations—that is, scaling up and down chromosome numbers of a species within a polyploid series—facilitate breeding polyploid asexual crops with their wild genetic resources endowment (Ortiz, 2004). Chromosome sets are manipulated with haploids, $2n$ gametes, and through interspecific-interploidy crosses. Analytical breeding schemes rely mainly on ploidy manipulations to "capture" diversity from exotic (wild or nonadapted germplasm) and use $2n$ gametes to incorporate this genetic diversity through unilateral (USP; $n \times 2n$ or $2n \times n$) or bilateral (BSP; $2n \times 2n$) polyploidization (Ortiz et al., 2009). Haploids are propagules with the gametophytic chromosome number (n) and $2n$ gametes possess the sporophytic chromosome number of the parental source.

Vegetative propagation is genetically a rather conservative strategy. Clonally propagated crops have individually limited genetic plasticity, and new variation from natural crosses in multiclonal fields and introgression from wild species are slow to arise under natural conditions. Consequently, most breeding programs consider recombination as the principal means of cultivar development. Modern biotechnology can be an important tool in the development of sustainable agricultural systems (Ferry et al., 2004). GM crops have the potential to contribute to increased productivity while conserving biodiversity and more sustainable agriculture and the environment. No set of biological tools has created greater expectations than those generated by achievements in the area of molecular, cellular, and other technologies broadly described as biotechnology. Biotechnology provides new tools for overcoming some of the problems for the genetic improvement of crops.

For example, biotechnology is useful when no resistance for a disease, pest, or virus is found in the germplasm of a crop. The transgenic approach of introducing foreign genes for these traits makes it possible to engineer plants with the traits of interest. The rapid and continued development of tools in biotechnology has introduced fundamental changes in crop improvement research and broadened the range of potential traits for genetic modification.

Although past achievements in conventional breeding of vegetatively propagated crops are impressive, it appears that an integrated approach combining both conventional and molecular breeding strategies is required to make further progress in crops that are characterized by polyploidy, a narrow genetic base, poor fertility, susceptibility to various pests and diseases, and the long durations its takes to breed elite cultivars (Lakshmanan et al., 2005). Two focus areas of biotechnology research in crop improvement include the development of molecular markers and molecular maps.

17.6.1 Molecular Markers

Molecular markers are discovered by establishing statistical associations between variations in particular DNA sequences and any traits that may be of interest (Reece and Haribabu, 2007). It is now generally accepted that molecular markers represent the most significant advance in breeding technology in the past few decades and are currently the most important application of molecular biology to plant breeding. There appears to be no resistance to the use of molecular marker technology in breeding as there is for GM organisms. Research in molecular marker technologies have blossomed in the last decade because they are essential if marker-assisted selection (MAS) in plant breeding programs is to be realized. Molecular markers can assist breeders in the process of evaluation and selection and shorten the timescale required for the production of improved cultivars. MAS eliminates the need for costly and time-consuming field trials to select superior individuals. Molecular markers enable breeders to detect the presence of multiple alleles that are associated with a single trait even when the individual alleles do not exert a detectable influence on the expression of the trait (Reece and Haribabu, 2007). Molecular markers can be used to facilitate decisions made during crossing, especially when they used to gain knowledge of the genetic diversity in parental clones. Therefore the search for target traits in many of the vegetatively propagated crops occupies the minds of many scientists. However, there remains a considerable need for parallel progress in the understanding of genotype-by-environment interaction and the influence of epigenetic mechanisms. Nevertheless, even with the current level of understanding, the application of DNA marker technology is likely to facilitate the attainment of new objectives in crop research and genetic improvement that have proven difficult to achieve using classical techniques. For example, highly precise MAS approaches require the development of high-density linkage maps. However, the generation of highly relevant and precise linkage maps is not routinely achievable in most crops.

DNA-based molecular markers have been used extensively for plant genome analysis, especially in characterizing genetic diversity, genome fingerprinting, genome mapping, gene localization, analysis of genome evolution, population genetics, taxonomy, and plant breeding (Sharma et al., 2005). All marker systems have different advantages and disadvantages in specific applications. Thus, it is important for molecular breeding programs to develop capacity in several assays in order that the most suitable system can be chosen and rapidly applied for any particular application. In addition, different DNA marker assays detect (and are therefore affected by) different types of genetic variation.

Improved molecular marker systems are required to enhance the adoption of MAS. Several factors are important when considering MAS, including ease of use, robustness, cost, and linkage to the trait of interest (de Koeyer et al., 2010). The ideal marker systems for polyploidy crops should be dosage sensitive and have the ability to distinguish heterozygous genotypes with multiple haplotypes within the target genomic region by the marker (de Koeyer et al., 2010).

New DNA technologies are constantly being developed especially for human genomic research, and some of them are being used in plant research. High-throughput technologies based on single nucleotide polymorphisms (SNPs) or small-scale insertions/deletions (indels) are efficient alternatives for traditional markers (restriction fragment length polymorphism [RFLP], random amplified polymorphic DNA [RAPD], amplified fragment length polymorphism [AFLP], and so forth) because of their greater abundance, high polymorphism, ease of measurement, and ability to reveal hidden polymorphisms where other methods fail (Dillon et al., 2007). SNPs also allow easy and unambiguous identification of alleles or haplotypes.

A good marker system for polyploid crops should be dosage sensitive and have the ability to distinguish heterozygous genotypes with multiple haplotypes (de Koeyer et al., 2010). High-resolution

DNA melting (HRM) analysis has been shown to have several advantages over other genotyping methods (Montgomery et al., 2007; Reed et al., 2007; Erali et al., 2008). The advantages include a short analysis time and the absence of post–polymerase chain reaction (PCR) sample processing or separation (de Koeyer et al., 2010). The three ways in which HRM can be used for genotyping and for variant screening are discussed in de Koeyer et al. (2010). HRM has been used in many crops and holds great potential for cultivar identification, especially in polyploids, mapping, polymorphism discovery, mapping candidate genes, and in combination with *quantitative trait loci* (QTL) or association studies for identifying genomic regions involved in important traits (de Koeyer et al., 2010).

17.6.1.1 Molecular Markers in Potato

More molecular markers are available in potato than for cassava. There are markers for late blight resistance (Bormann et al., 2004; Bisognin et al., 2005; Colton et al., 2006), resistance to potato virus Y (Hamalainen et al., 1997, 1998; Kasai et al., 2000; Flis et al., 2005; Gebhardt and Bellin, 2006), potato virus X (Gebhardt et al., 2006), and to the potato cyst nematode (Niewohner et al., 1995). Markers for MAS are limited to diploid material and include *Ns* for PVS resistance (Marczewski et al., 2002), Ry_{adg} for extreme resistance to PVY, *Gro1* for resistance to *Globera rostochiensis*, *Rx1* for extreme resistance to PVX, and *Sen 1* for resistance to potato wart (Gebhardt et al., 2006). Recently the marker for PVY resistance was validated as a fast and efficient way to select for PVY resistance (Ottoman et al., 2009). Markers tagging QTLs conferring resistance to *P. infestans*, *Verticillium dahlia*, and *V. albatrum* have been developed (Simko et al., 2007).

Recently a molecular marker for Verticillium wilt (VW) resistance has been identified in potato (Bae et al., 2008). Breeding for VW has been difficult because disease symptoms can easily be confused with natural senescence. Stem plating used to identify the disease is labor intensive and time consuming. Environmental factors influence disease expression, and great variation exists in pathogen population from stem to stem. Since the only effective control practice, soil fumigation, is expensive and has harmful environmental effects, host-plant resistance has been explored as a method of disease control (Pegg, 1974; Rowe et al., 1987). Most commercial cultivars in the United States are susceptible, although some exhibit moderate resistance (Rowe and Powelson, 2002). Major resistance genes have been identified for the disease. Inheritance of VW resistance in cultivated tetraploid potato appears to be polygenic and complex (Hunter et al., 1968; Pavek and Corsini, 1994) and most cultivars are not highly resistant to VW (Rowe and Powelson, 2002). The interesting aspect of this study was that tomato *Ve1* and *Ve2* gene sequence information was used to amplify candidate *Ve* gene orthologs in potato. The marker was tested for association with resistance in segregating populations and was found to be effective for the VW gene in potato.

17.6.1.2 Molecular Markers in Cassava

Molecular markers have made a great contribution to cassava breeding and genetics, in the assessment of genetic diversity, taxonomical studies, understanding the phylogenetic relationships in the genus, confirmation of ploidy, and the development of genetic maps (Fregene et al., 1997; Lokko et al., 2004). Very few markers have been identified for traits in cassava and are limited to markers for host-plant resistance and early yield (Jorge et al., 2000, 2001; Akano et al., 2002).

17.6.1.3 Molecular Markers in Sugarcane

The majority of useful traits in sugarcane are multigenic due to the polyploidy nature of the crop. Quantitative trait loci (QTL) for sucrose content and related traits have been identified (Ming et al., 2001; Ming, Del Monte, et al., 2002). Comparative analysis of QTL affecting plant height and flowering between sorghum and sugarcane identified QTL clusters in sugarcane (Ming, Del Monte, et al., 2002), highlighting the power of genome synteny. The conservation of chromosome organization across the grasses has the potential to locate and identify genes of interest in sugarcane. SNPs have been used to divide a set of 64 expressed sequence tags (ESTs) into two groups related to two genes family members of 6-phosphogluconate dehydrogenase (Grivet et al., 2001), to reveal the expression of an *Adh2* and two *Adh1* genes (Grivet et al., 2003) and the development of codominant cleaved, amplified polymorphic sequence markers (Quint et al., 2002) and to map several candidate genes and ESTs (McIntyre et al., 2005). Ming et al. (2001; Ming, Del Monte, et al., 2002) investigated the genetic basis of traits related to sugar content, plant height, and flowering in interspecific crosses between *S. officinarum* and *S. spontaneum*. Numerous QTLs were detected in a few genomic regions, suggesting that substantially fewer genes may actually be involved in the genetic control of these traits. Within an interspecific cross, wide phenotype segregation can provide a favorable setting for QTL detection. By comparison, a study of yield components (plant height, stalk diameter, stalk number, and Brix) in the selfed progeny of the modern cultivar 'R570' revealed numerous QTL with smaller individual effects (Hoarau et al., 2002). Similarly, Jordan et al. (2004) detected numerous small QTL for stalk number in sugarcane and found sorghum QTL for tillering in syntenic positions. The only major gene that has been localized so far in the sugarcane genome is a rust-resistance gene (Daugrois et al., 1996). This gene (called *Bru1* for brown rust) is currently the focus of a map-based cloning project (D'Hont et al., 2001; Asnaghi et al., 2004).

17.6.2 Molecular Maps

A molecular map is a construct that places the markers in order indicating the relative distances between them and assigning them to their linkage groups (Jones et al., 1997). Linkage maps are based on recombination frequencies, and the distance between points on a genetic map reflects the recombination frequencies between the points. Linkage maps provide valuable information about the genomic organization of a species. Maps provide a direct method for showing the positions of genes and other sequence features and allow selecting genes via detectable markers. Linkage maps for most crop species were initially developed using morphological markers, the first genetic markers used in biology. For example, Mendel studied visible traits such as seed and pod color, surface appearance of seeds and pods, and plant height. In the process of finding more markers, the first class of markers scored at the molecular level were isoenzymes. The advance of molecular biology provided a wide spectrum of technologies to assess different markers.

Two diverse approaches are used to facilitate the generation of appropriate linkage maps in polyploid species: (1) linkage mapping based on diploid relatives and extrapolation to the polyploid crop, and (2) polyploid mapping based on single-dose markers in populations derived from crosses between heterozygous tetraploid and diploid genotypes. The more markers there are on a map, the more likely it is that one will be closely linked to a trait of interest. Several different types of markers have been used to develop linkage map for crop species, including RFLP, RAPD, minisatellites or variable number tandem repeats, AFLP, and microsatellites or simple sequence repeats (SSR). The use of these markers for mapping is discussed adequately in many publications. SSR markers are advantageous to applied plant breeding because they are codominant, easily assayed, and detect high levels of polymorphism (Morgante and Olivieri, 1993). For these reasons SSR markers have become valuable markers to breeders for the purposes of genome and QTL mapping. In addition to their high polymorphism, SSR are usually single-locus sites, an important feature when dealing with allopolyploid species. SSR markers have thus become the marker class of choice for the molecular mapping of many crop species (Roa et al., 2000). While SSRs are standard PCR-based markers and can be considered as proven anchor-markers, their suitability for high-throughput mapping does not favorably compare to the new single nucleotide polymorphism (SNP) based genotyping techniques (Kilian et al., 2005). High-throughput genotyping technologies based on SNP or small-scale insertions/deletions (indels) could become efficient alternative techniques for traditional markers because of their greater abundance in the genome. However, development of massive SNP resources from the expressed portion of the genome, amenable to fully automated genotyping, is a difficult task for polyploidy species (Koebner and Summers, 2003; Somers et al., 2003). Diversity arrays technology (DArT), for which proof of concept was first reported by Jaccoud et al. (2001), is becoming increasingly adopted in many species. The technology combines a complexity reduction method (Wenzl et al., 2004) with hybridization-based polymorphism detection using high-throughput, solid-state platforms and has the potential to generate hundreds of high-quality genomic dominant markers with a cost- and time-competitive tradeoff (Kilian et al., 2005).

The development of high-resolution, easy-to-use genetic maps, coupled with sequencing and physical mapping, will revolutionize breeding of vegetatively propagated crops. Highly dense linkage maps are powerful tools for the localization of both qualitative and quantitative traits, for map-based cloning of genes, and to develop markers for marker-assisted breeding. They also provide information for understanding the biological basis of traits (Kriegner et al., 2000).

Constructing genetic maps in polyploids is challenging for several reasons, as outlined in Cervantes-Flores et al. (2008). Polyploids have multiple chromosomes of the same type and a large number of possible genotypes are expected in a segregating population due to the larger number of alleles combining in a particular event. This is especially true in autopolyploid species. The genotype of an individual is not always readily inferred through its marker phenotype. The type of ploidy (allopolyploidy or autopolyploidy) of many crops is unclear, making it difficult to determine patterns of inheritance (Wu et al., 1992; Ripol et al., 1999). In an autopolyploid species, corresponding chromosomes in the different genome copies are homologs and, therefore, can pair randomly between each other. In contrast, for allopolyploids chromosomes may originate from two or more different genomes and during meiosis will pair preferentially to their homologs from the same genome, and in lower frequency to a homeologous chromosome (Sybenga, 1996; Ramsey and Schemske, 2002). A commonly used approach to construct molecular genetic maps in polyploids is based on the use of single-dose fragments (SDF) (Wu et al., 1992). Since single-dose markers are markers present in one parent in a single copy, only half the progeny will carry the marker during gamete formation. Regardless of the ploidy of the genome, half the progeny will possess this fragment and half will not. SDFs combined with other mapping strategies such as testcross and pseudo-testcross approaches, have been used to construct linkage maps in several polyploid species, including potato, sugarcane, and sweet potato (Cervantes-Flores et al., 2008).

17.6.2.1 Potato

Potato has the densest genetic linkage map and one of the earliest cytogenetic maps among all plant species (van Os et al., 2006; Iovene et al., 2008). More than 40 major qualitative genes conferring resistance to important diseases and pests have been genetically mapped (Simko et al., 2007; Ottoman et al., 2009). To integrate these maps, bacterial artificial chromosome (BAC) clones are being hybridized to the pachytene chromosomes of potato (Iovene et al., 2008).

17.6.2.2 Cassava

The first genetic linkage map for cassava was constructed with predominantly RFLP markers and a full-sib intraspecific cross (Fregene et al., 1997). The map provided initial tools for genetic analysis of important traits of cassava (Jorge et al., 2000, 2001; Akano et al., 2002; Okogbenin and Fregene, 2002, 2003). A higher resolution map was developed using SSR markers (Okogbenin et al., 2006).

17.6.2.3 Sugarcane

The identification of single- (present on one chromosome only) and double-dose (present on two chromosomes) markers for mapping and QTL analysis was a breakthrough in sugarcane genetics. Despite the structural complexity of the sugarcane genome, genetic maps of sugarcane have been produced based on single-dose markers (Wu et al., 1992) for the two ancestral species *S. spontaneum* (Al-Janabi et al., 1993; Da Silva et al., 1993, 1995; Ming et al., 1998; Ming, Liu, et al., 2002) and *S. officinarum* (Mudge et al., 1996; Guimaraes et al., 1997; Ming et al., 1998; Ming, Liu, et al., 2002). Genetic maps have also been constructed for two modern sugarcane cultivars, 'Q165' (Aitken et al., 2005) and 'R570.' AFLPs and SSRs and resistance gene analogs (Rossi et al., 2003) were later mapped using a selfed 'R570' progeny. The AFLP-based map has more than 1,100 markers with a total length of 7,800 cM, which represents a coverage of about 46% of the anticipated genome size

(17,000 cM). A saturated map must be developed to be able to efficiently localize major genes or Mendelian factors involved in quantitative trait loci (QTL). Moreover, QTL mapping in sugarcane is a real challenge, since many alleles coexist at each locus due to the high polyploidy. At a particular locus, the effect of an allele should be perceptible only if it exceeds the average effect of all other segregating alleles in the background but not, as in diploids, if its effect simply exceeds that of a single alternative allele (D'Hont and Glaszmann, 2001; Hoarau et al., 2002). Two genetic maps of sugarcane were also constructed using AFLP, SSR, and RFLP markers (Raboin et al., 2006). A new, brown rust–resistant gene and one controlling stalk color were localized on the map.

17.7 GENETIC TRANSFORMATION

Genetic transformation has become an important tool for crop improvement and especially for vegetative propagated crops. Genetic transformation is defined as the transfer of any foreign gene(s) isolated from plants, viruses, bacteria, or animals into a new genetic background (Sharma et al., 2005). Successful genetic transformation in plants requires the production of normal, fertile plants expressing the newly inserted gene(s). The process of genetic transformation involves several distinct steps: identification of a useful gene, the cloning of the gene into a suitable plasmid vector, delivery of the vector into the plant cell (insertion and integration), followed by expression and inheritance of the foreign DNA encoding a polypeptide. Methods of gene insertion in plants can be achieved by direct gene transfer through particle bombardment or electroporation, or through biological vectors like a disarmed TI (tumor inducing)-plasmid of *A. tumefaciens*. However, this technology has not been realized in all crop plants because of the lack of suitable transformation methods. The major components for the development of transgenic plants include (1) the development of reliable tissue culture regeneration systems, (2) the preparation of gene constructs and transformation with suitable vectors, (3) efficient transformation techniques for the introduction of genes into the crop plants, (4) recovery and multiplication of transgenic plants, (5) molecular and genetic characterization of transgenic plants for stable and efficient gene expression, and (6) field evaluation of transgenic plants (Sharma et al., 2005).

17.7.1 Achievements and Prospects of Molecular Breeding in Potato, Cassava, and Sugarcane

Protection against diseases caused by fungi, bacteria, and viruses and losses due to insect herbivores are major challenges for crop improvement and are significant limiting factors in food production (Akhond and Machray, 2009). Current efforts to improve plant stress tolerance by genetic transformation have resulted in several achievements. Improvement in the nutritional qualities of our major plant foods is now within reach of crop improvement programs. Plants are also being used as biofactories for the production of pharmaceuticals, proteins, plastic, and so forth. The following section examines some of the achievements and prospects of molecular breeding in plant improvement programs. There is ample literature on these topics covering all three plants of interest. However, due to space restrictions, examples covering all plants are not discussed here.

17.7.1.1 Pest Resistance in Potato

A variety of genetic engineering strategies have been used to target insect pests in potato. One of the most widely used strategies is the use of the *cry* genes from *Bacillus thuringiensis*. The *cry* gene is part of a large family of homologous genes that encode proteins that are toxic to insects and their larvae (Davidson et al., 2002). Resistance to a number of insect pests has already been engineered in potato. They include:

1. Resistance to the Colorado potato beetle and potato tuber moth was derived by expression of one of the *cry* endotoxin proteins (Mohammed et al., 2000; Reed et al., 2001; Davidson et al., 2004).
2. Reduced feeding of the cotton bollworm larvae was observed in plants transformed with *cry1ab* gene (Chakrabarti et al., 2000).
3. Transformation of potato with a trypsin inhibitor from cowpea (Bell et al., 2001) and the lectin genes from snowdrop (Birch et al., 1999) also produced insect resistance in potato.
4. Genes that encode cysteine proteinases inhibitors (cystatins) confer resistance to the potato cyst nematode (*Globodera rostochiensis* and *G. pallida*) (Urwin et al., 2003).
5. The gene *chly* encoding for chicken lysozyme enzyme enhances resistance to blackleg and soft rot in potato (Serrano et al., 2000).

17.7.1.2 Virus Resistance

17.7.1.2.1 Cassava

Cassava mosaic disease (CMD) is a major constraint for cassava production in Africa, resulting in significant economic losses. The disease is caused by a complex of viruses named according to the regions in which they were discovered and includes the African, East African, and Southern African cassava mosaic virus. Breeding for resistance to CMD has been a major target in classical breeding programs. Several attempts have also been made since 2004 to genetically engineer virus resistance in cassava. Engineering African cassava mosaic disease (ACMV) resistance in cassava using sense and antisense RNA strategies produced lines with varying levels of resistance (Chellapan et al., 2004; Zhang et al., 2005). Since then, new approaches for developing transgenic cassava have been reported using RNA interference technology. RNA interference provides resistance against viruses via post-transcriptional and/or transcriptional gene silencing (PTGS and TGS) (Waterhouse and Fusaro, 2006). Vanderschuren et al. (2007, 2009) have developed two new ways of developing transgenic cassava for resistance to CMD. In the first instance they showed that expression of double-stranded RNA (dsRNA) homologous to virus sequences can interfere with RNA virus infection in plant cells by triggering RNA silencing (Vanderschuren et al., 2007). More recently, Vanderschuren et al. (2009) engineered a CMD-susceptible cassava cultivar to express a hairpin dsRNA homologous to a region of the replication-associated protein coding sequence (*Rep/AC1*). This sequence was selected because it is essential for replication of geminiviruses like CMD. Several of the transgenic cassava lines from these experiments showed resistance to CMD. Similar technologies could be used for other crops.

17.7.1.2.2 Potato

Research in developing transgenic potatoes for virus resistance began in the 1990s (Lamb and Hay, 1990; Kawchuk et al., 1991). Transformation of potato with the viral coat proteins of potato leaf roll virus conferred resistance to the virus (Kawchuk et al., 1990). Transformation with virally encoded proteinase gene sequences produced high levels of resistance to viral infection (Okamato et al., 1996). Replicase and coat protein genes from potato leaf roll luteovirus and potato Y potyvirus, respectively, provide resistance to these viruses (Duncan et al., 2002). RNA silencing technology has also been applied in potato to develop virus-resistant plants. Potato virus Y (PVY) is one of the most widespread viruses of potato. Missiou et al. (2004) generated transgenic potato that was resistant to potato virus Y.

17.7.1.3 Abiotic Stress Tolerance in Potato

Although oxygen is essential for aerobic life, reactive oxygen species (ROS) such as superoxide anion radicals ($-O_2^-$), hydroxyl radicals (–OH), and hydrogen peroxide (H_2O_2) are produced in all aerobic cells during metabolism (Asada, 1999; Foyer et al., 1994). Oxidative stress caused by ROS is one of the major factors that bring about damage in plants exposed to environmental stresses such as drought, salinity, extreme temperature, or strong light (Lim et al., 2007). The chloroplasts of plants are a rich source of ROS (Asada, 1999). Two enzymes—superoxide dismutase (SOD) and ascorbate peroxidise (APX)—are able to detoxify ROS in the chloroplasts. The removal of ROS suggests that plants will exhibit increased tolerance to environmental stresses. Tang et al. (2006) and Lim et al. (2007) developed transgenic potato and sweet potato, respectively, in which the genes for SOD and APX were expressed in the chloroplasts under the control of an oxidative stress-inducible promoter. The plants showed enhanced tolerance to high temperature (42°C for 20 h), oxidative, and chilling stress. These results suggest that the manipulation of the antioxidative mechanism of the chloroplasts may be applied in the development of transgenic crop plants with enhanced tolerance to multiple environmental stresses.

17.7.1.4 Improvement in Nutritional Qualities

Improvement in the nutritional qualities of crops is now within the reach of crop improvement programs through biotechnology. We provide a few examples of the role of biotechnology in improving the nutritional qualities of vegetative crops.

17.7.1.4.1 Generation of Cyanogen-Free Transgenic Cassava

Although cassava is used widely as a food crop, it contains toxic glucosides that are harmful to human health. The roots, leaves, and stems of cassava contain potentially toxic levels of cyanogenic glucosides (linamarin 95% and lotaustralin 5%). These cyanogens yield cyanide following enzymatic hydrolysis (Kakes, 1990; Koch et al., 1992; White et al., 1994). Currently, the cyanogen content of cassava is reduced to safe levels by maceration, soaking, rinsing, and frying (Siritunga and Sayre, 2003, 2004). However, many health-related disorders occur when cassava is not pretreated well before consumption. Chronic low-level cyanide exposure leads to the development of goiter and tropical ataxic neuropathy, while acute cyanide poisoning is associated with outbreaks of Konzo, a paralytic disorder and, in some cases, death (Rosling et al., 1992; Tylleskar et al., 1992). The development of acyanogenic cassava cultivars will be a safe and effective way of making this food crop safe and acceptable. Biotechnology has made it possible to identify the two genes that catalyze the first step in the production of linamarin and lotaustralin (Anderson et al., 2000). These genes were used in an antisense orientation to transform cultivar 'MCol 2215' of cassava. The transformed plants were shown to have reduced levels of linamarin in the leaves (94%) and roots (99%). Further studies in reducing cyanogens in cassava showed that transgenic cassava with a 92% reduction of in cyanogenic glucoside content in tubers and acyagenic leaves could be obtained by RNA interference that blocked the enzyme in the first committed step in the production of linamarin and lotaustralin (Jorgensen et al., 2005). These acyanogenic cassava plants represent a safe and more marketable product, demonstrating the value of biotechnology in making safer foods.

17.7.1.4.2 Nutritional Enhancement of Potato

Because the potato is an important highly nutritious vegetable and staple food, it has become an excellent target for biotechnological research with regards to nutritional improvement, food compositional analysis, and production of dextrans. Improving the nutritional content of potato would have a significant impact on vitamin and nutrient intake for millions of people who depend on the crop for food. Potato has been engineered for enhanced nutritional value of specific carotenoids. For example, overexpression of phytoene synthase resulted in 19-fold higher levels of lutein as well as higher β-carotene levels (Ducreux et al., 2005), zeaxanthin contents were increased up to 130-fold through disruption of zeaxanthin catabolism (Römer et al., 2002), and accumulation of astaxanthin was achieved through expression of a microbial β-carotene ketolase gene (Morris et al., 2006).

Vitamin E (tocopherol) is a powerful antioxidant essential for human health and is synthesized only by photosynthetic organisms. Humans and other animals cannot manufacture tocopherols and obtain them from their vegetarian diet. It is estimated that approximately 90% of children and adults in the United States do not consume the recommended amount of vitamin E (Ahuja et al.,

2004; Drewel et al., 2006). Efforts to enhance total tocopherol levels in plants have met with limited success but have raised important questions regarding the regulation of the tocopherol biosynthetic pathway. Studying the development and tissue-dependent regulation of vitamin E synthesis may provide more effective approaches for engineering enhanced vitamin E content in crop plants (Crowell et al., 2008).

Genetic and molecular approaches were used to determine if increased levels of tocopherols can be accumulated in potato tubers through metabolic engineering (Crowell et al., 2008). Since vitamin E synthesis has only been studied in leaves and seeds thus far, potato tuber is a good model system for studying the accumulation of vitamin E and the regulation of its synthesis in a below-ground, nonphotosynthetic plant organ. Two transgenes were constitutively overexpressed in potato: the *Arabidopsis thaliana p*-hydroxyphenylpyruvate dioxygenase (At-HPPD) and *A. thaliana* homogentisate phytyltransferase (At-HPT). Tocopherol levels in the transgenic plants were determined by high-performance liquid chromatography. In potato tubers, overexpression of At-HPPD resulted in a maximum 266% increase in x-tocopherol, and overexpression of At-HPT yielded a 106% increase. However, tubers from transgenic plants still accumulated approximately 10- and 100-fold less ct-tocopherol than leaves or seeds, respectively. The results indicate that physiological and regulatory constraints may be the most limiting factors for tocopherol accumulation in potato tubers and perhaps in other plants. Studying regulation and induction of tocopherol biosynthesis should reveal approaches to more effectively engineer crops with enhanced tocopherol content (Crowell et al., 2008).

17.7.2 Vegetatively Propagated Plants as Biofactories

Bioengineering of plants to produce various useful products has been in progress for over 20 years. Many crops, including sugarcane and potato, are considered to have potential to be used as biofactories for the production of pharmaceuticals and other useful products. In the following section we examine how sugarcane and potato are being used for this purpose.

17.7.2.1 Sugarcane

Several efforts are underway to develop sugarcane as a biofactory system. In Australia, researchers have developed transgenic sugarcane producing polyhydroxybutyrate, a thermoplastic, and are also working on other higher-value products (Brumbley et al., 2004). In the United States, efforts are underway to produce pharmaceutical-grade proteins for human therapeutic use from sugarcane (Holland-Moritz, 2003). Other research to produce high-value proteins in sugarcane is ongoing in Brazil, South Africa, and elsewhere. Wang et al. (2005) outlined the advantages of sugarcane for transgene/product containment that could position the crop as a "secure" platform for production of pharmaceuticals. Commercial sugarcane is propagated vegetatively. In many places commercial cultivars do not normally flower, so there is little chance of pollen drift or production of viable seed. If viable seed were produced, it would not become mixed with seed supplies because commercial fields are planted with vegetative pieces. Sucrose, the food commodity derived from sugarcane, is sold as a refined crystal that is essentially free of protein, rather than a whole fruit or vegetable. In the unlikely event that sugarcane producing a pharmaceutical protein was mixed into the food supply, the food product (refined sucrose) would remain unaffected. Considering these benefits, sugarcane was tested for transgenic production of human granulocyte macrophage colony stimulating factor (GM-CSF). GM-CSF is a cytokine that is produced in low concentrations by many cell types throughout the human body (Metcalf, 1991). The purified protein is used for treatment of neutropenia and aplastic anemia and is administered to bone marrow transplant patients to reduce infection risk by accelerating the response of neutrophils (Dale et al., 1995; Shin et al., 2003). Recombinant human GM-CSF protein has been produced in mammalian cells (Lee et al., 1985; Wong et al., 1985a, 1985b; Okamoto et al., 1990), insect cells (Chiou and Wu, 1990), yeast (Ernst et al., 1987; Balland et al., 1998), bacteria such as *Escherichia coli* (Libby et al., 1987), plant cell cultures such as tobacco (James et al., 2000) and rice (Shin et al., 2003), and seeds of tobacco (Ganz et al., 1996; Sardana et al., 2002) and rice.

Accumulation of GM-CSF protein ranged from undetectable to 0.02% of total soluble protein in transgenic sugarcane plants. It was further shown that human bone marrow cells (TF-1), which require GM-CSF for cell division, proliferated when growth media was supplemented with the extracts from transgenic sugarcane. The sugarcane-produced protein had essentially identical activity levels as commercially produced GM-CSF. These results suggest that sugarcane may provide a highly secure system for biofactory of pharmaceutical proteins.

Sugarcane has also been used as a model system for the production of industrials such as poly-3-hydroxybutyrate (PHB). Brumbley et al. (2003) engineered the genes encoding the enzymes for PHB production in sugarcane and showed that the leaves accumulated PHB up to 1.2% of their dry weight without affecting plant growth. Transgenic sugarcane has also been used for the production of *p*-hydroxy-benzoic acid that reached up to 7% of dry weight of leaves without any adverse affect on the growth of the plants (McQualter et al., 2004).

17.7.2.2 Production of Dextran in Transgenic Potato

There is great interest in using plants for the production of novel polymers (Kok-Jacon et al., 2005). Potato has been used for the production of fructan (Gerrits, 2000) and silk (Scheller et al., 2001) indicating that potato, like sugarcane, can be used as bioreactors for the production of novel polymers. Dextrans are important extracellular glucans. Commercially, dextran is produced by *Leuconostoc mesenteroides* NRRL B-512F via fermentation at a large industrial scale since 1948 (Groenwall and Ingelman, 1948). Dextrans are used in chromatographic media, as a soil conditioner (Murphy and Whistler, 1973), and as biodegradable hydrogels (Hennink and van Nostrum, 2002). Dextran-based gels are used for specific targeting of drugs to the colon, where they are degraded by a secreted bacterial dextranase. Kok-Jacon et al. (2005) were successful in producing transgenic potato that produced dextrans concentrations that were almost double that of untransformed plants, providing further evidence of the power of biotechnology in plants.

17.7.2.3 Improved Cassava Starch

The demand for starch for industrial purposes is on the rise since it is used in paper, textile, medical, and other applications. Starch is modified in many different ways to meet the criteria for various applications. Cassava starch consists of two glucan polymers: amylose and amylopectin. Amylose free-starches have different physicochemical properties compared to those that contain amylase and they are used for manufacturing different products (Raemakers et al., 2005). For example, amylose-free potato starch showed a higher granule melting temperature, less retrogradation, and better adhesive properties that that containing amylase (Visser et al., 1997). The presence or absence of amylase is controlled by the action of one enzyme, granule bound starch synthase I (GBSSI) (Raemakers et al., 2005). Since conventional breeding of cassava is difficult, the transgenic approach was successful in developing a genotype that produces amylase-free starch (Raemakers et al., 2005).

17.7.3 Risks Associated with Transgenic Crops

The rapid progress of transgenic biotechnology has accelerated the production of transgenic or GM crops. While GM crops are considered to have great benefits in solving the problems of world food security, GM technology is also purported to have serious biosafety concerns that have sparked heated debates. The major issues concerning GM crops include the environmental consequences of transgene flow from GM to non-GM cultivars and to its wild or weedy relatives. Movement of genes from one population to another could occur via pollination between individuals of different populations, seed dispersal between different populations, and through the dispersal of vegetative organs between different populations.

The release of GM crops for commercial purposes has raised concerns including food, feed, and human health safety caused by GM products, environmental safety, labeling of GM products, and detection of possible transgene or derived protein in agricultural products, socioeconomical and ethics concerns due to application of GM products and technology, regulatory procedures for GM-related issues, general public perception or acceptance of GM products, and risk assessment systems in relation to the environmental release and cultivation of GM products (Bao-Rong Lu, www.icgeb.org/~bsafesrv/pdffiles/Bao-Rong.pdf, accessed 23 April 2010). Consequently, regulatory systems have been and are being established in many countries to assess plants produced by genetic manipulation prior to commercial release. In the following section we examine some of the risks associated with transgenic crops and the research that is being undertaken to address these risks.

17.7.3.1 Risk of Transgenic Escape

Gene flow and introgression between cultivars and to wild and weedy relatives have always taken place in most of the important crops in the world (Ellstrand et al., 1999). Consequently, the possibility of transgene flow from transgenic crop plants to wild relatives is a matter of international concern (Ellstrand et al., 1999; Rieger et al., 2002). Generic studies are being carried out to ascertain whether transgenes can move from GM crops to plants related to the crop that grow either outside of cultivation, close to those crops, or as weeds within the crop (Bonnet et al., 2008). The chances of hybridization with wild relatives of crops' species vary with the crop. Some plants, such as the potato, are cultivated in regions that are far away from the center of origin and the likelihood of transgene escape to wild relatives is remote. However, the situation is different when transgenic potato are developed and grown within the center of origin of the crop. Therefore each crop has to be studied to assess its potential for outcrossing with wild relatives and transfer of the transgene (Green et al., 2005). For example, to assess the risk of a transgene escaping from sugarcane to related species, Bonnet et al. (2008) assessed the literature and reported that there has been a demonstrated transfer of DNA from *Zea mays*, *Sorghum*, *Erianthus*, and *Miscanthus* and between *Saccharum* species.

Current scientific literature has many similar studies for other crop species that will not be addressed in this chapter.

17.7.3.2 Risks to Other Organisms

In 2005 insect-resistant crops expressing *Bacillus thuringiensis* (*Bt*) toxins represented approximately 29% of the estimated global area of transgenic crops of about 90 million hectares (James, 2005). Insect-resistant transgenic plants have been suggested to have unpredictable effects on the biodiversity of the agro-ecosystem, including potential effects on insects' natural enemies, beneficial in control of crop pests. Although carnivorous as adults, many of these predators may also consume plant tissues, in particular plant pollen and nectar. The coleoptera are important in terms of agro-ecological research because of the large number of species in this order and for their role as biological control agents. Therefore any detrimental impact on this group of insects would be highly undesirable.

The first evidence that transgenic *Bt*-corn pollen caused significant mortality of the Monarch butterfly appeared in 2000 (Jesse and Obrycki, 2000). Laboratory experiments showed that Monarch caterpillars fed on milkweed leaves dusted with *Bt*-corn pollen ate less, grew more slowly, and suffered a higher mortality rate than those fed on leaves with normal pollen or with no pollen at all. Since then, other studies have shown that *Bt* toxins do not cause any detrimental effects on some beneficial coleopterans.

Commercial potato cultivars are highly susceptible to damage by the Colorado potato beetle (CPB) (*Leptinotarsa decemlineata*). Conventional breeding of potato has not been successful in producing a cultivar with resistance to this major pest. It is estimated that in 2005, 3.2 million pounds of insecticide were used on 1.2 million acres of potato grown in the United States. The pesticides used against the CPB are broad spectrum, killing not only the target pest but most of its natural enemies (Ferry et al., 2007). The effects of transgenic potato expressing the coleopteran-specific *Bacillus thuringiensis*–endotoxin Cry3A (*Bt* Cry3A) on the ladybird beetle *Harmonia axyridis* and the carabid beetle *Nebria brevicollis* were investigated via the bitrophic interaction of the adult ladybird with potato flowers and the tritrophic interaction of the carabid consuming a nontarget potato pest. Immunoassays confirmed accumulation of the transgene product in potato leaves and floral tissues at levels of up to 0.01% (pollen) and 0.0285% (anthers) of total soluble protein. Although *H. axyridis* and *N. brevicollis* belonged to the targeted insect order, no significant effects were observed on the survival or overall body mass change of either beetle. Furthermore, Bt Cry3A had no detrimental effects on reproductive fitness of either beetle species, either in terms of fecundity or subsequent egg viability. Behavioral analysis revealed no significant impact of Bt Cry3A on beetle activity or locomotor behavior. Despite the potential for detrimental effects, in that the ladybird beetles belong to the targeted order of insects and binding of Cry3A toxin to the gut was demonstrated, Cry3A expressed in transgenic potato plants had no overall significant acute effects on survival, body mass change, fecundity, or egg viability of *H. axyridis*.

Additionally *L. oleracea* was used as a model nontarget pest to assess the effects of tritrophic prey-mediated exposure to insecticidal proteins expressed in transgenic plants. Despite the potential for detrimental effects, in that carabids belong to the targeted order of insects, Cry3A expressed in transgenic potato plants had no overall significant acute effects on survival, body mass change, fecundity, or egg viability of *N. brevicollis*. In conclusion, this study shows that the cultivation of transgenic potato expressing the *Bt* Cry3A toxin presents a low risk to coleopteran insects other than the targeted chrysomelid larvae, due to the high specificity of the toxin.

17.7.3.3 Safety of Biotech Foods

There has been much concern and debate about the safety of GM food for humans. Hence, rigorous risk safety assessment has been recommended prior to commercial cultivation and market approval of GM crops and products. Shepherd et al. (2006) have outlined the procedures for assessing the safety of GM crops that are summarized in the following section. The first step in the risk assessment strategy is to identify appropriate methodologies and approaches to compare the GM crop or product with its non-GM counterparts. This comparison is the starting point of the safety assessment, which then focuses on any intended or unintended differences identified. The concept of substantial equivalence is based on the idea that an existing organism used as food or feed with a history of safe use can serve as a comparator when assessing the safety of genetically modified food/feed (OECD, 1993). An important part of this process is to evaluate any compositional changes that might occur in the GM crop that might influence its nutritional composition or toxic properties (Kuiper and Kleter, 2003; Howlett et al., 2003). It is important that such comparisons should only be made with GM and non-GM counterparts grown under the same regimes and environments. Where substantial equivalence does not occur, this does not necessarily identify a hazard. Where a trait or traits are introduced with the intention of modifying composition significantly and where the degree of equivalence cannot be considered substantial, then the safety assessment of characteristics other than those derived from the introduced trait(s) becomes of greater importance. It should be emphasized that unintended effects, although known to occur in GM crops, also happen frequently in conventional plant breeding via point mutations as well as through chromosomal recombination mechanisms (Mohan Jain, 2001). Furthermore, unintended effects that occur in conventional plant breeding have traditionally been identified and eliminated by removing lines that show inferior characteristics (such as reduced yield or altered appearance) to established cultivars (Cellini et al., 2004, and references therein). GM crops must meet the same criteria.

In substantial equivalence testing, the GM crop is grown side by side with the parental cultivar or other controls in randomized plots, usually at more than one location. Then selected components are measured in GM and control plants, and tested statistically for significant differences. Data from these field trials may also be compared with historical data for conventional cultivars to indicate the safe range for each component. In practice, the compounds selected are the important nutrients plus specific antinutrients and toxicants known for that crop.

Targeted analysis of specific compounds that make an important contribution to the nutritional value or safety of the crop species in question is the currently accepted approach to assess compositional equivalence (OECD, 1993). However, more unbiased profiling technologies such as metabolomics, proteomics, and transcriptomics are considered as emerging technologies that would extend the breadth of comparative analyses, reduce uncertainty, and identify the need for further risk assessment (Catchpole et al., 2005). Should new technologies be applied, then the expectation is that all approaches are properly validated and that statistical analyses have been performed to the highest standard (Kuiper et al., 2001).

Targeted compositional analysis was carried out on transgenic potato tubers of cultivars 'Record' or 'Desirée' to assess the potential for unintended effects caused by the genetic modification process. The range of transgenic lines analyzed included those modified in primary carbohydrate metabolism, polyamine biosynthesis, and glycoprotein processing. Controls included wild-type tubers, tubers produced from plants regenerated through tissue culture (including a callus phase), and tubers derived from transformation with the empty vector—that is, no specific target gene included (with the exception of the kanamycin resistance gene as a selectable marker). Metabolite analysis included soluble carbohydrates, glycoalkaloids, vitamin C, total nitrogen, and fatty acids. Trypsin inhibitor activity was also assayed. Targeted compositional analysis revealed no consistent differences between GM clones and respective controls. No construct specifically induced unintended effects. Statistically significant differences between wild-type controls and specific GM lines did occur but appeared to be random and not associated with any specific construct. Such significant differences were also found between wild types and both tissue-culture-derived tubers and tubers derived from transformation with the empty vector. This raises the possibility that somaclonal variation (known to occur significantly in potato, depending on genotype) may be responsible for an unknown proportion of any differences observed between specific GM clones and the wild type (Shepherd et al., 2006).

17.7.4 Marker-Free Transgenic Plants

Plant transformation is based on the ability to integrate foreign DNA into host-plant genomes and on the regeneration efficiency of transformed cells into shoots or embryos (Putcha, 2003). The transformation efficiency for many crops is low and selectable markers genes are used to identify the transgenic plants. Two types of marker genes are currently being used: a selectable marker gene (encoding antibiotic resistance, herbicide tolerance, and so on) and a marker gene used for visual screening of potential transgenics (Akhond and Machray, 2009). Environmentalists and consumer organizations have raised concerns about the use of the first type of selectable markers. The concerns include the transfer of herbicide resistance genes to weeds by outcrossing (Dale, 2002). In addition, the presence of resistance genes against antibiotics in food products might theoretically lead to the spread of these resistances via the intestinal bacteria in human populations (Puchta, 2003). These concerns have hindered the testing and widespread use of transgenic crops in many countries, especially where transgenic crops can potentially increase food productivity. Therefore studies to avoid markers genes or to eliminate them after use have been conducted, and a growing number of methods are under development for the elimination of selectable markers genes. Several methods developed for genetic modification of plants without foreign genes are described by Akhond and Machray (2009). Native genes and regulatory elements can also be introduced into plants without using selectable markers by using plant transfer DNAs (P-DNAs) derived from a crossable donor plant of the same or closely related species (Rommens et al., 2005, 2007). This approach referred to as the cisgenic or intragenic approach has been used in the genetic modification of potato (Rommens et al., 2004). Requests have been initiated to alter the legislation concerning transgenic and cisgenic plants (Havertkort et al., 2008). Transgenic plants contain genes from plants or other species that could not be introduced through introgression (crossing). Cisgenic plants contain indigenous genes from crossable species. Cisgenes with their native promoters introduced into a crop cultivar do not introduce new phenotypic traits into a species, as they were there already. They do not introduce new fitness traits, so putatively will not influence the environment or the risks in food or feed in another way than with traditional breeding. Cisgenesis may even be safer than conventional breeding because it prevents introduction of genes via linkage drag, which could lead to all kinds of unwanted traits (Haverkort et al., 2008). It is hopeful that the use of such marker-free techniques will make genetically modified crops more acceptable to the public.

17.8 CONCLUSIONS

Vegetatively propagated crops are staple foods in many of the poorer countries of the world. The world's population is headed for 9.2 billion people by 2050. The challenge of doubling food production in the next 40 years has become a daunting task. We have shown that most vegetatively propagated crops have complex genetic systems, and conventional breeding alone will not achieve the development of new improved cultivars. Molecular breeding can play an important role in realizing the full potential of difficult-to-breed vegetatively propagated crops. However, there appear to be many hurdles. Development of efficient genotype-independent transformations systems is the ultimate goal for the production of transgenic crops. This remains elusive due to varying morphogenic responses of some crops. The development of MAS has not reached the stage where it has become routine in breeding programs. While newer and faster techniques for developing molecular markers keep emerging, the so-called orphan crops lag far behind in this MAS.

Although the development of high-density maps for crops such as potato, rice, and so forth is forging ahead, there is very little research support for development of maps of other crops that feed the majority of the poor of this world. In this regard researchers of such crops can take advantage of concepts such as genome snyteny. Comparative mapping of related species has shown that colinearity and synteny exist in the arrangement of genes. The degree of genome conservation observed between species provides the possibility of transferring genetic information and mapping resource from well-characterized genomes to less studied and more complex genomes (Asnaghi et al., 2000). For example, knowledge of sugarcane, sorghum, maize, and rice genetic maps enables researchers to identify chromosome segments corresponding to the region that harbors the rust-resistance gene in sugarcane (Asnaghi et al., 2000).

This chapter provides evidence of the role of molecular breeding with regards to the development of crops with disease and pest resistance. Biotechnology is playing an important role in improving the nutritional qualities of some crops that are also being used as biofactories to produce useful products for man. One of the major concerns of GM crops is the safety it poses for human health and for the environment. It is encouraging to see that research to address these problems is ongoing and strict guidelines are being developed for GM foods.

Genetically modified crops have been commercial on a substantial scale for over 10 years. There is evidence to show that the technology has reduced pesticide spraying and decreased the environmental impact associated with herbicide and insecticide use. GM technology has reduced the release of greenhouse gases and has produced net economic benefits at the farm level. Despite these benefits, the debate on the adoption of GM foods still continues. Such debates can only act to stimulate new research such as the development of marker-free transgenic plants and research on cisgenics.

REFERENCES

Ahuja, J.K.C., J.D. Goldman, and A.J. Moshfegh. 2004. Current status of vitamin E nutriture. *Ann. NY Acad. Sci.* 1031:387–390.

Aitken, K.S., P.A. Jackson, and C.L. McIntyre. 2005. A combination of AFLP and SSR markers provides extensive map coverage and identification of homo(eo)logous linkage groups in a sugarcane cultivar. *Theor. Appl. Genet.* 110:789–801.

Akano, A.O., A.G.O. Dixon, C. Mba, E. Barrera, and M. Fregene. 2002. Genetic mapping of a dominant gene conferring resistance to cassava mosaic disease. *Theor. Appl. Genet.* 105:521–525.

Akhond, M.A.Y. and G.C. Machray. 2009. Biotech crops: Technologies, achievements and prospects. *Euphytica* 166:47–59.

Al-Janabi, S.M., R.J. Honeycutt, M. McClelland, and M. Sobral. 1993. A genetic linkage map of *Saccharum spontaneum* L. 'SES208.' *Genetics* 134:1249–1260.

Altman, A. 1999. The plant and agricultural biotechnology revolution: Where do we go from here? In: *Plant biotechnology and in vitro biology in the 21st century*, A. Altman, M. Ziv and S. Izhar, eds., 1–7. Dordrecht, The Netherlands: Kluwer Academic Publishers.

Alwala, S., C.A. Kimbeng, K.A. Gravois, and K.P. Bischoff. 2006. Trap a new tool for sugarcane breeding: Comparison with AFLP and coefficient of parentage. *J. Amer. Soc. Sugarcane Technol.* 26:62–87.

Anderson, M.D., P.K. Busk, I. Svendsen, and B.J. Moller. 2000. Cytochromes P-450 from cassava (*Manihot esculenta* Crantz) catalyzing the first steps in the biosynthesis of the cyanogenic glucosides linamarin and lotaustralin. *J. Biol. Chem.* 275:1966–1975.

Arvinth, S., R.K. Selvakesavan, N. Subramonian, and M.N. Premachandran. 2009. Transmission and expression of transgenes in progeny of sugarcane clones with *cry1Ab* and *aprotinin* genes. *Sugar Tech.* 11:292–295.

Asada, K. 1999. The water-water cycle in chloroplasts: Scavenging of active oxygen and dissipation of excess photons. *Annu. Rev. Plant Physiol. Plant Mol. Biol.* 50:601–639.

Asnaghi, C., F. Paulet, C. Kaye, L. Grivet, J.C. Glaszman, and A. D'Hont. 2000. Application of synteny across the *Poaceae* to determine the map location of a rust resistance gene of sugarcane. *Theor. Appl. Genet.* 10:962–969.

Asnaghi, C., D. Roques, S. Rusel, C. Kaye, J.Y. Hoarau, H. Telismart, et al. 2004. Targeted mapping of a sugarcane rust resistance gene (*Bru1*) using bulk segregant analysis and AFLP markers. *Theor. Appl. Genet.* 108:759–764.

Avasthy, P.N. 1983. Insect pest management for sugarcane in India. In: *Sugarcane pest management in India*, M. Balasubramanian and A.R. Solayappan, eds., 71–77. Madras: Tamil Nadu Coop Sugar Federation.

Bae, J.J., D. Halterman, and S. Jansky. 2008. Development of a molecular marker associated with Verticillium wilt resistance in diploid interspecific potato hybrids. *Mol. Breed.* 22:61–69.

Balland, A., D.A. Krasts, K.L. Hoch, M.J. Gerhart, K.E. Stremler, and S.M. Waugh. 1998. Characterization of the microheterogeneities of pixy32 1, a genetically engineered granulocytemacrophage colony-stimulating factor/interleukin-3 fusion protein expressed in yeast. *Eur. J. Biochem.* 25(3):812–820.

Bao-Rong Lu. www.icgeb.org/~bsafesrv/pdffiles/Bao-Rong.pdf (accessed 23 April 2010).

Bell, H.A., E.C. Fitches, R.E. Down, L. Ford, G.C. Morris, J.P. Edwards, et al. 2001. Effect of dietary cowpea trypsin inhibitor (CpTI) on the growth and development of the tomato moth *Lacanobia oleracea* (Lepidoptera: Noctuidae) and on the success of the gregarious ectoparasitoid *Eulophus pennicornis* (Hymeroptera: Eulopdidae). *Pest Manage. Sci.* 57:57–65.

Birch, A.N.E., I.E. Geoghegan, M.E.N. Majerus, J.W. McNicol, C.A. Hackett, A.M.R. Gatehouse, et al. 1999. Tri-trophic interactions involving pest aphids, predatory 2-spot ladybirds and transgenic potatoes expressing snowdrop lectin for aphid resistance. *Mol. Breed.* 5:75–83.

Bisognin, D.A., D.S. Douches, L. Buszka, G. Bryan, and D. Wang. 2005. Mapping late blight resistance in *Solanum microdontum* Bitter. *Crop Sci.* 45:340–345.

Bonnel, E. 2008. Potato breeding: A challenge, as ever. *Potato Res.* 51:327–332.

Bonnet, G.D., E. Nowak, J.J. Olivares-Villegas, N. Berding, T. Morgan, and S.K. Aitken. 2008. Identifying the risks of transgene escape from sugarcane crops to related species, with particular reference to *Saccharum spontaneum* in Australia. *Trop. Plant Biol.* 1:58–71.

Bormann, C.A., A.M. Rickert, R.A. Ruiz, J. Paal, J. Lubeck, J. Strahwald, et al. 2004. Tagging quantitative trait loci for maturity-corrected late blight resistance in tetraploid potato with PCR-based candidate gene markers. *Mol. Plant Microbe Interact.* 17:1126–1138.

Bradshaw, J.E. and G.R. Mackay. 1994. Breeding strategies for clonally propagated potatoes. In: *Potato genetics*, J.E. Bradshaw and G.R. Mackay, eds., 467–497. Wallingford: CABI International.

Brumbley, S.M., L.A. Petrasovits, P.A. Bonaventura, M.J. O'Shea, M.P. Purnell, and L.K. Nielsen. 2003. Production of polyhydroxyalkanoates in sugarcane. *Proc. Int. Soc. Sugar Cane Technol. Mol. Biol. Workshop*, Montpellier, France 4:31.

Brumbley, S.M., M.P. Purnell, L.A. Petrasovits, M.G. O'Shea, and K.L. Nielsen. 2004. Development of sugarcane as a biofactory/biopolymers. Paper presented at the Plant & Animal Genomes XII Conference, January 10–14, 2004, San Diego, CA.

Bull, T.A. and K.T. Glasziou. 1963. The evolutionary significance of sugar accumulation in *Saccharum. Aust. J. Biologic. Sci.* 16:737–742.

Camadro, E.L., D. Carputo, and S.J. Peloquin. 2004. Substitute for genome differentiation in tuber-bearing *Solanum*: Interspecific pollen–pistil incompatibility, nuclear-cytoplasmic male sterility and endosperm. *Theor. Appl. Genet.* 109:1369–1376.

Catchpole, G.S., M. Backmann, D.P. Enot, M. Mondhe, B. Zywicki, J. Taylor, et al. 2005. Hierarchical metabolomics demonstrates substantial compositional similarity between genetically modified and conventional potato crops. *Proc. Natl. Acad. Sci.* (USA) 102:14458–14462.

Ceballos, H., C.A. Iglesias, J.C. Pérez, and A.G.O. Dixon. 2004. Cassava breeding: Opportunities and challenges. *Plant Mol. Biol.* 56:503–515.

Cellini, F., A. Chesson, I. Colquhoun, A. Constable, H.V. Davies, K.H. Engel, et al. 2004. Unintended effects and their detection in genetically modified crops. *Food Chem. Toxicol.* 42:1089–1125.

Cervantes-Flores, J.C., G.C. Yencho, A. Kriegner, K.V. Pecota, M.A. Faulk, R.O.M. Mwanga, et al. 2008. Development of a genetic linkage map and identification of homologous linkage groups in sweet potato using multiple-dose AFLP markers. *Mol. Breed.* 21:511–532.

Chakrabarti, S.K., A.D. Mandakar, A. Shukla, D. Pattanayak, P.S. Naik, R.P. Sharma, et al. 2000. *Bacillus thuringiensis cry1Ab* gene confers resistance to potato against *Helicoverpa armigera* (Hubner). *Potato Res.* 43:143–152.

Chellapan, P., V.M. Munyaradzi, V. Ramachandran, N.J. Taylor, and C.M. Fauquet. 2004. Broad spectrum resistance to ssDNA associated with transgene-induced gene silencing in cassava. *Plant Mol. Biol.* 56:601–611.

Chiou, C-J. and M-C, Wu. 1990. Expression of human granulocyte-macrophage colony-stimulating factor gene in insect cells by a baculovirus vector. *FEBS Lett.* 259(2):249–253.

Chrispeels, J.M. and D.E. Sadava. 2003. *Plants, genes and crop biotechnology*. 2nd ed. Sudbury, MA: Jones and Bartlett Publishers.

Colton, L.M., H.I. Groza, S.M. Wielgus, and J. Jiang. 2006. Marker-assisted selection for the broad-spectrum potato late blight resistance conferred by gene RB derived from a wild potato species. *Crop Sci.* 46:589–594.

Crowell, E.F., J.M. McGrath, and D.S. Douches. 2008. Accumultion of vitamin E in potato (*Solanum tuberosum*) tubers. *Transgenic Res.* 17:205–217.

Dale, D.C., W.C. Liles, W.R. Summer, and S. Nelson. 1995. Review: Granulocyte colony-stimulating factor—Role and relationships in infectious diseases. *J. Infect. Dis.* 172(4):1061–1075.

Dale, P.J., B. Clark, and E.M.G. Fontes. 2002. Potential for the environmental impact of transgenic crops. *Nat. Biotech.* 20:567–574.

Da Silva, J.A.G., W.L. Burnquist, and S.D. Tanksley. 1993. RFLP linkage map and genome analysis of *Saccharum spontaneum*. *Genome* 36:782–791.

Da Silva, J., R.J. Honeycutt, W.L. Burnquist, S.M. Al-Janabi, M.E. Sorrels, S.D. Tanksley, et al. 1995. *Saccharum spontaneum* L 'SES208' genetic linkage map combining RFLP- and PCR-based markers. *Mol. Breed.* 1:165–179.

Daugrois, J.H., L. Grivet, D. Roques, J.Y. Hoarau, H. Lombard, J.C. Glaszmann, et al. 1996. A putative major gene for rust resistance linked with a RFLP marker in sugarcane cultivar 'R570.' *Theor. Appl. Genet.* 92:1059–1064.

Davidson, M., J. Jacobs, J. Reader, R. Butler, C.M. Frater, N.P. Markwick, et al. 2002. Development and evaluation of potatoes transgenic for a *cry1Ac9* gene conferring resistance to potato tuber moth. *J. Amer. Soc. Hort. Sci.* 127(4):590–596.

Davidson, M.M., R.C. Butler, S.D. Wratten, and A.J. Conner. 2004. Resistance of potatoes transgenic for a *cry1Ac9* gene, to *Phthorimaea operculella* (Lepidoptera: Gelechiidae) over field seasons and between plant organs. *Assn. Appl. Biol.* 145(3):271–277.

Davies, H.V. 1998. Prospects for manipulating carbohydrate metabolism in potato tuber. *Aspects Appl. Biol.* 52:245–254.

De Koeyer, D., K. Douglass, A. Murphy, S. Whitney, L. Nolan, Y. Song, et al. 2010. Application of high-resolution DNA melting for genotyping and variant scanning of diploid and autotetraploid potato. *Mol. Breed.* 25:67–90.

D'Hont, A. and J.C. Glaszmann. 2001. Sugarcane genome analysis with molecular markers, a first decade of research. *Proc. Int. Soc. Sugarcane Technol.* 24:556–559.

D'Hont, A. and M. Layssac. 1998. Analysis of cultivars genome structure by molecular cytogenetics and the study of introgression mechanisms. France: Centre de coopération internationale en recherche agronomique pour le développement, Annual Crops Department (CIRAD-CA).

D'Hont, A., L. Grivet, P. Feldman, S. Rao, N. Berding, and J.C. Glazmann. 1996. Characterization of the double genome structure of modern sugarcane cultivars (*Saccharum* spp.) by molecular cytogenetics. *Mol. Gen. Genet.* 250:405–413.

D'Hont, A., O. Grasmeur, L.M. Raboin, F. Paulet, D. Begum, R. Wing, et al. 2001. Chromosome walking towards a major resistance gene for common rust of sugarcane. *Proc. Int. Soc. Sugarcane Technol.* 24:315–317.

Dillon, S.L., F.M. Shapter, R.J. Henry, G. Cordeiro, L. Izquierdo, and L. Slade Lee. 2007. Domestication to crop improvement: Genetic resources for sorghum and *Saccharum* (Andropogoneae). *Ann. Bot.* 100:975–989.

Drewel, B.T., D.W. Giraud, S.R. Davy, and J.A. Driskell. 2006. Less than adequate vitamin E status observed in a group of preschool boys and girls living in the United States. *J. Nutr. Biochem.* 17(2):132–138.

Ducreux, L.J.M., W.L. Morris, P.E. Hedley, T. Shepherd, H.V. Davies, S. Milam, et al. 2005. Metabolic engineering of high carotenoid potato tubers containing enhanced levels of beta-carotene and lutein. *J. Exp. Bot.* 56:81–89.

Duncan, D.R., D. Hammond, J. Zalewski et al., 2002. Field performance of transgenic potato, with resistance to Colorado potato beetle and viruses. *HortScience* 37:275–276.

Ellstrand, N.C., H.C. Prentice, and J.F. Hancock. 1999. Gene flow and introgression from domesticated plants into their wild relatives. *Annu. Rev. Ecol. Syst.* 30:539–563.

Erali, M., K.V. Voelkerding, and C.T. Wittwer. 2008. High resolution melting applications for clinical laboratory medicine. *Exp. Mol. Pathol.* 85:50–58.

Ernst, J., J. Mermod, J. Delamarter, R. Mattaliano, and P. Moonen. 1987. O-glycosylation and novel processing events during the secretion of alpha-factor/gm-csf fusion by *Saccharomyces cerevisiae. Biotechnol.* (NY) 5:831–834.

Ferry, N., M.G. Edwards, J.A. Gatehouse, and A.M.R. Gatehouse. 2004. Plant-insect interactions: Molecular approaches to insect resistance. *Curr. Opin. Biotechnol.* 15:155–161.

Ferry, N., E.A. Mulligan, M.E.N. Majerus, and A.M.R. Gatehouse. 2007. Bitrophic and tritrophic effects of *Bt* Cry3A transgenic potato on beneficial, non-target, beetles. *Transgenic Res.* 16:795–812.

Flis, B., J. Hermig, D. Strzelczyk-Zyta, C. Gebhardt, and W. Marezewshi. 2005. The *Ry-*f_{sto} gene from *Solanum stoloniferum* for extreme resistance to potato virus Y maps to potato chromosome XII and is diagnosed by PCR marker $GP122_{718}$ in PVY resistant potato cultivars. *Mol. Breed.* 15:95–101.

Foyer, C.H., P. Descourvières, and K.J. Kunert. 1994. Protection against oxygen radicals: An important defense mechanism studied in transgenic plants. *Plant Cell Environ.* 17:507–523.

Fregene, M., E. Okogbenin, C. Mba, F. Angel, M.C. Suarez, J. Guttierez, et al. 1997. Genome mapping in cassava improvement: Challenges, achievements and opportunities. *Euphytica* 120:159–165.

Ganz, P., A. Dudani, E. Tackaberry, R.K. Sardana, C. Sauder, and I. Altosaar. 1996. Expression of human blood proteins in transgenic plants: The cytokine gm-csf as a model protein. In: *Transgenic plants: A production system for industrial and pharmaceutical proteins*, M.R.L. Owen and J. Pen, eds., 281–297. New York: John Wiley and Sons.

Gebhardt, C. and D. Bellin. 2006. Marker-assisted combination of major genes for pathogen resistance in potato. *Theor. Appl. Genet.* 112:1458–1464.

Gebhardt, C. and J.P.T. Valkonen. 2001. Organization of genes controlling disease resistance in the potato genome. *Annu. Rev. Phytopathol.* 39:79–102.

Gebhardt, C., D. Bellin, H. Henselewski, W. Lehmann, J. Schwarzfischer, and J.P.T. Valkonen. 2006. Marker-assisted combination of major genes for pathogen resistance in potato. *Theor. Appl. Genet.* 112:1458–1464.

Gerrits, N. 2000. Tuber-specific fructan synthesis in potato amyloplasts. Ph.D. diss., Utrecht University, The Netherlands.

Green, J., M.T. Fearnehough, and H.J. Atkinson. 2005. Development of biosafe genetically modified *Solanum tuberosum* (potato) cultivars for growth within a centre of origin of the crop. *Mol. Breed.* 16:285–293.

Grivet, L. and P. Arruda. 2002. Sugarcane genomics: Depicting the complex genome of an important tropical crop. *Curr. Opin. Plant Biol.* 2:122–127.

Grivet, L., J.C. Glaszmann, and P. Arruda. 2001. Sequence polymorphism from EST data in sugarcane: A fine analysis of 6-phosphogluconate dehydrogenase genes. *Genet. Molec. Biol.* 24:161–167.

Grivet, L., J-C. Glaszmann, M. Vincentz, F. Da Sliva, and P. Arruda. 2003. ESTs as a source for sequence polymorphism discovery in sugarcane: Example of the *Adh* genes. *Theor. Appl. Genet.* 106:190–197.

Groenwall, A.J. and B.J.A. Ingelman.1948. Manufacture of infusion and injection fluids. US Patent 2: 437–518.

Guimaraes, C.T., G.R. Sills, and B.W.S. Sobral. 1997. Comparative mapping of Andropogoneae: *Saccharum* L (sugarcane) and its relation to sorghum and maize. *Proc. Natl. Acad. Sci. (USA)* 94:14261–14266.

Hämaläinen, J.H., K.N. Watanabe, J.P.T. Valkonen, A. Arihara, R.L. Plaisted, F. Pehu, et al. 1997. Mapping and marker-assisted selection for a gene for extreme resistance to potato virus Y. *Theor. Appl. Genet.* 94:192–197.

Hämaläinen, J.H., V.A. Sorri, K.N. Watanbe, C. Gebhardt, and J.P.T. Valkonen. 1998. Molecular examination of a chromosome region that controls resistance to potato Y and A potyviruses in potato. *Theor. Appl. Genet.* 96:1036–1043.

Haverkort, A.J., P.M. Boonekamp, R. Hutten, E. Jacobsen, L.A.P. Lotz, G.J.T. Kessel, et al. 2008. Societal costs of late blight in potato and prospects of durable resistance through cisgenic modification. *Potato Res.* 51:47–57.

Hawkes, J.G. 1990. *The potato: Evolution, biodiversity and genetic resources*. London: Belhaven Press.

Heiser, C.B. 1981. *Seed to civilization: The story of food*, 2nd ed. San Francisco: W.H. Freeman and Co.

Hennink, W.E. and C.F. van Nostrum. 2002. Novel cross-linking methods to design hydrogels. *Adv. Drug Deliv. Rev.* 54:13–36.

Hershey, C.H. and D.L. Jennings. 1992. Progress in breeding cassava for adaptation to stress. *Plant Breed. Abstr.* 62:823–83.

Hershey, C. and C. Ocampo. 1989. New marker genes found in cassava. *Cassava Newslett.* 13:1–5.

Hijmans, R.J. 2001. Global distribution of the potato crop. *Am. J. Potato Res.* 78:403–412.

Hoarau, J.Y., L. Grivet, B. Offmann, L.M. Raboin, J.P. Diorflar, J. Payet, et al. 2002. Genetic dissection of a modern sugarcane cultivar (*Saccharum* spp.) II. Detection of QTLs for yield components. *Theor. Appl. Genet* 105:1027–1037.

Holland-Moritz, P. 2003. Sugarcane-derived therapeutic proteins avoid contamination issues. *Drug Discov. Dev.* (September).

Howlett, J., D.G. Edwards, A. Cockburn, P. Hepburn, J. Kleiner, D. Knorr, et al. 2003. The safety assessment of novel foods and concepts to determine their safety in use. *Int. J. Food Sci. Nutr.* 54(Supplement):S1–S32.

Hunter, D.E., H.M. Darling, F.J. Stevensen, and C.E. Cunningham. 1968. Inheritance of resistance to Verticillium wilt in Wisconsin. *A. Potato J.* 45:72–78.

Iglesias, C.A. and C. Hershey. 1994. Cassava breeding at CIAT: Heritability estimates and genetic progress in the 1980's. In: *Tropical root crops in a developing economy*, F. Ofori and S.K. Hahn, eds., 149–163. Wageningen, Netherlands: ISTRC/ISHS.

Ikoigbo, B. 1980. Nutritional implications of projects giving high priority to the production of staples of low nutritive quality. The case of cassava in the humid tropics of West Africa. *Food Nutr. Bull.* (United Nations University, Tokyo), 2:1–10.

Iovene, M., S.M. Wielgus, P.W. Simon, C.R. Buell, and J. Jiang. 2008. Chromatin structure and physical mapping of chromosome 6 of potato and comparative analyses with tomato. *Genetics* 180:1307–1317.

Jaccoud, D., K. Peng, D. Feinstein, and A. Kilian. 2001. Diversity arrays: A solid state technology for sequence information independent genotyping. *Nucleic Acids Res.* 29:e25.

James, C. 2005. Executive summary of global status of commercialized biotech/GM crops: 2005. ISAAA Briefs No. 34. Ithaca, NY: ISAAA.

James, C. and A.F. Krattiger. 1996. Global review of the field testing and commercialization of transgenic plants, 1986 to 1995: The first decade of crop biotechnology. ISAAA Briefs No. 1. Ithaca, NY: ISAAA.

James, E.A., C. Wang, Z. Wang, R. Reeves, J.H. Shin, N.S. Magnuson, et al. 2000. Production and characterization of biologically active human gm-csf secreted by genetically modified plant cells. *Protein Expr. Purif.* 19(I):131–138.

Jenkins, G.H. 1966. *Introduction to cane sugar technology*. New York: Elsevier Publishing Co.

Jenning, D.L. and C. Iglesias. 2002. Breeding for crop improvement. In: *Cassava: Biology, production and utilization*, R.J. Hollocks and J.M. Thresh, eds. 149–166. Oxon: CABI.

Jesse, L.C.H. and J.J. Obrycki. 2000. Field deposition of *Bt* transgenic corn pollen: Lethal effects on the monarch butterfly. *Oecologia* 125:241–248.

Johnson, A.A.T. and R.E. Veilleux. 2003. Integration of transgenes into sexual polyploidization schemes for potato (*Solanum tuberosum* L.). *Euphytica* 133:125–138.

Jones, R.A.C. 1985. Further studies on resistance-breaking strains of potato virus X. *Plant Pathol.* 34:182–189.

Jones, N., H. Ougham, and H. Thomas. 1997. Markers and mapping: we are all geneticists now. *New Phytologist* 137:165–177.

Jordan, D.R., R.E. Casu, P. Besse, B.C. Carroll, N. Berding, and C.L. McIntyre. 2004. Markers associated with stalk number and suckering in sugarcane co-locate with tillering and rhizomatousness QTLs in sorghum. *Genome* 47:988–993.

Jorge, V., M. Fregene, M.C. Duque, M.W. Bonierbale, J. Tohme, and V. Verdier. 2000. Genetic mapping of resistance to bacterial blight disease in cassava (*Manihot esculenta* Crantz). *Theor. Appl. Genet.* 101:865–872.

Jorge, V., M. Fregene, C.M. Velez, M.C. Duque, J. Tohme, and V. Verdier. 2001. QTL analysis of field resistance to *Xanthomonas axonopodis* pv *manihotis* in cassava. *Theor. Appl. Genet.* 102:564–571.

Jorgensen, K., S. Bak, P.K. Busk, C. Sorensen, C.K. Olsen, J. Puonto-Kaerlas, and B.L. Moller 2005. Cassava plants with depleted cyanogenic glucoside content in leaves and tubers, distribution of cyanogenic glucosides, their site for synthesis and transport, and blockage of the biosynthesis by RNA interference technology. *Plant Physiol.* 139:363–374.

Kakes, P. 1990. Properties and functions of the cyanogenic system in higher plants. *Euphytica* 48:25–43.

Kasai, K.Y., Y. Morikawa, V.A. Sorri, J.P.T. Valkonen, C. Gebhardt, and K.N. Watanabe. 2000. Development of SCAR markers to the PVY resistance gene *Ryadg* based on a common feature of plant disease resistance genes. *Genome* 43:1–5.

Kawano, K. 2003. Thirty years of cassava breeding for productivity: Biological and social factors for success. *Crop Sci.* 43:1325–1335.

Kawchuk, L., R.R. Martin, and J. McPherson. 1990. Resistance in transgenic plants expressing the potato leaf roll luteovirus coat protein gene. *Mol. Plant-Microbe Interac.* 3:301–307.

Kawchuk, L., R.R. Martin, and J. McPherson. 1991. Sense and antisense RNA-mediated resistance to potato leaf roll virus in Russet Burbank potato plants. *Mol. Plant-Microbe Interac.* 4:247–253.

Kilian, A., E. Huttner, P. Wenzl, D. Jaccoud, J. Carling, V. Caig, et al. 2005. The fast and the cheap: SNP and DArT-based whole genome profiling for crop improvement. In: *Proceedings of the international congress "In the wake of the double helix: From the green revolution to the gene revolution,"* 27–31 May 2003, 443–461. Bologna, Italy: Avenue Media.

Koch, B., V. Nielsen, B. Halkier, C. Olsen, and B. Moller. 1992. The biosynthesis of cyanogenic gylcosides in seedlings of cassava (*Manihot esculenta* Crantz). *Arch. Biochem. Biophys.* 292:141–150.

Koebner, R.M. and R.W. Summers. 2003. 21st century wheat breeding: Selection in plots or detection in plates? *Trends Biotechnol.* 21:59–63.

Kok-Jacon, J.A., J-P. Vincken, C.J. Luc, M. Suurs, D. Wang, S. Liu, and R.G.F. Visser. 2005. Production of dextran in transgenic potato plants. *Transgenic Res.* 14:385–395.

Kriegner, A., J.C. Cervantes, K. Burg, R.O. Mwanga, and D.P. Zhang. 2000. A genetic linkage map of sweet potato (*Ipomoea batatas* (L.) Lam.) based on AFLP markers. *CIP Program Report 1999–2000*, 303–313.

Kuiper, H.A. and G.A. Kleter. 2003. The scientific basis for risk assessment and regulation of genetically modified foods. *Trends Food Sci. Technol.* 14:277–293.

Kuiper, H.A., G.A. Kleter, H.P.J.M. Noteborn, and E.J. Kook. 2001. Assessment of the food safety issues related to genetically modified foods. *Plant J.* 27:503–528.

Lakshmanan, P., R.J. Geijskes, K.S. Aitken, C.L.P. Grof, G.D. Bonnet, and G.R. Smith. 2005. Sugarcane biotechnology: The challenges and opportunities. *In Vitro Cell. Dev. Biol. Plant* 41:345–363.

Lamb, J.W. and R.T. Hay. 1990. Ribozymes that cleave potato leafroll virus RNA within the coat protein and polymerase genes. *J. Gen. Virol.* 71:257–2264.

Lee, F., T. Yokota, T. Otsuka, L. Gemmell, N. Larson, J. Luh, et al. 1985. Isolation of cDNA for a human granulocytemacrophage colony-stimulating factor by functional expression in mammalian cells. *Proc. Natl. Acad. Sci. (USA)* 82(13):4360–4364.

Libby, R., G. Braedt, S. Kronheim, C. March, D. Urdal, T. Chiaverotti, et al. 1987. Expression and purification of native human granulocyte-macrophage colony-stimulating factor from an *Escherichia coli* secretion vector. *DNA* 6(3):221–229.

Lim, S., Y-H. Kim, S-H. Kim, S-Y. Kwon, H-S Lee, J-S. Kim, et al. 2007. Enhanced tolerance of transgenic sweet potato plants that express both CuZnSOD and APX in chloroplast to methyl viologen-mediated oxidative stress and chilling. *Mol. Breed.* 19:227–239.

Lokko, Y., M. Gedil, and A. Dixon. 2004. QTLs associated with resistance to the cassava mosaic disease: New directions for a diverse planet. In: Proceedings of the 4th International Crop Science Congress, 26 September–1 October 2004, Brisbane, Australia. http://www.cropscience.org.au/icsc2004/poster/3/7/4/1154_lokko.htm (accessed 28 July 2010).

Magoon, M.L., Krishnan, R. and K.V. Bai. 1969. Morphology of the pachytene chromosomes and meiosis in *Manihot esculenta* Crantz. *Cytol*ogy 34:612–626.

Marczewski, W., J. Hennig, and C. Gebhardt. 2002. The *Potato virus* S resistance gene *Ns* maps to potato chromosome VIII. *Theor. Appl. Genet.* 105:564–567.

Mariotti, J.A. 2002. Selection for sugarcane yield and quality components in subtropical climates. *Sugarcane Int.* March/April:22–26.

McIntyre, C.L., R.E. Casu, J. Drenth, D. Knight, V.A. Whan, B.J. Croft, et al. 2005. Resistance gene analogues in sugarcane and sorghum and their association with quantitative trait for rust resistance. *Genome* 48:391–400.

McQualter, R.B., B.F. Chong, K. Meyer, D.E. Van Dyk, M.G. O'Shea, N.J. Walton, et al. 2004. Initial evaluation of sugarcane as a production platform for *p*-hydroxybenzoic acid. *Plant Biotechnol J.* 2:1–13.

Menossi, M., M.C. Silva-Filho, M. Vincentz, M.-A. Van-Sluys, and G.M. Souza. 2008. Sugarcane functional genomics: Gene discovery for agronomic trait development. *Int. J. Plant Genom.* Article ID 458732, 11 pages DOI:10.1155/2008/458732.

Metcalf, D.1991. Control of granulocytes and macrophages: Molecular, cellular, and clinical aspects. *Science* 254:529–533.

Ming, R., S-C. Liu, Y-R. Lin, J. DaSilva, W.Wilson, D. Braga, A. Deynze, et al. 1998. Detailed alignment of *Saccharum* and *Sorghum* chromosomes: Comparative organization of closely related diploid and polyploidy genomes. *Genetics* 150:1663–1682.

Ming, R., S.C. Urn, P.H. Moore, J.B. Irvine, and A.H. Paterson. 2001. QTL analysis in a complex autopolyploid: Genetic control of sugar content in sugarcane. *Genome Res.* 11:2075–2084.

Ming, R., T.A. Del Monte, B. Hernandez, P.H. Moore, J.B. Irvine, and A.H. Paterson. 2002. Comparative analysis of QTLs affecting plant height and flowering among closely-related diploid and polyploid genomes. *Genome* 45:794–803.

Ming, R., S.C. Liu, J.E. Bowers, P.H. Moore, J.E. Irvine, and A.H. Paterson. 2002. Construction of a consensus genetic map from two interspecific crosses. *Crop Sci.* 42:570–583.

Missiou, A., K. Kalanidis, A. Boutla, S. Tzortzakaki, M. Tabler, and M. Tsagris. 2004. Generation of transgenic potato plants highly resistant to potato virus Y (PVY) through RNA silencing. *Molec. Breed.* 14:185–197.

Mohammed, A., D.S. Douches, W. Pett, E. Grafius, J. Coombs, W. Liswidowati, et al. 2000. Evaluation of potato tuber moth (Lepidoptera: Gelechiidae) resistance in tubers of Bt-cry5 transgenic potato lines. *J. Econ. Entomol.* 93:472–476.

Mohan Jain, S. 2001. Tissue culture-derived variation in crop improvement. *Euphytica* 118:153–166.

Montgomery, J., C.T. Wittwer, R. Palais, and L. Zhou. 2007. Simultaneous mutation scanning and genotyping by high-resolution DNA melting. *Nat. Protoc.* 2:59–66.

Morgante, M. and A.M. Olivieri. 1993. PCR-amplified microsatellites as markers in plant genetics. *Plant J.* 3:175–182.

Morris, W.L., L.J.M. Ducreux, P.D. Fraser, S. Millam, and M.A Taylor. 2006. Engineering ketocarotenoid biosynthesis in potato tubers. *Metab. Eng.* 8:253–263.

Mudge, J., W.R. Andersen, R.L. Kehrer, and J.D. Fairbanks. 1996. A RAPD genetic map of *Saccharum officinarum. Crop Sci.* 36:1362–1366.

Murphy, P.T. and R.L. Whistler. 1973. Dextrans. In: *Industrial gums*, R.L. Whistler, ed., 513–542. New York: Academic Press.

Niewohner, J., F. Salamini, and C. Gebhardt. 1995. Development of PCR assays diagnostic for RFLP marker alleles closely linked to alleles *Gro1* and *H1*, conferring resistance to the root cyst-nematode *Globodera rostochiensis* in potato. *Mol. Breed.* 1:65–78.

OECD. 1993. Safety evaluation of foods derived by modern biotechnology: Concepts and principles. Organisation for Economic Co-operation and Development (OECD), Paris. http://www.oecd.org/dsti/sti/s_t/biotech/prod/MODERN.pdf

Okamoto, M., C. Nakayama, M. Nakai, and H. Yanagi. 1990. Amplification and high-level expression of a cDNA for human granulocyte-macrophage colony-stimulating factor in human lymphoblastoid Namalwa cells. *Biotechnology* (NY) 8(6):550–553.

Okamato, D., M. Nielson, M. Albrechtsen, and B. Borkhardt. 1996. General resistance against potato virus Y introduced into a commercial potato cultivar by genetic transformation with PVYN coat protein gene. *Potato Res.* 39:271–282.

Okogbenin, E. and M. Fregene. 2002. Genetic and QTL mapping of early root bulking in an F_1 mapping population of non-inbred parents in cassava (*Manihot esculenta* Crantz). *Theor. Appl. Genet.* 106:58–66.

Okogbenin, E. and M. Fregene. 2003. Genetic mapping of QTLs affecting productivity and plant architecture in a full-sib cross from non-inbred parents in Cassava (*Manihot esculenta* Crantz). *Theor. Appl. Genet.* 107:1452–1462.

Okogbenin, E., J. Marin, and M. Fregene. 2006. An SSR-based molecular genetic map of cassava. *Euphytica* 147:433–440.

Ortiz, R. 2004. Breeding clones. In: *Encyclopedia of plant and crop science*, R.M. Goodman, ed., 174–178. New York: Marcel Dekker, Inc.

Ortiz, R., P. Simon, S. Jansky, and D. Stelly. 2009. Ploidy manipulation of the gametophyte, endosperm, and sporophyte in nature and for crop improvement—A tribute to Prof. Stanley J. Peloquin (1921–2008). *Ann. Bot.*104:795–807.

Ottoman, R.J., D.C. Hane, C.R. Brown, S. Yilma, S.R. James, A.R. Mosley, et al. 2009. Validation and implementation of marker-assisted selection (MAS) for PVY resistance (Ry_{adg} gene) in a tetraploid potato breeding program. *Am. J. Potato Res.* 86:304–314.

Pavek, J.J. and D.L. Corsini. 1994. Inheritance of resistance to warm-growing-season fungal diseases. In: *Potato genetics*, J.E. Bradshaw and G.R. Mackay, eds, 403–410. Wallingford: CABI.

Pegg, G.F. 1974. *Verticillium* diseases. *Rev. Plant Pathol.* 53:157–182.

Puchta, H. 2003. Marker-free transgenic plants. *Plant Cell Tissue Organ Cult.* 74:123–134.

Purseglove, J.N. 1979. *Tropical crops: Monocotyledons*. London: Longman Group Ltd.

Quint, M., R. Mihaijevic, C.M. Dutsle, M.L. Xu, A.E. Meichinger, and T. Lubberstedt. 2002. Development of RGA-CAPS markers and genetic mapping of candidate genes for sugarcane mosaic virus resistance in maize. *Theor. Appl. Genet.* 105:355–363.

Raboin, L.M., K.M. Oliveira, L. Lecunif, H. Telismart, D. Roques, M. Butterfield, et al. 2006. Genetic mapping in sugarcane, a high polyploid, using bi-parental progeny: Identification of a gene controlling stalk colour and a new rust resistance gene. *Theor. Appl. Genet.* 112:1382–1391.

Raemakers, K., M. Schreuder, L. Suurs, H. Furrer-Verhorst, J-P. Vincken, N. de Vetten, et al. 2005. Improved cassava starch by antisense inhibition of granule-bound starch synthase I. *Mol. Breed.* 16:163–172.

Ramsey, J. and D.W. Schemske. 2002. Neopolyploidy in flowering plants. *Annu. Rev. Ecol. Syst.* 33:589–639.

Reece, J.D. and E. Haribabu. 2007. Genes to feed the world: The weakest link? *Food Policy* 32(4):459–479.

Reed, G.H., J.O. Kent, and C.T. Wittwer. 2007. High-resolution DNA melting analysis for simple and efficient molecular diagnostics. *Pharmacogenomics* 8:597–608.

Reed, G.L., A.S. Jensen, J. Riebe, G. Head, and J.J. Duan. 2001. Transgenic *Bt* potato and conventional insecticides for Colorado potato beetle management: Comparative efficacy and non-target impacts. *Entomol. Exp. Appl.* 100:89–100.

Rieger, M.A., M. Lamond, C. Preston, S.B. Powles, and R.R.T. Roush. 2002. Pollen mediated movement of herbicide resistance between commercial canola fields. *Science* 296:2386–2388.

Ripol, M.I., G.A. Churchill, J.A.G. DaSilva, and M. Sorrells. 1999. Statistical aspects of genetic mapping in autopolyploids. *Gene* 235:31–41.

Roa, A.C., P. Chavarriaga-Aguirre, M.C. Duque, M.M. Maya, M.W. Bonierbale, C. Iglesias, et al. 2000. Cross-species amplification of cassava (*Manihot esculenta*) (*Euphorbiaceae*) microsatellites: Allelic polymorphism and degree of relationship. *Am. J. Bot.* 87:1647–1655.

Römer, S., J. Lubeck, F. Kauder, S. Steiger, C. Adomat, and G. Sandmann. 2002. Genetic engineering of a zeaxanthin-rich potato by antisense inactivation and co-suppression of carotenoid epoxidation. *Metab. Eng.* 4:263–272.

Rommens, C.M., J.M. Humara, J.S. Ye, H. Yan, C. Richael, L. Zhang, et al. 2004. Crop improvement through modification of the plant's own DNA. *Plant Physiol.* 135:421–431.

Rommens, C.M., O. Bougri, H. Yan, J.M. Humara, J. Owen, K. Swords, et al. 2005. Plant-derived transfer DNAs. *Plant Physiol.* 139:1338–1349.

Rommens, C.M., M.A. Haring, K. Swords, H.V. Davies, and W.R. Belknap. 2007. The intragenic approach as a new extension to traditional plant breeding. *Trends Plant Sci.* 2(9):397–403.

Rosling, H., N. Mlingi, T. Tylleskar, and M. Banea. 1992. Causal mechanisms behind human diseases induced by cyanide exposure from cassava. In: *Proceedings of the first international scientific meeting of the Cassava Biotechnology Network*, 336–375, Cartagena, Cali, Colombia.

Rossi, M., P.G. Araujo, F. Paulet, O. Garsmeur, V.M. Dias, H. Chen, et al. 2003. Genomic distribution and characterization of EST-derived resistance gene analogs (RGAs) in sugarcane. *Mol. Gen. Genet.* 269:406–409.

Rott, P., R.A. Bailey, J.C. Comstock, B.J. Croft, and A.S. Saumtally. 2000. *A guide to sugarcane diseases.* Montpellier, France: La librairie du Cirad.

Rowe, R.C. and M.R. Powelson. 2002. Potato early dying: Management challenges in a changing production environment. *Plant Dis.* 86:1184–1193.

Rowe, R.C., J.R. Davis, M.L. Powelson, and D.I. Rouse 1987. Potato early dying: Causal agents and management strategies. *Plant Dis.* 71:482–489.

Sardana, R.K., Z. Alli, A. Dudani, E. Tackaberry, M. Panahi, M. Narayanan, et al. 2002. Biological activity of human granulocyte-macrophage colony stimulating factor is maintained in a fusion with seed glutelin peptide. *Transgenic Res* 5:521–531.

Scheller, J., K-H. Guhrs, F. Grosse, and U. Conrad. 2001. Production of spider silk proteins in tobacco and potato. *Nature Biotechnol.* 19:573–577.

Scott, G.J.M., M.W. Rosegrant, and C. Ringler. 2000. *Roots and tubers for the 21st century: Trends, projections and policy options*. Food, Agriculture and Environment Discussion Paper 31, Washington, DC: International Food Policy Research Institute (IFPRI).

Serrano, C., P. Arce-Johnson, H. Torres, M. Gebauer, M. Moreno, X. Jordana, et al. 2000. Expression of the chicken lysozyme gene in potato enhances resistance to infection by *Erwinia carotovora* subsp. *atroseptica*. *Am. J. Potato Res.* 77:191–199.

Sharma, K.K., P. Bhatnagar-Matur, and T.A. Thorpe. 2005. Genetic transformation technology: Status and problems. *In Vitro Cell Dev. Biol.* 41:102–112.

Shepherd, L.V.T., J.W. McNicol, R. Razzo, M.A. Taylor and H.V. Davies. 2006. Assessing the potential of unintended effects in genetically modified potatoes perturbed in metabolic and developmental processes: Targeted analysis of key nutrients and anti-nutrients. *Transgenic Res.* 15:409–425.

Shin, Y.J., S.Y. Hong, T.H. Kwon, Y.S. Jang, and M.S. Yang. 2003. High level of expression of recombinant human granulocyte-macrophage colony stimulating factor in transgenic rice cell suspension culture. *Biotechnol. Bioeng.* 82(7):778–783.

Simko, I., S. Jansky, S. Stephenson, and D. Spooner. 2007. Genetics of resistance to pests and diseases. In: *Potato biology and biotechnology advances and perspectives*, D. Vreugdenhil, J. Bradshaw, C. Gebhardt, F. Govers, D.K.L. Mackerron, M.A. Taylor, and H.A. Ross, eds., 117–155. Oxford: Elsevier.

Sinha, S.K. and M.S. Swaminathan. 1984. New parameters and selection criteria in plant breeding. In: *Crop breeding, a contemporary basis*, P.B. Vose and S.G. Blixt, eds., 1–31. Cambridge: Oxford.

Siritunga, D. and R. Sayre. 2003. Generation of cyanogen-free transgenic cassava. *Planta* 217:367–373.

Siritunga, D. and R. Sayre. 2004. Engineering cyanogen synthesis and turnover in cassava (*Manihot esculenta*). *Plant Mol. Biol.* 56:661–669.

Somers, D.J., G. Fedak, and M. Savard. 2003. Molecular mapping of novel genes controlling Fusarium head blight resistance and deoxynivalenol accumulation in spring wheat. *Genome* 46:555–564.

Sonnewald, U., M.R. Hajirezaei, S. Biemelt, and M. Mueller. 2003. Designer tubers for production of novel compounds. In: *Congress proceedings, BCPC International Congress: Crop Science and Technology.* U.K. Bracknell, ed., 123–132. Glasgow: BCPC.

Sybenga, J. 1996. Chromosome pairing affinity and quadrivalent formation in polyploids: Do segmental allopolyploids exist? *Genome* 39:1176–1184.

Tang, L., S-Y. Kwon, S-H. Kim, J-S. Kim, J.S. Choi, K.Y.Cho, C.K. Sung, S-S. Kwak, and H-S. Lee. 2006. Enhanced tolerance of transgenic potato plants expressing both superoxide dismutase and ascorbate peroxidise in chloroplasts against oxidative stress and high temperature. *Plant Cell Rep.* 25:1380–1386.

Taylor, N., P. Chavarriaga, K. Raemakers, D. Siritunga, and P. Zhang. 2004. Development and application of transgenic technologies in cassava. *Plant Mol. Biol.* 56:671–688.

Tylleskar, T., M. Banea, N. Bikangi, R. Cooke, N. Poulter, and H. Rosling. 1992. Cassava cyanogens and konzo, an upper motor neuron disease found in Africa. *Lancet* 339:208–211.

Urwin, P.E., J. Green, and H.J. Atkinson. 2003. Expression of a plant cystatin confers partial resistance to *Globodera*, full resistance is achieved by pyramiding a cystatin with natural resistance. *Mol. Breed.* 12:263–269.

Vanderschuren, H., R. Akbergenov, M.M. Pooggin, T. Hohn, W. Gruissem, and P. Zhang. 2007. Transgenic cassava resistance to African cassava mosaic virus is enhanced by viral DNA-A bidirectional promoter-derived siRNAs. *Plant Mol. Biol.* 64:549–557.

Vanderschuren, H., A. Adler, P. Zhang, and W. Gruissem. 2009. Dose-dependent RNAi-mediated geminivirus resistance in the tropical root crop cassava. *Plant Mol. Biol.* 70:265–272.

van Os, H., S. Andrzejewski, E. Bakker, I. Barrena, G.J. Bryan, B. Caromel, et al. 2006. Construction of a 10,000-marker ultradense genetic recombination map of potato: Providing a framework for accelerated gene isolation and a genome wide physical map. *Genetics* 173:1075–1087.

Visser, R.G.F., Luc C.J.M. Suurs, P.A.M. Steeneken, and E. Jacobsen. 1997. Some Physicochemical Properties of Amylose-free Potato Starch. *Starch* 49:443–448.

Wang, M-L., C. Goldstein, W. Su, P.H. Moore, and H.H. Albert. 2005. Production of biologically active GM-CSF in sugarcane: A secure biofactory. *Transgenic Res.* 14:167–178.

Waterhouse, P.M. and A.F. Fusaro. 2006. Viruses face a double defense by plant small RNAs. *Science* 313:54–55.

Wenzl, P., J. Carling, D. Kudrna, D. Jaccoud, E. Huttner, A. Kleinhofs, and A. Kilian. 2004. Diversity array technology (DArT) for whole-genome profiling of barley. *Proc. Natl. Acad. Sci. (USA)* 101:9915–9920.

White, W., J. McMahon, and R. Sayre. 1994. Regulation of cyanogenesis in cassava. *Acta Hort.* 375:69–77.

Wong, G.G., J.S. Witek, P.A. Temple, K.M. Wilkens, A.C. Leary, D.P. Luxenberg, et al. 1985a. Human gm-csf: Molecular cloning of the complementary DNA and purification of the natural and recombinant proteins. *Science* 228:810–815.

Wong, G.G., J.S. Witek-Giannotti, P.A. Temple, K.M. Wilkens, A.C. Leary, D.P. Luxenberg, et al. 1985b. Isolation of cDNAs encoding human and gibbon gm-csf. *Prog. Clin. Biol. Res.* 191:351–366.

Wrigley, G. 1982. *Tropical agriculture: The development of production*, 4th ed. New York: Longman Inc.

Wu, K.K., W. Burnquist, M.E. Sorrels, T.L. Tew, P.H. Moore, and S.D. Tanksley. 1992. The detection and estimation of linkage in polyploids using single dose restriction fragments. *Theor. Appl. Genet.* 83:294–300.

Zhang, P., H. Vanderschuren, H. Futterer, and W. Gruissem. 2005. Resistance to cassava mosaic disease in transgenic cassava expressing antisense RNAs targeting virus replication. *Plant Biotechnol. J.* 3:385–397.

18 Future Prospects

Rodomiro Ortiz, Michael Pillay, and Abdou Tenkouano

The preceding chapters of this book deal with several subjects: botany, evolution, genetic resources, cytogenetics, genetics, plant health management, genetic enhancement, postharvest processing, micropropagation, and delivery of new cultivars to farming systems. The last chapter provides an update on molecular breeding of asexual crops from which the next genetic enhancement schemes can be sought. This brief chapter builds on the overviews given by the specialists in each of the contributing chapters by summarizing their major findings and providing an outlook on future research.

The most significant advances in *Musa* botany in recent years are on root systems and bunch morphology (Chapter 1). Research on the former was driven by the needs of defining an ideotype for breeding better root systems in *Musa* at large, but especially for plantains. Experiments during the second half of the 1990s in the degraded forest of West Africa and in the East African highlands led to a better understanding of the relationships between root and shoot traits, the variability in root system sizes, the biophysical effects on root development, and devising alternative methods for root evaluation. The study of bunch morphology of East African Highland bananas (Lujugira-Mutika) provided means for clustering them into four morphologically distinct clone sets. The sets bring together clones that share traits important to farmers and consumers, though they were not previously used for grouping this diverse *Musa* germplasm. Further research should focus on finding DNA markers linked to important underground and shoot traits for their further selection. DNA markers can also be used for gaining insights onto the evolution of bananas and the role that somatic mutations play on it, which will need to go hand in hand with plant morphology research to establish sound evolutionary pathways and relationships as a result of both human and natural selection.

In the last two decades, molecular markers have been extensively used for characterization and classification of *Musa* germplasm collections (Chapter 2). Nonetheless, these advances brought by molecular taxonomy will need to consider both complementary and contrasting morphology data, which will be required to resolve the relationships and phylogeny in *Musa*. Researchers on this subject should use numerical taxonomy to bring together morphological and molecular data, which will be supplemented by cytogenetic methods. Such an integrated approach will allow unraveling relationships among *Musa* species and cultivars, which are complicated by interspecific hybridization, heterozygosity, and polyploidy, that are common in this genus. Priority research with DNA markers should focus on *M. balbisiana* to understand its evolution and systematics, and on the parental contributions of *M. acuminata* subspecies to the origin of triploid banana and plantain cultivars. A better understanding and knowledge on *Musa* origin, domestication, and evolution will facilitate the genetic enhancement of this crop.

The third chapter deals with the genetic resources of wild-seeded bananas. Recent research on *Musa* wild species gave priority to establishing the boundary between sections and among species therein. Today, *Australimusa*, *Callimusa*, *Musa* (which includes the edible bananas), *Rhodochlamys*, and *Ingentimusa* are the five known sections of *Musa*. Their species are a great reservoir of important resistance traits that can be further used for breeding new edible bananas. Further research with DNA markers and geo-referenced systems may assist in understanding their demographic history and dispersal mechanisms, and how they account for *Musa* genetic diversity among populations and across geographical regions.

Chapters 4 and 5 provide state-of-the-art knowledge of banana cytogenetics and genetics. *Musa* species are among the best examples of organisms whose complex cytogenetic structure ensued through extensive chromosomal rearrangements. Since the 1990s a wealth of knowledge, facilitated by the reemergence of breeding programs and new tools, particularly ensuing from advances in molecular biology, became available for genome identification, ploidy analysis, and trait inheritance. *In situ* hybridization, DNA markers, and flow cytometry were the tools used for molecular cytogenetic research in the last 20 years. The molecular characterization of the *Musa* genome and the development of physical maps are therefore suggested topics for further cytogenetic research. The prospects of elucidating genetic mechanisms controlling important traits will undoubtedly accelerate banana germplasm enhancement in the "omics" era. The sequencing of the banana genome will bring new knowledge in breeding this crop. With the aid of bioinformatics, inter- and intragenome comparisons may overcome the lack of high-resolution physical and genetic maps in banana.

Several pathogens (Chapter 6) and pests (Chapter 7) affect the crop and cause yield losses. Sigatoka leaf spots, Fusarium wilt, banana weevil, and parasitic nematodes are of worldwide importance, while new emerging diseases such as banana Xanthomonas wilt or banana bract mosaic are regional constraints for growing bananas in East and Central Africa, and Asia, respectively. Host-plant resistance remains the cornerstone for banana health management, which along with cultural practices and biological control provides the means for an integrated approach to overcome many biotic stresses affecting the crop. Identifying sources of host-plant resistance to the main pathogens and pests in *Musa* germplasm for further use in banana breeding will remain an important research thrust in forthcoming years.

Reproductive biology, which is central to the species breeding system and for success of banana breeding schemes, is the subject of Chapter 8. Little research has appeared in the literature on *Musa* in recent years, though the knowledge of sexual plant reproduction of other species progressed considerably in the same period. Most of the recent research was on West African plantains and East African Highland bananas. The studies were on male and female fertility, as well as on $2n$ gametes, hybridization, seed set, and germination, which are very important traits for the genetic enhancement of the crop. New research should address why the low seed set occurs in cultivars and how to improve seed germination rates, which remain among the main bottlenecks for banana crossbreeding.

Not surprisingly, five chapters are devoted to the core of this book: the genetic enhancement of banana. They provide comprehensive overviews on crossbreeding techniques (Chapter 9), the use of mutations in cultivar development (Chapter 10), biotechnology approaches (both transgenics and genomics led) for breeding (Chapter 11), genotype-by-environment interactions (Chapter 12), and quality improvement and biofortification (Chapter 13). Although significant progress was made in the past two decades, a few hybrid cultivars (mostly tetraploids) are grown in significant acreages by banana farmers, and a handful of field experiments are known for transgenic bananas. The focus of banana breeding seems to be gradually shifting from addressing existing constraints to assessing the risk potential of emerging threats and preparing to respond to them, particularly under a changing climate due to global warming that may increase both abiotic and biotic stress incidences and severity in the crop. Emphasis for banana genetic enhancement should be given to preemptive breeding—particularly through broadening approaches—to deal with new strains of major pathogens and pests, as well as other emerging constraints. Nutritional quality should be added as a priority target trait, along with host-plant resistance and stabilizing yield traits, for banana breeding. Likewise palatability (according to consumers' preferences) and value-adding traits during postharvest processing (Chapter 14) cannot be ignored in banana breeding undertakings. High-yielding and stabilizing traits will allow surplus harvests in small landholders' fields. They can sell such an excess of fresh produce to nearby or other markets to improve their incomes. The ongoing analysis and sequencing of the banana genome, identification of its genes and their expression, recombination, and diversity will provide more effective and efficient means for the genetic enhancement of this crop. Genomics-led and evolutionary breeding approaches can indeed pave the way for banana breeding-by-design.

Large-scale propagation methods of true-to-type materials (Chapter 15) are needed for both banana breeding and delivery systems. More research should focus on low-cost methods for generating new breeding materials and for delivering outstanding clones to the farmers. The management of the genotype-by-environment interaction remains crucial in order to have efficient selection schemes and reach end users, whereas multilocation testing of selected genotypes should be mandatory prior to release of improved cultivars (Chapter 16). The farmers' cultural and economic contexts will also dictate their cultivar choice for adoption. Biophysical and socioeconomic factors need to elucidate how pest-resistant cultivars can be introduced into the farmers' cropping system through association with their own landraces and other crops. Likewise policy research cannot be ignored because the level of adoption of new banana cultivars appears to be associated with an induced model of adoption where farmers are able to assess the yield before embarking on the cultivation of the new cultivars and are thereafter supported with educational materials and incentives for adopting the new technology. Governmental intervention could be often needed through policy initiatives that should go beyond the *ad hoc* responses triggered by pest outbreaks.

Molecular breeding in banana has not kept pace with progress that is occurring in some of the other major food crops. Chapter 17 highlights some of the progress in using biotechnology to improve some traits in potato, cassava, and sugarcane. Breeding of these vegetatively propagated crops faces similar constraints as those faced by banana breeders—yet greater progress has been made in the development of molecular markers for disease and pest resistance and abiotic stresses, and in constructing molecular maps in the three crops addressed in this chapter. This chapter provides valuable insights for researchers wishing to improve banana and plantains.

Index

F

G

H

N

O

P